W9-ADS-306

THE LEPIDOPTERA

FORM, FUNCTION AND DIVERSITY

To my colleagues and my wife for their help and forbearance

THE LEPIDOPTERA
FORM, FUNCTION AND DIVERSITY

by

Malcolm J. Scoble

Natural History Museum Publications

OXFORD UNIVERSITY PRESS
1992

Oxford University Press, Walton Street, Oxford OX2 6DP
Oxford New York Toronto
Delhi Bombay Calcutta Madras Karachi
Petaling Java Singapore Hong Kong Tokyo
Nairobi Dar Es Salaam Cape Town
Melbourne Auckland

and associated companies in
Berlin & Ibadan

Oxford is a trade mark of Oxford University Press

Published in the United States
by Oxford University Press, New York

A catalogue record for this book is available from the British Library

Library of Congress Cataloging-in-Publication Data
Scoble, M. J.
   The lepidoptera: form, function, and diversity/by Malcolm J. Scoble
   p. cm.
   ISBN 0-19-854031-0 : $60.00
   1. Lepidoptera.   2. Lepidoptera-Classification.   I. Title.
QL542.S39   1992   92-4297
595.78-dc20   CIP

ISBN 0-19-854031-0

Typeset in Palatino by Spottiswoode Ballantyne, Colchester, Essex
Printed by St Edmundsbury Press, Bury St Edmunds, Suffolk

CONTENTS

PART II

PART III: MAJOR TAXA

(vii)

PREFACE

This book, on the form, function, and diversity of Lepidoptera, was stimulated because no general volume has appeared on the order as a whole for a number of years. Nevertheless, this century has seen several earlier continental works on the group. Hering (1926a) and Portier (1949) provided treatments on the form, function and habits of Lepidoptera, in German and French respectively. Two works dealing with both form and function, but which also review the families of Lepidoptera, are Zerny & Beier (1936), in German, and Bourgogne (1951), in French. The present work is in the same general mode as those of Zerny & Beier and of Bourgogne, but is restricted to the outer form of the insect and includes information on the 'soft' anatomy only in passing.

Around the time at which the manuscript of this book was being completed, Dr I.F.B. Common (1990) produced a major study of Australian moths. Despite the geographical emphasis, the widespread distribution of most families of Lepidoptera means that much of the information in Ian Common's book is applicable broadly. Although there exists some overlap between the two works, the present volume deals with all families, including butterflies and those groups occurring outside Australia, but it naturally lacks the special emphasis and detail on the Australian Lepidoptera fauna presented in the latter book. Moreover, a much greater proportion of the present work has been assigned to dealing with the form, function, and habits of the Lepidoptera.

By dealing with the Lepidoptera on a global basis, and with half the volume concerning general aspects of the group's biology, this book also departs from the pattern adopted in the previous volumes of this series (on Hymenoptera and on Hemiptera).

This book is intended for all those interested in the general biology and diversity of Lepidoptera. It is aimed not only at lepidopterists, although I hope that they will find a single volume useful and convenient, but also at those who teach or study whole-animal biology and biological diversity - particularly at degree level. Moreover, Lepidoptera are used extensively as study organisms by ecologists, conservationists, and environmentalists, and a major consideration in the construction of this work has been to provide these biologists with a general guide to the order as a whole so that the species they use may be placed in context. I hope, also, that amateur entomologists, who have provided so much empirical information about lepidopteran biology, will find this book useful.

ACKNOWLEDGEMENTS

In preparing this book I have benefited enormously from the knowledge of my colleagues. They have given freely of their advice, provided me with sightings of unpublished manuscripts, supplied photographs or, most time consuming of all, read various drafts of the text in part or as a whole. My debt to them is enormous. Therefore, I thank warmly Philip Ackery, Hans Bänziger, Mike Bascombe, Clive Betts, Michael Boppré, John Bradley, David Carter, Mark Cook, Don Davis, Phil DeVries, Mark Epstein, Kevin Gaston, David Goodger, Jeremy Holloway, Martin Honey, Marianne Horak, Ian Kitching, Niels Kristensen, Marcus Matthews, Jim Miller, Joël Minet, Laurence Mound, Wolfgang Nässig, Ebbe Nielsen, Linda Pitkin, Ken Preston-Mafham (Premaphotos Wildlife), Gaden Robinson, Klaus Sattler, Michael Shaffer, Kevin Tuck, Michael Tweedie, Dick Vane-Wright, Erik van Nieukerken, Peter and Sondra Ward (Natural Science Photos), Alan Watson, and Jason Weintraub. For the extensive comments they made on earlier versions of the text I am especially grateful to Don Davis, Jeremy Holloway, Ian Kitching, Niels Kristensen, Gaden Robinson, and Dick Vane-Wright.

Many of the black and white photographs in this book were prepared by members of the Photographic Unit and the Electron Microscope Unit of The Natural History Museum, London (BMNH). I am very grateful to all those concerned, particularly to Phil Hurst who prepared nearly all the illustrations for Part III, Peter York, for various photomicrographs, and Susan Barnes and Chris Jones for scanning electron micrographs. I thank Shell U.K., Ltd., for their readiness to permit me to publish two of their colour transparencies. The line drawings were executed mostly by Geoffrey Kibby and Malcolm Kerley. Their blend of entomological knowledge and artistic expertise is acknowledged with gratitude. Several of these illustrations were drawn, as originals, and many were based on existing works. Finance for most of the line drawings was given by a donor who wishes to remain anonymous. This support is valued deeply and acknowledged with grateful thanks.

I thank also Myra Givans and Lynn Millhouse, of the Publication Section at The Natural History Museum, and the staff of Oxford University Press for their editorial advice. The index was prepared by Mark Cook.

This work was commenced at Oxford several years ago and I am very grateful to Sir Richard Southwood for his support for this volume.

Finally, but not least, I thank my wife, Theresa, who not only helped me better understand some important entomological literature written in German, but endured, with understanding, the long period over which this book was compiled.

PART I

1

INTRODUCTION

It is impossible to do justice in a book of modest length to a group of organisms as large, and with so complex a biology, as the Lepidoptera; so this volume is a summary and a selective one at that. My emphasis is on the general form and function of Lepidoptera with an eye to phylogeny. Aspects of form and function are conspicuous mainly in the external anatomy, consequently the 'soft' parts, such as the gut and the reproductive system, are mentioned only in passing.

The text falls into three parts, of which the second is short and concerns the environmental importance of Lepidoptera. In Part I, the form and function approach is at its most obvious since we are dealing with basic lepidopteran morphology and the way in which various structures are used in the life of these insects. In Part III, a summary of the structure and composition of each superfamily is presented together with remarks on higher classification, and here, where possible, general biological information is noted, in particular any peculiarities of form and habits. The text might have been organised differently under chapters concerned with topics such as feeding, sensation, colour pattern, and migration, but, to emphasize the criteria of form and function, I have discussed the biological importance of various morphological parts in broadly anatomical sequence.

So in Part I, the adult head, thorax, and abdomen are first discussed, each in a separate chapter. After dealing with the basic morphology of each of the three body divisions, I expand on the function of various structures. Thus the chapter on the adult head deals with feeding and sensation because of the cephalic location of the mouthparts and the major sense organs. Both haustellate and mandibulate conditions of the mouthparts are considered, for although functional mandibles are rarely present in adult Lepidoptera they occur in the most primitive members of the order. After a discussion of basic structure and mode of operation of the proboscis of higher Lepidoptera, the adaptability of this organ is demonstrated by reference to its capacity, in certain species, to pierce the skin of fruit and the skin of mammals. From here, the chapter turns to the kind of food that the proboscis has permitted moths and butterflies to exploit; effectively this is a glimpse at the natural history of adult feeding. Since most of the prominent sensory organs (e.g., eyes and antennae) are concentrated on the head, this chapter provides a convenient point at which to examine sensation.

Chapter 3 deals with the thorax. The morphology of this region is particularly complex. The thorax bears wings and legs - organs of locomotion. Wings permit flight, and flight allows migration - a phenomenon occurring in certain Lepidoptera. Migration is introduced from a general point of view before it is illustrated by three lepidopteran examples showing that it may be a response to different stimuli in different species. A study of wings illustrates extremely well the observation of G.C. Williams that structures may often have a primary function but a secondary effect (Williams, 1966). In the case of wings, the distinction is clear: the primary function of wings is for flight, but the broad surface area of these structures and the varied colour patterns they display has given rise to extensive secondary functions relating to defence, courtship, and thermoregulation. In the discussion of the involved biology of wing colour, the basic formation of colour pattern is explained. Further complexity is

1

added by the position of the wings at rest, and it is noted that wing position and wing pattern are, not surprisingly, related.

The abdomen, that most obviously segmented part of insects, bears the genitalia, those structures of great interest to lepidopteran taxonomists. The basic structure of both male and female genitalia is described, although apart from their obvious function relatively little is known about the precise mechanisms of their particular parts. It is the morphology of the female genitalia that has been so important in the basic (subordinal) taxonomic divisions of the Lepidoptera. Male genitalia have been found, on the whole, to be of more taxonomic use at lower ranks, particularly those of species and genus. Despite the undoubted taxonomic importance of genitalia, the pregenital abdomen provides structures of special interest. This is true particularly of the abdominal base and its connection with the thorax. The articulation itself has provided characters important in phylogenetic reconstruction, but also it is either at the base of the metathorax or at the base (anterior end) of the abdomen that tympanal hearing organs are found in many moths, organs considered in more detail in Chapter 6.

The section on immature Lepidoptera (Chapter 5) follows the same general principles as those on the adult, but less extensive work has been carried out on these stages. Of these early stages, the larvae (or caterpillars) have been best studied. Much of Chapter 5 is concentrated upon the adaptive radiation of caterpillars - the way in which the basically simple body form is adapted to a variety of life styles, including feeding within plant tissue, external consumption of plants, and carnivory. The importance of eggs, larvae, and pupae in lepidopteran classification is outlined.

Some organ systems are not associated with any one body region; examples are those concerned with the reception and production of sound, and with the dissemination of scent. Therefore, hearing organs, stridulatory structures, and the various scent brushes and glands are discussed in a separate chapter (Chapter 6). Special emphasis is given to the importance of hearing in adult Lepidoptera since the impact that the evolution of tympanal organs has had in the lives of nocturnal moths (and most moths are nocturnal) in their defence against bats is very great. At the same time, the roles of hearing and sound production in courtship rather than defence has almost certainly been underestimated. More recent work on the behaviour and physiology of some species suggests that the use of sound production and hearing in courtship is more extensive than acknowledged hitherto. With regard to scent, interest in insect pheromones has grown particularly because an understanding of this subject is of potential significance for controlling pests. In this book, pheromones are discussed briefly with emphasis on the kinds of structures and habits involved in the dissemination of lepidopteran scent, and in the roles scent plays in the lives of Lepidoptera. In particular, I discuss the distribution of pheromone glands on the bodies of these insects, the function of scent in courtship, and in male-male interactions - a subject of growing interest.

Morphological terms used in Part I (and Part III) have been defined or explained where deemed necessary, the page on which any definition is given appearing in the index. There now exists a modern and comprehensive entomological glossary (Nichols et al., 1989) to which the reader is referred for an extensive coverage.

In Part II are summarized the issues that seem to be the essence of the environmental significance of Lepidoptera. Although it touches on ecological issues, Chapter 7 is emphatically *not* an 'ecology of the Lepidoptera'. Rather, I first note the potential effect that Lepidoptera have on the environment. Second, I consider the diversity of plants consumed, the potential that plants offer as food, and the barriers that Lepidoptera have successfully surmounted in their radiation on plants. Finally, I turn to the way in which I think we should view the natural role of Lepidoptera. The importance of Lepidoptera to man is considered in this section, but chiefly in the broad context of the role of these insects in the natural economy. Nevertheless, the value of moths and butterflies as indicators of environmental change is considered, and there is a short section on silk, a natural product of Lepidoptera that has been used commercially for centuries, and another on the importance of Lepidoptera as a source of food.

Part III is a guide to the major taxa of Lepidoptera. Forty-one superfamilies are recognized in this work, a number that is by no means definitive, but simply one that represents a synthesis derived from the literature incorporating, as far as possible, recent findings or views. My aim has been to provide a summary of structure, diversity, and general comments on the biology of each superfamily, with similar treatments for families and important subfamilial taxa as far as possible. Inevitably, there is imbalance. The standard of work on different groups, whether involving accuracy or comprehensiveness, is very variable, and the text reflects this variation. I discuss phylogenetic relationships wherever possible, but we are a long way from achieving satisfactory schemes within many superfamilies to say nothing of relationships between the superfamilies.

Composed of five chapters, Part III is introduced (Chapter 8) by a brief discussion of the historical background of lepidopteran classification, and some of the problems with our existing schemes. The superfamilies are treated in four chapters, reflecting the way in which lepidopterists often think of, or refer to, large sections of the Lepidoptera - even if these groupings are not monophyletic divisions of the order. The divisions are grades - the members of which share primitive characters. For example, primitive moths (Chapter 9) are all homoneurous and include but a small fraction of the members of the order. Considering the small number of species, they receive an apparently disproportionate treatment, but this approach is justified by their structural diversity and phylogenetic significance. Within the group occur not only species with functional mandibles, but those displaying the earliest signs of the characteristic lepidopteran proboscis, and also those that have developed a proboscis with intrinsic musculature. Size and life history also vary considerably from the very small, weakly flying Micropterigidae with larvae living mainly among liverworts, through several families of leaf miners, to Hepialoidea, a group including some of the largest members of the Lepidoptera with much more powerful flight, and larvae that typically tunnel in the soil or in roots. Chapter 10, on the early Heteroneura, deals with those groups the larvae of which are chiefly leaf-miners. The lower Ditrysia, together with the primitive Lepidoptera and the early Heteroneura, are traditionally called 'Microlepidoptera', an assemblage which, although not monophyletic, is deeply ingrained in the minds of lepidopterists, and remains a term in frequent and widespread use. Lower Ditrysia (Chapter 11) include groups with larvae that are typically internal tissue feeders (e.g., leaf-miners and gall-dwellers), or more weakly concealed feeders (such as leaf-rollers or -tiers). The higher Ditrysia (Chapter 12) include the 'Macrolepidoptera'. Whether or not the assemblage is monophyletic, the superfamilies included undoubtedly form a recognisable grouping of large Lepidoptera with larvae that are predominantly external feeders. They are divided into relatively few superfamilies compared with the microlepidopterans, and are dominated, numerically, by two families - Noctuidae and Geometridae.

The decision as to what to illustrate in Part III was particularly difficult. Even if space was unlimited, the variation of wing pattern, wing shape and wing size within many lepidopteran families makes it impossible to cover diversity. But one halftone illustration has been presented for almost every family. In the case of larger families, more than one species has been illustrated each representing a major grouping. Usually these groupings are subfamilies, but in certain instances they are more inclusive (e.g., as in trifine Noctuidae).

2

THE ADULT HEAD:
FEEDING AND SENSATION

Cephalization occurred early in animal evolution, and the organs of feeding and many of those prominently involved in sensation tend to be concentrated on the head. In Lepidoptera, as in all other insects, the head is a composite - probably of six segments, although the number is controversial (Matsuda, 1965). Its segmented nature, largely obscured postembryonically, is apparent in the mouthparts, which represent paired, segmental appendages.

Although the mouthparts of most adult Lepidoptera are of the haustellate variety with a long sucking proboscis, in the most primitive members of the order they are mandibulate. The most prominent sensory structures of the head include a pair of antennae and a pair of compound eyes. Ocelli may be visible or not visible, as may some less familiar sensory structures termed chaetosemata, the functions of which are unknown.

Head capsule

The head capsule of adult Lepidoptera (Denis & Bitsch, 1973) is well-sclerotized, which is to be expected in a structure that bears the mouthparts and their associated musculature. It is filled largely by the brain, and the sucking pump with its associated muscle bundles.

The lepidopteran head has been divided into a number of general areas (Matsuda, 1965), but these areas cannot be defined precisely in terms of their development. They are used by lepidopterists mainly as topographical guides in descriptions (Figs 1-3). The sclerites that form the 'shell' of the insect head are divided by seams known as sutures. These primary divisions are difficult to distinguish from sulci, which are secondary, and strengthening, thickenings. Because of the uncertainty concerning the identity of sutures and sulci, all dividing lines on the head of the adult are generally referred to as sulci - a convention adopted here. There is general agreement about the homology of sulci both within the order and often between orders. Within the Lepidoptera, the strength of sulci varies as does their presence or absence individually. For example, in the primitive family Micropterigidae a relatively full lepidopteran complement is present, whereas in Mnesarchaeidae the sulci are virtually lost (Kristensen, 1968a). Sulci have not been studied comprehensively within the order and few generalizations can be made about their taxonomic significance. Only the laterofacial sulcus persists consistently in higher Lepidoptera (Kristensen, 1968a).

The head is usually covered with scales which may be hair-like or lamellate. In many groups they form a tuft on the vertex and the frons or on the vertex alone. In the tufted condition the head is said to be rough-scaled (Fig. 4) in contrast with the situation where the scales are appressed to the head capsule and said to be smooth-scaled (Fig. 5). Amongst microlepidopterans the smooth-scaled condition is found more often in the less primitive families, but the character is of limited diagnostic importance.

4

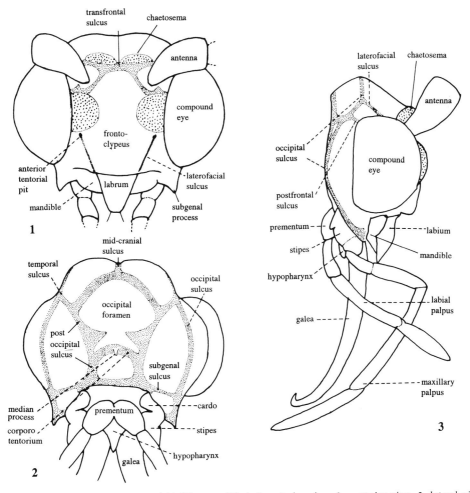

Figs 1–3. *Head of* Neopseustis meyricki (*Neopseustidae*). *1, anterior view; 2, posterior view; 3, lateral view.* *(After Kristensen, 1968a.)*

Internal skeleton

Like internal skeletons in general, that of the lepidopteran head functions in support and movement. The major component is composed of the tentorium, which gives rigidity to the head and provides a base for the insertion of various muscles. The tentorium consists of a pair of apodemes (the anterior tentorial arms), which run from the front of the head to the back of the capsule where they meet a pair of short posterior tentorial arms, united by a median bar-like sclerite termed the corporotentorium, or tentorial bridge (Fig. 2). The arms may be strengthened in Lepidoptera that pierce fruit (p. 16). The points on the front of the head from which the anterior tentorial arms arise are indicated by a pair of anterior tentorial pits (Fig. 1). The position of these pits varies, but they are usually found well below the antennae where the clypeus meets the genae. A pair of dorsal tentorial arms may also arise from the

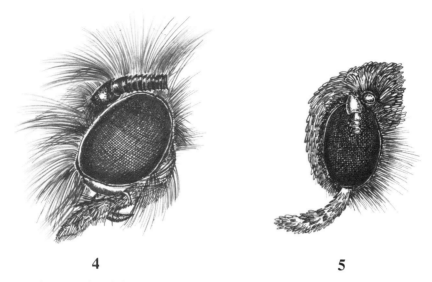

4   **5**

Figs 4, 5. *Heads. 4, 'rough-scaled',* Monopis pavlovski (*Tineidae*); *5, 'smooth-scaled,* Glyphipteryx antidota
(*Glyphipterigidae*) (*note prominent external ocellus*). (*Drawn by Malcolm Kerley.*)

anterior arms, particularly in primitive Lepidoptera. The dorsal arms either almost
reach the roof of the head-capsule or they may be represented merely by little more
than swellings of the main arms. In higher moths they are generally reduced to
swellings, but there are exceptions such as in some papilionid butterflies (Ehrlich,
1958b, Miller, 1987c).

The occipital foramen (Fig. 2) is divided horizontally by the posterior arms plus the
corporotentorium. A median process arises posteriorly from the corporotentorium (Fig.
2) onto which the ventrolongitudinal muscles of the cervix are inserted. This process is
unique to the Lepidoptera (Kristensen, 1984b).

Mouthparts

The dominant feature of the mouthparts of most Lepidoptera is the long coilable
proboscis (Fig. 12) - a structure formed from the union of the two extended galeae of
the maxillae. The proboscis is coiled under the head when the insect is at rest, and it is
extended for imbibing fluids. The structure is unique to the Glossata, a taxon including
all but a small fraction of the Lepidoptera. Although the occurrence of vestigial
mandibles is widespread in adult Lepidoptera fully developed and functional mandibles are rare. In contrast, lepidopteran larvae (caterpillars) are mandibulate and feed,
typically, on plant tissue. Thus larvae and adults have quite distinct methods of feeding
and each, therefore, has a different ecological impact.

Maxillary and labial palpi (Fig. 3) vary in size and number of segments. In higher
moths, the former are reduced while the latter are often well developed. Both pairs of
palpi are usually covered with scales, often densely so. The labrum, or upper lip, is
reduced in most Lepidoptera to a pair of bristled, lateral lobes termed pilifers. A very
small median lobe may also remain. In some hawkmoths (Sphingidae), pilifers are
involved in hearing (Chapter 6). These structures developed early in lepidopteran
evolution, they are probably not homologous with the lateral bundles of bristles found
on the labrum of *Micropterix* (Micropterigidae) (Kristensen, 1984b).

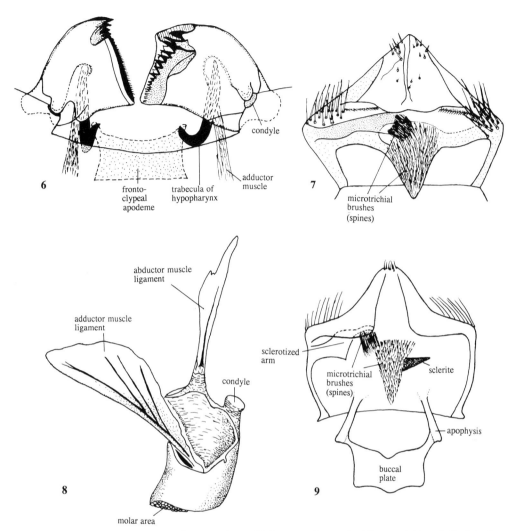

Figs 6–9. *Head structure, showing elements of mandibulate mouthparts in Micropterigidae. 6, 7, Sabatinca incongruella: 6, mandibles; 7, epipharynx; 8, 9, Micropterix calthella: 8, base of right mandible; 9, epipharynx. (6, 7, after Tillyard, 1923; 8, 9, after Hannemann, 1956a.)*

The mandibulate condition

Although mandibles are present as functionless lobes in many adult Lepidoptera, they are thought to be used for chewing in just two families of primitive moths, the widespread Micropterigidae and the South American Heterobathmiidae. Large mandibles with distinct cranial articulations and powerful muscles occur in Agathiphagidae (Kauri moths, found in the SW Pacific region and in northern Queensland). However there are no incisor elements, suggesting that the moths may not feed (Kristensen, 1984b). Only micropterigid moths have actually been observed using their mandibles, but Heterobathmiidae were seen to visit flowers of *Nothofagus* (Fagales), so mandibulate feeding is likely (Kristensen & Nielsen, 1983). Micropterigid moths are typically diurnal and visit flowers to feed on pollen grains. Some species of *Sabatinca* (Micropterigidae) feed on spores of ferns (G.W. Gibbs *in* Kristensen, 1984b).

The mandibles of Micropterigidae (Figs 6, 8) are used in conjunction with an armature on the epipharynx (the inner surface of the labrum) (Figs 7, 9) and hypopharynx (see below). Studies of the structure and associated armature of mandibles were reported for *Sabatinca* from New Zealand (Tillyard, 1923) and the European *Micropterix calthella* (Hannemann, 1956a) (both Micropterigidae), and the Patagonian heterobathmiid genus *Heterobathmia* (Kristensen & Nielsen, 1979). The grinding of pollen grains or spores involves the mandibles, stiff spines on the epipharynx (Figs 7, 9), and a depression in the hypopharynx, known as the 'triturating basket' (Tillyard, 1923) or the 'infrabuccal pouch'.

Labrum and epipharynx

In contrast with most Lepidoptera, the labrum is well developed in Micropterigidae, Agathiphagidae, and Heterobathmiidae (as it is in the primitive glossatan family Eriocraniidae where the mandibles in the adult are not functional).

The epipharyngeal (ventral) surface of the labrum bears asymmetrical structures with brushes composed of powerful spines (Figs 7, 9). These spines are long microtrichia (minute cuticular outgrowths) and are found in both Micropterigidae and Heterobathmiidae. They are absent from Agathiphagidae (Kristensen, 1984b; Dumbleton, 1952). The epipharyngeal armature is not unique to micropterigids and heterobathmiids since similar structures have been found in a xyelid sawfly (Hymenoptera) (Kristensen, 1984b). If, as is quite possible, it has been independently derived in Hymenoptera and Lepidoptera, then its presence in Lepidoptera is a ground plan condition of the order with a secondary loss occurring early in lepidopteran evolution.

The epipharyngeal sclerotizations vary somewhat in their degree of development. In the most fully developed arrangement (e.g., *Micropterix calthella*, Fig. 9), there is a pair of asymmetrical sclerites and two brushes of microtrichia of unequal size. On the right side is found a rod-like arm and on the left side a more complex sclerite. Of the two brushes, the larger is medial and the smaller lies close to the sclerotized arm of the right side of the epipharynx. The epipharyngeal armature varies in degree of development in those Lepidoptera possessing it; but generally, where there is a well-developed sclerotized arm on the right side the medial brush is also strong. In *Heterobathmia* both sclerites are present, as is a median brush of strong microtrichia arising from the larger (left side) sclerite and the surrounding area (Kristensen & Nielsen, 1979).

The epipharyngeal brushes are assumed to work in conjunction with the mandibles since they lie immediately over the area of closure of their cutting and grinding surfaces (Tillyard, 1923). Although the labrum is well developed and trilobed in *Agathiphaga* (Agathiphagidae), no epipharyngeal armature exists (Dumbleton, 1952).

The necessary mobility of the labrum and the epipharyngeal sclerotizations is provided by three muscles: one fronto-labral, one clypeo-epipharyngeal, and one labro-epipharyngeal (Hannemann, 1956a).

Mandibles

The existence of functional mandibles is atypical for Lepidoptera. Essentially, lepidopteran mandibles (e.g., Figs 6, 8) are like those of other pterygote insects. Unlike the condition occurring in the lowest apterygotes ('below' the lepismatoid assemblage), where mandibular articulation with the head-capsule is single, in pterygotes the articulation is double. One of these points of articulation is a rounded protuberance (condyle), while the other is a groove or cavity (ginglymus). Well-developed abductor and particularly adductor muscles, found widely in mandibulate pterygotes, are also present. They arise from different areas of the head capsule and their attachment to the base of the mandibles is shown in Fig. 8.

The armature of the mandibles of Micropterigidae and Heterobathmiidae varies between species, and the cutting and grinding components of mandibles in individual moths are not mirror images. Both mandibles bear 'teeth' (an 'incisor' area) and possess a grinding 'molar' area. In *Sabatinca incongruella* (Fig. 6), the biting surface of the left mandible is mainly molariform, but it bears teeth at the apex of the outer angle. On the right mandible, in contrast, a series of teeth runs along the occlusal surface dividing it into incisor and molar areas. At the apex of the right mandible are three large teeth at the edge of a cavity. While grinding, the teeth at the apex of the left mandible work into this cavity (Tillyard, 1923).

The armature of the mandibles of *Heterobathmia* is also asymmetrical, but rather less so than that of *Sabatinca incongruella*. Both molar and incisor areas are present, but whereas in Micropterigidae teeth are concentrated at the apex of the left mandible, in *Heterobathmia* they run proximally for most of its length (Kristensen & Nielsen, 1979).

Maxillae

The maxilla of Micropterigidae (Fig. 10) is composed of a basal cardo and stipes, a five-segmented maxillary palpus, a small galeal lobe, and a lacinia. The galea is destined to become greatly extended and united with its partner on the other maxilla to form the characteristic lepidopteran proboscis (p. 11). According to Kristensen & Nielsen

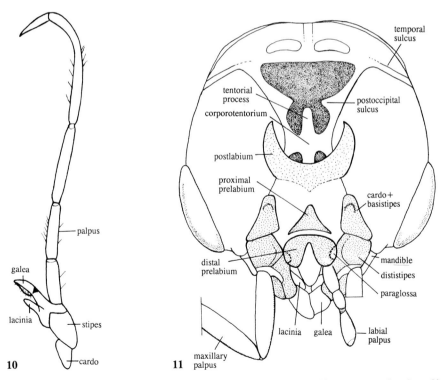

Figs 10, 11. *Head structure. 10, maxilla of* Micropterix aruncella *(Micropterigidae); 11, posterior view of head of* Heterobathmia pseuderiocrania *(Heterobathmiidae) showing postlabium and labium. (10, after Tillyard, 1923; 11, after Kristensen & Nielsen, 1979.)*

(1979), in Micropterigidae the most basal component of the maxilla is actually composed of both the cardo and that part of the stipes called the basistipes. The basistipes is divided by a costa from the dististipes. The costa is absent from Heterobathmiidae.

Hypopharynx

The hypopharynx of Micropterigidae, which lies above the labium, is a broad, tongue-shaped lobe strongly modified for grinding pollen. It bears an infrabuccal pouch or depression with numerous strong spines, and is often termed a 'triturating basket' (Tillyard, 1923). Curved, sclerotized supports called trabeculae (Tillyard) or suspensoria (Fig. 6) (Hannemann, 1956a) are present on each side of the hypopharynx. The apodeme on which the adductor muscle of the mandible is attached is invaginated from an area immediately adjacent to the trabecula so that the hypopharynx acts together with the mandibles in triturating pollen. A similar triturating basket has also been found in Heterobathmiidae, but connections between the mandibles and trabeculae were not definitely observed (Kristensen & Nielsen, 1979). Anterior to the infrabuccal pouch, the hypopharynx also bears a diamond-shaped patch of spines. The salivary ducts unite to form a common duct opening on the lower surface of the hypopharynx.

Labium

In Micropterigidae and Heterobathmiidae the labium is composed of three sclerites (Fig. 11). The postlabial sclerite lies slightly below the corporotentorium, and the prelabium includes a proximal and a distal sclerite (Kristensen & Nielsen, 1979). The labial palpi extend from the distal sclerite. Both a ligula (fused glossae) and paraglossae are present.

To summarize: the presence of mandibles in Lepidoptera is primitive, but the reduction of the double articulation, the biting surfaces and the structures associated with the mandibles are specializations. Those Lepidoptera with well-developed and well-articulated mandibles are among the most primitive members of the order, yet they exhibit considerable morphological specializations in their mode of feeding.

Feeding in mandibulate moths

The mode of action of the pollen-grinding mechanism in Micropterigidae was described by Tillyard (1923) and Hannemann (1956a). Hannemann studied both the functional morphology of the various components and their mode of action in living moths feeding at flowers.

The maxillary palpi, each bearing a claw-like terminal segment (e.g., as in Fig. 10), gather pollen from anthers of flowers by alternating raking motions carried out at the considerable rate of about 15 per second. Pollen is scraped to the mouth by the palpi, whilst the mobile paraglossae of the labium probably help retain pollen in the buccal cavity.

Pollen is ground by the combined action of the epipharyngeal spines, the mandibles, and the triturating, spinose pouch on the hypopharynx. The movements of the epipharynx, mandibles, and hypopharynx are co-ordinated. The labrum is raised by the fronto-labralis muscle to coincide with mandibular abduction, while at adduction the labrum moves against the mandibles. The actions of the clypeo-epipharyngealis and labro-epipharyngealis lead to an outward movement of the spines on the epipharynx, clearing pollen from the mandibles.

The connection between the mandibular adductors and the hypopharynx is such that the movements of these structures are synchronized. At adduction the hypopharynx is brought towards the mandibles and at abduction it moves away. The hypopharynx not only moves towards the mandibles thus helping to grind pollen, but also contracts so adding to the grinding process. During adduction the mandibles move at a slight angle to each other, and this further enhances grinding. Food is ground with saliva and forced into the mouth. The observation that few coarse particles of food were encountered in pharyngeal sections suggests that the grinding process is effective (Hannemann, 1956a).

The haustellate condition

The proboscis is a characteristic structure of the vast majority of the Lepidoptera. Coiled under the head at rest (Fig. 12), it is extended to probe flowers for nectar, or other substrates for other substances. The organ is of a complex and remarkable design, and its precise mode of operation is controversial. Its function is to imbibe fluid,

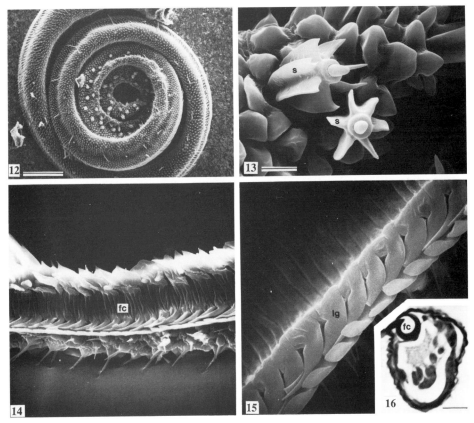

Figs 12–16. *Head structure, proboscis. 12, 13, Nola pustulata (Noctuidae): 12, proboscis coiled (scale, 100 μm); 13, sensilla styloconica (s) near tip of proboscis (scale, 5 μm); 14, 15, Ethmia bipunctella (Ethmiidae): 14, galea, showing food canal (fc); 15, linking plates (legulae, lg) of galeae; 16, transverse section of galea of Neopseustis (Neopseustidae) showing food canal (fc) (scale 20 μm). (12, 13, courtesy of Dr J.S. Miller, after Miller, 1991; 14, 15, courtesy of Dr K. Sattler/EM Unit, BMNH; 16, courtesy of Dr N.P. Kristensen and Dr E.S. Nielsen, after Kristensen & Nielsen, 1981.)*

particularly nectar and water, but some moths are able to pierce the skin of fruit or of mammals and their modified proboscis, which is used for the uptake of juice or blood, is described below. Others feed on saliva and the lachrymal and nasal secretions from cattle or other animals, and some probe at urine, and general wet patches, or faeces where they obtain certain salts or, possibly, nitrogen.

Several groups of insects bear sucking mouthparts. In Lepidoptera, uniquely, the proboscis is formed by the union of the greatly extended galeae. The medial surface of each galea is strongly convex, so that when united with its opposite member a channel, or food canal, is formed up which liquids can pass.

The proboscis of Eriocraniidae

In extant Lepidoptera a proboscis first appears in Eriocraniidae, the most primitive family of the Glossata. A study of the proboscis of Eriocraniidae by Kristensen (1968b) was based on an examination of *Heringocrania* and *Eriocrania*.

The proboscis (Fig. 17) is short and, when the moth is at rest, remains coiled under the head. The proximal part of each galea is sclerotized entirely, but the rest of the structure only dorsally. The galeae are linked both by dorsal and ventral projections, but the cohesion is weak as was demonstrated when the proboscis 'split' apart in moths that pressed their tongues against a glass petri dish while sucking water (Kristensen, 1968b). The wall of the food groove is composed of a series of slightly overlapping plates. Two sets of extrinsic muscles (a levator and a flexor) arise on the stipes but, unlike the condition in other Lepidoptera, no intrinsic muscles are present. In higher Lepidoptera, extrinsic muscles are involved in recoiling the proboscis. Presumably, coiling in Eriocraniidae depends on elasticity, a property considered important, although insufficient in itself, in accounting for coiling in higher Lepidoptera (Bänziger, 1971). Chemoreceptors are found at the apex of the galeae and a few are present in the walls of the food canal. The base of the proboscis leads into the stipes, itself a hollow structure.

The functional nature of the proboscis of these primitive moths is evident from observations of them feeding on sap from damaged leaf-buds (Kristensen, 1984b) as well as water. However, Eriocraniidae do not seem to visit flowers.

Intrinsic proboscis musculature is also absent from Acanthopteroctetidae and Lophocoronidae (Kristensen, 1986).

Proboscis structure in higher Lepidoptera

The morphology and function of the proboscis of *Pieris brassicae* was detailed by Eastham & Eassa (1955), in a study from which a great many generalizations about the organ within the order as a whole have been made. In higher Lepidoptera, the structure of the proboscis is much more complex than in Eriocraniidae. However, like the eriocraniid proboscis, that of higher Lepidoptera is composed of two hollow, united galeae, which form a channel (Fig. 14) approximately circular in cross-section (Figs 18, 19).

In transverse section (Figs 18, 19), the outer wall of each galea is formed by a series of exocuticular ribs embedded in flexible endocuticle. This arrangement gives the proboscis its annulated appearance. These ribs are C-shaped and do not extend to the exocuticular lining of the food canal, an arrangement allowing greater flexibility, which assists the coiling and uncoiling process. The lining of the food canal is composed of exocuticular bars, which themselves consist of tightly packed laminae. Dorsally and ventrally, the food canal is firmly closed by an interlocking system of plates (Figs 14, 15, 19).

Dorsally, the exocuticular bars fuse so that in each galea there is a longitudinal elastic bar running the entire length of the proboscis. The gap between the dorsal bar of each galea is bridged by a series of lance-shaped, cuticular plates (legulae of Davis, 1986). These plates, which arise from both galeae, overlap to provide a dorsal linkage (Figs 15, 19). Ventral linkage is stronger than dorsal linkage. It is effected by a series of interlocking hooks. The hooks are very tightly packed along each galea, and ventral to each hook is a tooth. Hooks and teeth interlock (Fig. 19) to form an extremely firm connection, yet one which allows each galea to slide along the other. The ability to slide is of great importance in the functioning of the mouthparts of fruit and skin piercing moths, which operate by rapid alternating thrusts of each galea (p. 16). Galeal linkage is sufficiently strong to prevent the structures becoming forced apart as hydrostatic pressure rises in each galeal cavity, or when the proboscis becomes distorted through muscle action.

The lumen of each galea is divided by two longitudinal septa (Fig. 18), representing inturnings of the galeal wall. These septa divide the lumen into dorsal, lateral, and medio-ventral chambers. A trachea runs within the dorsal chamber, the galeal nerve and the primary oblique muscles are situated within the lateral chamber and, where present, the secondary oblique muscles occur in the median ventral chamber. The septa run from near the base of the proboscis to near its apex, but they are absent from the extreme tip.

Proboscis function also depends on the interaction of various muscles. Stipital muscles are responsible for closing the aperture between the cavities of the head and the stipes. In higher Lepidoptera both extrinsic and intrinsic galeal muscles occur. Of the extrinsic muscles, the retractors pull the base of the galeae backwards bringing the food canal into the cibarial region, while the elevators raise the base of the proboscis prior to feeding. The intrinsic muscles include primary and secondary oblique muscles, with the latter occurring only at the 'knee bend' of the proboscis - a point about one third the length of the proboscis from its base. The primary oblique muscles are found within the haemocoel throughout the coilable length of the proboscis. They are broadly attached to the outer wall and taper to a point on the ventral wall close to the septum. The muscles function to bring the dorsal and ventral galeal walls towards each other.

The lepidopteran proboscis bears numerous sensilla of two main kinds (Börner, 1939; Miller, 1991) - sensilla basiconica, with a broad base and a distal peg, and sensilla styloconica (Fig. 13). The former occur over the entire length of the proboscis, while the latter are restricted to the distal third of the organ. Sensilla styloconica may be fluted or winged (Fig. 13), or they may occur without flutes or with reduced flutes. Both kinds of proboscis sensilla are widespread in the order although no comprehensive survey has been carried out. Although sensilla basiconica are present on the galeae of *Micropterix* (summary in Miller, 1991), fluted styloconic sensilla are absent. Their presence may be a specialization for Lepidoptera above Micropterigidae (Miller, 1991). Loss of fluting appears to have occurred in several different instances (Kitching, 1987; Miller, 1991).

Proboscis structure in primitive Glossata

All Lepidoptera from Neopseustidae onwards have intrinsic musculature in the proboscis and are collectively termed Myoglossata (Kristensen & Nielsen, 1981a). However, the arrangement of proboscis muscles in the primitive families Neopseustidae, Opostegidae, Nepticulidae, Incurvariidae, and Tischeriidae are more simple than in higher Lepidoptera (Kristensen & Nielsen, 1981a). In primitive moths, one or a few longitudinal muscle fibres run from the stipes to the apical region close to the ventral wall of the lumen of each galea.

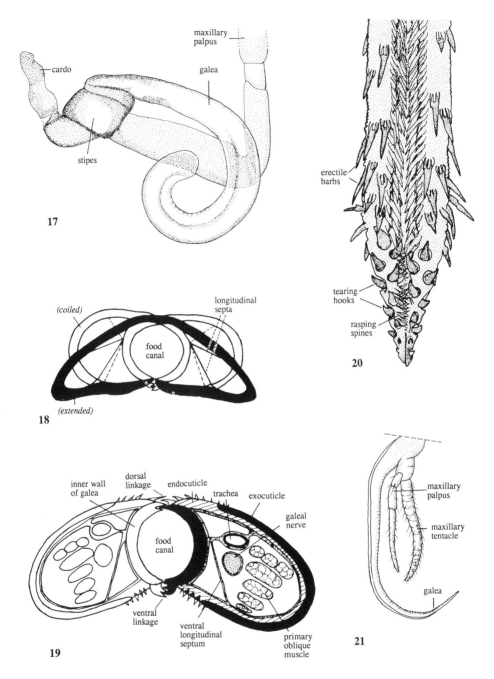

Figs 17–21. *Maxilla structure. 17, maxilla of* Eriocrania semipurpurella (Eriocraniidae) *showing short proboscis (palpus truncated); 18, 19, transverse sections of proboscis (diagrammatic) in* Pieris brassicae (Pieridae): *18, coiled (not shaded; longitudinal septa shown as broken line) and extended (shaded); 19, transverse section of proboscis (extended) at middle of length, with detail; 20, apex of proboscis of the blood-sucking moth* Calyptra eustrigata (Noctuidae); *21, maxillary tentacle of* Parategeticula pollenifera (Prodoxidae). (*17, drawn from a photograph by Kristensen, 1968; 18, 19, after Eastham & Eassa, 1955; 20, after Bänziger, 1980; 21, after Davis, 1967.*)

In Neopseustidae there are two notable modifications in the structure of the proboscis (Kristensen & Nielsen, 1981b; Davis & Nielsen, 1980). First, a closed channel exists within *each* galea so that the proboscis bears a double food canal (Fig. 16). Second, the galeae are linked dorsally by curved microtrichia and special 'zip-scales'. Zip-scales are apparently not homologous with the dorsal linking plates found in higher Lepidoptera; they form a single row of conspicuous, apically serrated scales, running along each galea just below the food groove. The scales interlock with their opposite members on the other galea.

The function of this remarkable double food canal is unclear. An apparent disadvantage is that the cross sectional area of a single food groove of a given perimeter length is double the total cross sectional area of two small canals with a combined perimeter length equivalent to that of the single canal. However, the capillary force in a food canal of smaller radius is correspondingly greater than in the larger canal (Kristensen & Nielsen, 1981b).

The coiling mechanism

The mechanism by which the proboscis is coiled and uncoiled is controversial. In *Pieris brassicae* (Pieridae), Eastham & Eassa (1955) rejected a favoured view that extension was caused by the direct pumping of blood from head to proboscis. Instead, they proposed that extension was achieved indirectly by contraction of the primary oblique muscles of the galeae on the hydrostatic skeleton of the proboscis. Hydrostatic pressure was thought to be maintained within the proboscis by closing the stipital valve, the resultant turgidity permitting the oblique muscles to have a single uniform effect on the galeal walls. The effect of oblique muscle contraction was thought to change the shape, in transverse section, of the proboscis from relatively flat to arched (Figs 18, 19) It was considered that this change in shape would lead, automatically, to proboscis extension.

This explanation was questioned by Bänziger (1971) who observed, in a series of experiments on *P. brassicae* and representatives of seven other families, that it is indeed blood pressure, rather than indirect muscle action that brings about proboscis extension. Blood, forced into the stipes, raises stipital blood pressure. Since the stipes is closed by one valve where it meets the head cavity and another where it adjoins the galeal lumen, pressure is transmitted to the proboscis when the stipital valve to the proboscis is opened. Whereas Eastham & Eassa thought that uncoiling or extension resulted indirectly due to *contraction* of the primary oblique muscles aided by closed, turgid galeal haemocoela, Bänziger demonstrated that the oblique muscles were actually *relaxed*.

Eastham & Eassa (1955) considered that the proboscis was *coiled* solely by its elasticity. In particular, there is a continous, dorsal cuticular bar in each galea, which, when a galea is divided dorsoventrally, causes the dorsal half to coil up. Although this resilin bar is *largely* responsible for coiling, it appears (Bänziger, 1971) that the final 'tightening' of the coil is achieved by contraction of the oblique muscles. Nevertheless, in moths without intrinsic musculature (e.g., Eriocraniidae), coiling must result from elasticity alone.

Coiling appears to be aided by secretions of single-celled glands situated along the length of the dorsal chamber of each galea. Eastham & Eassa suggested that their secretions lubricate the plates of the dorsal linkage.

The *secondary* oblique muscles, found at the knee-bend of the proboscis, were said by Eastham & Eassa to act antagonistically to the primary oblique muscles so flattening the proboscis in cross section at this one point. They considered that the effect of such antagonistic contraction was to retain a coiled or semi-coiled state just at this point, allowing the proboscis considerable side-to-side movement to facilitate probing.

Although contraction of the primary oblique muscles is now thought to be involved in coiling rather than in extension of the proboscis, contraction of the secondaries probably acts in the formation of the knee-bend by distorting the normally circular cross section of the proboscis at this point.

The sucking pump

The sucking pump is derived from both the cibarium and the pharynx. It is particularly well developed in those Lepidoptera that suck copious quantities of fluid.

The cibarium is that part of the preoral cavity (the cavity formed by the mouthparts) enclosed by the epipharynx, i.e., the inner wall of the clypeolabrum and part of the hypopharynx. The hypopharynx is continuous with the buccopharyngeal region behind it. In higher Lepidoptera the free (anterior) part of the hypopharynx is lost (Eastham & Eassa, 1955), although it is distinct but reduced in primitive groups. The hypopharynx posteriorly forms a cibarial plate constituting the floor of the sucking pump. The roof of the pump is membranous and flexible and has an associated complex of muscles that contract the pump. Movement of the roof permits changes in the volume of the pump causing fluid to be sucked in or forced out.

Essentially there are three groups of muscles (Eastham & Eassa, 1955) involved in the working of the sucking pump - those that control the flow of fluid into and out of the pump, those that compress the pump, and those that dilate it. When fluid is sucked in, the anterior sphincter is released and the posterior sphincter is contracted so that the pump is open anteriorly and closed posteriorly. Contraction of the dilators raises the roof of the pump increasing its volume and drawing in fluid. By contraction of the anterior sphincter and relaxation of the posterior sphincter the pump is closed anteriorly and opened posteriorly. Contraction of the compressor muscles and relaxation of the dilators forces fluid into the oesophagus as the volume of the pump decreases.

Piercing: sap and blood sucking in adult Lepidoptera

The juice of fruit and the blood of mammals are among the variety of fluids imbibed by adult Lepidoptera. While many species suck fluid from damaged fruit or from wounds, some are able to pierce the skin of both fruit, animals, or both. All moths with a piercing proboscis belong to the Noctuidae (Catocalinae, including Ophiderinae, see Part 3).

The proboscis of piercing moths, whether it penetrates the skin of fruit or mammals, differs considerably from that of non-piercers (Bänziger, 1970). It is thicker and therefore stronger, and whereas the organ tapers towards the apex in non-piercers, it does not do so in piercers. In piercers, the proboscis terminates suddenly to form a heavily sclerotized point, and the apical section of the proboscis bears erectile barbs and tearing hooks (Fig. 20) (called galeal barbs by Johannsmeier, 1976). These structures are probably highly modified sensilla basiconica (I.J. Kitching pers. comm.) In non-piercers they are not typically present; only simple bristles and some rasping spines or hairs usually occur. Furthermore, the ability of the galeae to slide along each other in opposite directions (the antiparallel movement of Bänziger) seems to be confined to piercing Lepidoptera. This antiparallel movement of the galeae is achieved by the action of cranio-stipital muscles (a galeal retractor and protractor). Four cranial-stipital muscles occur in piercing moths whereas in other higher Lepidoptera there is a maximum of three (Bänziger, 1970). Wu & Chou (1985) found that in piercers the tentorial arms are strong and arched whereas in non-piercers they are relatively weakly developed and not arched. The cephalic musculature studied by these authors (the retractor of the proboscis and the cranial adductor and cibarial dilator) was found to be best developed in moths that pierce intact fruit, less so in those that pierce intact and damaged fruit, and least developed in those that feed only on damaged fruit.

In the skin-piercing and blood-sucking moth *Calyptra eustrigata* the tearing hooks are stronger and larger than in the fruit piercing moth *C. thalictri*, and the walls and tip of the proboscis are more heavily sclerotized. The piercing mechanism is somewhat more complex in *C. eustrigata*, which is not surprising considering the great resistance posed by the elasticity of mammalian skin.

The piercing mechanism, which is confined to males, was described by Bänziger (1980) who carried out observations on moths feeding on cattle and large, wild mammals, and also on his own finger (Figs 22, 23). The moth forces the tip of the proboscis into cracks, creases, or sores in the skin. Initially, the antiparallel motion of the galeae, together with a roughly circular rotation and twisting of the proboscis and a change of direction of thrust, forces the tip of the proboscis into an unevenness in the skin. The turgid proboscis takes up a slightly S-shaped position, which permits a much stronger thrust to be transmitted than if it was arched. Terminally, the proboscis bears several strong tearing hooks, the points of which project backwards assisting the proboscis to bore into the skin. With its tip pressed against or into the skin, the proboscis bends from side to side in a series of high speed movements, which simulate the motion of a spindle. This action is accompanied by a twisting motion so that the proboscis describes a roughly circular motion. Since, as a result of the antiparallel

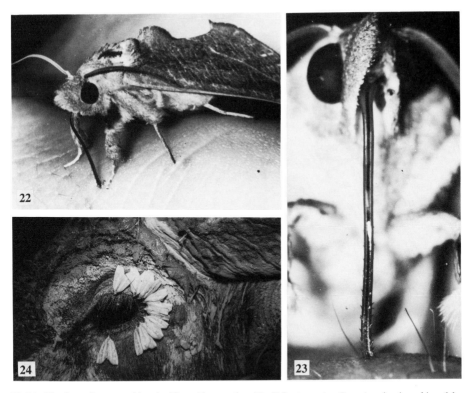

Figs 22–24. *Blood- and tear-sucking in Noctuidae moths. 22,* Calyptra minuticornis *piercing skin of human finger; 23,* C. eustrigata *piercing skin of Malayan Tapir (note erectile barbs); 24,* Lobocraspis griseifusa *sucking tears from eye of a Water Buffalo in Thailand. (All courtesy of Dr H. Bänziger, 22, after Bänziger, 1979; 23, after Bänziger, 1980; 24, after Bänziger, 1983.)*

movement, only one galea is forced against the skin, the thrust transmitted is increased and eventually leads to penetration. Once penetration has occurred, the side to side movement of the proboscis is effectively neutralized as the moth turns its head from side to side. The antiparallel motion of the galeae continues as the proboscis is forced deep into the tissue. As one galea is extended, the other suffers a retractive force, but actual retraction is prevented by the erectile barbs on the proboscis. When the proboscis is turgid, the blood pressure causes the barbs, which are normally directed forward, to become directed backwards. As the retractive force is applied, the tissue becomes lacerated by the barbs, resulting in a flow of blood. This flow is increased with partial withdrawal and repenetration. Feeding may last for several minutes. The blood pressure drops when the proboscis is withdrawn causing the erectile barbs to return to their forward pointing position.

The erectile barbs and oscillatory twisting of the proboscis in *C. eustrigata* seems to be unique within insects. Not only are different components involved, but whereas the mouthparts of other insects are rigid, the proboscis of Lepidoptera is coiled at rest and made turgid by hydrostatic pressure (Bänziger, 1970). The effectiveness of the various mechanisms is evident from the fact that the moth can pierce the healthy skin of large ungulates and elephants.

Fruit-feeding moths may be divided into those that actively pierce fruit and those that suck exuded juices or probe the surface of wounds. The differences between the proboscis of piercers and non-piercers has already been described, but fruit-piercers themselves have also been divided by Bänziger (1982) into primary and secondary piercers. Primary piercers penetrate the intact skin of fruit, while secondary piercers insert their proboscis into holes made by primary piercers or into breaks in the skin made by other insects or other agents. However, the categorization of a moth as a primary or secondary piercer must be made with reference to the fruit. Few primary piercers of soft-skinned fruit are able to penetrate hard-skinned fruit. Only nine out of 86 species of fruit-feeding moths from Thailand were able to pierce the hard-skinned longan fruit (Bänziger, 1982) and, in general, most piercers penetrate only fruits with soft-skin. A primary piercer on soft-skinned fruit may be a secondary piercer on hard-skinned or thick-skinned fruit. The ability to pierce may also vary with the condition of the fruit; for instance the proboscis of *Ophiusa tirhaca* (Noctuidae) is capable of piercing the skin of a ripe but not an unripe peach (Bänziger, 1982).

In contrast with piercers, non-piercers simply imbibe sap that is freely available from damaged fruit. Nevertheless, sap exudation may be increased in some species by the action of rasping exposed pulp with bristles on the proboscis.

Fruit-piercing moths are found in most parts of the world, but they are particularly well represented in the tropics and subtropics. Most of the attention focussed on fruit-piercing moths has been generated as a result of damage caused to orchard crops. Primary piercers pave the way for secondary agents such as fungi, and insects such as *Drosophila* and small beetles, which lay their eggs in the fruit and whose larvae feed on the pulp. Since primary piercers prefer to feed on intact fruit, ignoring that which rots on the ground, damage to fruit crops can be extensive.

Norris (1936), who reviewed the literature on fruit-piercers, noted that nearly all fruit-piercing moths were noctuids. However, there are examples of damage by other groups. The nymphalid butterfly *Crenis boisduvali* was once observed to ruin an entire crop of apples in east Africa. Hundreds of individuals of *Charaxes neanthes* and *C. zoolina* were seen attacking oranges, often with seven or eight butterflies on a single fruit.

The basic lepidopteran proboscis is evidently preadapted for piercing. A prerequisite for the piercing habit was a strengthening of the tongue, an adequate lacerating apparatus, and some mechano-behavioural modifications such as the antiparallel movement of the galeae and the oscillatory motion of both the proboscis and the head.

Those moths that pierce hard or thick-skinned fruit or mammalian skin have the most highly developed mechanical and behavioural modifications.

Food substrates and what they provide

In many Lepidoptera the proboscis is absent or vestigial so the adults do not feed. However, in most the organ is well developed. Despite numerous observations of adult butterflies and moths, many basic questions about feeding have yet to be investigated. For example, precisely what substances do these insects seek when they feed?; what are the substances that stimulate feeding?, and what effect does adult feeding have on such characteristics as longevity, fecundity, and flight activity?

Most information on adult feeding has been gained from general field observations. More critical assessments, together with simple experiments, were made by Norris (1936), Downes (1973), and Bänziger (e.g., 1970), and detailed experiments were undertaken by Arms et al. (1974). In this section some of the work undertaken by these authors, and by others, is considered to illustrate the form and function of the proboscis, and the sort of substances that this organ allows Lepidoptera to obtain. But, until we have more answers to the kind of questions posed above, strict categorizations of food sources based on fundamental requirements of the insects (such as precisely what substances are sought) will remain elusive. The relatively few Lepidoptera that have been studied means that we should be wary of making sweeping generalizations from limited investigations. The only conclusions that can be drawn at this stage are first that adult Lepidoptera derive a variety of substances, rarely only one item, from most substrates upon which they feed, and second that different species probably have different requirements. So it may be inaccurate to suggest simply that all Lepidoptera feeding at puddles and damp sand seek the same nutrient.

From observations and experiment, it appears that four substances are most frequently sought by Lepidoptera. These are water, sugars, salts, and amino acids. They are potentially and variously derived from the many substrates on which adult Lepidoptera have been seen to feed. Adler (1982) listed the following sources other than floral or extrafloral nectaries: mud puddles, soil, dung, urine, crushed bodies of conspecifics, moist ashes from a campfire, carrion, saliva, exposed heads of basking turtles, soap suds, lachrymal secretions and pus, perspiration, plain salt, blood, frog-hopper larval secretion, aphid honeydew, nectar gland secretion of lycaenid larvae, rotten fruit, sound fruit, cocoa seeds, rotten seeds, fermented milk, rotten cheese, borage plants, tree sap, red wine, ink, honey, and pollen steeped in nectar, damp walls, floors, and ceilings of abandoned, crumbling cement structures, owl pellets, wet rocks, certain woods in the form of prepared and exposed timber, bird droppings, fungi, and certain withered plants from which pyrrolizidine alkaloids are obtained (pp. 81, 168).

Since this section is primarily about form and function in the lepidopteran proboscis, and as uncertainty exists as to which basic substances are sought by butterflies and moths, feeding in adult haustellate Lepidoptera is discussed under headings of habits and substrates rather than substances sought.

Water

The proboscis may have evolved in response to the need for early Lepidoptera to replenish water lost in dry conditions (Common, 1973). The only record of a primitive (homoneurous) moth drinking water was reported by Kristensen (1968b), who observed members of *Eriocrania* (Eriocraniidae) drinking from water droplets in the laboratory. He pointed out that since Eriocraniidae fly during the warmest parts of the day, water loss through the spiracles may well be significant for these moths. The variety of Lepidoptera that visit and probe at damp or wet substrates is considerable (Norris, 1936; Downes, 1973; Adler, 1982). The habit is well known among large Lepidoptera, but five Gracillariidae and a species of *Bucculatrix* (Bucculatricidae) were also observed puddling (Adler, 1982).

Longevity and fecundity were roughly halved in *Ephestia caudatella* if the insects were deprived of water, although in *E. kuniella* these factors were only slightly affected (Norris, 1934). Although there is little doubt that adult Lepidoptera often need water, the extent to which this is a general requirement is unknown. It is not reasonable to assume that individuals seen drinking in nature from wet or moist substrates are primarily deriving water even though very considerable amounts are sometimes imbibed (e.g., Downes, 1973; Adler, 1982). As Norris (1936) pointed out, 'water drinking in the Lepidoptera is inextricably confused with that of their attraction to the dung and perspiration of animals and human beings. There is some reason to believe that practically all water-drinking is primarily due to such attraction.' Norris was led to this view because at least one water-drinking assemblage of a variety of Lepidoptera was feeding at soil from which human perspiration was traced. She also noted that butterflies in the alps usually drink at roads or paths even when the hillsides are running with water, and that individuals frequently feed at dung. Numerous observations have been made of Lepidoptera discharging water from the end of the abdomen as they drink. *Gluphisia septentrionis* (Notodontidae) even squirts fluid in rhythmic jets over a distance of 30 cm (Adler, 1982). Evidently, something as well as, or other than, water is sought by the insects in such situations.

Decaying organic matter

There are numerous examples of visits to wet or damp substrates that have been contaminated with animal excreta or remains (Norris, 1936; Downes, 1973; Adler, 1982). Amongst some of the sources recorded by Norris are perspiration-soaked garments, water that had run over bat guano emerging from a cave, and a watering place of buffaloes. In a study of the habits of diurnal (mainly butterflies, but including two species of pyralid moths) and nocturnal adult Lepidoptera feeding at margins of puddles at a site in Ontario, Canada it was suggested (Downes, 1973) that the primary function of drinking was not simply to obtain water since free water, in the form of dew, was available widely while the insects were probing at the margins of the puddles. Instead, as the puddles in question were supplied, from time to time, by run off from a nearby path, it was considered that contaminants, such as organic debris or solutes in the run-off might be that to which the lepidopterans were actually attracted. When dung and carrion was added to the puddles, both the number of individuals visiting and the rate at which they probed increased greatly. Individuals feeding at the dung and particularly at the carrion were far less easy to disturb, an observation emphasizing the intentness of their feeding. Further support for the view that the insects were primarily deriving substances other than water was gained by offering some butterflies dry organic material (frog skin and muscle). Probing was observed, but in this situation butterflies exuded a drop of saliva from the end of the proboscis to moisten the dry substrate before it was imbibed. This cycle was repeated many times.

Attraction to the puddles seemed usually to be 'accidental', a response, perhaps, to certain odours. However, bird droppings appeared to provide a visual stimulus to some butterflies. Sulphur butterflies landed close to other individuals, and it is probably a visual stimulus that explains the familiar aggregations of butterflies feeding at puddles and damp sand in warmer parts of the world.

Observations by Downes (1973) were extended to night time. As with diurnal Lepidoptera, the nocturnal visitors were less easily disturbed when probing organic matter. Most records of moths definitely probing were of Geometridae and Pyralidae, but a few Noctuidae, Pterophoridae, and Tortricidae were also noted. About 96 percent of the visitors, whether diurnal or nocturnal, were males. Records were made only when probing was clearly observed.

From this study, it was suggested that the purpose of puddling might be to obtain salts, or some trace substance, needed (particularly by males) for flight (Downes, 1973). Proteins or amino acids were considered as unlikely both because of the absence of reports of proteases in the lepidopteran gut, and because it is not apparent that males require nitrogen more than females.

A study by Arms *et al.* (1974) provided direct information on the precise substance that is being sought, although since the work was confined to the tiger swallowtail butterfly (*Papilio glaucus*) in Ithaca, New York, it remains to be seen just how far the results of this study apply to other Lepidoptera. Controlled experiments were carried out in which wild butterflies were offered trays of sand soaked with solutions of various substrates. Extended puddling visits were restricted almost entirely to those trays containing sodium ions. Anions, for example chloride, phosphate (PO_4), nitrate (NO_3), did not seem to induce puddling behaviour. Furthermore, the higher the concentration of sodium ions the greater was the attraction of butterflies to the trays. The authors concluded that the threshold concentration required to stimulate puddling was below $10^{-3}M$.

There was no support for the view that amino acids stimulated puddling as it was not induced by presenting sand soaked with amino acid solutions. Nevertheless, the use of radioactive samples of amino acids demonstrated that amino acids taken up incidentally during puddling may be incorporated into body proteins. The question of whether protein synthesis was involved in this incorporation was not determined. Underlining the importance of sodium to herbivorous insects, Arms *et al.* suggested that the relatively low concentration of sodium in land plants may act as a defence against grazers by denying them an adequate supply of the ion. Most herbivorous insects require sodium ions in the sodium-dependent pump in their excretory system. Even those with a potassium-pump require sodium for their neuromuscular system.

This study supports the earlier view, based on simple observations, that it is salt that stimulates puddling behaviour and that salt is the primary substance obtained by puddling. It also demonstrated that the key substance required by the butterflies is the sodium ion.

Overripe fruit is another frequently encountered source of decaying organic matter. This substrate is utilised by several Lepidoptera, particularly Noctuidae in temperate countries (e.g., Norris, 1936). Nymphalid butterflies of the genus *Charaxes* are also well-known to feed on ripe figs and grapes, and a favoured method of collecting these fast flying insects is by the use of banana bait in special traps hung in trees.

Many, perhaps all, adult Danainae and Ithomiinae (Nymphalidae) obtain pheromone precursors from dead, dry plants belonging to the families Boraginaceae and Asteraceae (p. 168). The compounds, pyrrolizidine alkaloids or PAs, are secondary plant substances used in plant defence. Their uptake by certain Lepidoptera may have originally evolved to confer chemical protection to the insects. The use of PAs as pheromone precursors has presumably occurred secondarily (p. 168). The habit of seeking and taking up secondary plant substances for uses other than primary metabolism, or merely for foodplant recognition, has been termed 'pharmacophagy' (Boppré, 1984a).

Exuding sap of plants, and honey dew

Although they frequently feed at sugary fluids, Noctuidae have rarely been recorded at sap (Norris, 1936), although *Catocala* may be an exception. Many nymphalid butterflies, including *Charaxes*, have been observed feeding at wounds of trees. Sometimes (see Norris) the butterflies become intoxicated and incapable of flight.

Related to the uptake of sap is the habit of feeding on honeydew, a substance that has attracted, on occasion, enormous numbers of Noctuidae. Thorpe (1928) found *Yponomeuta cagnagella* (Yponomeutidae) feeding on buds infested with aphids. Honeydew was exuded by the aphids when they were touched by the proboscis of the moth, and the substance was immediately imbibed, but Thorpe considered that this prodding was accidental. Lycaenid butterflies in particular feed on honeydew, and at least two species stroke aphids with their proboscis.

Honey

The Death's Head Hawkmoth (*Acherontia atropos*) enters bee hives piercing the operculum over honey cells to obtain honey. *Laothoe populi* and *Sphinx ligustri* are also said to 'steal' honey (Norris, 1936) but, at least in the case of *populi*, this action seems impossible given its reduced, non-functional proboscis.

Blood

Blood-sucking (p. 16) was considered (Bänziger, 1980) to have developed from fruit-feeding. The mouthparts of moths that pierce the skin of fruit are similar to those in which animal skin is pierced. Both kinds differ from those of moths that merely imbibe fluid without piercing, whether the fluid is nectar, lachrymal secretion, or the juice exuded from damaged fruit. Furthermore, both fruit- and animal skin-piercers exhibit similar feeding mechanisms. Just as fruit-piercers occasionally withdraw and insert their proboscis while feeding to induce the flow of sap, this habit is retained in the animal tissue piercers even when blood has been exuded from the wound. Both blood and fruit supply a moth with protein and sugar; however, whereas fruit juice may be seasonally limited, blood is potentially available at all times (Bänziger, 1980). Feeding on blood may increase longevity of moths (Johannsmeier, 1976).

From Bänziger's hypothesis one may conclude that two independently derived trends have developed from the generalized nectar-feeding habit. In one, the (polyphyletic) trend has been towards lachryphagy via feeding on animal excreta. In the other, blood-sucking is derived from fruit-piercing.

In contrast to Bänziger's interpretation, Downes (1973) considered that fruit-piercing and blood-sucking developed independently because, he argued, while fruit-piercing provides Lepidoptera with sugar, eye-frequenting and blood-sucking offers nutrients of a different kind. Downes considered that the sugar-feeding habit is widespread and primitive in Lepidoptera - fruit-feeding is merely a less frequently encountered and more specialized way of obtaining sugar than nectar-feeding. Blood-sucking was regarded by Downes as the most specialized habit developed from what he called collectively the 'animal excreta' feeding habit, under which is included mud-, dung- and carrion-feeding, and eye-frequenting. Perhaps the most effective evidence to support the view that blood-feeding is derived from fruit-piercing is taxonomic. The blood-sucking moths (*Calyptra eustrigata* and some others, Bänziger, 1979, 1989b) belong to the *Othreis - Calyptra* group, most members of which are probably fruit-piercers.

Eye-exudates

Some adult Lepidoptera imbibe tears from the eyes of horses, cattle, and certain other mammals (including humans) (e.g., Shannon, 1928; Bänziger, 1973, 1983, 1988a,b,c, 1989a). Families with such lachryphagous representatives include Pyralidae, Geometridae, Thyrididae, Notodontidae, Noctuidae, and Sphingidae, but the species involved appear to be confined to tropical America, Africa, and Asia.

The only moth known to feed exclusively on eye-fluids is *Lobocraspis grisefusa* (Noctuidae) (Fig. 24); other zoophilous moths feed partly, or largely, on other exudates such as blood, perspiration, sebum, and urine. Tears contain not only water and salts, but also protein, and *L. grisefusa* is possibly unique among Lepidoptera in secreting a proteinase in its gut (Bänziger, 1973). The proboscis of *grisefusa* resembles that of those moths that suck juices from damaged fruit. It is flexible, with a soft, blunt tip and non-erectile sensilla.

Several lachryphagous moths also drink perspiration from the hides of animals, as well as droplets of blood egested onto a host's skin by blood-sucking mosquitoes. Thus different substances appear to be sought by different species of moths. Moreover, members of a particular species may not seek the same substances throughout the year; and requirements may depend not only on season, but also on habitat. For example, whereas lachryphagy may predominate over imbibition of skin exudates in dry areas, where the need for water is greater, the opposite situation may prevail in humid regions (Bänziger, 1983).

Lachryphagy in Lepidoptera seems to be confined to nocturnal groups, possibly because a night-time approach occurs when the host animals are drowsy, and therefore less alert (Bänziger, 1988b).

Although the proboscis is relatively soft and flexible compared with that of most Lepidoptera, eye-frequenting moths evidently cause some discomfort to host animals. But, apart from discomfort, it has been suggested that these insects may act as vectors for forms of ophthalmia in cattle (Büttiker & Whellan, 1966).

Flowers

Many Lepidoptera visit flowers (see also Chapter 7). Observations on flower visiting have been made mainly on diurnal species, particularly butterflies, both because most flowers are day-openers and because of the relative ease with which observations may be made by day. Nevertheless, night-opening flowers are also visited by some moths, a habit exploited by collectors 'sugaring' or 'treacling' the bark of trees with a viscous mixture of treacle and beer. Flowers provide Lepidoptera with both nectar and pollen, although it is the former that is chiefly sought.

Nectar. Nectar is essentially a solution of various sugars in water although other substances such as amino acids, proteins, certain enzymes, and secondary plant substances such as pyrrolizidine alkaloids also occur. The most important nutrients to Lepidoptera seem to be sugars of a low molecular weight (mono- and disaccharides rather than oligosaccharides) and, to a lesser extent, amino acids. Nectar-feeding animals fall into three categories - low-energy demanders, high-energy demanders, and those that feed at flowers with nectar that satisfies the needs of both (Watt *et al.*, 1974). Most Lepidoptera fall into the first category, but hawkmoths (Sphingidae) exemplify the second.

The variable nature of sugar requirements of different species of Lepidoptera were demonstrated experimentally by Norris (1934, 1936). She found (Norris, 1934) that in both *Ephestia cautella* and *E. kueniella*, longevity was considerably increased if individuals were fed on a solution of cane sugar, although fecundity was no greater than if the moths drank only distilled water. But sugar was found to increase the fecundity in the pierid butterfly *Pieris rapae* (Norris, 1936).

Sugar-uptake in butterflies depends on several factors. Significant among these are the attraction of the insects to the flowers, the concentration of the nectar, and the molecular weight of the sugars in the nectar. *Colias* butterflies (Pieridae) were found to visit flowers with dilute nectar in which a high proportion of the constituent sugars were represented by monosaccharides (Watt *et al.*, 1974). This finding was considered to be consistent with the relatively high water requirements of small bodied insects, in which a high surface to volume ratio leads to relatively high evaporation. Uptake of dilute nectar may allow the butterfly to visit flowers continuously without stopping to drink at another source. One objection to this explanation is that hawkmoths, which are high-energy demanders, take nectars that are no less dilute, an observation suggesting that perhaps nectar viscosity poses a mechanical problem for the proboscis (Watt *et al.*, 1974). However, butterflies fed on a sucrose solution often die, apparently because the sucrose crystallizes in the proboscis and therefore blocks it. When glucose is used the problem does not occur (I.J. Kitching, pers. comm.), so the problem, although mechanical, may not simply be a matter of viscosity.

Butterflies generally feed on nectars with a sugar concentration of 15-25 percent (Watt *et al.*, 1974). According to a model by Kingsolver & Daniel (1979), the optimum for net rate of energy gain is found in nectar with a sugar concentration of 20-25 percent, so butterflies actually do seem to feed on nectars that optimize their energy gain. Indeed, butterfly-pollinated flowers produce nectar of this very concentration. The optimum for net rate of *energy gain* was found to be independent of both metabolic rate and the size and shape of the proboscis over the ranges found in butterflies, even though (according to Kingsolver & Daniel) the optimal rate of *extraction* of the nectar does depend on size and shape of the proboscis. Heyneman (1983) stressed that the maximum sugar flux for sucrose solutions of a concentration between 20 and 25 percent is 'relatively insensitive to several parameters in [Kingsolver & Daniel's] analysis'. These parameters include geometry and size of the proboscis to an extent beyond the range found in butterflies, the mechanism inducing nectar flow, and the energy cost of the process of ingesting sugar solutions.

Both the optimum rate of flow of fluids in sucking tubes and the concentration of nutrients are governed by the relationship of the viscosity of the fluid to the nutrient concentration. The length of time of flower visiting by *Colias* seems to lie within the estimated time needed for extraction of the nectar. The compressive strength of the chitin forming the proboscis is not apparently a mechanical limitation.

According to a model of sugar intake, or energy flux, of nectars by nectar-feeding pollinators (Heyneman, 1983), the efficiency by which sugar is taken up (energy flux) is maximal (maximum flux concentration) at particular sugar concentrations. Since viscosity rises rapidly with concentration, but as high viscosity sharply reduces intake by nectar-feeders, a high concentration of sugars may lower energy intake rate. For flower nectar the maximum flux concentration was predicted to be about 26 percent, and for pure sucrose solutions the predicted figure was 22 percent. More concentrated nectars were predicted to be exploited by nectar feeders with a high energy cost in travelling to nectar sources, but low costs for feeding. Relatively dilute nectars (i.e., those close to the maximum flux concentration of 26 percent) are exploited by nectar feeders that minimize feeding times (to avoid predation or reduce hovering energy costs).

Butterflies and moths appear to pollinate flowers with nectar concentrations near the maximum flux concentration (Heyneman, 1983). The exploitation of these relatively dilute nectars may be explained in terms of minimizing feeding time to reduce the chance of predation. Nectars of so-called 'butterfly flowers' (see Chapter 7) tend to be relatively rich in non-sugar constituents such as amino acids. Such 'impurities' increase viscosity, and therefore decrease nectar flux. Measurement of sucrose concentration alone is therefore a potentially inaccurate source of information for comparison of energy flux of nectars. The nectars of flowers pollinated by hawkmoths (and also bats and hummingbirds) tend to have relatively low levels of non sugar constituents, so these insects, which have feeding habits with a high energy cost, visit flowers with nectars that are 'dilute' but 'pure'. They therefore accept a nectar somewhat lower in energy potential but one that requires the expenditure of less energy (i.e., a shorter time) for its imbibition.

However, the view that Lepidoptera feed on relatively dilute nectars was challenged by the discovery that a European skipper (*Thymelicus lineola*: Hesperiidae) feeds on floral nectars ranging from 40-65 percent sucrose, a figure considerably above previous estimates (Pivnick & McNeil, 1985). In fact, under laboratory conditions, butterflies were tested with various sucrose solutions and it was found that volume intake did not decrease with increased concentration nearly as rapidly as predicted by previous models. At higher temperatures intake rates were higher, a phenomenon explained by a lower viscosity of the sucrose solution and an increase in power output of the cibarial pump of the insect. A maximum rate of sucrose intake was found to occur at 40 percent sucrose. Thus a new model was proposed (Pivnick & McNeil, 1985) differing from previous ones largely in that it took account of the point that although the power output of the cibarial pump remains relatively constant, the pressure drop created by the pump, which induces suction, varies greatly. This change in the model results in an optimal rate of sucrose intake at a concentration of about 40 percent. The model applies to a wide range of shapes and sizes of proboscis and types of suction pump. Insects, including Lepidoptera, presumably prefer nectar with sucrose concentrations corresponding to optimal rate of sucrose uptake, particularly where feeding costs per unit time are high or predation is a serious problem. But if energy costs are high in transport then even higher concentrations would be expected to be preferred, and more dilute nectars selected at times when insects are water-stressed.

Although nectar is primarily a source of sugar for Lepidoptera, amino acids are also available. Preliminary work (Watt et al., 1974) suggested that *Colias* uses nectar as an 'easily soluble source of general nitrogen' rather than a source of particular amino acids. The nectars of plants visited by the butterflies were either those with amino acids low in carbon relative to nitrogen or those having amino acids with side chains enriched with nitrogen.

Pollen. Although the proboscis obviously functions for uptake of fluids, Gilbert (1972) discovered that butterflies of the predominantly neotropical genus *Heliconius* feed on pollen. Both sexes gather pollen, but the loads carried by females are consistently greater. The exploitation of pollen by these butterflies has a marked impact on their general biology. Gilbert demonstrated effectively that pollen was actively sought and not simply picked up by chance during nectar-feeding. For example, butterflies were seen to scrape the tip of the proboscis over floral anthers. Fourteen species of *Heliconius* were observed to carry pollen loads on the tongue. Since the proboscis can only suck liquid foods, dry pollen loads are processed so that they can be taken up by the butterfly. A liquid (probably nectar) is exuded from the tip of the proboscis and mixed with the pollen load, which is carried ventrally between the base of the proboscis and the head. The pollen mass is agitated for several hours as the butterfly coils and uncoils its proboscis.

Proteins and free amino acids are known to be released from pollen when it is incubated in a sucrose solution. This process probably occurs with these butterflies since wet pollen loads taken from living specimens were found to have lost their protein and amino acids (except for proteinases), the solution presumably having been imbibed by the insect. Gilbert demonstrated that amino acids are directly incorporated into developing eggs. He also found that pollen-fed butterflies produced, on average, five times as many eggs per day during pollen-feeding, and five times as many eggs for five days after pollen-feeding had ceased. *Heliconius* butterflies are long-lived, and produce eggs over a long period at a low rate. Such a strategy disperses offspring in time, which improves their chance of survival against pre-imaginal parasitoids and predators. It was also pointed out that the uptake of proteins and amino acids by adults reduces the foraging time required for larvae to obtain the necessary protein for egg production. Again, this reduces the chances of parasitization and predation. The corresponding increase in foraging time required by the adults is appropriate to their aposematic defence strategy (*Heliconius* butterflies are distasteful) in which exposure is obviously adaptive. Protein uptake has the further adaptive function in *Heliconius* of increasing longevity.

Gilbert noted an analogous situation of protein and amino acid uptake in the Malayan blood sucking moth *Calyptra eustrigata* (p. 17) (and in mosquitoes), and also the potential gain of these substances by Lepidoptera that feed on other nitrogen-bearing substances such as dung and even nectar. However, it should be noted that only males appear to suck blood (Bänziger, 1975), and male Lepidoptera predominate in feeding at other nitrogenous substrates.

Other modifications

In the Yucca Moths (Davis, 1967) (*Tegeticula*, Prodoxidae: Incurvarioidea) each maxilla of the female is strongly modified to form a maxillary tentacle (Fig. 21). The morphological derivation of the tentacles is uncertain, but the structures may represent the extended palpifers (lateral components of the stipes). The tentacles are used by the female to gather pollen from flowers of the South American *Yucca*, and to force a ball of pollen into the pistil of a flower in the ovary of which it has laid an egg. The resulting fertilization of the plant ensures development of the seeds on which the larvae of *Tegeticula* feed. This association is of importance to both plant and moth; the plant relies on the moth for pollination and the moth relies on the seeds of the plant as the food-source for its larva.

Sensation

Compound eyes

The compound eyes are prominent, paired structures situated on the side and front of the head (Figs 1-3, 11), which are capable of perceiving movement and form. Externally, each is covered by a chitinous layer divided into hexagonal facets. Each facet subtends, and forms the outer part of, an ommatidium (Figs 25, 26). Each ommatidium is composed of a dioptic component distally (i.e., towards the exterior), and a receptive component, the 'retinula', proximally.

External appearance

Lepidopteran eyes are approximately round, oval, or somewhat pear-shaped. In Lycaenidae they are emarginate, this emargination corresponding with the ingress of the base of the antenna.

The size of the compound eye varies both in absolute terms and in relation to the proportion of the head it occupies. Several indexes have been proposed for the comparison of relative size differences between taxa. That of Powell (1973) is the ratio of the vertical diameter of the eye and the distance from the midpoint between the antennal base to the bottom of the frons-clypeus. The inter-ocular index of Davis (1975a) is the ratio between the vertical diameter of the compound eye and the inter-ocular distance measured at a point on the frons midway between the base of the antennal sockets and the anterior tentorial pits. Kristensen & Nielsen (1979) found their supraocular index useful to describe the proportion of the eye to the head in lower Lepidoptera. This index is the ratio of the height of the head-capsule above the compound eye to the total height of the head-capsule from the top of the epicranium to the tip of the subgenal process. Yagi & Koyama (1963) measured actual surface area of the compound eye of 84 species of Lepidoptera (20 representatives of each) distributed among 25 families. They found that, with few exceptions, eyes of male Lepidoptera are larger than females of the same species. A further widespread observation is that eyes in nocturnal Lepidoptera tend to be large while eyes in diurnal Lepidoptera tend to be relatively small.

The hexagonal facets of the cornea vary in size and number. The largest facet found by Yagi & Koyama (1963) was that of the satyrine butterfly *Melanitis* (Nymphalidae), and the smallest was found in *Stigmella* (Nepticulidae), a genus of minute moths the larvae of which are leaf-miners. In butterflies, facet size is generally relatively larger in terms of the surface of the eye than it is in moths.

On the whole, the larger the eye, the greater is the number of facets (Yagi & Koyama, 1963). The hawkmoth *Herse convolvuli* was recorded as having 27000 facets in an eye, whereas in a species of *Stigmella* there were only 200. Facet size tends to be reduced gradually towards the periphery of the eye, a trend more marked in large than in small eyes.

Interfacetal hairs are present in many Lepidoptera, particularly Papilionoidea. Each hair is the product of a trichogen cell and arises from the corners of the facets. The density and length of the hairs vary considerably both between the eyes of individuals and between species. The function of the hairs is unknown, but Yagi & Koyama (1963) noted a correlation between hair-density and habitat in butterflies. Density is generally lowest in butterflies inhabiting open fields, higher in those living at the edge of woodland, and highest in woodland or forest species. Interfacetal hairs are also prominent in Noctuidae and Lasiocampidae, and in many Acrolophidae (Hasbrouck, 1964). Yagi & Koyama found no hairs on the eyes of Saturniidae, Lymantriidae, or Bombycidae, nor on those of microlepidopterans, but Davis (1978) noted a few microtrichia irregularly scattered over the surface of the eye of certain Eriocranioidea.

Structure

The compound eye of Lepidoptera (Yagi & Koyama, 1963), like that of other insects, is composed of numerous ommatidia. Each ommatidium (Figs 25, 26) is indicated on the surface of the eye by a cuticular facet, below which lies the light sensitive component. The dioptic (distal) part of each ommatidium is composed of a cornea and a crystalline cone. The receptive (proximal) component is known as the 'retinula' and includes sensory cells, which continue as nerve fibres. The ommatidium is surrounded by pigment cells.

In the dioptic part of each ommatidium, the cornea is a transparent, hexagonal, cuticular structure composed of epicuticle, exocuticle, and endocuticle. Being curved, it acts as a lens. The degree of curvature and the thickness of the corneal lens varies. In general, the lenses of diurnal Lepidoptera are thicker than those of nocturnal species, although they are exceptionally thin in Papilionoidea. The corneal lens is a product of the iris pigment cells. Immediately below the lens a transparent corneal process, thought to be secreted by the crystalline cone cell, is usually found. The size of the process is generally related to the size of the cone. If the cone is long the process is absent or very small, but if it is of medium size the process is moderately developed. In butterflies, which have short cones, the process is well-developed, but it is absent in Hesperiidae. The corneal processes jointly form the corneagen layer.

Below the corneagen layer lie four crystalline cone cells, which secrete the corneal processes distally and the crystalline cone proximally. When the corneagen layer is absent the cells of the crystalline cone secrete the cornea. The cone, like the cornea, is

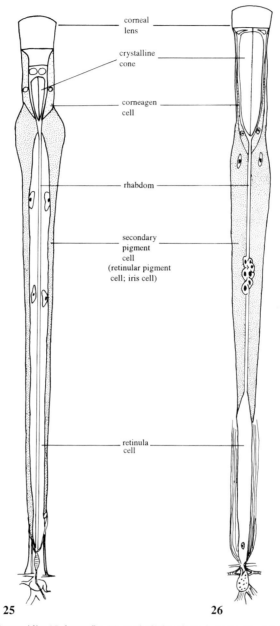

corneal
lens

crystalline
cone

corneagen
cell

rhabdom

secondary
pigment
cell
(retinular pigment
cell; iris cell)

retinula
cell

25   26

Figs 25, 26. *Ommatidia: 25, butterfly; 26, moth (light adapted). (After Yagi & Koyama, 1963.)*

laminated. Lepidoptera have eucone eyes, that is, a true crystalline cone is secreted by the cone cells. (In insects with exocone eyes there is a cone of extracellular, cuticular origin.) The cone has generally been considered to be merely an optical spacer, functioning to provide a gap sufficient to accommodate the focal length of the lens system of each ommatidium. However, studies by Nilsson *et al.* (1984) on the eye of the Australian butterfly *Heteronympha merope* (Nymphalidae) show that the cone is actually an extremely powerful lens (see below). The length of the crystalline cone is related to the thickness of the corneal lens; if the cornea is thick the cone is short and vice versa. Shorter cones are usually found in diurnal species whilst in nocturnal species the cones are usually long.

Two corneagen cells, which are strongly pigmented, surround the crystalline cone. In some species the pigment of the corneagen cells migrates according to light intensity (see below).

The receptive component of the ommatidium (Figs 25, 26) comprises a number of retinula sense cells situated below the crystalline cone and the corneagen cells. In Lepidoptera, there are usually seven or eight of these cells but there may be up to 15. They are continuous proximally with a nerve fibre. Internally they secrete an optical rod, the rhabdom, which runs through the centre of the retinula. That part of the rhabdom secreted by each retinula sense cell is called the rhabdomere. Two types of retinula sense cells are found in Lepidoptera. In papilionoid butterflies and most diurnal moths the cells are not constricted along their length (Fig. 25), but in most crepuscular and nocturnal moths the distal part of each cell is constricted so that the cell is broader proximally. The relative lengths of the constricted part and the thickness of the unconstricted section vary within Lepidoptera. These two types of retinula sense cell are found, respectively, in eyes forming an apposition image and eyes forming a superposition image (see below).

Surrounding the retinula sense cells and the corneagen cells, are six heavily pigmented secondary pigment cells (known also as retinula pigment cells), which isolate each ommatidium from its neighbour. The cells run from the cornea to the basement membrane on which each ommatidium rests. Yagi & Koyama (1963) noted three main types of secondary pigment cells (their iris cells), which they described as shortened, separated, and lengthened. In the shortened type (characteristic of nocturnal moths) the length of the secondary iris cell is the same as the length of the constricted part of a retinula sense cell; the proximal part is drawn out into a fine thread-like component. In the separated type (characteristic of Hesperiidae and occurring in some Sphingidae) each secondary pigment cell is effectively divided into a distal part, which surrounds the crystalline cone, and a proximal section surrounding the thickened part of the retinula sense cell. The two parts are joined by a very fine, drawn out component. The lengthened type is found in papilionoid butterflies, each cell running the whole length of the ommatidium.

Function

Functionally, insect eyes have been divided traditionally into those producing an apposition image and those producing a superposition image, which has led to them being termed apposition or superposition eyes. In the typical apposition eye an inverted image is formed in each ommatidium, the corneal lens focussing rays onto the distal tip of the rhabdom. The distal tip of the rhabdom therefore lies at the focal length of the lens. Each ommatidium is screened from its neighbours by pigment in the primary and secondary iris cells, so that each acts as an isolated optical unit. In contrast, a superposition eye is known to produce an image derived from light entering several ommatidia 'super imposing' on each other to form an image deep within the retina. The image is erect and afocal, the rays converging within the crystalline cone. The image is formed from rays from several ommatidia since the ommatidia are not screened from each other. Instead, the pigment is concentrated distally in the region of the pigment cells surrounding the crystalline cone.

In many nocturnal moths the superposition eye may become light-adapted by dispersion of pigment into the proximal part of the pigment cells thereby screening each ommatidium (Fig. 26). Thus the condition in each ommatidium in light-adapted superposition eyes of nocturnal moths resembles that in apposition eyes.

Apposition eyes are the most widespread kind of compound eyes in insects. The pigment in the secondary pigment cells does not migrate to any significant extent so optical isolation of each ommatidium is permanent. It follows that there is little light-dark adaptation in apposition eyes, and only the light immediately above a facet, not that from its neighbours, reaches the rhabdom. In the apposition eye each ommatidium is isolated and its rhabdomes are typically long. Although apposition eyes occur in Lepidoptera in Papilionoidea, they are afocal differing significantly in this and other ways from simple apposition eyes (see below).

The superposition eye, optically more flexible than the *simple* form of the apposition eye, is found typically in nocturnal Lepidoptera which, as described above, may be light or dark adapted. In some superposition eyes, pigment is fixed in the concentrated proximal position so superposition images are always formed. Thus superposition images are formed when a so-called 'clear-zone' is found in the primary and secondary cells surrounding the retinula sense cells. This situation pertains whether this clear-zone is fixed, or whether it occurs when pigment migrates in intense light. In Bombycoidea the clear-zone condition is achieved by longitudinal movement of the pigment alone, whereas in Pyraloidea, Gelechioidea, Tineoidea, and most Noctuoidea movement of retinula cell bodies and iris pigment is responsible. Virtually no pigment movement occurs in Hesperiidae, although the members of this family have superposition, clear-zone, eyes which are permanently dark adapted (Walcott, 1975).

In superposition eyes, both the cornea and the crystalline cone act as lenses. However, whereas the cornea acts as a normal lens by bending light by spherical refraction, the crystalline cone acts as a 'lens cylinder'. A lens cylinder bends light by continuous refraction along its length - the situation occurring in the crystalline cone, which is not homogeneous in its composition but has a graded refractive index. The principle of the lens cylinder was proposed by Exner (1891) to explain the method by which light rays pass through the eyes of moths and of *Lampyris* (the glow worm) (Land, 1985). It has been confirmed only recently.

Effectively, the superposition eye, with its two lenses, acts as if it were a telescope (Land, 1985). The first 'lens' is represented by the distal part of the lens cylinder, which focuses light halfway along the cylinder as an inverted image. The second 'lens' is equivalent to the proximal half of the lens cylinder. In fact the light is refracted partly by the surface of the cornea and partly by the refractive gradient of the crystalline cone. Thus, although 'the optical elements in superposition eyes are not simple lenses [they] behave as though they were two-lens, unity magnification telescopes.', (Land, 1985: 272).

Recently, the traditional view of the apposition eye of butterflies was challenged by Nilsson *et al.* (1984) who found that the crystalline cone, far from being an optical spacer as traditionally thought, actually acts as an extremely powerful lens. The focusing properties of the cones probably result from a concentric protein gradient. Studies were carried out on the Australian nymphalid butterfly *Heteronympha meropa*, but the findings seem to apply throughout Papilionoidea (but not Hesperioidea, see above) (Nilsson, 1989). Nilsson *et al.* (1984) found that sections of crystalline cone act as converging lenses, forming inverted images. By measuring the magnification of sections cut from along the cone they found that most of the power lay at its proximal tip, that is the part just distal to the rhabdom. The optical properties of the ommatidium can be simulated as a series of thin lenses. Nilsson *et al.* deduced that rays are focused by the cornea and the middle region of the cone. As the rays emerge from the tip of the cone they are collimated (made parallel) and then enter the rhabdom. The last 8 μm of the cone has a power of 0.2 megadioptres, apparently representing the most powerful lens known to man. The optical properties of the ommatidium can be modelled by a spherical surface (equivalent to the cornea), and two thin lenses (equivalent to the effect of the cone). This model represents the optical form of a telescope, a situation similar to that occurring in superposition eyes.

The afocal apposition eye is of a far superior design to the simple focal apposition system in terms of light capture and image resolution. But the advantages can only accrue in diurnal organisms (Nilsson, 1989). So the classical view of the butterfly apposition eye as a simple focal system is invalid. Although anatomically the eye cannot be distinguished from the form of the simple apposition eye, optically it resembles the afocal superposition eye typical of nocturnal moths. However, in contrast to the dark-adapted state of moth eyes the image is apposition rather than superposition. But whereas the butterfly apposition eye is probably merely a variant of the superposition eye of moths, the *focal* apposition eye of bees is fundamentally different (Nilsson *et al.*, 1984).

Ocelli

Ocelli are simple eyes with a single lens (Fig. 27). Although structures of a roughly similar structure are found in larvae, their function and developmental pathway differ (Land, 1985). Since the structures do not appear to be homologous in adults and larvae they have been given different names. Simple eyes in larvae are termed *stemmata*.

In adult insects there are usually three ocelli, which are positioned on top of the head. One ocellus occurs above each eye (lateral ocelli), the third is situated anteriomedially. In Lepidoptera, the anteriomedial ocellus is absent. This loss represents a groundplan specialization of the order (Kristensen, 1984b) for a medial ocellus is present in Trichoptera, the sister group of the Lepidoptera. Lateral ocelli are present in the groundplan, but are frequently said to be lost within the order. In fact, external ocelli have been overlooked on many occasions because of their small size in many groups, their colour and, sometimes, their atypical position (Dickens & Eaton, 1973). In

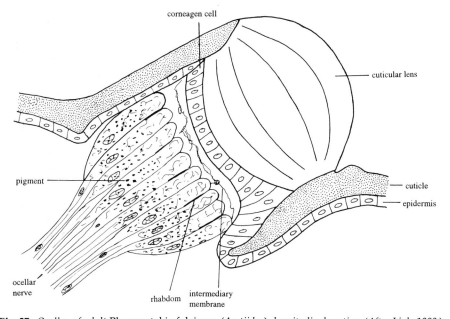

Fig. 27. *Ocellus of adult* Phragmatobia fulginosa (*Arctiidae*), *longitudinal section. (After Link, 1909.)*

those Lepidoptera studied by Dickens & Eaton and previously said to be anocellate, small external ocelli were in fact present. Small corneal lenses were recognized, and in Sphingidae nerve branches (arising from 'internal ocelli') were traced to the external ocellus. Moreover, what appear to be retinular cells underlie the external ocellus. Those Lepidoptera studied included representatives of Sphingidae, Saturniidae, Arctiidae, and all families of butterflies (including Hesperiidae). Therefore, ocelli are possibly never strictly absent in Lepidoptera and should more accurately be regarded as reduced until demonstrated to the contrary. The phrases 'external ocelli present' or 'external ocelli absent' are used, as appropriate, in Part 3 of this book.

In a typical ocellus (Link, 1909; Ehnbom, 1948; Land, 1985) a thick, and strongly biconvex cuticular lens is present together with an extended retina (Fig. 27). There are fewer retinular pigment cells than in an ommatidium. The sensory cells vary in number (Bourgogne, 1951). Some variation occurs in ocelli in the order. For example, in lower Lepidoptera the cornea is not thickened, unlike the situation in higher Lepidoptera (except Zygaenidae and Sesiidae) (Ehnbom, 1948), and in Eriocraniidae there occurs a thick, cellular 'vitreous body', formed by the corneagen cells and situated under the cornea, which extends along the posterior wall of the capsule (Kristensen, 1968b).

Although stemmata of larvae are well-focussed, the ocelli of adult insects are apparently defocussed, and their most likely function is orientational (Land, 1985). Because ocelli are situated on top of the head, they could be used to sense any change in an insect's concept of the position of the horizon when a change occurs in the organism's orientation through tilt or pitch. A system that does not focus will avoid distractions, such as leaves, from the immediate surroundings. Some support for this view comes from the fact that there is a very well developed nervous connection from each ocellus that runs directly to the optomotor system. Such a connection would allow rapid corrections in orientation to be made. Moreover, the dorsal ocelli are positioned so that each has a field of view of 150° or more. Nevertheless, as Land points out, such a sensory function could be carried out by compound eyes, and there seems to be no function for which ocelli are uniquely fitted.

Antennae

Insect antennae are mobile structures arising between the eyes. Embryologically, they represent the appendages of the second head segment. Their function is primarily sensory, and they are correspondingly well supplied with various sensilla.

General structure and appearance

Lepidopteran antennae (e.g., Bourgogne, 1951; and, with particular reference to butterflies, Jordan, 1898), like those of other insects, comprise a basal segment, known as the scape, followed by a further segment, the pedicel, and a flagellum formed of many 'segments'. The segments are more appropriately called units or flagellomeres (Fig. 28), since there are no intrinsic flagellar muscles in Lepidoptera (or other true insects). Antennae are moved by extrinsic muscles from the tentorial arms to the scape, and by muscles arising within the scape and inserting on the pedicel.

The scape is covered with scales, and may be flattened and expanded to form an 'eye-cap' (e.g., as in Nepticuloidea and cemiostomine Lyonetiidae). A pecten, comprising a series of bristles, often arises from the anterior edge of the scape. The flagellum is usually only partially covered with scales: usually two rings are formed around each flagellar unit, or two bands may be found on the dorsal side alone. In moths, scales are usually absent from the ventral surface. Sensory areas of the antennae are not covered by scales, and frequently the ventral surface of the flagellum is naked. Sensory areas of

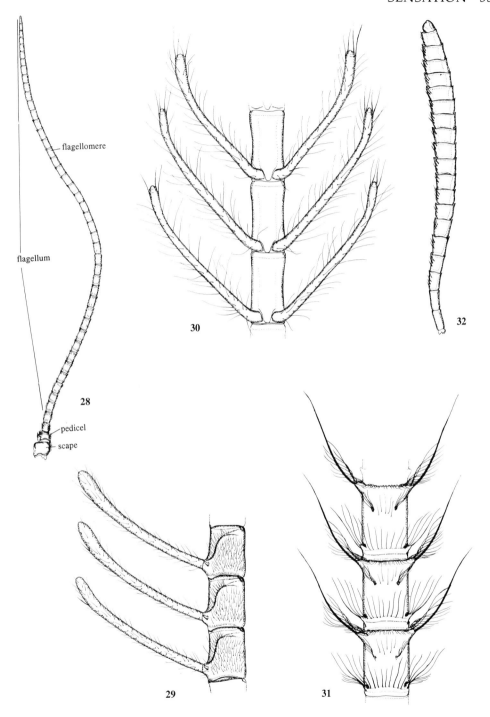

Figs 28–32. *Antennae showing main types of flagellomeres in Lepidoptera. 28, filiform (whole antenna); 29, unipectinate; 30, bipectinate; 31, fasciculate; 32, clubbed. (Drawings by Malcolm Kerley.)*

clavate antennae are usually confined to the club, which is unscaled. The flagellomeres of butterflies are frequently grooved.

Major variation in antennal morphology occurs in the number of flagellomeres, the shape of each flagellomere and the arrangement of sensilla. The main shapes are moniliform, filiform, lamellate, and prismatic. A flagellomere may be pectinate, that is extended into a ramus or two or more rami. Depending on the number of rami per flagellomere, an antenna is described as unipectinate (Fig. 29), bipectinate (Fig. 30), tri- or quadripectinate. Conspicuous hair-like sensilla often occur on the flagellomeres, besides those that need higher magnification for their study (see below). Their presence and distribution are used in the description of antennal type. Antennae may be ciliated-setose, when the sensilla are short and numerous, ciliated when they are longer and less numerous, and fasciculate (Fig. 31) when they arise in bunches. The flagellum as a whole may be dilated apically or expanded into a distinct club (Fig. 32) - sometimes extended into a short apical projection called an apiculus. Sexual dimorphism is common, with sensilla and sensilla-bearing structures best developed in the male, an occurrence fitting the conventional view that, on the whole, male Lepidoptera are attracted to females by pheromones.

An antennal character probably unique to Lepidoptera (Kristensen, 1984b) is the presence of a small intercalary sclerite situated in the membrane between the scape and the pedicel. The sclerite is usually roughly triangular, and may be reduced. It has been recorded extensively from primitive Lepidoptera (Kristensen, 1984b), but there is no review of its presence, secondary reduction or loss throughout the order. In most Hepialoidea (Exoporia) the intercalary sclerite is elongated and extends into a pocket in the scape.

Sensilla

Antennae bear many sensilla of various kinds (e.g., Figs 33-38). They may be distributed fairly evenly over the ventral unscaled areas, or restricted to particular areas (Jordan, 1898).

There have been two approaches to the classification of insect sensilla, a classical one, based on the form of the cuticular parts and, recently, one based on function. Both systems are in use today. That based on form is of limited taxonomic value, but has the advantage of being usable by workers equipped with a light microscope. However, this essentially utilitarian classification has its restrictions when attempting to constuct a natural classification since sensilla that look alike under the compound microscope

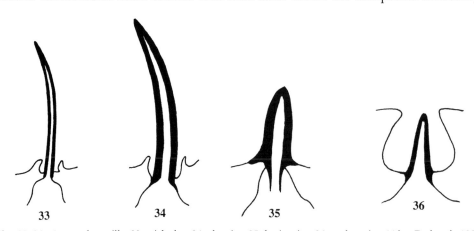

Figs 33-36. *Antennal sensilla. 33, trichodea; 34, chaetica; 35, basiconica; 36, coeloconica. (After Zacharuk, 1985.)*

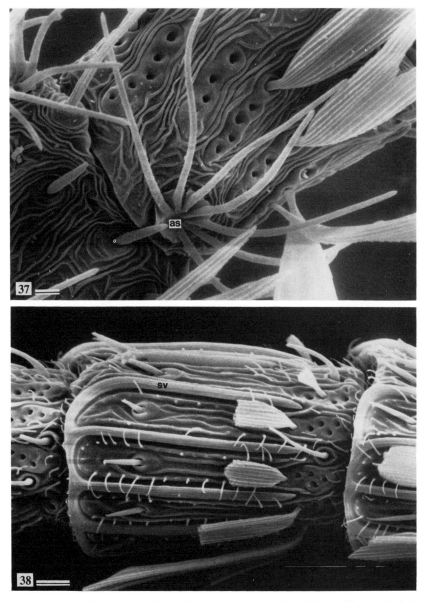

Figs 37, 38. *Antennal sensilla. 37, ascoid sensillum (as) on Opostegidae; 38, sensillum vesiculocladum (sv indicates one branch) on Nepticulidae. (Courtesy of Dr E.J. van Nieukerken; after Nieukerken & Dopp, 1987.) Scales, 5 μm.*

may not be homologous. A particular problem with the traditional classification of sensilla is that the number of categories continues to increase as more, superficially different, kinds of sensilla are discovered. That such sensilla cannot be accommodated by the existing scheme demonstrates the unpredictive quality of the classification. The problem with the more recent, functional, scheme is that relatively few sensilla have been studied to the depth required for functional interpretation; hence the system is not comprehensive. The basis of the modern classification reviewed by Zacharuk (1985) rests on the presence or absence, or number, of pores on a sensillum and the functional implications of the pores. An aporous sensillum does not respond to chemical stimuli, one that is multiporous responds strongly, and a uniporous sensillum responds less than a multiporous sensillum.

At present lepidopterists use the old, but comprehensive, system of classification for sensilla. On antennae the best known forms include sensilla trichodea, sensilla chaetica, sensilla basiconica, and sensilla coeloconica (Figs 33-36). However, these different kinds appear to respond either to mechanical or chemical stimuli, or to both.

Three other kinds of sensilla found on the antenna are notable for their conspicuous form. On the flagellum of Micropterigidae occur large multibranched sensilla similar to the 'ascoids' found in psychodid Diptera (Kristensen & Nielsen, 1979). Such structures have also been observed on the flagellum of Opostegidae (Nepticuloidea) (Fig. 37) (Nieukerken & Dop, 1987; Davis, 1989). Ascoids are essentially branched sensilla trichodea.

Sensilla auricillica (shoehorn sensilla of Callahan, 1975) are assumed to be present throughout Lepidoptera (Kristensen & Nielsen, 1979), although these chemoreceptors are absent from Micropterigidae and Agathiphagidae, and are considered to have been secondarily lost in several other families. Their surface is furrowed and they bear numerous apertures. The variation exhibited by these sensilla within the Lepidoptera casts some doubt on their homology.

A peculiar sensillum, apparently homologous with the 'ascoids' in Opostegidae, was described from the Nepticulidae by Nieukerken & Dop (1987), who termed it a sensillum vesiculocladum - meaning a small, branched blister (Fig. 38). These branched, tubular sensilla, which are raised a little from the surface of the flagellomere, are found on all flagellomeres in males except the last, but only on some units of the female. Typically, a main branch runs partly along the distal rim of a flagellomere, and sends out five branches running at right angles. There are two sensilla vesiculoclada on each nepticulid flagellomere. Grooves run along these tube-like sensilla within which are numerous pores, and many microtrichia lie alongside the arms of the sensilla. Considerable variation is found between and within nepticulid genera in the form taken by the sensilla vesiculoclada. The 5-branched arrangement appears to be the primitive condition. The multiporous nature of these sensilla, and the observation that they are found more extensively in males rather than in females, suggests that they are olfactory, and presumably represent the major pheromone receptors in Nepticulidae (Nieukerken & Dop, 1987).

Johnston's Organ

Johnston's organ is a chordotonal organ (see also Chapter 6) found between the pedicel and the first flagellomere of all hexapods with the exception of Collembola and Diplura. It is composed of a group of scolopidia (sensors that respond to stretch), and functions to perceive movement of the flagellum.

Chaetosemata

Chaetosemata (singular: chaetosema), known also as Jordan's or Eltringham's organs, are raised, cuticular patches found on the head of many Lepidoptera, from which arise bristles or narrow scales (Fig. 39). They are usually visible on dried specimens where their bristled appearance generally distinguishes them from the surrounding scaled area. The structures were first described by Jordan (1923b) who reviewed their occurrence within the order. When present, each chaetosema is usually found near the eye and behind the antenna. Jordan illustrated and discussed two main arrangements. In *Sematura* (Sematuridae) each chaetosema is a discrete, rounded swelling with numerous bristles, but without scales. In the other main arrangement, exemplified by *Micronia* (Uraniidae), each chaetosema is roughly triangular with the apex of each meeting mesally on top of the head. Both bristles and small scales arise from this type of chaetosema. According to Jordan, each kind of structure may be found within the same family and, furthermore, intergradations occur between the two. The organs are often present or absent within a single family.

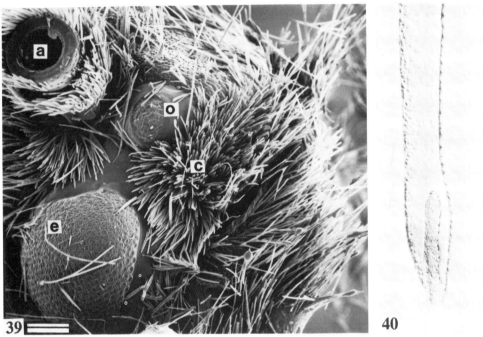

Figs 39, 40. *Sensory structures on head. 39, chaetosema (c) of* Janseola titaea *(Heterogynidae) (a, base of antenna; e, compound eye; o, ocellus; scale, 100 μm); 40, organ of vom Rath on labial palpus of* Macrosoma tipulata *(Hedylidae). (39, EM Unit; 40, Photographic Unit, BMNH.)*

Chaetosemata are found in all butterflies and in many families of moths (Jordan, 1923b). Kristensen & Nielsen (1979) recorded their presence in the primitive genus *Heterobathmia* (Heterobathmiidae), the organs being identified by the presence of piliform scales in distinct groups on the head. In this genus a chaetosema occurs above and slightly medial to each antenna, and a narrow, elongate chaetosema runs on each side from the top of the head laterally down to the level of the bottom of its opposite number. The structures have not been recorded from the most primitive Lepidoptera, but they occur in primitive Glossata such as Eriocraniidae (Kristensen, 1968b) and Neopseustidae (Kristensen, 1968a). In Hepialoidea the structures are absent, although there is a patch of hair-like bristles behind the compound eye on the primitive African genus *Afrotheora* (Nielsen & Scoble, 1986). Chaetosemata are found throughout the monotrysian Heteroneura. The presence or absence of the organs within the Ditrysia is noted, where known, under the appropriate sections in the taxonomic part of this work. Amongst the lower Ditrysia chaetosemata are often absent (e.g., from Tineoidea, Yponomeutoidea, Gelechioidea, Cossoidea), but also frequently present (e.g., most

Tortricoidea and chalcosiine Zygaenidae). In Pyralidae they may be present or absent, in Noctuoidae and Bombycoidea they are absent, while in Geometridae they are present. In Hesperiidae there are often two pairs (Jordan, 1923b), one pair on the occiput and one member of the other pair just in front of each antenna.

Chaetosemata were considered to be sensory structures by Jordan (1923b), a supposition indirectly confirmed by Eltringham (1925), who observed a nerve leading to a chaetosema of *Sematura empedocles* (Sematuridae), and by Ehnbom (1948), who examined *Zygaena* and *Ino* (Zygaenidae). This species was selected because the large size of the organ facilitated the preparation of serial sections. Histological examination of chaetosemata seems confined to Eltringham's work on *Sematura empedocles*, so that other so-called chaetosemata are assumed to be sensory largely on the basis of his observations. Whether all bristle-bearing, raised cuticular patches on the head are homologous with those investigated by Eltringham remains, to some extent, an open question. Nevertheless, few would doubt the homology of structures that strongly resemble them on even superficial examination: comparison with more diffuse patches are less convincing.

Organ of vom Rath

The distal (third) segment of the labial palpi of Lepidoptera usually bears a flask-shaped or pit-like invagination (Fig. 40). The structure was first recorded by Rath (1887) who also observed cone-like sensilla at the base of the pit and a single, large sensory cell associated with them. The organ is apparently unique to Lepidoptera (Kristensen, 1984b), an observation supporting the monophyly of the order. It is present in most primitive moths, and although found extensively in the Lepidoptera is not uniformly present. In Mnesarchaeidae, for example, there is no deep invagination although sensilla are present terminally (Kristensen, 1968a), and the same is true of Neopseustidae (Kristensen, 1968a; Davis, 1975c). Shape and depth of the invagination vary within the Lepidoptera. The large aperture of the invagination of at least one palaephatid moth (*Metaphatus ochraceus*) permits eversion of the organ so exposing the sensilla (Davis, 1986).

Two adjacent subapical apertures of the invaginations on each palpus were found in *Micropterix calthella* (Micropterigidae) (Chauvin & Faucheaux, 1981), the larger of which represents the organ of vom Rath and contains about 15 coeloconic sensilla. The second, smaller cavity includes a striated 3-pored heart-shaped organ.

The function of vom Rath's organ remains largely speculative. However, in *Amerila* (Arctiidae) the sensilla within these apical pits were shown to demonstrate a strong response to carbon dioxide by Bogner *et al.* (1986) in a study using microelectrodes. Despite exposing the organs to a variety of substances, only carbon dioxide at atmospheric concentrations elicited a marked response. Pilot studies on *Achaea* (Noctuidae) and *Pieris brassicae* (Pieridae) gave a similar response. Under continuous stimulation, a stable plateau response was exhibited (i.e., there was no physiological adaptation) so precise monitoring of environmental carbon dioxide levels would be permitted in principle.

The biological function of the response to carbon dioxide remains unknown. Sensitivity to the substance found in haematophagous arthropods and in root feeding insects might be an adaptation to host-seeking. Bogner *et al.* (1986) suggested that concentrations of carbon dioxide, resulting from conspecific insects or from foodplants, might be of significance to Lepidoptera. Living insects and fresh leaves confined in syringes for some time elicited strong responses from the palp receptors.

3

THE ADULT THORAX -
A STUDY IN FUNCTION AND EFFECT

The thorax of Lepidoptera, as in other insects, is the site of the locomotory appendages. After a short account of thoracic structure, this chapter deals with the form and function of the organs of locomotion - particularly the wings.

Williams (1966) made an important distinction, applicable to much of evolutionary process, between what he called *function* and *effect*. The difference is particularly well illustrated by the biology of lepidopteran wings. The primary function of lepidopteran wings is for flight. However, flight is by no means the only function of wings; secondarily, wings have been modified for a number of purposes such as stridulation and, particularly in Lepidoptera, for the display of diverse colour patterns for advertisement and camouflage. Such features are examples of what Williams calls *effects*.

Body

The thorax is divided into the prothorax, the mesothorax, and the metathorax. Each of these segments bear a pair of legs. The meso- and metathorax each have a pair of wings, which are sometimes reduced to mere vestiges.

The structure of the lepidopteran thorax is complex, and its comparative morphology was reviewed by Shepard (1930) and Matsuda (1970). Comments on the ground plan were made by Kristensen (1984b).

The form of the lepidopteran thorax is best understood by considering the basic modifications that have occurred in the evolution of the insect thorax in general. It is assumed that primitively both terga and sterna were each formed by simple segmental sclerites, themselves separated by intersegmental sclerites. The complexity of the lepidopteran thoracic terga (nota) and sterna is the result of secondary sclerotization of this basic plan. The intersegmental sclerites become associated with the nota of the thoracic segments. The postnotum represents an intersegmental sclerite that has become fused with the notum anterior to it, while the acrotergite represents an intersegmental sclerite that has become associated with the notum posterior to it.

The *pleural* sclerites are believed by many to have been derived from the subcoxa of the legs that was incorporated into the trunk wall. The pleuron is divided by the pleural suture into the anterior episternum and the posterior epimeron. In Lepidoptera, the upper part of the episternum (the anepisternum) is separated from the lower part (the infraepisternum) by the anepisternal cleft (Fig. 42). The infraepisternum is itself divided into the preepisternum and the katepisternum by the precoxal suture. In many Lepidoptera, but not in lower members of the group, (and in some other orders) occurs a secondary development of the preepisternum termed a hypopteron or prepectus. The anterior margin of the preepisternum has been mistaken sometimes for the paracoxal suture causing the hypopteron to be misidentified as the preepisternum (Matsuda, 1970). The complex sclerites of the wing bases are derived from both pleural and notal elements.

The primary sclerotizations of the *sternum* are the segmental sclerite (the eusternum) and the intersegmental sclerite. The latter is produced internally as a spine - the spinasternum. From the eusternum arise a pair of internal apophyses, the sternal apophyses ('furca'). In higher insects these diverge inside the thorax from a common stem to form the characteristic furca.

In Lepidoptera, as in most other insects, considerable modification of the basic pattern described above is evident. The prothorax is the smallest and simplest of the three thoracic segments, and the mesothorax is nearly always the best developed.

Cervix

In Lepidoptera, as generally in insects, the head is joined to the thorax by a membranous cervix within which are embedded a pair of laterocervicalia or lateral cervical sclerites (Fig. 41). (In some primitive insects there are additional sclerites.) Each sclerite, which represents a modified preepisternum (Matsuda, 1970), joins the occipital condyle on the corresponding side of the head and posteriorly the pleura of the prothorax. The cervix appears to be a composite structure derived from both cephalic and thoracic elements.

Anteriorly near the apex of each laterocervicale is a small process bearing a proprioceptive hair-plate, an occurrence possibly representing an independent special-ization in Lepidoptera (Kristensen & Nielsen, 1979). However, although such a hair-plate is absent from Trichoptera and Mecoptera it is found in some other insects, and therefore may represent a primitive character that has been lost in those two orders (Kristensen, 1984b). The plate is present in *Heterobathmia* (Heterobathmiidae), weakly developed in Micropterigidae, well developed in Agathiphagidae, and well developed or weakly developed in the Eriocraniidae (Kristensen & Nielsen, 1979).

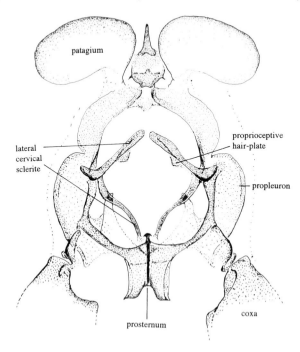

Fig. 41. *Prothorax of* Sarcinodes *sp. (Geometridae) showing laterocervicalia and patagia. (Drawn by Geoffrey Kibby.)*

Each laterocervicale is approximately triangular in primitive moths, but L-shaped in more advanced members of the order. In Trichoptera and at least in primitive Lepidoptera, the lower, posterior corner is produced towards the sternum of the prothorax (prosternum) (Kristensen & Nielsen, 1979).

Prothorax

The prothorax (Fig. 41) is larger in primitive Lepidoptera than in more advanced members of the group (where it is reduced to a collar-like structure). In Micropterigidae, the pronotum bears a pair of setose warts on the median dorsal plate of the pronotum, a condition assumed to represent a unique character of the amphiesmenopteran groundplan. These warts are homologues of the patagia found in higher Lepidoptera. Patagia are membranous, flap-like structures (Fig. 41), of variable shape (Schultz, 1914). Sometimes they are stalked, and in other cases lost altogether. In some, a parapatagium lies posterior to each patagium. The so-called lateral plate of the prothorax has been treated as of both notal and pleural derivation (see Matsuda, 1970) but, from the arrangement in certain primitive Lepidoptera, it is apparently of notal origin. Whereas in the propleuron of *Micropterix* the katepisternum is divided from the anepisternum by a membranous cleft, in higher Lepidoptera this division is weak or absent, and the propleuron is more heavily sclerotized. The trochantin, an articular sclerite between the coxa and the episternum, is absent from the prothorax in most members of the order.

Another prothoracic structure probably unique to the Lepidoptera is the 'free' condition of the profurcal arm in the endoskeleton. The arm arises from the bridge between the prosternum and the lower posterior corner of the propleuron (Kristensen & Nielsen, 1979; Kristensen, 1984b). It is displaced laterally (i.e., arises near the propleural corner) in Neolepidoptera (Exoporia and Heteroneura) (Kristensen, 1978b; 1984b).

Brock (1971) stated that the fusion of the anteriolateral angles of the prosternum with the propleuron is a condition confined to the Heteroneura and does not occur in Trichoptera, Micropterigidae, Eriocraniidae and Hepialidae. No such fusion occurs in Opostegidae (Nepticuloidea), and in Nepticulidae (Nepticuloidea) the sclerite is weakly developed. However, the absence or weak development of this character in the Nepticuloidea is arguably a secondary reduction.

The spina may be universally present in the prosternum in Lepidoptera (Matsuda, 1970).

Pterothorax

The pterothorax (Figs 42, 162) is composed of the mesothorax and the metathorax, segments that are more complex structures than the prothorax reflecting the extensive modifications to accommodate wings, legs and their associated musculature. Of the two pterothoracic segments the mesothorax is the largest in Lepidoptera, although in some primitive groups of the order the size difference is not marked.

The anterior part of the mesonotum is bent strongly down in both Lepidoptera and Trichoptera to form an attachment site for the longitudinal dorsal muscles. A prescutum is not generally distinctly delimited from the very large and prominent scutum. The scutum is demarcated from the scutellum, which is smaller than the scutum, by the scutoscutellar suture. The scutum may bear a median, longitudinal suture.

On each side of the anteriolateral section of the mesonotum arises a large mobile structure called the tegula (plural tegulae) (Hering, 1958) (Fig. 43). Enlarged tegulae are characteristic of Lepidoptera. Each tegula is supported by a subtegula (i.e., a detached prescutal plate) and a tergopleural apodeme (Sharplin, 1963b; Matsuda, 1970). Four possible functions of tegulae were suggested by Hering (1958). Firstly, it appears that the tegula may play a role analogous to that of an axillary sclerite (p. 48). Secondly,

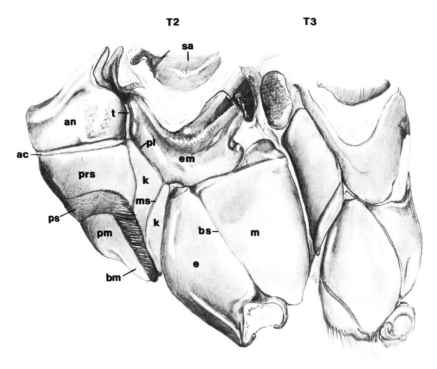

Fig. 42. *Pterothorax (T2, mesothorax; T3, metathorax), lateral view of lower section, of* Cossus cossus (*Cossidae*). *ac, anapleural cleft; an, anepisternum; bm, basisternum; bs, basisternal suture; em, epimeron; k, katepisternum; m, meron; ms, 'marginopleural' suture; pl, pleural suture; pm, parepisternum; prs, preepisternum; ps, parepisternal suture; sa, subalare; t, tergopleural apodeme. (Drawn by Geoffrey Kibby.)*

the structure probably assists wing coupling in Lepidoptera with a frenular-retinacular system since, through pressure, the tegulae, or brush-like scales extending from it, may help the frenulum to remain locked within its retinaculum. The discovery (H.J. Hannemann *in* Hering, 1958) that a tegular nerve branches in the area where the tegula has maximum contact with the forewing led to a third suggestion that the organ may function in sensing air resonance. A fourth function, probably incidental, of the tegulae is their role in simple protection of the wing bases. Apart from the functions noted by Hering, in certain Galleriinae (Pyralidae) the tegulae act as sound producing organs (p. 153).

Laterally, the mesoscutum is produced into anterior, median, and posterior notal processes, which form articulation points at the base of the forewing. From the posterior fold of the scutellum, the tubular axillary cord extends along the posterior margin of the forewing.

The main sclerites of the mesopleuron are the episternum and the epimeron, which are divided by the pleural suture. The pleural suture is the external indication of the

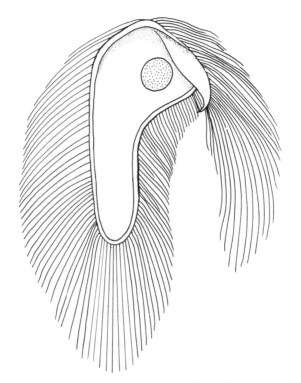

Fig. 43. *Tegula of* Catocala elocata (*Noctuidae*). (*After Hering, 1958.*)

pleural ridge. The episternum is divided into an anepisternum and an infra-episternum by the anapleural cleft (anepisternal suture of Shepard, 1930). In Papilionoidea, the anapleural cleft is displaced dorsally and in Nymphalidae it virtually disappears (Brock, 1971). The infraepisternum is strengthened by a suture system, which is somewhat variable within the order, and which needs further comparative study (N.P. Kristensen, pers. comm.). The following account is based on Brock (1971) and Matsuda (1970). A precoxal suture divides the infraepisternum into a preepisternum and a katepisternum. In Heteroneura, the preepisternum is itself divided by a cleft. The lower section of the preepisternum between this so-called parepisternal cleft and the basisternum is known as the parepisternum. The parepisternal cleft becomes sclerotized in more advanced Lepidoptera to form the parepisternal suture, and it migrates towards the anapleural cleft. The katepisternum is not always well defined, but when it is it takes the form of a narrow sclerite. In primitive Lepidoptera, the trochantin (a structure absent from most members of the order) lies between the katepisternum and the coxa. In many Lepidoptera a secondary sclerotization, the hypopteron or preepimeron is found between the episternum and the epimeron. This structure has been misinterpreted frequently as the preepisternum, but it is actually a modification of the mesepimeron.

Apart from the sclerotization of the parepisternal suture and its migration towards the anapleural cleft, two other main trends in the sternopleural region of the mesothorax were noted by Brock (1971). One of these involves the gradual disappearance of the free, upper section of the precoxal suture. The other relates to the suture termed 'marginopleural' by Shepard (1930) and Brock. This suture, which extends across the katepisternum, becomes elongated and strengthened. According to Brock it may be an extension of the internal fold of the pleural suture that has become deflected onto the episternum.

A 'tergopleural apodeme' arising from the upper section of the pleural suture of the mesothorax was described by Sharplin (1963b). The structure is a uniquely derived character of the Lepidoptera (Kristensen 1984b).

Compared with the mesonotum, the metanotum is much reduced, largely as a result of the reduction in size of the scutum. A reduction in metanotal length is a progressive trend through the Ditrysia, resulting, finally, in the most specialized macrolepidopteran condition where the mesal part of the anterior margin of the metascutum is concealed beneath the metascutellum (Brock, 1971). As in the mesonotum, a series of notal processes are present in the metanotum. These processes are not conspicuous, but the posterior notal process is well developed and may be partly or wholly detached from the metanotum (Matsuda, 1970). In Amphiesmenoptera (Trichoptera + Lepidoptera), the metapostnotum is separated from the metanotum by a thin membranous area, and is united with the first abdominal tergum. In Geometridae, a family with tympanal organs at the base of the abdomen, the postnotum is modified to form a characteristic accessory tympanum (p. 137) in all subfamilies other than Archiearinae. The accessory tympanum is an enlarged fenestra media, a membrane occurring in several other Lepidoptera, including Archiearinae (Minet, 1983).

The dorsal articulation of the metathorax with the abdomen is complex and further studies of this region particularly are needed to interpret homologies.

The sclerites of the meso- and metapleurosternum (Matsuda, 1970) are generally similar in position, but not in shape. In the metapleurosternum, the anapleural cleft is probably reduced or absent in most Lepidoptera, and the paracoxal suture is reduced or lost. In some lower Lepidoptera a weakly developed trochantin (a sclerotized remnant of the coxopodite, which is situated at the base of the leg) may be present, and a hypopteron occurs in many members of the order. The tympanum of those moths with metathoracic hearing organs is a modification of the membranous area dorsad of the metepimeron.

The meso- and metathoracic furcae are complex structures in higher Lepidoptera, with the metafurca particularly being of considerable taxonomic significance. The secondary arms of the furca of both pterothoracic segments in Trichoptera and Lepidoptera are fused with the posterior margin of the epimera of the corresponding

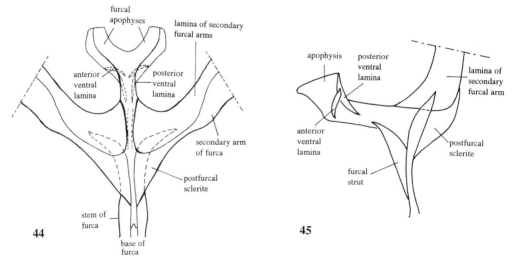

Figs 44, 45. *Metafurca of Lepidoptera. 44, posterior view; 45, lateral view. (After Brock, 1971.)*

segment. This is a specialized amphiesmenopteran character (Brock, 1971; Kristensen, 1984b). Essentially the mesothoracic furca, when viewed posteriorly, appears Y-shaped; it is the lateral arms that are fused with the epimeron. The furca arises from the discrimen (that inflected ridge dividing the sternum longitudinally).

The metathoracic furca is a relatively simple structure in the most primitive Lepidoptera, but secondary sclerotizations render it more complex in more advanced groups (Figs 44, 45). Three main trends in the development of the metathoracic furca were noted by Brock (1971). While in primitive groups the furcal stem is relatively long, in more advanced families the secondary furcal arms diverge closer to where the stem arises from the sternum. The development of a postfurcal sclerite in later groups emphasizes this shortening of the furcal stem. Anteriorly, the furcal stem of more advanced groups is supported by furcal struts formed from the lower, posterior portion of the posterior ventral laminae. In a number of groups, the apophyses of the metathoracic furca are fused to the secondary furcal arms (Brock, 1971; Davis, 1986).

The spina, which is the median apodemal process of the spinasternum, is absent from the mesothorax of most Lepidoptera; it is always absent from the metathorax in higher insects.

Legs

Each thoracic segment bears a pair of legs. A leg is divided (Figs 46-49) (proximally to distally) into a coxa, trochanter, femur, tibia, tarsus, and pretarsus.

The *coxa*, which articulates basally with the thoracic wall, is relatively immobile and takes the form of a truncated cone. In Lepidoptera (and many other higher insects) the coxa is divided by the basicoxal sulcus into a posterior meron and an anterior eucoxa (Fig. 42). A small *trochanter*, having a double condylar articulation with the coxa, links the coxa with the femur.

The *femur*, the stoutest part of the leg, is somewhat leaf-like in shape and articulates with the long and narrower *tibia*. The tibia of the prothoracic (fore-) leg typically bears

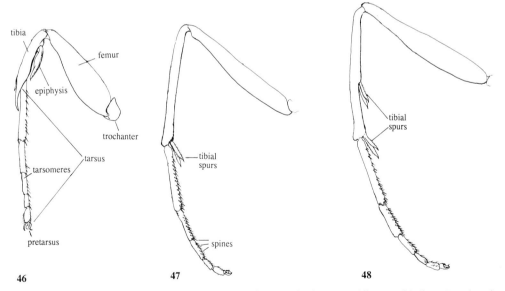

Figs 46–48. *Legs of* Oenochroma polyspila (*Geometridae*). *46, foreleg; 47, midleg; 48, hindleg.* (*Drawings by Geoffrey Kibby.*)

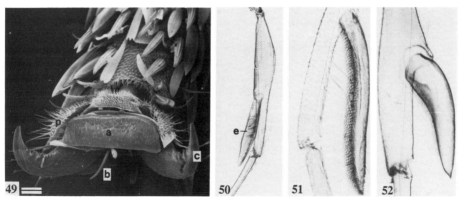

Figs 49–52. *Leg structure. 49, pretarsus of* Sarcinodes *sp.* (Geometridae) *(a, arolium; b, pseudempodial bristle; c, claw; p, pulvillus; scale, 45 μm). 50-52, epiphyses (e) of* Macrosoma *sp.* (Hedylidae): *50, M. semiermis; 51, M. napiaria; 52, M. tipulata. (49, EM Unit, BMNH; 50-52, Photographic Unit, BMNH, after Scoble, 1986.)*

a comb-like epiphysis (Figs 50-52), used for cleaning the antenna (e.g., Jander, 1966; O Dell *et al.*, 1982; Robbins, 1989) (see below) or the proboscis. It arises from the inner wall of the foretibia. The epiphysis is a structure unique to Lepidoptera (e.g., Kristensen, 1984b), which has frequently been secondarily lost. The foreleg lacks tibial spurs, although prominent apical spines are sometimes present, but usually the mesothoracic (mid-) leg bears a pair of apical spurs, and the metathoracic (hind-) leg bears a pair of medial and a pair of apical spurs. The number of tibial spurs is often expressed as a formula (for example and typically 0-2-4), a convention followed in Part III. The number of spurs sometimes deviates from that usually encountered.

The *tarsus* is composed of up to five *tarsomeres*. Usually the full number is present, but sometimes this figure is reduced. This reduction probably occurs by secondary fusion both because the five-segmented condition exists in the most primitive Lepidoptera, and because the original segmentation is often suggested by the pattern of scales on the tarsus. The tarsomeres differ from the true segments of the leg since they lack muscles; the tarsus is elevated and depressed by muscles originating in the tibia and inserting on the tarsus proximally.

The *pretarsus* (Fig. 49) is formed by a membranous base subtending a median lobe termed the arolium. Usually there occurs a medial, bristle-bearing 'pseudempodium', and frequently a pair of pulvilli arise one from each side of the pretarsal base. A pair of prominent claws typically arise from the pretarsus. Reductions in various elements of the pretarsus are common (e.g., reduction of the arolium and pulvilli in Papilionidae), and other modifications, such as bifid pretarsal claws in Pieridae, also occur.

Function

Legs function primarily for walking. Some Lepidoptera are gressorial, the forelegs being folded under the body while at rest and when the insect is walking. This condition occurs in both males and females of Hedylidae and Nymphalidae, and in male Riodininae (Lycaenidae). Reduction of foreleg length, or in various foreleg components, is widespread in butterflies, but occurs also in other Lepidoptera (e.g., some Geometridae). Although in the forelegs of male Lycaenidae the tarsomeres are fused and the pretarsal claws lost, these butterflies (except male Riodininae) use the forelegs for walking (Robbins, 1988b).

In many male Lepidoptera the legs are often the site of scale brushes or pencils, structures connected with the distribution of scent. Most of the categories of male scent organs (McColl, 1969, and see Chapter 6) are represented by organs present on the legs. Frequently encountered in Lepidoptera (e.g., many Noctuidae and Geometridae), is a brush concealed at rest within a groove in the tibia.

Legs of certain Lepidoptera are involved in stridulation, the midleg or hindleg bearing a grater, which rubs against a file situated on the forewing or hindwing (Chapter 6).

Both mechanoreceptor and chemoreceptor sensilla occur on the tarsi of Lepidoptera, a generalization based on studies of various butterflies (see McIver (1985) and Zacharuk (1985) for a summary and references on mechanoreceptors and chemoreceptors respectively). Probably the most important function for chemoreceptors on the tarsi of females is that of oviposition (Chew & Robbins, 1984). It appears that most chemoreceptors respond to a wide variety of secondary plant compounds, but there are some that are specific. The sensilla function either to inhibit or to stimulate oviposition.

Legs are used in cleaning the antennae (Jander, 1966; O Dell *et al.*, 1982; Robbins, 1989). In those Lepidoptera where the foreleg bears an epiphysis, particles are removed from the antenna as it is dragged along the inner surface of the epiphysis, this latter being covered with acanthae (minute cuticular spines - microtrichia). In Hesperiidae, the inner surface of the epiphysis bears a groove, devoid of acanthae, but this groove has not been observed in other Lepidoptera (Robbins, 1989).

Alternative means of cleaning the antennae occur in Lepidoptera lacking epiphyses. In some Pieridae, experiments using fluorescent powder suggested that the antennae are cleaned by drawing them across the tibiae and tarsi (Robbins, 1989). They do not appear to use the foretibial scale-brush for cleaning. Antennae of Lycaenidae (but not Riodininae) and Nymphalidae (including Libytheidae) are usually cleaned by the tibial scale-brush of the *midleg* (Jander, 1966; Robbins, 1989).

Wings

This section is concerned firstly with the basic structure of lepidopteran wings and the way in which the various components function, such as in wing-folding and wing-coupling. The use of wing venation in lepidopteran classification is also considered, and the primary and secondary values of wings to Lepidoptera are then discussed.

The primary function of wings is for flight. This complex subject is examined, before one of the behavioural consequences of the ability to fly - namely migration - is discussed. Apart from flight, an important secondary aspect of wings in Lepidoptera is their diversity of colour patterns. Indeed, pattern is one of the most notable features of the biology of the order. It is usually created from the numerous scales that cover the wings of nearly all species and which give this order of insects its name. The nature of colour involves a discussion both of pigments and of scale ultrastructure. The actual development of colour pattern in Lepidoptera is considered against the background of studies demonstrating an underlying and unifying explanation for what, at first sight, appears as a bewildering variety of forms. Colour pattern has many functions. Important amongst them is the role it plays in crypsis or advertisement, therefore it is closely linked to topics such as resting position, time of flight, distastefulness, mimicry, and courtship. Colour also plays an important part in thermoregulation in butterflies.

Diversity

Wings vary in shape from narrow to broad (see Figs 186-321). In many small moths both forewings and hindwings are narrow and delicately fringed with fine, hair-like scales, often called 'cilia'. By contrast, very broad wings are found among the

Bombycoidea, particularly Saturniidae and Brahmaeidae. Relatively narrow wings are found in Sphingidae, an arrangement assumed to be aerodynamically efficient for long-distance and powerful flight. In several unrelated groups (e.g., Zygaenidae, Papilionidae, and Saturniidae), the hindwings are sometimes, or often, extended into long tails. Wing reduction or loss, usually restricted to females, is found in less than one percent of lepidopteran species, but representation of these phenomena is widespread occurring in 25 families (Sattler, 1991b).

The level of reduction ranges from moderate to extreme, but there are four categories, divisions which are not discrete (review by Sattler, 1991b). In brachyptery the length of the wings is usually less than that of the abdomen, but some tubular veins are retained, at least in the forewing. Sustained flight is not possible. Stenoptery, in which the width of the forewing is greatly reduced and the hindwing often vestigial, is very rare. Stenopterous Lepidoptera cannot fly. Both pairs of wings of micropterous species are reduced to small lobes and lack tubular veins. Few strictly apterous Lepidoptera occur. Wing reduction is strongly related to environmental conditions. Those few species where males are affected inhabit coastal habitats or small oceanic islands, areas where wind conditions are such as to prevent individuals flying directionally towards potential mates. Jumping is a typical mode of progression in members of these species. Wing reduction in females is strongly related to the degree of egg maturation at eclosion. Well developed eggs leave little room for the bulky flight muscles, which become reduced. Reduction of flight muscles leads to flightlessness, a first stage in wing reduction. Tympanal organs, which occur mainly at the base of the thorax or of the abdomen, also tend to become reduced or lost. Females emerging from the pupa with well developed eggs have no need to sustain themselves while the eggs mature. As long as males can find them, and provided eclosion occurs on a suitable larval foodplant, these females can lay their eggs rapidly. Species exhibiting these characteristics are often active in the cold season (e.g., many Geometridae) where rapid oviposition, occurring before extreme conditions overtake individuals, is clearly advantageous.

Wing base

The wings are articulated with the thorax by way of various sclerites situated at their bases (Figs 53, 54). Articulation must permit an up and down movement of the wings, and the system also determines the degree to which the wings may be folded against the body when the insect is at rest.

The complex arrangement of sclerites at the base of the wings of Lepidoptera was discussed by Sharplin (1963b,c), who did much to standardize the terminology of the components of this region. She also investigated the use of the various parts during wing movement in flight and in wing folding. A range of Lepidoptera were studied, although observations on the most primitive taxa of moths were necessarily confined to Micropterigidae, Eriocraniidae, and Hepialidae - several of the additional groups having not been discovered at the time of her work.

Essentially, the sclerotized components involved include the notal processes of the thorax, the axillary sclerites, the plates associated with the bases of the wing veins, and the sclerites below the wing.

Forewing (Fig. 53)

Notal processes. Three articular processes arise from the alinotum (the wing-bearing plate of the dorsum of the meso- or metathorax). The anterior notal process is made up of the suralare and the adnotale, which are plates extending from the scutum. The

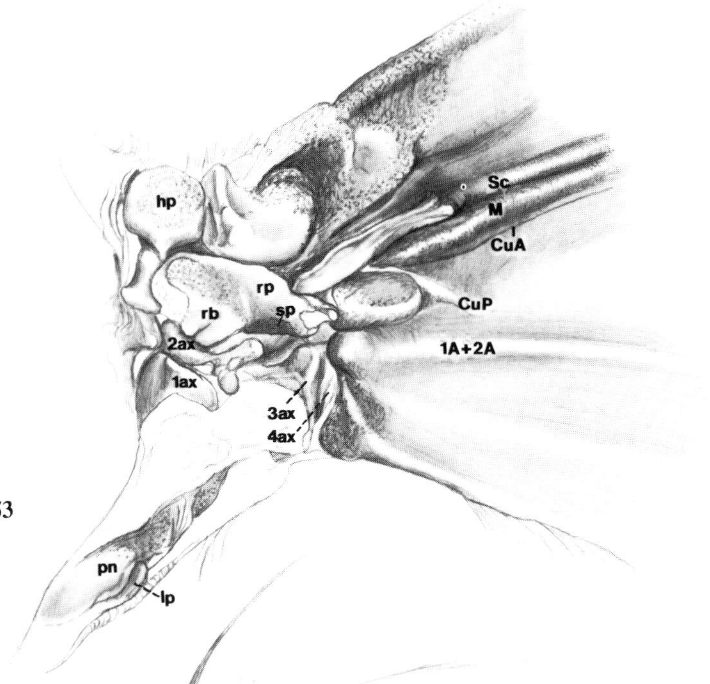

Figs 53, 54. *Wing bases of* Cossus cossus (Cossidae). *53, forewing; 54, hindwing. (1ax, 2ax, 3ax, 4ax, first to fourth axillary sclerites; ap, anal plate; hp, humeral plate; lp, lateral process of posterior notal wing process; ma, median arm; mp, median plate; pn, posterior notal wing process; rb, radial bridge; rp, radial plate; sp, scale plate.* (Drawings by Geoffrey Kibby.)

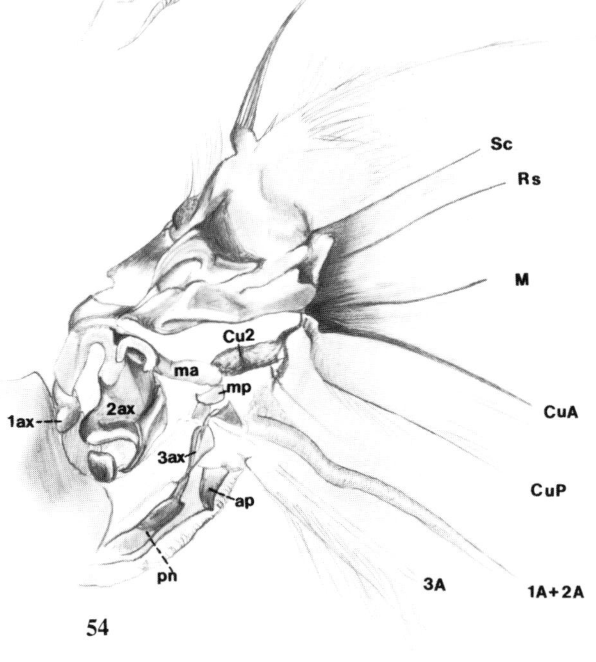

median notal process is smaller in primitive moths than in higher Lepidoptera, its size being related to that of the first axillary sclerite with which it articulates. The posterior notal process articulates with the third axillary sclerite.

Axillary sclerites. There are three, or sometimes four, axillary sclerites in Lepidoptera. The first axillary sclerite is connected to the scutum by 'ligaments' which, morphologically speaking, are cuticular areas with special staining properties. The sclerite connects with the base of the subcostal vein, the first notal process, the median notal process and, distally, with the second axillary sclerite. The second axillary sclerite extends dorso-ventrally through the thickness of the wing (see also Nielsen & Kristensen, 1989). In Trichoptera and Lepidoptera this sclerite has a unique double structure. The dorsal component of the second axillary sclerite connects with the radial vein, the first axillary sclerite and the first median plate. The ventral component is joined by 'ligaments' to the subalare and the pleural wing process. It forms a complex consisting of the original ventral component, fused with the radial plate by way of bending cuticle. The radial plate is a dorsal widening of the base of the radial vein. The ventral component is also suspended from the pleural wing process posteriorly by a 'ligament', and the pleural wing process itself is joined, by another 'ligament', to a projection from the ventral part of the radius. The third axillary sclerite articulates with the posterior notal process, and also with the first median plate. In Incurvarioidea and primitive Ditrysia a strip of bending cuticle cuts off the posterior region of the third axillary sclerite to form (functionally) a fourth axillary sclerite.

Median plates. The median plates are situated between the second and third axillary sclerites and the median vein. The first (i.e., proximal) median plate is fused to or connected with the third axillary sclerite. Near this junction the plate extends to the lower surface of the wing. It articulates with the dorsal component of the second axillary sclerite, and is joined to the ventral component of the second axillary sclerite by a 'ligament' or by bending cuticle. The second (i.e., distal) median plate is composed of normal cuticle in Micropterigoidea, Eriocranioidea, and Hepialoidea. It articulates with the distal end of the first median plate and is fused to the base of the median vein, when that vein is present. In Incurvarioidea and Nepticuloidea the base of the cubitus (Cu$_2$ of Sharplin) overlies the distal end of the second median plate. That part of the median plate overlain is soft. In many Ditrysia the whole of the plate is soft.

Sclerites below the wing. Anterior and posterior to the pleural wing processes are additional sclerites. Anteriorly, the two basalar sclerites (the first and second basalares) are associated with the mesepisternum. In primitive Lepidoptera they are joined to the upper part of the episternum from which they are hardly distinct. In Ditrysia, the first basalare is clearly distinguishable being either partly or entirely separated from the mesepisternum. The second basalare in Ditrysia is detached from the episternum and remains connected to the first basalare by a 'ligament' only. Posterior to the pleural wing process is the *subalare* or subalar sclerite, which lies above the epimeron (Fig. 42). Onto the basalare and the subalare respectively are inserted the anterior and posterior pleural muscles of the wing.

Hindwing (Fig. 54)

In primitive moths, the pterothoracic segments are approximately of equal size, and the wing bases are correspondingly similar. In higher Lepidoptera the metathorax is smaller than the mesothorax and the base of the hindwing is modified to some extent.

Unlike the situation in the forewing, there is no proximal enlargement of the first axillary sclerite in the hindwing of Ditrysia. The second axillary sclerite is a single sclerite, which joins the upper and lower wing membranes. In Micropterigoidea and Eriocranioidea the first median plate articulates with the second axillary sclerite as in the forewing, but in other Lepidoptera this articulation is lost. In non-ditrysians the third axillary sclerite is a simple plate. It articulates with the posterior notal wing process proximally, and with the first median plate anterio-distally. In Ditrysia there is an element of the third axillary sclerite on the ventral wing membrane near its point of articulation with the posterior notal process. In Cossoidea the posterior notal process articulates with ventral projections of the third axillary sclerite - a sclerite that resembles the fourth axillary sclerite of the forewing. In higher Ditrysia there is no functional fourth axillary sclerite. In non-ditrysians and in primitive ditrysians there is a small sclerite, known as the anal plate, on the dorsal wing membrane between the posterior notal wing process and the axillary cord (a strengthening ligament extending from the base of the wing along the proximal part of the anal edge of the wing).

There is probably only a single median plate in the hindwing of all Lepidoptera, including Micropterigidae and Eriocraniidae where it appears to be double as the result of a flexion line (Nielsen & Kristensen, 1989). The plate is partly or entirely composed of bending cuticle and is closely associated with the third axillary plate. The function of the median plate in Lepidoptera other than Micropterigoidea and Eriocranioidea is taken over by a small proximal and larger distal cubital plate. These plates are often fused in Bombycoidea. The base of the median plate is detached from the rest of the vein to form a *median arm* in all non-micropterigoid Lepidoptera. This arm projects from the posterior margin of the radial bridge to the second cubital plate.

Wing folding

Wing folding in Lepidoptera depends not only on the form and articulation of the basal sclerites, but also on the presence of a special kind of cuticle known as 'bending cuticle' (Sharplin, 1963a) (Fig. 55). Its study was stimulated by the observation that what appeared to be normal cuticle in the wing base, buckled to an unexpected degree during wing movement. In stained sections, the mesocuticle of this bending cuticle was shown to be divided into cones. Such cuticle can be both buckled and twisted. Although the exocuticle was not divided in this way, it was capable of being twisted far beyond the point at which normal cuticle splits. The way in which the cuticle is thought to bend is illustrated in Fig. 55, which demonstrates the function of the divided mesocuticle. In primitive Lepidoptera, the projections of the mesocuticle into the endocuticle take the form of irregular pegs varying in both length and diameter. In higher Lepidoptera the projections are cone-shaped and usually reach the inner margin of the endocuticle.

Wing folding (Sharplin, 1964) is achieved by a combination of the pivotal arrangement of the sclerites of the wing bases and the strategic location of bending cuticle (Figs 56, 57). The first and second axillary sclerites do not move in wing folding. The wing-folding muscle, which originates on the pleural ridge, inserts on the third axillary sclerite, and, in the forewing, the median plates transmit the contractive force to the bases of the veins. In the hindwing, the cubital plates fulfil the function of the median plates in the forewing. Bending cuticle permits the radial bridge to buckle during wing folding, which in turn allows the radial bridge to rotate around the ventral part of the second axillary sclerite. In moths with very tight wing folding, a patch of bending cuticle at the distal end of the radial plate allows the wing to be folded beyond its normal limits.

The ability to fold the wings over the body is secondarily lost in many Lepidoptera, notably butterflies and typical geometrids. (There are, however, several Geometridae in which wing folding is pronounced, see for example, in McFarland, 1988.) In Papilionoidea the median plates are reduced and are no longer connected, and the base of 'Cu$_2$' fails to meet the radial plate and only partly overlies the second median plate (Sharplin, 1963b).

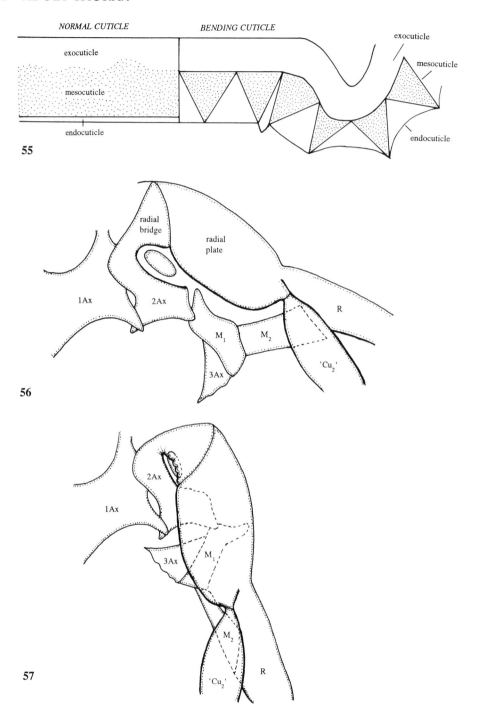

Figs 55–57. *Wing folding. 55, normal cuticle compared with bending cuticle; 56, 57, relative positions of basal wing sclerites during wing-folding (1Ax, 2Ax, 3Ax, first to third axillary sclerites): 56, wing not folded; 57, wing folded. (55, after Sharplin, 1963a; 56, 57, after Sharplin, 1964.)*

Tympanal organs

Tympanal organs occur in swollen veins at the base of the wings of some butterflies and in certain Thyrididae. These structures are discussed in Chapter 6.

Venation

Through the thin, two-layered, membranous wings of insects run several tubular veins. A nerve and a trachea are present within each of the major veins and, since the veins connect with the haemocoel, haemolymph may, in principle, circulate within them. Around the edge of the wing, runs a lacuna into which the veins open. In transverse section, the walls of the veins are seen to be relatively well sclerotized. Veins are derived from blood spaces (lacunae), which develop in the internal wing buds of the larvae of holometabolous insects such as Lepidoptera. These lacunae are invaded first by tracheoles (i.e., larval or provisional tracheoles) and later by outgrowths of large tracheae. The larval tracheoles degenerate and disappear in the pupal wings. (Since the pattern of venation in the pupa may differ from that of the adult, caution is necessary when using pupal veins to compare homologies of veins in different adults.)

Lepidopteran venation is relatively simple in the sense that there are few cross veins and, in many families, there has been considerable coalescence, loss, or both.

In the lowest Lepidoptera (up to, and including, the Hepialoidea, see Table 1), the venation of the forewing is similar, overall, to that of the hindwing: an arrangement termed the homoneurous condition (Figs 58, 59). All other Lepidoptera (more than 98 percent of the order) are heteroneurous (Figs 60, 61), that is the venation of the two wings forms a different pattern. The venation of Micropterigidae (Zeugloptera) (Fig. 58) is similar to that of primitive Trichoptera. The absence of vein M_4 in both forewing and hindwing of Micropterigidae led to the view that this provided a way in which Lepidoptera could be distinguished from Trichoptera. However, the discovery of M_4 in the forewing of Agathiphagidae (Common, 1973), and sometimes in the hindwing (Kristensen, 1984b) means that its absence can no longer be regarded as diagnostic of Lepidoptera (Kristensen, 1984b). Thus early fossil Lepidoptera are very difficult to recognize as such because frequently the structures most visible in this material are the wings and their veins (Kristensen, 1984b).

In the forewing, the costal vein (C) is simple and runs along the leading edge of the wing. In homoneurous moths the subcosta (Sc) is frequently branched in the forewing into Sc_1 and Sc_2. The radius (R_1) may also be forked. The radial sector (Rs, usually numbered R_{2-5}) is 4-branched. Three branches of the median (M) are typical although, as mentioned immediately above, four are present in Agathiphagidae. Anterior and posterior cubital veins (CuA and CuP) are present, the former being forked into CuA_1 and CuA_2. Up to three anal veins (A) are present. Vein 2A usually merges distally with 1A, and 3A (if present) usually unites with 2A before it meets 1A. As a result, a double 'loop', called the anal loop, is formed. Few crossveins are found. A humeral vein unites Sc with the costa at the base of the forewing in primitive homoneurans.

In the hindwing, vein Sc may run together with R_1 in Micropterigidae (as in Australian species, e.g., Fig. 58), but they remain separate in other homoneurans. In Micropterigidae R_1, all branches of Rs, M, and CuA are present as is vein CuP. The first signs of reduction in both forewing and hindwing are seen even within the homoneurous moths with Sc and R_1 being represented each by a single trunk, and a weakening of CuP.

In the primitive Heteroneura, the difference between the venation of the forewing and that of the hindwing is marked. In particular, in the hindwing, Sc and R_1 are fused to form $Sc+R_1$, and Rs is reduced to a single branch. A further occurrence in these

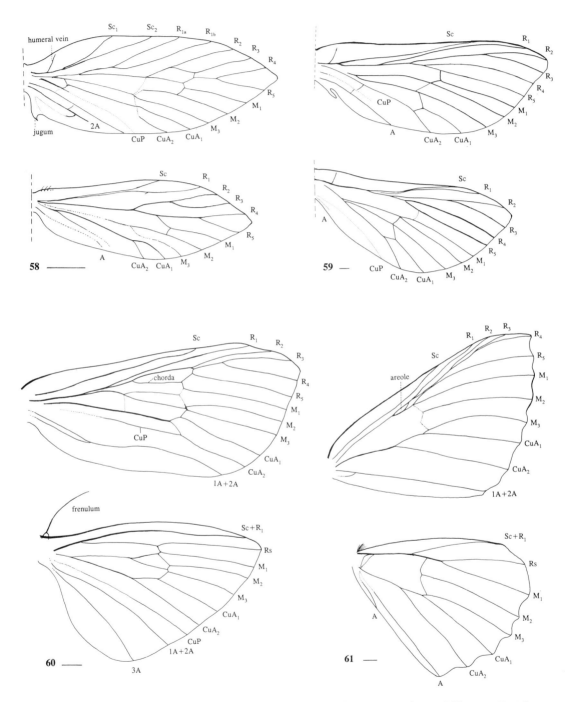

Figs 58–61. *Wing venation. 58, 59, homoneurous condition: 58,* Sabatinca *sp.* (Micropterigidae); *59,* Hepialus humuli *(*Hepialidae*); 60, 61, heteroneurous condition: 60,* Degia imparata *(*Cossidae*); 61,* Prognostola caustoscia *(*Geometridae*). (Drawings by Geoffrey Kibby.) Scales, 1 mm.*

more advanced Lepidoptera, and one that is present, with some exceptions, throughout the rest of the order, is the formation of a 'discal cell'. Although the stem of R_{4+5} (the chorda) divides this cell in the forewing of Cossidae (Fig. 60), as does M, both veins are absent in most Heteroneura, an arrangement making the discal cell very large (Fig. 61). In more specialized Lepidoptera, CuP weakens and disappears (Fig. 61).

In many 'higher' Lepidoptera crossveins or anastomoses (or both) in the $Sc-M_1$ region form one or more secondary cells (e.g., most Geometridae, Drepanidae, Anthelidae, and Noctuoidea).

The importance of venation in classification

The pattern of venation in Lepidoptera has been used extensively in the classification of the order and continues to be so. The nomenclature of veins in insects usually follows the system adopted by J.H. Comstock and J.G. Needham (summarized by Comstock, 1918), although a number of modifications have been adopted in the present work. In Part 3, and elsewhere in this book, with one exception I follow the recommendations of Wootton (1979), who made a plea for conservatism in the naming of insect veins and rejected the revised terminology of Hamilton (1972). The exception concerns the numbering of the branches of the radial sector (Rs). Wootton recommended naming these Rs_1, Rs_2, Rs_3, Rs_4 but lepidopterists are generally resistant to this convention preferring to name the branches R_2, R_3, R_4, R_5. This latter system is adopted, reluctantly, in this book.

The first comprehensive use of venation in lepidopteran classification was made by Herrich-Schäffer (1843-1856). This author numbered the veins from 1, which represented the anal veins (1a and 1b), through to number 12, representing the subcosta. Although this system of nomenclature is used less frequently today than the scheme devised by Comstock (1918), the influence of Herrich-Schäffer's study on venation (and other structures) remains considerable. Many of the families that we recognize today were established by Herrich-Schäffer and based, to a significant extent, on wing venation (e.g., the number of veins present and their branching patterns).

Position of wings at rest

Although the wings of Lepidoptera are variously positioned when these insects are at rest (e.g., Pl. 1:1-10, and Tweedie, 1988; Tweedie & Emmet, 1991), three basic resting postures, termed by McFarland (1979) tectiform, planiform and veliform, are observed. The most widespread of these is the *tectiform* condition (e.g., Pl. 1:1,2), the situation occurring typically in the most primitive Lepidoptera but also in many other families. Here, the wings are folded over the abdomen, often (and typically in the most primitive moths) in a slanting, 'roof-like' position. This position was called stegopterous by Graham (1950). In the *planiform* position (which occurs for example in many Geometridae), the wings are held away from the body and more or less flatly appressed to the substrate (e.g., Pl. 1:3,4). Many Lepidoptera, notably butterflies, hold their wings vertically so that the dorsal surfaces are touching (e.g., as in Pl. 1:9). This 'wings up' position is termed *veliform*. The veliform posture is seen not only in papilionoid butterflies but also, for example, in some Geometridae, many Hesperiidae, and in Callidulidae.

Oudemans' (1903) classification of resting positions is slightly different. The planiform position was regarded as a subdivision of the tectiform position, and the veliform position was distinguished from the situation where the wings are held only partially erect. Both Oudemans and Graham noted many variations in the tectiform (stegopterous) position.

Although still regarded as tectiform, the wings in many Lepidoptera are slightly less steeply angled (e.g., Pl. 1:2) than they are in the most primitive families such as Micropterigidae. Such an 'ordinary stegopterous' instead of 'primary stegopterous' position (Graham, 1950) is exemplified by most Tineoidea, Tortricoidea, Cossoidea, Castnioidea, Pyraloidea, Noctuoidea, and by certain Geometroidea. There are various other modifications of the three basic patterns, a well known one of which is typical of Lasiocampidae and some smerinthine Sphingidae where the hindwings project beyond the forewings (e.g., Pl. 1:6). In Pterophoridae, which hold their wings in a planiform position at rest, the hindwings lie largely beneath the forewings (Pl. 1:7). In some skippers (Hesperiidae) the forewings are elevated while the hindwings lie flat (Pl. 1:10). In Epipleminae (Uraniidae), the forewings are typically rolled when the insect is at rest (e.g., as in Pl. 1:8).

Correlated, to a fairly high degree, with the position of the wings at rest is the presence or absence of the wing-thorax coupling system (p. 61). This apparatus is present in those Lepidoptera where the wings are folded tightly against the body. In those groups in which the wings are not held in the tectiform position, or even in those where the wings are not folded tightly against the body (e.g., as in Bombycoidea), the microtrichial patches on the underside at the base of the forewing and on the metathorax are usually absent (Common, 1969a; Kuijten, 1974). Exceptions to the general rule exist. Alar and metathoracic patches persist even when the wings are not held in the tectiform position, for example in Alucitidae and in those Pyralidae that do not hold their wings close to the body. Similarly, the patches are absent from those Geometridae that fold their wings against the body, a situation occurring in many Ennominae.

The position of the wings at rest, wing colour pattern, and the resting posture of the insects are strongly integrated. The relationship between colour pattern and resting posture was examined in detail in macrolepidopterans by Oudemans (1903) in a work that has influenced studies on the subject ever since. Oudemans' main principle is that, in general, the colour of parts of the wings exposed to view when the insect is in a well-settled resting position contrasts markedly with that of areas which are concealed. The principle applies also to other parts of the body. Furthermore, the patterns on the wings in the resting position frequently form a fully integrated pattern. Such patterns are not integrated in cabinet-set specimens since rarely, if ever, do they reflect the true resting position. In most moths, the integrated (visible) colour pattern is cryptic. Concealed parts are usually drab and of a uniform colour rather than integrated into a cryptic pattern overall. There are, however, many exceptions. Hindwings, for example, which are typically concealed, may bear bright marks, such as eyespots. These markings may be revealed suddenly when the insect is disturbed (flash coloration). Examples of such integration are particularly well exemplified in quadrifine Noctuidae, Saturniidae, Sphingidae, and Geometridae where bands across each forewing meet to form a continuous band when the moth is at rest (e.g., as in *Petrophora cervinata*: Geometridae), or where at rest a continuous U-shaped band extends from the costal edge of one forewing through both hindwings to the costal edge of the other forewing (e.g., as in *Phrygionis*: Geometridae, Pl. 1:4).

Such integration may function to break up the typical moth outline, which birds may use as a search image. However, the purpose of much of the detail is uncertain.

Wing-coupling

It is a general assumption in entomological texts that while the wings of primitive insects tend to beat out of phase, those in more advanced groups beat together. This latter effect is achieved largely by various modifications that couple forewing with hindwing so that they function as single unit. There are exceptions to this general rule, including examples in Lepidoptera as has been shown by high-speed photography of moths and butterflies in flight. Nevertheless, wing-coupling mechanisms are frequently well developed in Lepidoptera, and within the order there are three basic kinds: jugal, frenulo-retinacular, and amplexiform.

Jugal

The jugum is a subtriangular (e.g., as in Micropterigidae: Zeugloptera) or finger-like (e.g., as in Hepialidae: Exoporia,) extension of the jugal lobe of the forewing (Figs 62, 63). It is demarcated from the rest of the wing by a thin, membranous line. Best developed in many Hepialidae (Fig. 63), it has been reported from several other families of primitive Lepidoptera. Precisely where the jugal lobe of the forewing becomes sufficiently large to be called a jugum is a matter of judgement. The jugum has sometimes been distinguished from the *fibula* because the two structures were thought to operate in different ways in wing-coupling. However, both the jugum and the fibula are almost certainly equivalent, rendering the term 'fibula' redundant.

Where present, the jugum or jugal lobe rests on the upper (dorsal) surface of the hindwing when the wings are extended, and probably fails to act as an effective wing-coupling mechanism. When the wings are closed it is folded under the forewing. The base of the jugum in primitive moths lies along the posterior distal margin of the third axillary sclerite of the wing base (Sharplin, 1963b). When the wings close, the underfolding of the jugum occurs automatically as the third axillary sclerite moves over the posterior notal process. This folding of the jugum or jugal area occurs in all monotrysian moths and in the lower ditrysian superfamilies Tineoidea (including the 'Psychoidea' of Sharplin), Tortricoidea, Zygaenoidea, and Pyraloidea. In Cossoidea, a jugal area is present, but only partly curled under the wing at rest. In other Ditrysia there is no trace of a jugum (Sharplin, 1963b). Underfolding was noted in certain families of Lepidoptera by Tillyard (1918) and by others, but the mechanism was not understood until the study of Sharplin.

The way in which the jugum functions has been misinterpreted in the past. Tillyard (1918) realized that in Trichoptera the jugal lobe simply rests upon part of the costa of the hindwing thus effecting some degree of coupling. But in Lepidoptera, Comstock (1918) and Tillyard considered that the jugum formed a notch in the forewing into which the costa of the hindwing fitted. The view that the jugum lies *under* the hindwing was probably based on observations of Hepialidae, since Tillyard argued that in that family the jugum operated in a way quite different from Micropterigidae. According to Tillyard, in Micropterigidae a series of costal bristles on the hindwing engages the notch formed by the jugum, the jugum itself being folded under the forewing. Tillyard claimed that Micropterigidae were actually *frenate* moths, the costal bristles representing what he regarded as a true frenulum, and termed the system 'jugo-frenate'.

The view that the jugum of Micropterigidae is indeed folded under the forewing was confirmed by Sharplin (1963b) whose interpretation is summarized above. However, this folding occurs only at rest; in flight the jugum is extended. The jugum does *not* act like a retinaculum for the costal spines (i.e., Tillyard's 'frenulum') (Philpott, 1924; Braun, 1924), and the term jugo-frenate is misleading since no comparison can justifiably be made between this arrangement and the frenulo-retinacular system in higher groups.

Philpott (1925) demonstrated that neither the jugum of Hepialidae nor of Micropterigidae extends under the hindwing as a 'finger-and thumb' coupling system. He noted that in some species of this family the jugum was so small as to be ineffective as a coupling device. In those where it was sufficiently long he suggested that previous workers may have been misled by giving undue weight to finding *some* set, cabinet

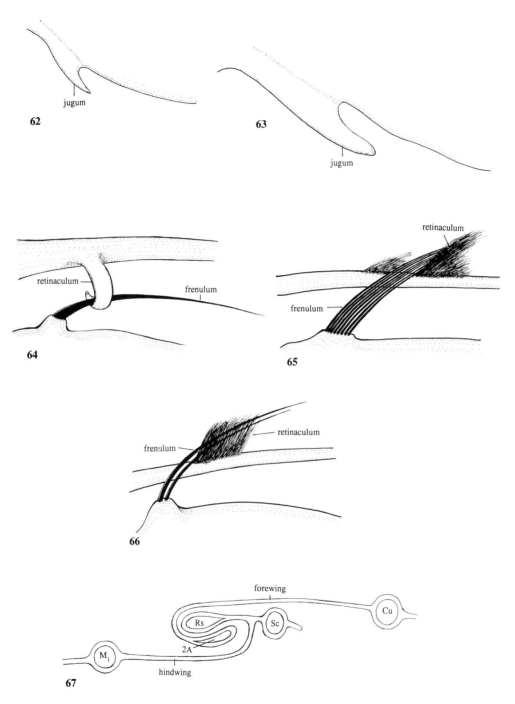

Figs 62–67. *Wing-coupling. 62, 63, jugal coupling: 62, jugum of* Sabatinca *(Zeugloptera, Micropterigidae); 63, jugum of* Hepialus humuli *(Exoporia, Hepialidae); 64-66, frenular-retinacular coupling: 64, 65,* Hippotion scrofa *(Sphingidae): 64, male, 65, female; 66,* Chrysodeixis eriosoma, *female (Noctuidae); 67, Sesiidae wing-coupling. (62, 63, drawn by Geoffrey Kibby; 64-66, after Tillyard, 1918; 67, after Heppner & Duckworth, 1981.)*

specimens showing the finger-and-thumb 'coupling' of the jugum with the costa. In *most* set specimens the jugum lay above the costa. Philpott based his interpretation on simple experimentation. By cutting the wings off a freshly killed specimen of *Porina cervinata* (Hepialidae), he found the jugum folded under the forewing. As the forewing of other freshly killed specimens were gradually pushed forward, the jugum gradually unfolded to project, eventually, from the anal edge. The jugum did not pass below the costa of the hindwing. In specimens that died with their wings up (instead of in the usual roof-like resting position) the jugum could not be seen from the underside, further suggesting that the normal position of the jugum was on top of the hindwing.

If the forewing and hindwing of jugate moths are not linked, then at upstroke the forewing must be able to rise without the hindwing. That this actually occurs was demonstrated by a series of photographs of male *Hepialus humuli* (Hepialidae) taken during their hovering courtship flights at dusk (Mallet, 1984), and of *Eriocrania haworthi* (Eriocraniidae) taken under laboratory conditions (Grodnitsky & Kozlov, 1985). These photographs show that at upbeat, the forewing rises prior to the hindwing. The hindwing meets the forewing before they descend together at downbeat. These authors noted that this out of phase beating was unlike the condition in Ditrysia. (The situation in some butterflies is noted below.)

Frenulo-retinacular (Figs 64-66, 68)

With many exceptions, the Heteroneura (i.e., most Lepidoptera) couple their wings by means of a frenulum and a retinaculum. This system also occurs in many non-Heteroneura. The most primitive group with this arrangement is the Acanthopterocte-tidae (Davis, 1978: 46). Essentially, a frenulo-retinacular system is one in which a single bristle, or a series of bristles, at the base of the hindwing interlocks with a hook, or a series of bristles, on the undersurface of the forewing. Such a mechanism effects wing-coupling. In males, the frenulum is composed of a single, but composite, spine at the base of the forewing (Figs 64, 68). In females, it may also comprise a single bristle (e.g., as in Sesiidae, many Tineidae, and some stictopterine and euteliine Noctuidae). However, usually it is represented by a series of bristles (2 to 9 according to Tillyard, 1918) (Figs 65, 66). In both sexes, the frenulum arises from the costal plate, a structure situated at the base of the hindwing, and engages the retinaculum on the undersurface of the wing. In males, the retinaculum is generally a membranous flap or hook, whilst in females it is usually a series of stiff hairs or scales.

The frenulum should not be confused with the series of costal bristles on the hindwing in Trichoptera and many primitive Lepidoptera, each member of which arises from a separate point along the base of the costa rather than the costal sclerite (Braun, 1919).

The position and form of the retinaculum among Lepidoptera varies not only between male and female, but also within each sex. Braun (1924), following Tillyard (1918), suggested that it may be subcostal or subdorsal. The subdorsal retinaculum was said to be associated with the base of the cubital veins. The position of the retinaculum actually varies so much that a division into subdorsal and subcostal is rather misleading.

In most males, the retinaculum arises from the base of the subcostal vein of the forewing, between Sc and C, or on a spur of Sc (Common, 1970). It usually takes the form of a membranous hook, being broad-based and rather short in the more primitive Lepidoptera, and narrow-based and long in more advanced groups (Fig. 68). In most Nepticulidae the retinaculum, although subcostal, comprises a series of strong hook-like scales into which the single-spined frenulum catches. No subcostal retinaculum exists in *Opostega* (Opostegidae) nor in some Cossidae (Braun, 1924). Although a subcostal retinaculum is absent from some male Megalopygidae, it is present in many others.

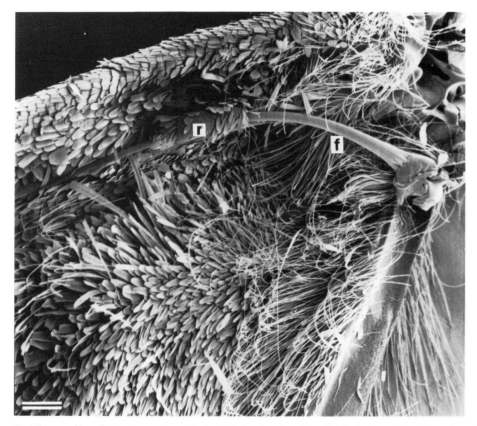

Fig. 68. *Wing coupling of* Agrotis segetum *(Noctuidae) showing frenulum (f) interlocking with retinaculum (r).* (EM Unit, BMNH.) Scale, 300 µm.

Besides the subcostal retinaculum, in some males a series of stiff bristles or hair-like subcubital scales also often occurs. In some Pyralidae these spines function, together with the subcostal retinaculum, to catch the frenulum.

A hook-like, subcostal retinaculum is usually confined to males, but in some groups (Sesiidae and some Lymantriidae) the structure is present in this form (although generally not fully developed) in females (Braun, 1924). The subcostal retinaculum of females is generally composed of a series of hooks or bristles. Sometimes the frenulum is held by no more than heavy scaling near the base of the subcostal vein.

Where present, the subcubital (subdorsal) retinaculum tends to play a more important role as a frenulum catch in females rather than males. It comprises a group of specialized scales. In Nepticulidae these take the form of a series of stiff scales.

In some groups, the difference between the wing coupling of males and females is marked. Male Nepticulidae have a single-spined frenulum, and a subcostal retinaculum is retained as well, whereas female nepticulids bear hooked costal spines on the hindwing that interlock with two rows of scales near the base of the cubitus. A weak subcostal retinaculum occurs in some species in addition to the subcubital, and in *Stigmella pteliaeella* (Nepticulidae) the subcostal retinaculum is absent.

Sesiidae have, in addition to the frenular-retinacular device, a unique coupling system (Fig. 67) in which Rs+M$_1$ of the hindwing is folded and curls around to interlock with the dorsal margin of the forewing, which is also rolled (Braun, 1924;

Heppner & Duckworth, 1981). In Gelechiidae the retinaculum of the females is subradial, comprising a row of strong scales.

The primary wing-coupling function of the frenulum and retinaculum is altered in the pyralid *Samcova* to become a stridulatory organ in which the retinaculum bears a file and the frenulum acts as a scraper (Chapter 6, and Fig. 169).

The effectiveness of the frenular-retinacular system in coupling wings is apparent from some of the high-speed flash photographs of Dalton (1975), particularly the sequence of six frames of *Noctua pronuba* (Noctuidae), which shows that the wings remain firmly coupled during the upbeat.

Amplexiform

In butterflies (with the exception of the male of *Euschemon* (Hesperiidae) and the Hedylidae) and Bombycoidea (except Sphingidae) there is no frenular-retinacular system for wing-coupling. Instead, the humeral area of the hindwing is expanded into a broad lobe forming an extensive overlap between forewing and hindwing. This provides an *amplexiform* wing-coupling device, although the wings are not locked. In fact, slow motion cine photography shows that the wings of at least some butterflies do not beat as one throughout the cycle of a single wing beat (C. Betts pers. comm.).

Forewing-thorax locking devices

In many Lepidoptera, when in the resting position, the forewings are coupled with the thorax (Common, 1969a; Kuijten, 1974). The mechanism by which this coupling is

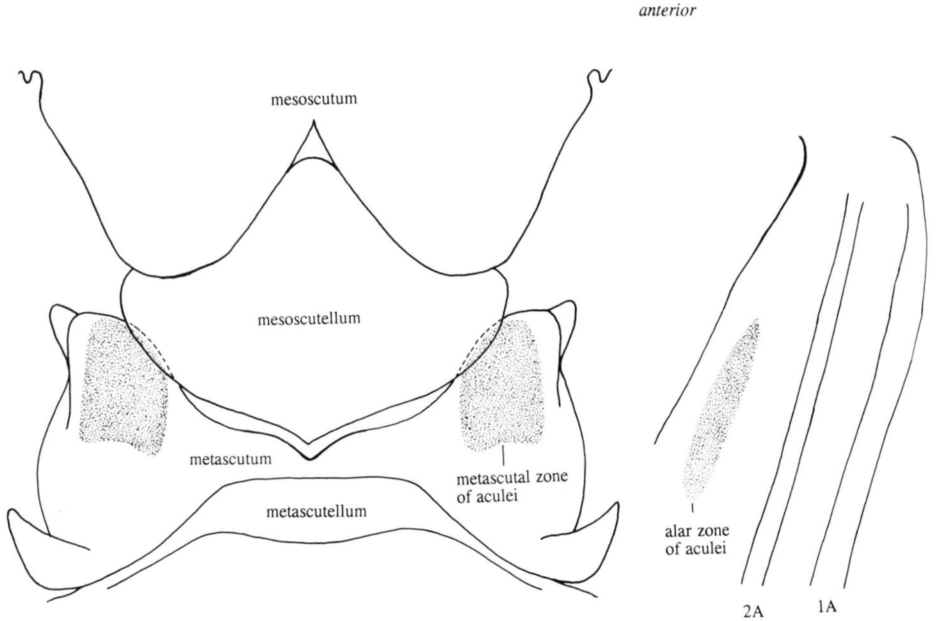

Figs 69, 70. *Forewing-thorax locking mechanism in* Xyleutes durvillei (Cossidae). *69, dorsal view of part of thorax showing spines (aculei or microtrichia) on metascutum; 70, forewing showing alar patch of spines on undersurface.* (*After Common, 1969a.*)

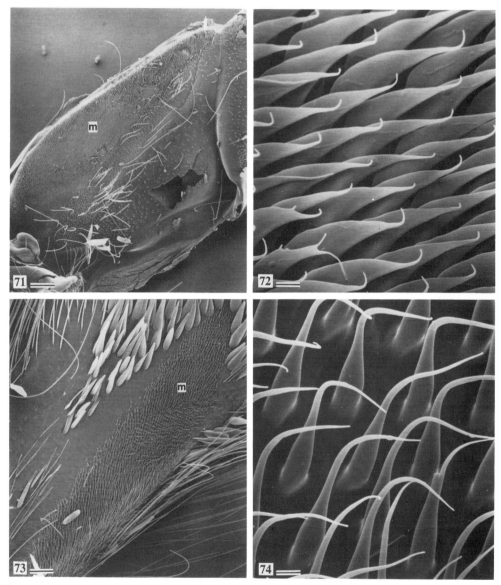

Figs 71–74. *Forewing-thorax locking device in* Orthosia stabilis (Noctuidae). *71, 72, thorax showing microtrichial spines (aculei) (m) on metascutum: 71, whole area (scale, 129 μm); 72, small area, enlarged (scale, 3.6 μm); 73, 74, microtrichial spines (m) on undersurface of forewing: 73, whole area (scale, 90 μm); 74, small area enlarged (scale, 3.6 μm). (EM Unit, BMNH.)*

primarily effected involves well-developed, spine-like aculei (microtrichia), confined to a patch on the metascutum, that interlock with aculei on a corresponding patch situated ventrally at the inner margin at the base of the forewing (Figs 69-74). The aculeate patch on the forewing has been referred to as the *Haftfeld*, spinarea (Minet, [1990a]), or holding area.

This wing-locking mechanism is found extensively in the most primitive moths and in the lower Ditrysia, although one species of Psychidae was found to lack it (Kuijten,

1974). The device is present in Pyralidae but appears to be absent from Thyrididae and Pterophoridae. It is absent from most macrolepidopterans, but occurs in most Noctuoidea. Within Drepanidae, it is present in Thyatirinae, but appears to be absent from Drepaninae and from the peculiar drepanid genus *Hypsidia* (Scoble & Edwards, 1988). Although found in Lasiocampidae, the mechanism is absent from other Bombycoidea. It is universally absent from butterflies (Papilionoidea, Hesperioidea, and Hedyloidea).

The locking device tends to occur in those Lepidoptera that keep their wings closely folded against the body. Where the wings are held away from the body (e.g., as, typically, in Geometridae) or vertically over it (e.g., as in Papilionoidea), the mechanism is lost. There are, however, exceptions.

The forewing-thoracic locking device may also function as a stridulatory mechanism in *Xyleutes* (Cossidae) (p. 151).

Besides the *Haftfeld*, another fairly discrete patch of aculei on the underside of the forewing occurs in many Ditrysia (Börner, 1939; Sattler, 1991a). It is situated on the humeral area and has been termed the *Achselkamm* (shoulder comb). A corresponding patch of aculei occurs on the mesepimeron. The *Achselkamm* functions (Sattler, 1991a) in a way similar to the *Haftfeld*, but locks the costa of the forewing to the side of the mesothorax.

The distribution of this other locking device within the Lepidoptera (Sattler, 1991a) appears to be a groundplan character of the Ditrysia. Although occurring widely it is absent from many families, even some which typically bear a *Haftfeld* (e.g., Pyralidae and Noctuidae).

Hindwing-thorax lock

In Micropterigidae, a hindwing-thorax lock occurs in addition to the one linking the forewing to the thorax (Kristensen, 1984a). The mechanism of the hindwing lock is similar: well-developed aculei forming a patch at the base of the underside of the hindwing interlock with aculei on the axillary cord (an extension of the posterior fold of the scutellum) of the metathorax.

Scales

The wings of Lepidoptera are scaled (e.g., as in Fig. 77), an attribute from which the order derives its name. Scales are considered here, under wings, because the effects they have on the appearance of the wing-surface are particularly pronounced. Nevertheless, scales clothe not only lepidopteran wings, but also the head (including antennae and palpi), other parts of the thorax (including the legs), and the abdomen (including parts of the external genitalia - particularly the valvae of the male). Morphologically, lepidopteran scales are macrotrichia, and thus homologous with the large hairs (and scales) that cover the wings of Trichoptera (caddisflies).

Although scale diversity is considerable (e.g., Kellogg, 1894; Downey & Allyn, 1975), scales share a common basic structure. A typical scale (Ghiradella, 1984) (Fig. 75) fits into a socket (Fig. 78) by way of a stalk or pedicel. The scale body or blade has an upper and a lower lamina. While the lower lamina (the surface that lies next to the wing membrane) is basically smooth, the upper lamina is complex. In surface view, the latter appears as a series of longitudinal ridges traversed by cross ribs so that the whole forms a lattice punctuated by a series of windows (Figs 75, 78). The ridges are composed of longitudinally overlapping lamellae. At least in Ditrysia, very fine vertical ribs (flutes) run down the sides of the ridges to the edges of the windows, or onto the

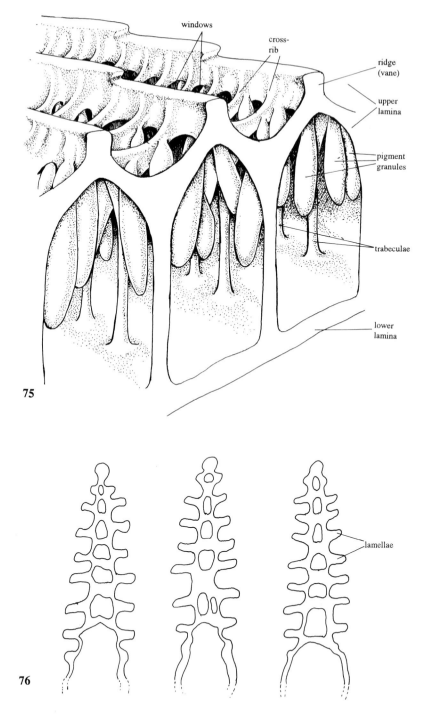

Figs 75, 76. *Scale structure. 75, structure of a typical scale; 76, 'Morpho-type' scale showing multilayer arrangement, formed by lamellae, on three vanes of a scale of* Eurema *(*Pieridae*). (After Ghiradella & Radigan, 1976.)*

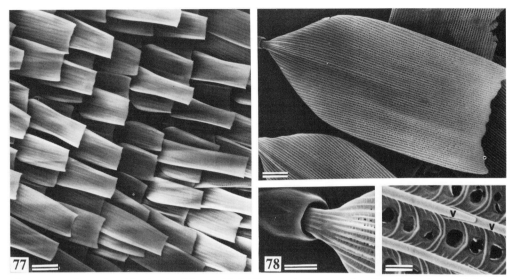

Figs 77, 78. *Wing scales of* Sarcinodes (*Geometridae*). *77, general view (scale-bar, 80 μm); 78, single scale (scale-bar, 20 μm), with detail of socket, lower left (scale-bar, 6 μm), and windows and overlapping ridges (vanes, v), lower right (scale-bar, 0.80 μm). (EM Unit, BMNH.)*

cross ribs. A series of supporting struts, termed trabeculae, run between the upper and lower laminae and also act as spacers.

On the wings of certain Danainae (Nymphalidae) occur androconial scales with a lattice on both upper and lower surfaces (Boppré & Vane-Wright, 1989). Each double-lattice scale arises from a secretory 'cushion' (probably a modified scale base). Not only may these scales function to give mechanical protection to the cushions, but the presence of a lattice on both surfaces would be expected to permit the transfer of secretion (Boppré & Vane-Wright, 1989).

Notable differences are found between scales of the wing surface of primitive moths (here meaning non-Glossata and Eriocraniidae) and other Lepidoptera (Kristensen, 1970). In primitive groups the scales are nearly always 'solid', that is they lack a lumen; in more advanced groups a lumen is present and they are said to be 'hollow'. Examination of the ultrastructure of wing-surface scales in Micropterigidae and Eriocraniidae demonstrated, however, the occasional occurrence of a narrow lumen (Kristensen, 1970). Although wing-surface scales are typically 'solid' in primitive moths, those of the wing margins, which are relatively long and slender and approximately circular in cross-section, have prominent lumens. In addition to being solid, wing-surface scales of primitive Lepidoptera are imperforate; there is no lumen into which 'windows' could open. The outer (obverse) surface of wing scales of primitive moths is densely set with transverse micro-ridges ('flutes'). In non-Glossata, these 'flutes' are overlain by what has been termed 'herring-bone crests' (summary in Kristensen, 1984b).

Scales also occur in insects other than Lepidoptera, including certain Trichoptera, the sister group of the order. In *Pseudoleptocerus chirindensis* (Trichoptera) the various types of scales resembled those of higher Lepidoptera being hollow and having internal trabeculae (Huxley & Barnard, 1988). Since the wing-scales of primitive Lepidoptera are typically solid, the hollow condition of those in Trichoptera are reasonably assumed to have evolved independently. Moreover, unlike the condition in ditrysian Lepidoptera, in Trichoptera the scale-ridges lack flutes.

Scales vary in shape. Downey & Allyn (1975) grouped them into three categories: piliform or hair-like, lamellar or blade-like, and 'other (variable form)'. Piliform scales are round or elliptical in cross section. The shapes of lamellar scales are often described in botanical terms applied to leaf-shape, such as ovate, obovate, or lanceolate. The pedicel (petiole of Ghiradella, 1984) may be short, failing to appear above the socket, or relatively long. In some cases the scale-blade lies at an angle to the pedicel, as in some androconia. The precise direction of the ridges on the upper lamina varies according to scale shape. Downey & Allyn (1975) summarized the various scale components, providing a terminology for wing-scale morphology in general. The vocabulary is large and outstrips our understanding of the precise function of most of the fine details of structures now visible as a result of scanning electron micrographs.

Scales and their sockets develop from a subgroup of epidermal cells (Nijhout, 1985). From the mother cell, the daughter cell nearest the wing surface divides to form the scale cell and the socket cell. (The inferior daughter cell degenerates.) The socket cell becomes wrapped around the former. The first signs of scale appearance occur around the middle stage of development. A finger-like process first projects, then flattens so that a spatulate head is attached to the scale-building cell by a narrow stalk. The extracellular cuticle then develops into the ridges and trabeculae described above.

Scales profoundly affect the natural history of the Lepidoptera. Possibly, the original function of lepidopteran scales was for insulation; refinements of scale design, number and arrangement have probably improved insulating qualities. In primitive Lepidoptera, the scales, which are 'solid', provide less efficient insulation than in more advanced members of the group where they have air-filled lumens. Furthermore, in more advanced Lepidoptera the density of scales appears to be higher, which improves insulation - both because of the greater number and also because of the chance of trapping layers of air between scale layers. Body scaling, especially that of the thorax, may be of particular importance given the need to maintain a high temperature during flight. However, some Tibetan and Patagonian Hepialidae are thinly scaled (G.S. Robinson, pers. comm.), and a high-altitude satyrine nymphalid butterfly from Venezuela has almost scaleless wings (R.I. Vane-Wright, pers. comm.). Since it is mainly scales that give lepidopteran wings their colour, and as dark wings may absorb more heat energy than pale wings, scales have a thermoregulatory function additional to insulation (p. 88).

Colours, and thus scales, play important roles in communication (p. 85) and protection (e.g., camouflage, warning, and mimicry) (p. 78). Physical protection from potential predators is provided to some myrmecophilous Lycaenidae by deciduous waxy scales that cover freshly emerged adults as they escape from the nests of ants.

As structures associated with scent glands and with the distribution of scent, scales are important in chemical communication, particularly in males (p. 161). Most simply, individual scales are associated with single epidermal scent glands; but scent organs have become enormously complex in many groups with scale tufts or brushes being developed.

Finally, scales play a role in increasing the lift:drag ratio in gliding flight (p. 67).

Flight

Lepidoptera fly mostly by flapping their wings, although sometimes there is a gliding component to their flight. Flapping flight occurs as either hovering, or as forward or backward motion. During hovering, or in certain kinds of forward flight, butterflies may use their wings rather like paddles to 'row' through the air (Betts & Wootton, 1988), but most lepidopteran flight involves other mechanisms.

Prior to the application of vortex theory to the aerodynamics of flapping wings, it was thought that insect flight could be explained predominantly in terms of conventional steady or quasi-steady state aerodynamics. However, this view assumed (the so-called 'quasi-steady' assumption) that instantaneous forces generated by a *moving* wing were explained in terms of a conventional *non-flapping* aerofoil like an aeroplane wing. But it was demonstrated (Ellington, 1984a, b) that the lift produced under steady or quasi-steady conditions would be insufficient to support body weight, at least during hovering flight; hence *unsteady* effects, in the form of vortices (air circulation or turbulence), must come into play. Likewise, the amount of lift actually generated during flight in some insects cannot be explained solely in terms of steady or quasi-steady state aerodynamics. In fast-forward flight, the likelihood of the quasi-steady assumption applying is much greater since flapping velocity (a function of wing velocity and the angle of attack of the wing) changes relatively little compared with the mean flight velocity. Deviations from the quasi-steady assumption are likely to increase as flight velocity falls, which is why unsteady mechanisms have been studied initially in hovering - when velocity is zero. But it appears (see Ellington, 1984a), that the quasi-steady assumption fails even to explain adequately fast-forward flight.

If the angle of attack of an aerofoil changes, a vortex is created forcing air downwards and creating lift. But lift will be maintained only if the angle of attack of the wing keeps changing. Since the wings of insects rotate at the top and bottom of every wing stroke, the flight of an insect can be considered in terms of constantly changing angles of attack of the wings, and lift generated by this mechanism is more effectively analysed in terms of shed vorticity (Ellington, 1984a, b).

Unsteady effects not only contribute to lift during flight but may also be employed to generate drastic increases in lift during take-off and possibly even manoeuvring. One such case has been demonstrated in *Pieris brassicae* (Pieridae). This butterfly is unusual in making use of a *vertical* (as opposed to a horizontal or inclined) stroke plane during take-off and hovering. The wings beat flatly at downstroke, but twist strongly during the upstroke so that the angle of attack approaches zero. At the start of the downstroke, the wings 'clap' together dorsally and 'fling' open. Air is sucked into the space between the wings as they move apart. The downward momentum of the air extends below the butterfly after the downstroke is complete and creates a 'vortex ring' as the mass of air is swirled through the surrounding air. The importance of the fling mechanism is that air circulations around the wing have been shown to produce sufficient lift for flight. A vertical stroke plane has also been observed in films of members of the papilionid genera *Pachliopta*, *Graphium*, and *Papilio* and in the fast forward flight of a species of the nymphalid genus *Precis*.

A modified form of the 'clap and fling' mechanism occurs in many Lepidoptera. Wings often clap only partially rather than completely, and then 'peel' apart rather than 'fling' apart at downstroke. The peel not only creates circulations around the wing comparable with those formed by the fling, but may also improve stability of the circulations (Ellington, 1984a, b). It has even been suggested that when the wings of Lepidoptera meet at the extremes of upstroke and downstroke, a funnel is formed through which jet-like forces may be produced (Bocharova-Messner & Aksyuk, 1981).

Another consideration in the aerodynamics of butterfly flight is the use of 'tailed' wings to smooth airflow and so optimize gliding performance (see discussion in Betts & Wootton, 1988). Additionally, these authors found that the forewings and hindwings of the butterflies studied, appeared to be separated during gliding, suggesting a slotting effect delaying stall at low speeds or at high angles of attack of the wings. Although said to be coupled by an amplexiform system (p. 61), the wings of butterflies are evidently separated during various stages of flight. In most moths, by contrast, the wings are coupled by a frenulum and retinaculum (p. 59). For this reason, perhaps, gliding may be less frequently expected in moths, or at least confined to higher speeds. The presence of scales on the wing has been said to increase lift in Lepidoptera by about 15 percent although the precise way in which scales influence the boundary layer of airflow adjacent to the wing is unknown (Nachtigall, 1976).

Flight behaviour can be predicted approximately from wing form (Betts & Wootton, 1988), at least in some butterflies. Although butterflies with fairly similar wing shape (e.g., *Troides* and *Papilio*) may exhibit markedly different flight patterns. Such differences may be explained by different morphological parameters not obvious from visual inspection of wing shape.

Betts & Wootton (who used a variety of morphological parameters) found that they could predict the basic flight modes of butterflies (fast forward, slow forward, hovering, and take-off and climb) from characterizations of wing-shapes. Yet there were also some unexpected observations. Predictions were made about flight characteristics expected for butterflies with short and broad wings, for those with long and slender wings, and for those in which the forewings are extended into narrow tips. Slow agile flight, expected in butterflies with short broad wings, was in fact observed in *Papilio rumanzovia* and *Pachliopta hector*. But although the flight speed of *P. hector* was *generally* low (hovering was frequently exhibited), occasionally the butterflies flew fast. Rapid, and unexpected, bursts of speed were also found in *Precis iphita* (Nymphalidae) despite its broad, short wings. However, in this instance, the high manoeuvrability and slow glides were predicted. Again, the great manoeuvrability of the skipper *Idmon* was expected, but *Idmon* (as with other skippers) is also capable of extremely fast flight.

Long slender wings are associated with prolonged flight. In fast fliers, like *Graphium sarpedon*, wing loading is high. Not only is *G. sarpedon* capable of fast flight, but it is also agile. The agility of *sarpedon* (a skilled hoverer and able to execute rapid manoeuvres) was predictable on the basis of the extension of the forewings into narrow tips.

Insect wings operate in a more complex fashion than do aeroplane wings not only because they flap, but also because they flex to a much greater degree. A far more appropriate analogue of an insect wing is a sail (Wootton, 1981; 1987). Both sails and insect wings are composed of flexible membranes ramified by various stiff supports controlling their deformability. When airflow imparts a propulsive force to a flexible membrane, that force is directed by the supports (e.g., the mast, boom, and gaff in sails, and veins in insect wings). The analogy extends to control. Unlike the situation in vertebrate wings where internal muscles extend along the arm, in insects the wings are controlled from the base - there are no intrinsic muscles. Sails are also controlled externally - by the crew.

In summary, the aerodynamics of even basic flight patterns is far from resolved let alone such complex manoeuvres as tight turns. Although quasi-steady mechanisms may play some part in explaining the aerodynamics of low-frequency and high-amplitude wing beats observed for many butterflies, unsteady mechanisms are undoubtedly required to explain other kinds of flight including the subtleties of basic flight modes.

Migration

Meaning and function

Migration is here taken to mean the act of moving from one place to another (e.g., Baker, 1978) in a direction that is usually seasonally determined and predictable. The purpose of the habit is generally viewed as a means of escaping unfavourable environmental conditions or for moving to an area favourable to the rapid growth of the population (e.g., Young, 1982).

Behavioural biologists tend to adopt a definition of migration more akin to dictionary meanings emphasizing the *action* of moving. By contrast, ecologists tend to view the habit as any ecologically significant displacement. This conceptual difference might be resolved by treating migration as a behavioural process with ecological consequences - a means by which ecologically significant displacements are achieved (Gatehouse, 1987).

Whereas migration is directed, *dispersal* is undirected (Douglas, 1986) but uncertainty exists as to where to draw the line between the two phenomena. In vertebrates it is relatively straightforward to distinguish migration from dispersal since whereas migration involves seasonal two-way movement of large numbers of animals, in dispersal there is a gradual spread from a natal area. Although there are some confusing instances, such as the periodic one-way emigration of lemming, there are sufficiently few anomalies to fit them into the simple migration/dispersal dichotomy. But in invertebrates this dichotomy may be a confusing and inhibitory oversimplification. In insects, certainly, dispersive movements are seen as more of a continuum between a short spread from one habitat to one nearby, through medium length movement, to the very long, and spectacular, mass flights of the Monarch butterfly (*Danaus plexippus*) of distances over 1500 miles. Care is needed not to overemphasize the spectacle of migration. Although some conspicuous mass movements are noted here, they should be considered against the background of other kinds of animal movements. Indeed, whereas migration was thought of as mass emigration and dispersal as a gradual spread resulting additively from short movements to find food or mates, the former is now viewed more as a regular feature in the lives of many species. An extreme point of view was held by Johnson (1969) who regarded migratory flight as 'the prime locomotory act in many, if not most, species of winged insects . . .', since migratory movements occur as one of the first activities of insects after emergence.

The behavioural concept of lepidopteran migration is exemplified by Williams (1958) who defined migration as 'a continued movement in a more or less definite direction, in which both movement and direction are under the control of the animal concerned.' Although noting that there is often a return flight to the original habitat, Williams did not insist on this as a part of the definition, although absence of return posed him with a conceptual difficulty since he thought (incorrectly) that it would lead to the loss of migratory behaviour.

Lepidoptera fall mainly into two of Johnson's three categories of migration (Johnson, 1969). In the first group, emigration occurs without return, usually by short-lived adults. In the second group (Johnson's third category) emigration takes place to hibernation or aestivation sites, and a return flight follows, after diapause of the adults, by the same individuals.

The first group is exemplified by the pierid butterfly *Ascia monuste* (Pieridae) (see below). The butterfly breeds throughout the year in Florida, but not in any particular localities. Adults migrate up to 100 miles along the coast to breed in other localities, although not every generation migrates. Somewhat intermediate in habits is the Painted Lady butterfly (*Vanessa Cynthia cardui*: Nymphalidae) individuals of which may travel beyond the breeding range and either die, or return to the breeding area. Migrants from North America and southern Europe may, for example, reach Scotland, which is beyond the breeding range.

The other main class of lepidopteran migrants include those that move to hibernation or aestivation sites and from which the same individuals return, at least part of the way, after diapause. The Monarch butterfly *Danaus plexippus* (Nymphalidae), which migrates from the northern U.S.A. to the highlands of Mexico, is one example. Another is the Australian Bogong moth (*Agrotis infusa*: Noctuidae), which flies from the plains of New South Wales and Victoria to summer aestivation sites in the Brindabella Range of the Australian Capital Territory before returning.

Johnson (1969) developed a *concept* of migration rather than a strict definition, and this seems appropriate for such an involved and diverse phenomenon that has arisen so many times not only within the animal kingdom as a whole, but also within the Lepidoptera. There are four main components to Johnson's concept of migration. First, migratory flights tend to start where adults emerge and end where they lay their eggs. Thus migration involves a transfer of adults from one breeding habitat to others. Second, migration involves *active* dispersal as opposed to accidental dispersal between breeding habitats. It is not always straightforward to distinguish between active and passive dispersal, but this is more a matter of limited methods and observation than a conceptual problem. Third, migration involves certain essential stages namely an exodus from a habitat with minimal chance of return, and a persistence of flight until a suitable breeding site is found. Most insects depend on wind currents to take them to suitable breeding sites or at least to their vicinity. Appetitive flight to foodplants is not considered part of migration. (An important way in which migratory flight may be distinguished from non-migratory (e.g., appetitive) flight is that the former is not distracted by environmental factors that would normally lead to feeding, mating, or egg-laying.) Fourth, migration, since it is underpinned by moving from one breeding habitat to another, clearly involves females in particular. The development of wings and flight muscles in migrants precedes the development of the ovaries. The development of both flight apparatus and ovaries is influenced by factors such as food, density and temperature.

Emphasis was given by Johnson to the so-called 'oogenesis flight syndrome'. The principle behind this idea is that an insect may become a migrant if the flight apparatus is developed before ovarian maturation. However, this view fails to explain the migration of males and is particularly inapplicable to butterflies (Baker, 1984). Furthermore, although the migratory habit in *Pieris* butterflies (Pieridae) peaks in pre-reproductives, migration in these insects occurs throughout life, and in the Monarch, individuals oviposit on their northerly return flight from Mexico after winter diapause. Nevertheless, in very many cases pre-reproductive migration by females does occur. But an alternative explanation for this situation is that selection maximizes the product of reproduction and flight instead of acting on one or the other. Rankin *et al.* (1986) suggested that juvenile hormone stimulates both migration *and* reproduction.

Migration was seen by Baker (1984) as a change in the 'mean expectation of migration', represented by $\bar{E}$. He viewed migration in the light of individual decision-making over the question of whether to stay in a particular habitat or to move. Individuals have different threshold values of $\bar{E}$. When numbers are too great for a particular habitat, then below-$\bar{E}$ individuals will leave until the habitat becomes suitable such that no more leave. Individuals that have left will look for habitats above $\bar{E}$. A picture of constant motion can be visualized with, at any given time, habitats in equilibrium or below or above equilibrium, and with individuals in transit between habitats. This picture is complicated by the ever changing nature of some habitats and hence the number of individuals that will find them above or below their level of tolerance. Although the tendency to migrate may increase with rising population density, the concept of individuals migrating from more transitory habitats places a different perspective on the subject. Baker emphasized an *individual's* view of habitat requirements and its response, rather than of populations or species.

In this respect, the concept of an individuals 'lifetime track' (e.g., Baker, 1984) is particularly apt and is a concept embracing all forms of movement from spectacular to local. The 'decision' as to whether to migrate or not is influenced by an individual's reproductive potential and the cost that such a migration may incur, for example the chance of not surviving, the energy cost, loss of opportunity for feeding, mating and oviposition.

There is more than one adaptive reason for migration. Williams (1958) took the view that the habit was related to seasonal variation in food supply associated with variable rainfall or variable temperature. Undoubtedly this is an extremely important factor, but others are also critical as in the Monarch where the butterflies cannot tolerate low temperatures in northerly breeding areas and avoid them by southerly movement (Calvert & Brower, 1986).

Southwood (1962) stated that 'the prime evolutionary advantage of migratory movement lies in its enabling a species to keep pace with the changes in the locations of its habitats.' The expectation of this hypothesis is that the migratory habit would be best developed in species with transient habitats and least well developed in those with permanent habitats. This view differs from suggestions that migration is a response primarily to overcrowding. Among the examples given by Southwood to support his hypothesis on arthropods in general, was that of the relationship between type of habitat and the migratory habit. Almost all migrants typically inhabit temporary habitats (e.g., the herbage of hedgerows, and arable land).

Southwood (1962) noted that most migratory Lepidoptera are found in semi-arid areas of the world where habitats are notably temporary. He quoted the work of E.P. Wiltshire on Middle East Lepidoptera who found that some species were more widespread because their migratory habit allowed them to move to and from ephemeral habitats.

Case histories

The three examples of lepidopteran migrants discussed below were selected because they have been particularly well studied, and because the migratory habit seems to have developed in response to different factors.

The Great Southern White butterfly (*Ascia monuste*) in Florida

Migrations of the Great Southern White, which occur along the coast of Florida, were studied over many years by detailed observations and marking experiments (Nielsen, 1961). The directional nature of migratory flight in *Ascia monuste* is suggested by certain flight characteristics. For example the butterflies fly in streams up to 15 meters wide, and close to the ground - rarely more than 3-4 meters above it. Furthermore, in windy conditions the insects fly on the leeward side of sand dunes, and a headwind may inhibit their flight; but in calm weather they fly directly over the dunes. Thus the flight path is essentially independent of wind. The butterflies probably orientate by the sun, presumably by means of polarized light. Visual stimuli, such as shoreline and roads, are also important.

The butterfly breeds along the coast of Florida, but breeding occurs throughout the year only in those colonies situated at the southern tip of Florida. The reproductive range of the butterfly extends a considerable distance up the east and west coasts and increases northwards with a few successive years of favourable conditions. Distribution of colonies is limited by lower winter temperatures.

Migrations occur during a period of about four months of the annual life-cycle, a time at which there is also increased breeding activity. For the rest of the year migratory behaviour does not take place. It appears (Nielsen, 1961) that migration in this species is triggered by crowding rather than by a shortage of food, for migrating butterflies were found to be well fed with nectar.

The Australian Bogong moth (*Agrotis infusa*)

In a few Lepidoptera, the same individuals both migrate and then return to their natal area. This situation is exemplified by the Australian Bogong moth (*Agrotis infusa*:

Noctuidae), a species studied by Common (1954). In spring, the moths migrate from the plains or lowlands of New South Wales and Victoria to caves and rock-crevices in the mountains of the Australian Capital Territory and the Bogong Mountains. These mountainous areas form northern extensions of the Australian Alps. Here in the mountains, the moths aestivate (although a proportion indulge in some flight activity) before moving back to the lowlands in autumn. The larvae feed on a variety of dicotyledonous annual plants that grow in the lowlands in winter, but in the summer the lowlands are dominated by grasses, which are apparently unacceptable.

No larvae are produced by the migrant moths (indeed the ovaries of migrating females remain undeveloped), and the same migrant adults that leave the lowlands in spring return to them in autumn - not a second generation. The moths, therefore, are not migrating to breeding grounds. The purpose of migration in the Bogong Moth may be to avoid laying eggs too early (Common, 1954), for if adults were to remain on the plains, climatic conditions would probably favour egg development well before the availability of foodplants. Evidence suggests that a brief facultative diapause does occur in adults emerging on the breeding ground in spring. This strategy, in which both diapause and migration are combined, contrasts with that usually observed in Lepidoptera where migration occurs to avoid diapause. The result, for the Bogong Moth, is that egg laying is delayed until the moths return to the lowland the following autumn, a time at which larval foodplants are developing.

Aestivation sites are usually found near the summit of mountains in crevices of caves among rock outcrops. The moths form a densely packed, single, imbricated layer (Pl. 1:12) in which each individual holds onto the rock with its foretarsi and the back of a neighbouring moth with its hind- and midtarsi. Such close packing may help the moths to resist desiccation (Common, 1954). Although most remain at rest throughout the day and the night, a small proportion of individuals frequently undergo random flights shortly after sunset. Typically, they return to the shelters after about an hour. A similar flight takes place just before sunrise so activity is related to changes in light intensity.

Aggregations of Bogong Moths in the mountains were recorded from early in the history of European settlement. The phenomenon was noted mainly because the high fat content of the aestivating moths made them attractive as a source of food for Aboriginal people.

The Monarch butterfly (*Danaus plexippus*)

The best known lepidopteran migrant is the Monarch butterfly, *Danaus plexippus* (Danainae), which engages in annual, long distance migrations particularly between North America and Mexico. The Monarch is not confined to North America, but has an extensive distribution (Ackery & Vane-Wright, 1984). Widespread in Central and South America, it also occurs in the West Indies and the Galapagos Islands, is established in Madeira and the Canary Islands, and has sometimes been recorded in western Europe. The species occurs on many Pacific Islands, and in Australia - particularly the eastern part of that continent. The migratory habit is not limited to the North American Monarchs, but is observed also in Australia and New Zealand. Not all Monarchs are migratory; some populations remain throughout the year in lower parts of southern North America and in Central America.

Monarchs occur in two populations in North America (Urquhart, 1987). In the western part of the continent, the butterflies inhabit river valleys in the Rocky Mountains. East of the Rockies they occur over a wide area. In autumn, western Monarchs migrate in a south westerly direction to overwintering sites on the Pacific coast between Bodega Bay in the north to Ventura in the south. Most of those from the east move south to the Neovolcanic Mountains in Central Mexico, but some colonize areas in Florida, Bermuda, the Bahamas, the Greater Antilles, the Lesser Antilles, and the Gulf of Mexico. In the following spring, the same butterflies leave their overwintering sites in Mexico and lay eggs on Asclepiadaceae foodplants in the southern States of North America. These eggs give rise to butterflies that continue the northwards migration. North America is colonized up to about the level of the Canadian border by a series of successive generations. Support for the view that at least most of the recolonizers are not the original overwintering butterflies comes from 'fingerprinting' studies by which Monarchs raised on foodplants from different areas can be identified by their cardenolides (Malcolm, Cockrell & Brower, *in press*). However, the possibility exists that some of the overwintering butterflies do form some component of the recolonization.

Migratory behaviour of North American Monarchs had been observed for many years (Urquhart, 1960) before the spectacular overwintering sites in the high altitude, transvolcanic belt of Central Mexico were discovered by K. Brugger in 1974 (Urquhart, 1976). But although small, winter-breeding colonies were known from Florida, their size failed to account for the numbers of Monarchs returning northwards in the spring (Brower, 1977).

The summer breeding grounds of the Monarch extend over the whole of the eastern half of the United States, the Great Mississippi Valley, and along the Pacific Coast. The migratory flight southwards in the autumn is persistent for it takes place not only with the wind but also against and across it. Migrating Monarchs frequently form temporary roosts at the end of each day; they do not fly at night. During migration, they feed regularly on the numerous meadow flowers in bloom across the continent. Such feeding is important (Brower, Walford, & Calvert, *in press*); autumn migrants starting their journey from central Texas and northern Mexico were found to have very large reserves of fat - far greater than those present in butterflies from the northern states. Furthermore, a significant fall in fat reserves by the end of the overwintering period demonstrated that reserves for winter survival are necessary. Fat content of females is enhanced towards the end of overwintering by transfer of nutrients in the spermatophores of males. Wells & Rogers (*in press*) show that in late winter the lipid content of females increases significantly whilst that of males is depleted.

Overwintering colonies have now been found up to altitudes of 10 000 to 11 000 feet (Calvert & Brower, 1986). Monarch roosts have turned out to be one of the great spectacles of the natural world with butterflies gathering in enormous aggregations (estimated to be of the magnitude of tens of millions) on trees within isolated patches of coniferous forests.

Development of overwintering aggregations (summary in Calvert & Brower, 1986) are preceded by the formation of many smaller, nuclear colonies. Flight activity seems to act as a signal for coalescence of the small groups into larger ones during November and December. Before January and February the butterflies settle on branches of trees, whereas later, when the final colonies are formed, they pack onto the trunks and form clusters sometimes extending the whole height of a tree. By roosting on trees rather than on the ground Monarchs avoid the lowest temperatures (which would often be lethal) and reduce the chance of predation by mice (Glendinning, *in press*), animals known to prey on the butterflies.

Monarchs do not remain entirely inactive at the roosts. During periodic dry and sunny periods tens of thousands of the butterflies fly from the colonies and drink at streams or other sources of moisture.

Why do Monarchs migrate to overwintering sites? One suggestion (Calvert & Brower, 1986) is that by so doing they avoid the lethally low temperatures of the northern summering areas yet avoid the low altitude areas of the tropics, the high temperatures of which would mean continual flight activity and the depletion of fat reserves required for breeding and the spring flight. Low temperatures are unquestionably fatal to larvae and newly formed pupae, which are killed by freezing temperatures (Urquhart, 1960). According to Urquhart, adults recovered from exposure to temperatures below freezing, although they succumbed over long periods of this treatment. Further studies (reported in Calvert & Brower, 1986) demonstrated that Monarch butterflies that are wetted (e.g., by dew) are killed or injured at temperatures slightly below freezing. (Dry butterflies survive lower temperatures.)

Migration and overwintering of Monarchs elsewhere. Monarchs that overwinter in California do so in smaller, but more numerous, sites than in Mexico and the southern United States. The iconoclastic suggestion (Wenner & Harris, *in press*) that Californian Monarchs, at least, do not migrate but exhibit instead a contraction and expansion of their range has been poorly received.

In Costa Rica, Monarchs (like many other butterflies) appear to migrate from areas of dry forest on the Pacific side of the country to the evergreen forest of the eastern side (Haber, *in press*).

In Australia, the Monarch is widely distributed over eastern parts of the continent, whereas in winter the butterflies are confined to two or three coastal areas. As in North America, Monarch migration seems to be related to climate (Smithers, 1977; James, 1984). Observations showed (James, 1984) reproductive dormancy to be induced by short periods of cool, overcast weather in late summer and autumn. Autumnal migratory activity was found to increase on warm sunny days with cold nights. Such activity decreased in late autumn and early winter when days became cool and nights cool to mild.

Colour and pattern

Colour plays an important part in the lives of Lepidoptera. In this section, pattern formation and the main functions of colour in the order are considered. The topics are discussed in this chapter because the structures most important in the display of colour pattern - the wings - are thoracic.

Form and diversity of colour

Colours in insects are structural or pigmental, although a combination of these types is also found. In Lepidoptera, scale-structure or the presence of pigment contained in the scales are largely responsible for colour in Lepidoptera. However, pigment has also been found in wing membranes (see below), and a cuticular thin-film multilayer system is responsible for producing the spectacular gold and silver colours in certain butterfly pupae.

'Structural' colours

Basically, colour is produced in three ways as a result of structural or physical properties of materials. It may be formed by the scattering of light, by interference, or by diffraction. A summary of the extensive literature is given by Grassé (1975). The various structural colours produced by lepidopteran scales are based on relatively minor modifications of the basic components, which are present in all scales, such as ridges, lamellae, microribs, and trabeculae (Ghiradella, 1984; Nijhout, 1991) (see p. 63).

Matt whites, common in Pieridae butterflies, are produced when light is *scattered* in all directions by corrugations and striations on scales. The white colour does not depend on uric acid in the wing scales (as some have suggested) because when this substance is chemically extracted the white colour remains. However, pterines, which may be derived from uric acid, do supplement structural whites. Pearly white found, for example, in *Argynnis* (Nymphalidae), is formed when light is scattered from lamellae formed by the upper and lower surfaces of overlapping scales. Tyndall blue, another colour formed by the scattering of light, is rare in insects. It occurs, apparently uniquely in Lepidoptera, in *Papilio zalmoxis* (Papilionidae) (Huxley, 1976). The Tyndall effect seems to be responsible for a major contribution to the blue colour of this butterfly. Essentially, Tyndall blue is formed when light is scattered by particles, including granules and air spaces, the dimensions of which correspond to the wavelength of blue light. Longer wavelengths tend to pass through. For Tyndall scattering to be visible, the background needs to be black so that light of other wavelengths is not reflected. In this way the blue is not masked. In *P. zalmoxis* the scales responsible for the blue colour are backed by a layer of black scales. The reflecting scales consist of a thick 'alveolar' layer of close-packed, vertical, air-filled tubes. This layer is attached to the longitudinal ridges above and, via trabeculae, to a thin basement lamella below. Apparently, the Tyndall effect is caused by the small size of the alveoli. In *zalmoxis*, Tyndall 'blue' also occurs in the hindwing, but the white or pale yellow patches can be explained by the presence of white rather than black underlying scales.

Interference colours are common in Lepidoptera. Scales are ideal for producing interference effects because their regularly spaced components are separated by distances comparable with the wavelengths of light. Incident light is split into its wavelengths by these periodic structures. When light is reflected the waves may be in phase or out of phase with each other. While waves that are out of phase can cancel each other, those that are *in* phase reinforce each other (constructive interference). When particular wavelengths interfere constructively, the reflected light appears as the colour associated with those wavelengths. In other words, surfaces appear coloured because only certain wavelengths of light are reflected from them. A simple example of interference colours is found in thin films, such as in soap bubbles or oil on water.

In lepidopteran scales, the most commonly encountered interference system is the multilayer arrangement. Reflecting structures are composed of alternating layers of material of high and low refractive index; the thicknesses of these layers are comparable with the wavelength of light. In Lepidoptera, air is often the medium with the low refractive index. A simple thin-film interference effect was proposed by Ghiradella (1984) for the moth *Coleophora apicalbella* (Coleophoridae), in which the gold colour is thought to be produced by the particularly close apposition of the upper and lower scale laminae with a very thin layer of air separating them.

Multiple layers are formed from various components of scales (reviews in Nijhout, 1985, 1991). In the so-called *Morpho*-type of scale, the vanes of the scale are extended into a series of ridges parallel to the surface of the scale (Fig. 76). Colours reflected depend on both the distance between the reflecting layers and the refractive index of the material. The *Morpho*-type scale occurs also in several other Nymphalidae and in certain Pieridae and Papilionidae.

In the *Urania*-type scale, the iridescence is effected by constructive interference of light reflected from stacks of sheet-like laminae within the body of the scale. These laminae are thought to be derived from the trabeculae of the scale. The change in colour occurring when the wing surface is viewed from different angles is brought about by what is effectively a change in the distance between the lamellae of individual scales. An oblique view will have the effect of reducing the inter-lamellar distance, whereas a view down into the surface of the wing results in an apparent increase of the distance. *Urania*-type scales occur also in *Chrysiridia* (Uraniidae), *Ornithoptera* and some species of *Papilio* (Papilionidae), *Aenea* and *Limenitis* (Nymphalidae), and some Lycaenidae. Differences in colour between *Papilio karna* and *P. palinurus* may be explained by differences in thickness of these internally situated multilayers rather than by distinction in refractive index (Huxley, 1975).

Thin-film iridescence is also caused when the microribs on the vanes of a scale become heightened, angled, and partially overlapped. This arrangement also acts as a multilayer system, and has been observed in certain members of Nymphalidae, Papilionidae, and Hesperiidae. The satin-like iridescence of *Battus philenor* (Papilionidae) and of some Nymphalidae is also brought about by modification of the microribs, which, in this instance, are closely spaced and extended across the body of the scale, but are not heightened.

White light is split by *diffraction* into its spectral colours by fine grooves or ridges separated by gaps corresponding to the wavelengths of light. Although scales might potentially provide a means for diffracting light because of their ridges (ribs), the phenomenon is not common in Lepidoptera. Where it exists, the iridescence is 'white' since the irregularity of the ridges produces overlap of the spectra. A diffraction grating is present in some species of *Pierella* (Nymphalidae) (see Laithwaite *et al.*, 1975). In *P. hyceta* there is an iridescent patch on the forewing from which the entire range of colours may be observed when viewed from slightly different angles. A three-dimensional, or volume, diffraction grating is considered to explain the uniform green iridescence over the underside of the wing in the Green Hairstreak butterfly (*Callophrys rubi*: Lycaenidae) (Morris, 1975). In this insect, light is diffracted by the lattice-like structure of each member of a mosaic of polygonal 'grains' found within the scale.

Pigment colours

Pigments may be formed either directly, entirely through the insect's own metabolism, or indirectly from plant chemicals. Although the precursor of a particular pigment may be derived from a plant, introduction of the final pigment may involve subsequent modification by the insect. However they are formed, pigmentary colours result from the absorption of light by pigment-containing chromophores lying within the scale lumen (Fig. 75) and the corresponding display of the colours of the non-absorbed wavelengths.

Six or seven different classes of pigments are currently recognized in Lepidoptera (Kayser, 1985; Nijhout, 1985, 1991; Vuillaume, 1975), but there are almost certainly others (M. Cook *pers. comm.*).

Carotenoids are red and yellow pigments belonging to the tetraterpenoids. They nearly always have a C_{40} skeleton. Two main groups of carotenoids occur - hydrocarbons or carotenes, and oxygen-containing carotenes or xanthophylls. Many different forms of these carotenoids are encountered in Lepidoptera. The substances occur in many parts of the body of adult and immature stages. Carotene precursors are required for the C_{40} skeleton, and Lepidoptera tend to derive these from xanthophylls rather than carotenes.

The tetrapyrrole group of pigments is represented in Lepidoptera by bile pigments, specifically biliverdin IXα and biliverdin IXγ. Bile pigments, generally blue or green, are open chain tetrapyrroles and are metabolized by the insects themselves. Biliverdin of the IXα-type is found in *Thaumetopoea* (Notodontidae). In other Lepidoptera bile pigments are of the IXγ-type. Of these, most occur as pterobilin of which there are examples in Papilionidae, Pieridae, the subfamilies Nymphalinae, Heliconiinae, and Acraeinae of the Nymphalidae, Saturniidae, Sphingidae, Lasiocampidae, and Lymantriidae. Derivatives of pterobilin are also found in Lepidoptera; phorcabilin in males of *Papilio phorcas* and in Saturniidae, and sarpedobilin in *Graphium sarpedon*, *G. weiskei* , and in some Saturniidae. Biliverdins are frequently combined with yellow carotenoids to form a green colour.

Ommochromes, red to brown pigments derived from tryptophan, were first discovered in insect eyes. As xanthommatin they occur in the eyes of Lepidoptera, and in the integument of both adults and larvae; as rhodommatin and ommatin they are found in the wings; and as ommins they are present in the eyes, testis, egg serosa, and ganglia of some species. In lepidopteran larvae, xanthommatin is the only ommochrome found in the epidermis.

Papiliochromes, which occur in the wings of Papilionidae, have a developmental pathway related to that of ommochromes. Several types have been discovered, but papiliochrome II has been most intensively studied. The substance is a complex of L-kynurenine, L-noradrenaline, and α-alanine, and its structure has been studied in *Papilio xuthus*. The presence of papiliochrome II in other Papilionidae is deduced from the presence of kynurenine pigments, which can be extracted from the wings with hot water. Papiliochrome II is considered to be the pale yellow pigment in a number of species of *Papilio*, in *Luehdorfia japonica* and in *Sericinus telamon*. Papiliochrome M is deep yellow and occurs in *Papilio machaon* and *Luehdorfia japonica*, and papiliochrome R is the reddish brown pigment found in the eye-spot of the hindwings of *P. demoleus* and *P. machaon*.

Pteridines occur in the form of pterin, particularly in Pieridae. Their chemical base is synthesized from a purine precursor. Pteridines are distributed widely in insects, and play a part in metabolism. Most are white, but in certain forms they are responsible for some pale yellows and reds. Apart from pterin in pierid wings, other pigmental pterines are present in lepidopteran eyes. In butterfly wings, they occur as elliptical, crystalline pterinosomes. Apparently only a little of the pterin in the wings represents stored excretory products; most wing pterin seems to be synthesized at that site. As wing pigments, pterines have been found only in pierid and heliconiine Nymphalidae.

Although widespread in plants, flavonoids are rare in animals and cannot be synthesized by them. Among insects, they have been found only in some butterflies, and the substances are probably derived from the hostplant by the larvae. In lepidopteran wings they range from white to yellow. Flavonoids are common in dismorphiine Pieridae, and present, but not common, in Papilionidae, Satyrinae, Riodininae, Sphingidae, Agaristinae, Noctuinae, and Castniidae.

Melanins are difficult to extract and purify, are heterogeneous in composition, and their structure is not fully understood. Although all black and most brown pigments found in lepidopteran wings are generally called melanins, these pigments have been identified with certainty in few Lepidoptera, including in the integument of the larvae of two species of moths.

Pigments are usually confined to the scales where they hang as granules from the cross-ribs and trabeculae in the lumen of the scale (Ghiradella, 1984). However in some Lepidoptera, pigment is found between the wing membranes - for example in *Graphium* where intermembranous blue-green and yellow-green are found, which can form an important part of the wing pattern (Allyn, *et al.*, 1981). Modifications occur to the scale layer overlying the areas containing the wing pigment (Allyn *et al.*, 1982). The scales may be transparent and closely appressed to the wing membrane so allowing the pigment to be seen through them. In other instances the scales may be reduced to slender hair-like structures, or they may be absent or reduced in number.

The functions of colour in Lepidoptera

Colour plays an important part in protection and communication. It protects some Lepidoptera by concealment or camouflage; others by the display of aposematic signals, which warn potential predators of toxicity or distastefulness; and yet others by mimetic resemblance to toxic or distasteful species.

In communication, colour functions in courtship and aggressive interaction (agonistic behaviour). A less obvious role exists in thermoregulation where, for example, degree of melanization of the wing bases is important in the absorption of radiant energy and its transfer to the body.

The functions of colour are considered below under the subheadings Protection, Communication, and Thermoregulation, but such divisions mask a compromise. Conspicuous colour patterns may be suitable for sexual communication, but conspicuousness will have to be compromised or occur in some other way if concealment from predators is to be effected. So colour pattern is more appropriately considered as a compromise between the constraints of pattern development and the functions of colour.

Protection

Crypsis. In its most obvious form, protection is gained by concealment, and numerous adult Lepidoptera use colour and pattern to blend with their background. In Geometridae, the wings of many species bear wavy lines or other markings and are held in the planiform (outstretched) position (p. 55). Often the hindwing is similarly marked to the forewing so that at rest the two wings form a continuous pattern, the narrow bands appearing to run on from forewing to hindwing. Dramatic examples of crypsis in Lepidoptera are seen in those species resembling dead leaves (e.g., *Othreis* and its relatives: Noctuidae, and *Kallima*: Nymphalidae).

The classic example of crypsis in Lepidoptera is the phenomenon of industrial melanism, a subject investigated most extensively in the Peppered moth *Biston betularia* by H.B.D. Kettlewell (see Kettlewell, 1973). The traditional story is well known, and need be stated only briefly. The species exists mostly in two forms, which are discontinuous in colour pattern. One form (*typica*) is pale, the other (the melanic, f. *carbonaria*) is dark. (A further intermediate form - *insularia*, also occurs.) Following industrialisation of parts of Britain there was a marked decline in the lichen-covering of trees, so that the trunks and branches became much darker, a marked change from their previous pale colour. Lichens are extremely sensitive to increases in levels of sulphur dioxide, a pollutant that was released into the air from factory chimneys. Correlated with industrialization was an increase in the number of melanic individuals over those of f. *typica*. It was suggested that the increase was a response to predation, particularly by birds: the dark form, better camouflaged against the lichenless trees than the pale form, was spotted, and eaten, less often. With the decline in industrial pollution, brought about primarily as a result of the Clean Air Act in 1956, the process has apparently reversed, and in a locality at Merseyside f. *typica* has increased in proportion.

Attractive as this explanation appears, there are many unresolved problems. A major difficulty lies in the very limited direct evidence for predation by birds (or other visual predators). It has been claimed (Mikkola, 1984) that Peppered moths rest on the undersurfaces of horizontal branches in the canopy, not tree trunks. During work on this moth for a period of over 25 years, Clarke *et al.* (1985) reported that only one specimen had been seen on a tree trunk, but these authors did not believe that Peppered moths rested in the canopy in their own study area. However, they concluded that the rise in frequency of the melanic form was still probably the result of industrial pollution, because, for example, implementation of the Clean Air Act was followed by an increase in f. *typica*. But the precise reason for that change is not fully understood. Although SO_2 levels have fallen in West Kirby, foliose lichens are virtually absent. Nevertheless, there has been a great increase in numbers of *typica*. One suggestion is that the reduction in smoke levels may have caused a decrease in the number of f. *carbonaria* because of a *general* lightening of the background. Other factors have to be considered. Colour pattern of the moths may depend on date of emergence, and the persistence of f. *typica* in industrial areas may be explained in part by immigration.

Choice of background by the moths is a form of behaviour probably under genetic control (Steward, 1985). Experiments show that while individuals of f. *carbonaria* prefer a dark background, dark specimens of the intermediate f. *insularia*, which are phenotypically much closer to those of f. *carbonaria*, tend to rest on a similar (pale) background to that selected by f. *typica*. However, background choice experiments carried out by Mikkola (1984) on f. *typica* and f. *carbonaria* failed to demonstrate any difference in choice between dark and pale backgrounds. These experiments were carried out using lichen-covered branches sprayed with black or white paint whereas earlier studies (Kettlewell, 1955) relied on less natural, smooth surfaces. Moreover, the results of previous release-recapture experiments were questioned by Mikkola on the grounds that the experimental moths were released shortly before sunset, whereupon they immediately alighted on tree trunks to roost instead of searching out their normal resting place.

Probably the best supported explanation of the polymorphism in the Peppered Moth (Berry, 1990 and references therein) is that many factors, not only visual predation, determine the melanic frequencies observed (e.g., the height of the bryophyte layer on trees, January temperature, and sulphur dioxide concentration). However, it is generally accepted that frequencies can be explained in terms of Darwinian selection.

Aposematic coloration. Distasteful or toxic organisms are frequently brightly coloured, and such aposematic colours are thought to signal unpalatability to potential predators. Most Lepidoptera are cryptically coloured and nocturnal (see above), but many are diurnal (true butterflies, skippers, and some moths) and bear what are considered to be warning colours. Some examples of such aposematically coloured Lepidoptera are given below together with a summary of the defensive chemicals thought to give these insects their unpalatable character.

Many Lepidoptera utilize (directly or indirectly) protective chemicals of plants for their own defence. Aposematic colour and then mimicry probably evolved against the background of this chemical defence (Turner, 1984). Unpalatable Lepidoptera are neither all diurnal, nor all aposematically coloured; and nor are all diurnal Lepidoptera unpalatable or aposematically coloured. Nevertheless, unpalatable Lepidoptera are often brightly coloured and diurnal; and the most toxic or distasteful are diurnal (Rothschild, 1985).

Warning colours in animals (e.g., Cott, 1940) are typically found in a combination of two or three such as: red, black, and white; yellow, black, and white; red and black; orange and black; yellow and black; and white and black. These are colours, or a combination of colours, that are conspicuous against most natural backgrounds. In adult Lepidoptera, warning colours occur particularly on the wings, and they are often also present on the abdomen, usually in the form of contrasting bands. With other aposematic animals, Lepidoptera share certain associated characteristics. Many are freely exposed (a prerequisite for effective advertisement) and slow-moving - the 'easy' flight of danaine butterflies provides an example. A number are gregarious, which probably increases the speed at which predators learn the pattern. Examples among the Lepidoptera include danaine, heliconiine, and acraeine butterflies (Cott, 1940) (although defence does not appear to be the primary function of aggregations in Danainae), and zygaenid and some arctiid moths (Rothschild, 1985).

To be effective, warning colours need to be both bold and limited in number: patterns and colours are most effective if they are learned easily and remembered by predators (Cott, 1940). Lepidoptera with aposematic colours may have a tough integument; an example well known to butterfly collectors is the robust nature of danaines. The advantage of such a characteristic is that members of the group can withstand a certain amount of damage by predators that have either not learned of their distastefulness, or that need to be reminded of it.

The effectiveness of bright colours and chemical defence against lepidopteran predators is uncertain. Little is known about the invertebrate predators of Lepidoptera (Brower, 1984) but, among vertebrates, the importance of birds is evident from both direct observation and from the study of beak marks on wings of butterflies (see reviews of Brower, 1984; Rothschild, 1985). Unpalatable species are more often beak-marked in collections than palatable ones, suggesting that beak-marked butterflies are frequently released after capture. Beak-marked wings have been further distinguished from beak-torn wings based on the principle that the former indicate that unpalatable butterflies will be released while the palatable butterflies are likely to escape only by breaking away from the beaks of birds.

Evidence to show that aposematically coloured Lepidoptera derive protection from predators is extensive (see review and references for butterflies in Brower, 1984). Nevertheless, protection is never complete, and there is always an 'evolutionary arms race' between predator and prey - a new strategy by the predator eliciting an evolutionary response from the prey and so on. For instance, orioles that feed on Monarch butterflies in overwintering colonies in Mexico have been observed to slit the abdomens of the butterflies and reject those with a high concentration of cardenolides. Notwithstanding the extreme avoidance of noxious butterflies by blue jays (e.g., Brower & Brower, 1964), birds will generally eat distasteful Lepidoptera if they are sufficiently hungry. The acceptance of Lepidoptera by predators is more appropriately viewed in terms of a 'palatability spectrum' rather than of absolute acceptance or rejection. Factors that affect this spectrum depend on such criteria as the physiological state, food requirements, and previous experience of the predator, as well as the degree of palatability or toxicity of the insect.

Brower (1984) divided defensive chemicals of butterflies into two main classes. Class I chemicals are noxious; they may be distasteful or may cause an adverse reaction to a predator, such as inducing vomiting. Class II chemicals are innocuous; they are not toxic, and cause no adverse reaction to a predator. The defensive properties of class II chemicals seem to arise because, by conditioning, a predator may associate them with physical unpalatability (e.g., being tough or spiny) or a prior experience of attacking prey that proved to be noxious. Class I chemicals possibly also act as class II chemicals when their concentration is high enough to be sensed but too low to be noxious.

Substances involved in chemical defence in adult Lepidoptera are either derived from secondary plant substances, or are synthesized by the insect either in the adult stage, or by the larva or pupa and then carried through to the adult. Both kinds may be represented in a given species.

The main secondary plant substances were divided by Harborne (1982) into three major groups - nitrogenous compounds, terpenoids, and phenolics. The distribution of toxic secondary plant substances in British macrolepidopterans was summarized by Rothschild (1985: table III). Many defensive chemicals have been extracted, so only a few are mentioned here to illustrate the diversity found. Nitrogenous compounds are exemplified by cardiac glycosides and pyrrolizidine alkaloids (PAs), substances that often co-occur in a few Danainae. While cardiac glycosides are undoubtedly Class I chemicals, being distasteful and emetic, it is unclear as to whether PAs act as Class I or as Class II chemicals or as both (Brower, 1984). PAs do not act as deterrents themselves because, although potentially lethal, they do not act immediately. Diterpenes belong to the terpenols and are found, for instance, in *Hyles euphorbiae* (Sphingidae), which derives the substance from *Euphorbia* as a larva.

Among those substances synthesized by the Lepidoptera themselves rather than derived from the foodplant is one of the most toxic of chemicals, namely hydrogen cyanide (HCN). HCN has been extracted from the crushed tissue of various stages in the life-cycle of zygaenid moths. Other self-synthesized substances involved in defence are acetylcholine, histamine, and those with acetylcholine, histamine, and adrenaline-like activities (Rothschild, 1985: table IV). By contrast, a further group of substances, pyrazines, are probably both synthesized and derived from plants.

The best known examples of noxious Lepidoptera that are aposematically coloured are found among Papilionoidea, Arctiidae, and Zygaenidae. There are, of course, many others examples outside these groups.

Both cardiac glycosides and PAs are found in danaine butterflies, the orange and black colours of which are typically aposematic. Cardenolides are sequestered by the larvae, which feed on members of the Asclepiadaceae (milkweed family), and make all of the stages in the life-history noxious. Cardenolides are both distasteful, and, if consumed, also have an emetic effect on predators. In the Monarch butterfly, the highest concentrations of cardenolides occur in the wings, but the abdomen contains the most highly emetic substances. In females both concentration and emetic potential is considerably higher than in males (Brower & Glazier, 1975).

Brower (1984) argued that the emetic properties of cardenolides can be explained without invoking group selection. Birds can discriminate emetic from non-emetic individuals by taste without killing the butterfly. Since a high concentration of distasteful cardenolides are found in the wings, and as birds tend to catch butterflies by their wings, the insects may be released relatively unscathed. Furthermore, cardenolides are not distributed regularly within the body of an individual or between the sexes, which implies that there is a physiological cost in the incorporation of these chemicals. This may explain the presence of non-emetic and emetic individuals in a population, the cost of the defence being balanced against the benefit of protection. However, it would surely appear necessary to invoke group selection in those Lepidoptera with emetic, but without distasteful substances.

PAs are known to function in courtship (Chapter 6), but their role as defensive chemicals in Danainae (Brower, 1984) is less clear. In butterflies, these substances are probably obtained by the adults from nectars or from withered vegetation of species belonging to Boraginaceae, Asteraceae, and Fabaceae. In some Arctiidae moths (e.g., *Utetheisa* and *Creatonotos*) PAs are derived by larvae from the foodplants (e.g., Boppré & Schneider, 1989). It is not clear whether PAs behave as Class I or as Class II chemicals (see above). However, since *Creatonotos* larvae may feed on plants that do not contain PAs without impairing mating success in the adult, these substances may provide some kind of chemical defence. In Danainae, they may act synergistically with cardenolides (Brower, 1984).

Brown (1984) found that certain Ithomiinae butterflies (close relatives of Danainae) are defended by PAs against the tropical orb spider *Nephila clavipes*. Ithomiinae also obtain PAs as adults, and spiders eat only freshly emerged butterflies, which lack these compounds. It appears that only males directly obtain PAs but that, because much of the PA component of males accumulates in spermatophores, female ithomiines probably derive their PAs by mating. This suggestion may explain why female ithomiines may mate several times.

Both PAs and cardenolides are also found in Arctiidae moths. The chemicals may be derived from a variety of plants, for example from members of the Boraginaceae, Papilionaceae, Asteraceae, Asclepiadaceae, and Apocynaceae. Arctiidae include many brightly coloured, diurnal species, with a great range of defensive chemicals. (Nocturnal species have been found to emit warning sounds in a number of cases, p. 158.) In *Arctia caja*, the PAs and cardiac glycosides in the adult are separated from the larval foodplants, while acetylcholine, histamine, and α,α-dimethylalcrylylcholine are synthesized by the insect itself (Rothschild *et al.*, 1979). The moth is red, black, and white and thus exhibits typical warning colours. Aplin & Rothschild (1972) suggested that whereas the Monarch butterfly shifted to asclepiad hostplants thus acquiring cardenolides, and while zygaenids secrete HCN permitting them to feed on cyanogenic clovers and vetches, *A. caja* and *Tyria jacobaeae* were perhaps protected by their warning colours and self-synthesized toxins allowing them to survive on a wide range of toxic plants. Arctiidae from different geographical areas do in fact feed on a wide range of plants containing PAs and cardiac glycosides. However, care is needed in generalizing about the acquisition of unpalatable substances by adult Lepidoptera from their larvae feeding on poisonous hostplants (Ackery, 1988).

Zygaenidae, which include the burnet and forester moths, also exhibit striking warning colours such as red, orange, black, and white in various combinations. (The larvae are also often aposematically coloured.) Furthermore, they often feed in aggregations, which adds to their effective advertisement. HCN is released from crushed tissue of the moths (and other stages of the life cycle), and it appears that the substance is synthesized by the insect via cyanoglucosides. The pliant, elastic cuticle of the moths helps recovery from attack by predators. Besides HCN, histamine and acetylcholine are also released as a result of damage. A further chemical defence is gained from the presence of pyrazines, substances that may be derived from food-plants as well as synthesized by the insect.

It was suggested (Turner, 1984), that Lepidoptera sequestered poisons from their foodplants prior to the evolution of conspicuous colours. Poisoning a predator, as opposed to being distasteful to it, obviously fails to protect the organism eaten. However, by the conditioning of the predator a selective advantage may be conferred on its relatives (kin selection). The evolution of distastefulness, bright (warning) colours, and finally gregariousness, provides defence for the individual. This view of a gradual development of aposematic protection implies a shift from kin selection to individual selection. Both probably operate in the Lepidoptera today.

Deception. MIMICRY. Mimicry is a complex subject that has grown around some basic observations and interpretations by the older naturalists, the best known of whom were H.W. Bates and F. Müller. Since the time of Bates and Müller there has been considerable debate as to how mimicry evolved, and how it is maintained. Mimicry is considered in this chapter because of the fundamental relevance of the colour and pattern of lepidopteran wings.

Although only Batesian and Müllerian forms of mimicry have been studied in detail (Turner, 1977), they should be seen in the context of mimicry as a whole. Mimetic resemblance may be classified into a number of theoretical compartments (Vane-Wright, 1976). These divisions are based on the dynamic interactions between three components, the 'mimic', the 'model', and the 'operator'. In this tripartite system, first clearly proposed by W. Wickler, the mimic and the model are seen as signallers, and the operator as the signal-receiver. The dynamics possible with these components involve the biological role of the operator to the model, the biological role of the operator to the mimic, and the result of the interactions of the operator on the model-mimic relationship. These three interactions may be positive or negative in each case so that there exist, in principle, 2^3 (i.e., 8) different categories of mimetic relationships.

The three components (model, mimic, and operator) may belong to one species, three different species, or to two species. In the case of two species, the model and operator may belong to the same species, the model and mimic may belong to the same species, or the mimic and operator may belong to the same species. There are thus 5 categories. By combining the 8 interactions with the 5 categories of species composition an 8×5 matrix can be drawn up giving 40 theoretical compartments of which 21 are represented in the natural world (Vane-Wright, 1976). It is in this perspective that Batesian and Müllerian mimicry should be seen. Under this view, the

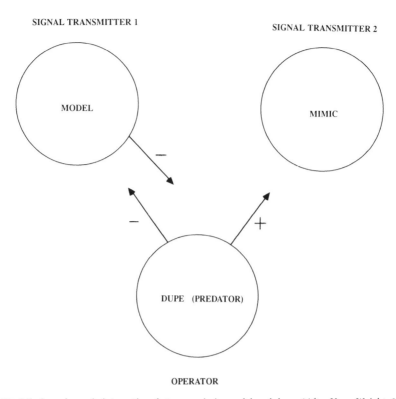

SIGNAL TRANSMITTER 1

SIGNAL TRANSMITTER 2

MODEL

MIMIC

–

–   +

DUPE   (PREDATOR)

OPERATOR

Fig. 79. *Mimicry: dynamic interactions between mimic, model and dupe. (After Vane-Wright, 1976.)*

concept of mimicry is of a *dynamic* system in which 'the terms model and mimic describe functional entities in a schematized biological interaction: they are not fixed properties of individuals' (Vane-Wright, 1981).

The dynamics of a typical case of mimicry (Fig. 79) involves a predator being duped by a palatable mimic resembling an unpalatable model. The response of the dupe (operator) is positive with regard to the mimic (it regards it as inedible), but negative (disadvantageous) to the model since if the predator *does* start to eat mimics it will associate the pattern with palatable prey and start killing models.

The two dominant forms of mimicry in the Lepidoptera may be summarized thus. In Batesian mimicry one or more palatable forms mimic an unpalatable model, whereas in Müllerian mimicry one or more unpalatable forms exhibit the same pattern. In the former, the model must be sufficiently numerous for the predator to associate its pattern with unpalatability. A relatively abundant mimic would serve only to advertise its own palatability, and would thus also endanger the model. In the latter, the effect of sharing the same pattern means that the predator needs to recognize one, rather than two or more patterns to associate it with unpalatability.

The purpose of this brief statement of a complex biological phenomenon is to put lepidopteran mimicry in perspective in the natural history of the order. Hence, rather than simply list the common examples that can be found in many entomological texts, two points of general interest from the discussion of Turner (1984) are summarized.

Batesian and Müllerian mimicry should be viewed not as disjunct phenomena, but rather as two phases of what has been termed the palatability spectrum (Turner, 1984). On the unpalatable side of the neutral point on this gradient from complete palatability to powerful unpalatability, the mimics will be Müllerian; on the palatable side the mimics will be Batesian. Each type of mimicry probably evolves in two stages. In Batesian mimicry the first stage involves a major genetic shift in which a pattern fairly close to the model develops. Although the resemblance may be rather weak, the degree of similarity of the mimic must clearly be sufficiently close to the model to confuse predators. If it is not sufficiently close, the model may advertise its own palatability.

The second stage involves a refinement through a gradual shift towards a near perfect resemblance between mimic and model by way of small genetic changes. Thus the basic (stage 1) similarity between *Battus philenor* and its mimic *Papilio polyxenes* is produced mainly by a single gene that changes the mimic from yellow to black. The stage of refinement (stage 2) involves details of the yellow spotting.

A similar two-stage evolution probably occurs in Müllerian mimics, in the first stage of which the least unpalatable form will converge on the most unpalatable. Although the first stage is similar in both Batesian and Müllerian situations, there are marked differences in the second step. In Batesian mimicry the development of a palatable mimic reduces the fitness of the model, so that a shift in the pattern of the model away from the converging mimic will be favoured. In Müllerian mimicry, the second stage involves a convergence of *both* forms on an intermediate pattern.

A consequence of Batesian mimicry is that in successful mimicry the pattern of the mimic must evolve to keep up with that of the model. Turner makes two points. One is that the advantage of being a mimic must be greater than the disadvantage of being a model. The other is that computer simulation has shown that a Batesian mimic is likely to shift more rapidly towards a model than are two Müllerian species likely to converge.

FALSE HEADS. In many Lycaenidae, the hindwings are patterned and otherwise modified such that the posterior end of the butterfly appears to bear a false head (Pl. 1:9) (review by Robbins, 1980). The modifications involve the presence of a dark, head-like pattern and an extension of each hindwing into 'tails', which have been though to resemble antennae. The tips of these pseudoantennae may be white, making them more conspicuous than the true antennae, and strong lines along both the forewing and hindwing may guide a vertebrate's eye to the false head of the insect.

Lycaenidae often move their hindwings back and forth in opposite directions when held vertically over the body in the typical resting position. Suggestions that this motion may enhance the deception of the morphological modifications by causing the 'tails' to move like probing antennae are questionable (Robbins, 1980). Doubts arise from observations that some Lycaenidae without false heads move their wings, and that wing motions are sporadic.

False heads defend the butterflies, probably by deflecting the attacks of predators (particularly birds and lizards) away from the true head rather than by either alarming predators or confusing them by the presence of two heads. Support for the false head hypothesis (Robbins, 1980) is indefinite but suggestive. First, wings break most easily at the anal angle of the hindwing, where the false head is situated, hence a butterfly has a better chance of breaking away from a predator if seized by this area. Second, Lycaenidae with false head markings have a higher incidence of predator damage on the hindwings than do those without such markings. Third, lizards have been observed to attack Lycaenidae at the posterior part of these butterflies, and the predators were successful only if part of a butterfly's body was seized (Someren, 1922).

The false-head hypothesis, proposed to explain markings on Lycaenidae, is probably a particular example of a more general phenomenon. Eyespots are found widely among Lepidoptera, occurring frequently on the wings and, on a number of occasions, on the bodies of caterpillars. Possibly (Blest, 1957a), larger eyespots alarm potential predators (see below) while small spots deflect attacks of predators from more vulnerable parts of the insect's body. But this division between the function of large and small eyespots is not sharp. Experimental evidence supporting the view that small eyespots actually do direct the attacks of small birds, was derived from controlled experiments in which Yellow Buntings were offered mealworms (highly edible) artificially marked with small eye-like spots (Blest, 1957a). However, the deflection response of the bird was found to be only transient, which limits the degree of protection gained.

ALARM PATTERNS. Eyespots on wings (e.g., Pl. 1:5) appear to function not only to deflect the attacks of predators from more vulnerable areas of the body (see above), but also to intimidate them. The most effective eyespots are those resembling most closely the eyes of large avian predators (as in *Caligo* and *Precis*: Nymphalidae). Eyespots have been shown experimentally to intimidate small insectivorous birds possibly because they resemble the eyes of their own predators (Blest, 1957a). The presence of such eyespots on lepidopteran wings is usually associated with movements to display them (Blest, 1957a). For example, *Inachis io* (Nymphalidae) opens its wings from the normal closed resting position displaying the forewing spots, then protracts the forewing so that the spots on the hindwing are exhibited. The butterfly also orientates itself so that the spots are clearly visible to the predator. The displays of certain Saturniidae and Sphingidae, with or without eyespots, possibly evolved from movements made by moths preparatory to flight (Blest, 1957b).

Communication

Nocturnal Lepidoptera communicate mainly by scent whereas diurnal species include a major visual component in their repertoire. Detailed studies of the use of colour in inter- and intra-sexual communication have been confined to a few species of butterflies, and generalizations are essentially extrapolations from these works. Therefore, although many moths are diurnal and may perhaps use colour in sexual communication, this section is necessarily restricted to work on butterflies.

Colour is particularly important in the initial attraction of male butterflies to females and in distance signalling between males (e.g., Silberglied, 1984). Signals based on ultraviolet (UV) reflection or absorption has been demonstrated to be of great significance in inter- and intra-sexual communication in some butterflies, and is undoubtedly important in many others (Silberglied & Taylor, 1978; Silberglied, 1979). Therefore colour plays both an intra-specific and an inter-specific role in butterflies.

Butterfly eyes respond to a much broader spectral sensitivity (300 nm in the UV through 700 nm in the red) than do vertebrate eyes (although no information is available for Hedylidae, see Chapter 12). This breadth means that butterflies can make use of a private system of communication invisible to vertebrates, as well as 'visible' colour for protection against vertebrate predators (Silberglied, 1984). There are at least two colour receptors in Papilionoidea and Hesperioidea, and perhaps more than three in some species of butterflies and moths (see Silberglied, 1984), so butterflies have good colour vision.

Male butterflies are attracted to females by visual stimuli - colour (including UV reflectance or absorption), movement and, to a lesser extent, shape. Experiments with cardboard dummies demonstrated that males are most attracted to colours displayed by their conspecific females. Despite their great variety, the details of colour patterns on butterfly wings appear to be insignificant in courtship, a finding that conflicts with the view commonly held.

Not only is colour (as perceived by vertebrates) an attractant, but so also is 'colour' in the UV. Males of *Colias eurytheme* and *C. philodice* (Pieridae) are attracted to UV-*absorbing* females. By contrast, UV-*reflecting* dummies were more attractive to males of

Pieris napi, P. bryoniae, and *P. rapae crucivora* than those that absorb UV. Whereas in *Colias* the females are UV-absorbing (males may reflect or absorb UV), *Pieris* females are UV-reflecting relative to their males (Silberglied & Taylor, 1978). In particular, at least in *Colias*, males are attracted mainly by the visual stimuli associated with the wings rather than by the body, and by the hindwings more than the forewings. Studies have shown also that hindwing colour must be of the correct hue and of low saturation and intensity. Silberglied and Taylor studied only ventral wing surfaces, which are the surfaces visible when the butterflies are at rest. After the initial visual response of males to females, odour takes over as the stimulus as the stages in courtship progress.

Vertebrate-visible colour of males is apparently of little or no significance in mate selection by females (e.g., Silberglied, 1984). This statement is based on very few studies, work on the topic being limited by the technical difficulties of obtaining a supply of sexually active males and virgin females. However, in some species, UV reflection certainly is of importance. Females of *Colias eurytheme* rejected significant numbers of males experimentally changed from their normal condition as UV-reflectors to UV-absorbers. By contrast, females of *C. philodice* were not prevented from accepting their naturally UV-absorbing males when these males were artificially made to reflect UV. Visual stimuli (vertebrate-visible or UV) seem to play no role in mate choice. Females of *C. eurytheme* seem to respond to the UV absorption to distinguish between their conspecific males and those of *C. philodice* (Silberglied & Taylor, 1978).

It has been suggested (Silberglied, 1977; 1984) that the main purpose of the brilliant vertebrate-visible coloration of so many butterflies is explained by their use in distance-signalling between males. The evidence is circumstantial, but the explanation would account for the relatively low variability of male colours and the absence of mimicry restricted to the male sex. Moreover, male butterflies are well known to chase others out of a particular area or to defend perches from which they pursue females (see Silberglied, 1977). This idea contrasts with the view more usually held, that bright colours were assumed to be selected by females in mate choice.

UV reflectance may also be important in male-male interaction. The UV-reflecting dorsal surfaces of *Colias eurytheme* are displayed to other males during flight. Furthermore, at the approach of other males, males in copula have been observed to open their wings so exhibiting their UV-reflecting patches. This habit has the effect of deterring the flying male from further approach.

Thermoregulation

Thermoregulation is considered here because it principally involves control of thoracic temperature over fluctuations in environmental temperatures. Although the process is usually associated with adequately raising the temperature of muscles for flight, overheating must also be avoided. The most obvious need for thermoregulation is in arctic or alpine environments where very low temperatures are encountered, or in areas with cold winters where adult Lepidoptera are active. But some form of temperature control probably occurs in all Lepidoptera. Extremes of temperature can be survived only by hibernation and diapause. But hibernation permits such extremes to be avoided so it contrasts with thermoregulation, where there is some form of control.

Physiological, behavioural, and morphological mechanisms are involved in thermoregulation. For example, at low temperatures the temperature of the thoracic muscles may be raised by the physiological process of muscular thermogenesis ('shivering', wing vibration). At higher temperatures orientation to the sun to maximise solar heat energy is important. Morphological adaptations, such as the 'hairiness' of the thorax and the bases of the wings, also play a significant part in absorption and retention of heat.

Butterflies are usually ectothermic (in the sense that they gain heat from the environment), relying on solar heat energy to increase body temperature, while moths are endothermic (in the sense that they gain heat from biochemical processes inside the insect) and engage in muscular thermogenesis to raise body temperature (summary and references in May, 1985). But although 'shivering' is not usually associated with butterflies, there are several exceptions (summary and references in Heinrich, 1981). For example, in the Mourning Cloak butterfly (*Nymphalis antiope*), 'shivering' causes the thoracic temperature to rise by 8-11°C in a few minutes permitting effective use of the leg muscles so that the insect may force its way out of cramped overwintering conditions (Douglas, 1986).

Because ambient temperatures are usually too low, 'shivering' (summaries and references in Kammer, 1981; Heinrich, 1981; May, 1985) permits large moths to raise their thoracic temperature to a level at which take-off is possible. 'Shivering' involves isotonic contraction of the flight muscles, so the amplitude of wing movement is very small. Essentially, the wings do not beat. By not beating, possible damage to the wings is avoided, predators are less likely to be attracted to the moth, and loss of heat by convection is reduced. Neurophysiological studies suggest that 'shivering' has evolved independently many times within the Lepidoptera, including within single families. The effectiveness of such pre-flight warming is considerable: thoracic temperature rises in a linear fashion, so heat production must increase exponentially to account for thoracic heat storage and heat loss during the process. While the temperature of the thorax may rise up to 10° per minute, that of the abdomen rises little so the increase is virtually confined to the flight muscles.

Diurnal Lepidoptera (particularly butterflies) bask in sunshine making considerable use of radiant energy to raise thoracic temperature to the level necessary for flight (summaries and references in Casey, 1981; May, 1985; Shreeve, 1990). It appears that reduced flying time over the whole life of female butterflies leads to reduced fecundity (e.g., Kingsolver, 1988), so basking is evidently adaptive. Butterflies tend to be most active when their thoracic temperature lies between around 28-30°C. Different basking strategies are used by different butterflies, but important factors common to all of them are the position of the wings to the sun and colour of the wings. In 'dorsal basking' the wings are held fully open so that their dorsal surfaces are warmed. In 'lateral basking' the wings are closed over the body and the butterfly tilts so that only their undersides receive sunshine. Uniquely in Pierinae (Pieridae), the upper surfaces of the wings are used to reflect radiation onto the body ('reflectance basking') (Kingsolver, 1988).

The body itself directly absorbs a significant portion of the heat energy gained from sunlight. However, the importance of the wings in raising body temperature is demonstrated by the fact that when butterfly wings are shaded, the excess body temperature (i.e., that gained by basking) is reduced by about 30 percent (Wasserthal, 1975). But only the basal third of the wings (about 15 percent of their surface area) is involved, and the pattern on the rest appears to be irrelevant in the absorption of heat either directly (by conduction) or indirectly (by convection) transmitted to the body (Wasserthal, 1975; Kingsolver, 1988). Heat transmission by means of conduction, which seems to occur in lateral baskers but not significantly in dorsal baskers, is limited to the base of the wings because of the low thermal conductivity of the wings (Kingsolver, 1988). These findings resolved the problem in earlier work (e.g., Vielmetter, 1954; 1958), which recognized the importance of wings in thermoregulation, but failed to address the question of whether wings functioned directly in heat absorption, or whether they influenced body temperature indirectly by altering airflow around the body. They also disputed the results of Kammer & Bracchi (1973) (working on the Monarch butterfly - *Danaus plexippus*) that the wings were of little importance in absorption of radiant heat.

In dorsal baskers the wings influence body temperature by causing an accumulation of warm air under the wing bases (Wasserthal, 1975). In neither dorsal nor lateral baskers does haemolymph circulation play a significant role in the transfer of radiant heat energy from wings to body. The increase in temperature of the body due to the wings in lateral baskers is high because large areas of the wings are in contact with the body during basking.

Several studies have shown (see Kingsolver, 1988) that heat absorption rises with intensity of melanization at the base of the wing, and that degree of melanization is related to environmental conditions. In colder regions, melanization is greater and where there is seasonal variation in temperature within an area, multivoltine species differ in their degree of melanization.

The close relationship between the orientation of the wings and degree of melanization is seen within Pieridae in which there are two different thermoregulatory mechanisms (Kingsolver, 1988). *Colias* (Coliadinae) butterflies are lateral baskers; the underside of the hindwing is exposed to sunlight. By contrast, in *Pieris* (Pierinae) butterflies the dorsal surfaces of the wings are involved in basking and the wings are held partly open. Furthermore, while the bases of the wings are employed in heat absorption, the middle sections in *Pieris* are used to *reflect* radiation onto the body (Figs 80, 81). Since melanization causes heat to be absorbed, the areas of the wing involved in reflecting heat are pale. The relative extent of the dark and pale areas on the wings beyond their bases would be expected, on theoretical grounds, to determine the angle at which the wings are held during basking; the more extensive the melanization of the marginal areas, the further apart the wings would be expected to be held. Such different wing angles are indeed observed within *Pieris*. Moreover, when the amount of black pigment was increased experimentally in the subgenus *Artogeia*, wing angle during basking was increased significantly.

When body temperatures of butterflies reach 40-42°C overheating occurs. At these temperatures the insects adopt a heat-avoidance temperature in which the body and wings are orientated parallel to the sun's rays.

Wing scales were found (Wasserthal, 1975) to be responsible for about 14 percent of body temperature generated by basking. The dense covering of hair-scales over the thorax and wing bases increases heat absorption and reduces heat loss; and the striations on the scales (p. 63) may also increase absorption (Douglas, 1986). A pilose covering of the wing bases, thorax, and abdomen has evolved independently in arctic and alpine Lepidoptera.

Considerable amounts of heat energy are generated during flight itself. Heat loss is related to body size (inversely), weather, and mode of flight (e.g., rate and amplitude of wing beat). At high wing-beat frequency overheating is a hazard, particularly if the ambient temperature is high. Blood circulation plays an important role in cooling.

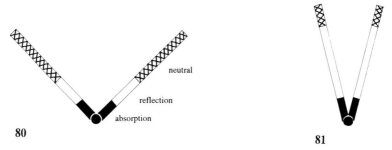

Figs 80, 81. *Melanization and wing angle in* Pieris (Pieridae) *(diagrammatic). In 80, wing angle is large and solar reflecting areas are small. In 81, wing angle is reduced and solar reflecting areas are increased. (After Kingsolver, 1988.)*

The development of pattern

The great diversity of pattern in lepidopteran wings (at least among macrolepidopterans) may be accounted for by a relatively simple developmental system (Nijhout, 1978, 1981, 1985, 1991). Essentially, the wing-patterns of butterflies and other macrolepidopterans can be classified into a few fundamental elements, such as eyespots ('ocelli') and bands. These elements are probably developed by means of a chemical signal termed a morphogen, thought to be secreted by a small cluster of wing cells ('foci'). Spreading from any one source or focus, the morphogen induces pigments to be synthesized in individual scales.

Ultimately, the overall colour and pattern of lepidopteran wings depends on the colour and distribution of individual scales. Each scale usually carries a single pigment, and most Lepidoptera have only three basic colours - perhaps up to five in the most colourful butterflies. The enormous range of non-structural colours is accounted for by the different concentration of pigment in particular scales, and by the mixing of different coloured scales to give varying effects. At least occasionally, colour in individual scales is formed from more than one pigment (e.g., green in *Nessaea* (Nymphalidae), Vane-Wright, 1979).

Earlier this century, wing patterns were classified into a few basic elements. The full complement of these elements was expressed as a 'groundplan' (Fig. 82), an arrangement resembling, and expressly derived from, patterns typically found in nymphalid butterflies. Nijhout referred to this arrangement as the *Schwanwitsch/Süffert 'nymphalid groundplan'*, in recognition of those workers independently responsible for its construction. Although termed a groundplan, the arrangement is purely hypothetical and not intended to imply an ancestral or primitive condition. From the components of the groundplan most patterns displayed by butterflies and other macrolepidopterans can potentially be derived. The elements of the groundplan occur against a *background* of one or more colour fields. Such fields, which typically occupy large areas of the wing, are often disrupted by components of wing pattern.

Wing pattern elements frequently appear as serially repeated basic components. A row of ocelli, for instance, is made up of one ocellus in each wing-'cell', a cell, in this sense, being defined as an area bounded by two wing veins. Indeed, Nijhout constructed a groundplan additional to that of Schwanwitsch and Süffert, not for the entire wing but for a wing-'cell'. It is at the level of the wing-'cell' that the determinants of pattern act. Just as with wing pattern in its entirety, a relatively few sources can comfortably account for the diversity of wing-'cell' patterns observed.

Basic patterns generated by the sources include, for example, a border 'ocellus' and an intervenous stripe. Interaction between sources probably affects the shape of the basic elements, and a further basis of variation is the superposition of basic elements. Whereas overall wing pattern was constructed from a comparative study of wing pattern alone, the derivation of wing-'cell' pattern has been explained in terms of morphogen theory and from the results of cautery experiments on pupal wings prior to pattern formation.

According to Nijhout, the factors determining pattern are: the foci on the wing-'cell' midline; the characteristics of the signal or signals emitted from the foci (e.g., they may differ in strength and timing); the way in which the signals interact; the function of the wing veins (e.g., as sinks for signals emitted from foci); and the way in which signals are interpreted by cells that build scales. Pattern-determination proceeds in three steps. First, the position and properties of sources and sinks are determined; second, the sources are activated and gradients are established for the substances or signals produced by the sources; and third, the gradients are interpreted by scale-building cells so that a particular pigment is determined for each scale.

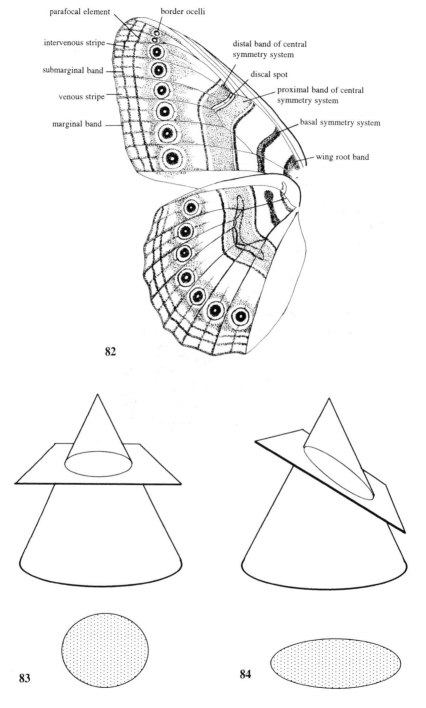

Figs 82-84. *82, the nymphalid 'groundplan' wing pattern; 83, 84, concentration gradient and wing pattern: 83, projection of conic section forming a circular pattern; 84, projection of conic section at an inclined plane forming an elongated pattern. (82, after Schwanwitsch, 1924; 83, 84, modified after Nijhout, 1978.)*

Wing patterns are usually determined early in the pupal stage before pigmentation is visible, for if the area of a potential eyespot on the wing is cauterized no such spot will develop when pigments are formed. It is assumed that cauterization destroys the cells of the focus responsible for secreting the morphogen. The morphogen is thought to act by switching on or off genes responsible for pigment formation. Concentration of the morphogen gradually reduces as it diffuses from the source. The effect of reduction in morphogen concentration may induce a corresponding change in pigment concentration.

The effect of the morphogen can be visualized (Nijhout, 1985) in terms of a conical concentration gradient (Fig. 83). The focus is represented by the apex of the cone. Morphogen concentration is highest at the apex, and reduces, along a gradient, from the apex. Under this model, particular scale-producing cells are thought to respond to particular concentrations of morphogen. Cells producing, for example, black scales might be induced to do so at one concentration; those producing blue scales at another. Circular patterns (spots or rings) are explained by imagining that the cells producing the pigment are induced to do so at a certain concentration. The concentration is represented by a transverse section across the cone (Fig. 83).

Central and basal symmetry systems (Fig. 82) are derived from processes similar to those responsible for the formation of ocelli. Microcautery experiments have shown that the central symmetry system of *Ephestia kuhniella* (Pyralidae) appears to be derived from two or three sources. Bands are formed by the merger of pigments derived from the same kinds of sources that form ocelli. Scallop-shaped bands are produced when rows of foci form rings that merge with each other.

However, many arrangements, for instance straight-edged bands, sagittate shapes, and ellipses, cannot be explained from circular diffusion patterns. Nijhout pointed out that any deviation from a circular eyespot must mean either that the morphogen fails to diffuse at the same rate and in all directions from the source, or that the cells are not influenced by the morphogen in the same way. It is believed that differential response of the cells is of prime importance. In terms of Nijhout's model, just as a circular eyespot may be considered the result of a transverse section of the conic projection, so an ellipse would be formed if the plane of the section sloped (Fig. 84). It has been proposed that sagittate shapes may be induced by the interaction of a second interacting with the first. Although this explanation is entirely hypothetical, Nijhout pointed out that the interaction of two gradients could potentially account for great numbers of variants found on lepidopteran wings.

4

THE ADULT ABDOMEN: SEGMENTATION, AND THE GENITALIA

The abdomen, which is the most obviously segmental of the three sections of the insect body, is associated particularly with visceral functions such as digestion, excretion, and respiration. In this chapter, special emphasis is given to the reproductive structures and their function. Emphasis is placed on the male and female genitalia not least because they are of particular importance in our understanding of the phylogeny of the Lepidoptera. The structure of the female genitalia is also discussed with reference to the way in which the spermatozoa are transferred from the storage receptacle to the egg. Egg-laying is also considered in relation to the form of the female genitalia.

Pregenital abdomen

The segments that precede those modified to form the genitalia are of a similar, annular structure although basally the sterna are modified. Each segment is composed of a tergum, sternum, and two pleura, and the segments are separated by intersegmental membranes. The pleural membranes of females are enlarged permitting expansion as the ovaries develop and increase in size.

Abdominal base

The greatest complexity in the pregenital segments occurs at the base of the abdomen where there are extensive modifications, particularly to the sternum. The modifications are chiefly the result of the articulation of the thorax with the abdomen and, in several groups, the presence of abdominal tympanal organs. The basal, sternal sclerites of the primitive Lepidoptera were discussed by Kristensen & Nielsen (1980), and the abdominal terga and sterna of *Micropterix* (Micropterigidae) were studied in detail by Kristensen (1984a).

The first abdominal sternum (A1) is much reduced even in the most primitive Lepidoptera, and it is absent from all Heteroneura. In *Micropterix* sternum A1 is a small distinctive structure with its corners extended into pointed arms (Fig. 85). A further pair of processes arise paramedially from the posterior margin of the sclerite. Sternum A2 is more like a typical annular abdominal sternum. Its most conspicuous modification is the presence of a pair of membranous windows (fenestrae), which almost divide the anterior rim of the sclerite from the remainder. In Agathiphagidae, Heterobathmiidae, Eriocraniidae, and Neopseustidae the form of sterna A1 and A2 are basically the same as in Micropterigidae. In Acanthopteroctetidae the anterior rim of sternum A2 is completely detached from the rest of that sclerite and associated with sternum A1 (Fig. 86). In Hepialidae sclerotizations of sternum A1 are often absent, as are fenestrae on sternum A2. Sternum A2 of some primitive Heteroneura is represented by two sclerites, one smaller and anterior and the other larger and posterior. In all Heteron-

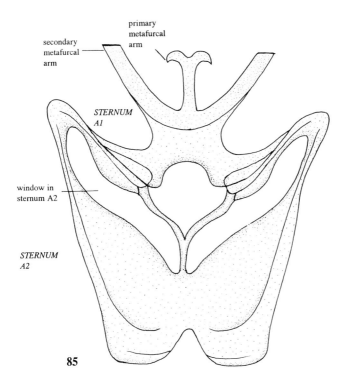

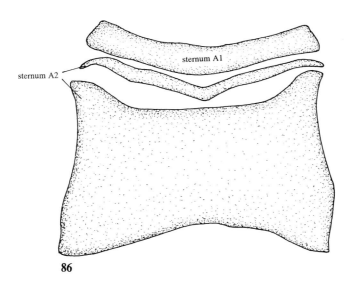

Figs 85, 86. *Base of abdomen* (*ventral*). *85,* Micropterix calthella (*Micropterigidae*); *86,* Acanthopteroctetes bimaculata (*Acanthopteroctetidae*). (*85, after Kristensen 1984a; 86, after Davis 1978.*)

eura, sternum A1 is absent (Figs 87-89), and this loss is a specialised character of the group. Within the Heteroneura sternum A2 is positioned more anteriorly, and in Ditrysia it lies below, or partly below, tergum A1.

In Ditrysia (Figs 88, 89), sternum A2 bears a pair of apodemes extending from its anterior margin into the thoracic lumen. The presence or absence of a pair of sclerotized sternal thickenings (here termed venulae, following Minet (1983), a term equivalent to 'sternal rods' of Brock (1971)) was said to divide the Ditrysia into those with a 'Tineoid type' arrangement (with venulae) and those with a 'Tortricoid type' arrangement (without venulae) (Brock, 1971; Heppner, 1977). In fact, the development of apodemes and venulae is too variable within the groups to permit accurate separation (Kyrki, 1983a).

A more consistent division of the Ditrysia is between those Lepidoptera in which the anterior corners of sternum A2 are elongated into anteriolateral processes where they

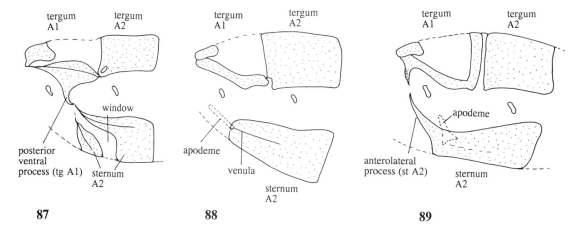

Figs 87-89. *Base of abdomen (lateral). 87, Incurvariidae; 88, 'tineid' arrangement; 89, 'tortricid' arrangement. (After Kyrki 1983a.)*

meet ventral processes from tergum A1 (tortricid type), and those without these processes (tineid type) (Kyrki, 1983a). In Ditrysia, the tergo-sternal connection is pre-spiracular (Fig. 89), while in non-Ditrysia it is post-spiracular (Fig. 87). A probable lepidopteran specialization is the presence of a process, on each lateral margin of tergum A1, which extends downwards and posteriorly and articulates with the anterior corners of sternum A2 (Kristensen & Nielsen, 1979; Kristensen, 1984b). Although this process is frequently reduced, it is of general occurrence in non-Ditrysia (Börner, 1939).

In *Micropterix*, tergum A1 is sclerotized only at the margins, the desclerotization of this tergum possibly representing a lepidopteran groundplan specialization (Kristensen, 1984b). The dorsolongitudinal muscles of the abdomen are apparently three-layered in pterygotes; in *Micropterix* the loss of the outer layer of these three muscles is correlated with the desclerotization of tergum A1. The acrotergite is formed largely from a pair of paramedial lobes that appear as extensions of the anterior sclerotized margin of tergum A1. Morphologically, these lobes are actually derivatives of the metapostnotum (Kristensen, 1984a).

The presence of tympanal organs at the base of the ventral abdomen in pyralids, geometrids, and some other families is discussed below (Chapter 6).

Other modifications

Glands on sternum A5

The presence, in many primitive Lepidoptera, of a pair of glands opening on sternum A5 (Fig. 90) is a specialized character of Amphiesmenoptera (Kristensen, 1984b), but these structures have become independently reduced in many groups of both Trichoptera and Lepidoptera. The only detailed study of their morphology in Lepidoptera has been carried out for Micropterigidae (Kristensen, 1984a). An efferent duct runs internally from the opening of each gland to a sac-like reservoir, the contents of which are expelled by contraction of a network of surrounding muscle fibres. The contents are released at the aperture of the gland by a muscle inserting on the efferent duct. In the southern hemisphere group of Micropterigidae (see Chapter 9) the opening of each

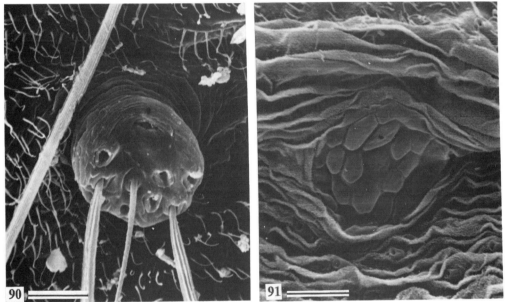

Figs 90, 91. *Abdominal structures in Micropterigidae. 90, sternum A5 gland in* Sabatinca chalcophanes; *91, dorsal tuberculate plate on segment A2 of* Micropterix calthella. (*Figures courtesy of Dr N.P. Kristensen; after Kristensen, 1984a.*) *Scales, 20 μm.*

gland is found on a protrusion from the sternal wall. Scales, presumably for dissemination of the contents of the reservoir, are present on the protuberance.

The glands are thought to contain pheromones or defensive repellents (Kristensen, 1984a). They are present (although not universally) in both sexes of Micropterigidae (Fig. 90) and Eriocraniidae, but only in males of Agathiphagidae (Davis, 1975c).

Besides the sternal A5 glands, there are paired windows on sternum A4 of female Eriocraniidae, Neopseustidae, and Nepticulidae (Davis, 1978). At least in Eriocraniidae, these windows seem only to indicate the presence of reservoirs, the contents of which are released from the orifices on sternum A5.

Tuberculate plates

On the pleura of certain primitive moths, exist structures called 'tuberculate plates' (Fig. 91) (Kristensen, 1978a). First noted on pleura A2-6 of the hepialoid genus *Neotheora*, the plates were also found to occur in a corresponding position in

Micropterix. In *Micropterix* the dorsal plates mark the site of attachment of chordotonal organs, and in *Eriocrania* (Eriocraniidae) the ligaments that open the spiracles are attached to them. In *Afrotheora* (Hepialoidea), plates of a similar texture are present along the anterior margin of terga A2-7 (Nielsen & Scoble, 1986). 'Tuberculate plates' also occur in primitive Ditrysia (Nielsen, 1978). These structures generally demarcate insertions of dorsoventral muscles that are usually broken down after adult eclosion (Kristensen, 1984a).

Further modifications of the pregenital abdomen confined to individual groups, such as the cteniophore of certain Notodontidae and the pleural lobes of Yponomeutoidea, are noted under the appropriate sections in Part 3.

Genitalia

Male

The male genitalia of Lepidoptera (Figs 92, 93) are complex structures, the various components of which are not fully homologized. In particular, it has been difficult to assign several of the sclerotizations to particular segments from which they have been derived. The value of the male genitalia in the classification of the order is well known, and these structures have been used in innumerable taxonomic revisions. Species in Lepidoptera are often largely 'defined' from related species by the subtle differences between the male genitalia. A bewildering variety of terms has been applied to the various parts, and an invaluable attempt at their homology, together with a summary of the structure of the organs, was provided by Klots (1970).

Proximally, the framework of the genitalia is provided by modification of segment A9 and part of A10. Typically, tergum A9 (the tegumen) is fairly large and hood-like, and is known as the tegumen. The anteriolateral corners of the tegumen usually extend to meet the lateral arms of sternum A9, termed the vinculum, and takes the form of a narrow U-shaped band, usually extending medioventrally as a sac-like structure of variable shape called the saccus. The union of the lateral arms of the vinculum and the lateral extensions of the tegumen forms a ring-like structure. Sometimes an unbroken ring is formed, a situation thought to exist in the lepidopteran ground plan (Kristensen, 1984c) and to have developed independently in a number of more advanced members of the order.

The major part of tergum A10 forms the uncus, although an element of it is incorporated into the tegumen, a structure that is therefore compound. Often, the uncus is a triangular sclerite, with the base running along the caudal edge of the tegumen. Frequently, the apex of the uncus is divided, and in some instances the entire structure is lost. At the base of the uncus, on each side, may occur a socius. Socii are soft, hairy lobes of varying sizes. Sometimes prominent, at other times they are very short and difficult to see. Their derivation is uncertain, but they are generally considered to be modifications of tergum A10, or of the intersegmental membrane between A9 and A10, or of the cerci (appendages of abdominal segment A11). The socii may not be homologous in different groups. It was tentatively suggested (Kristensen, 1984c) that socii may be derived from the venter of A10.

The most prominent modification of sternum A10 is the gnathos. This structure, when single, is actually formed by a fusion of components from each side. Sometimes the two parts do not fuse, and in other instances the gnathos may be entirely absent.

The phallus, penis or, as it is generally termed by taxonomists, the aedeagus, and the paired clasping organs, the valvae, are derived from the primary phallic lobes. The primary phallic lobes develop in the late larval or early pupal stage in Lepidoptera (e.g., Matsuda, 1976). Each lobe divides into an inner mesomere and an outer

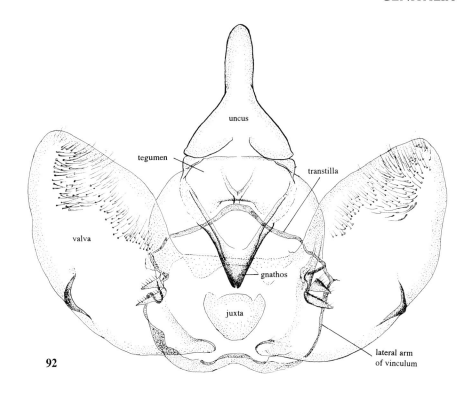

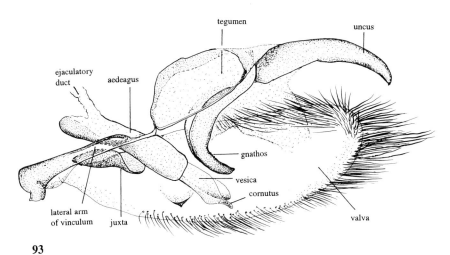

Figs 92, 93. *Male genitalia of* Arhodia lasiocamparia *(Geometridae): 92, ventral view (aedeagus removed); 93, lateral view. (Drawings by Geoffrey Kibby.)*

paramere. The paramere becomes the valva and represents, if correctly interpreted, a true segmental appendage - the coxopodite. The aedeagus, the distal portion of the phallus, is formed through development and fusion of the mesomeres. The basal part of the phallus is termed the phallobase. The aedeagus becomes invaginated from its caudal end during development to form an endophallus (the vesica of taxonomists), which in turn fuses with the caudal part of the ductus ejaculatorius. The vesica is eversible, and usually bears sclerotized spines, or other structures, called cornuti. These help to provide a grip within the bursa copulatrix of the female during mating.

The end of the abdomen is closed by a membrane, the diaphragma, which is perforated by the aedeagus. Three functional regions occur in the diaphragma (Klots, 1970): the fultura superior (dorsal), the fultura inferior (ventral), and the anellus (central), which folds around the aedeagus. However, there are no clear divisions between these regions. There are several sclerotizations of the diaphragma, the homologies of which are uncertain. A ventral sclerotization of the fultura inferior acting as a support for the aedeagus, and found widely in Lepidoptera, is known as the juxta. The homology and derivation of the juxta is not entirely clear, but studies on primitive Lepidoptera suggest that at least part of the structure in higher Lepidoptera is very likely to be homologous with sclerites of corresponding position in Micropterigidae and Eriocraniidae (Birket-Smith & Kristensen, 1974) and Heterobathmiidae (Kristensen & Nielsen, 1979).

The valvae function primarily to clasp the female during mating. They range from small to large, and may be simple or complex. Although great variety exists in the form of the valvae, Sibatani *et al.* (1954) attempted to homologize the various areas of these structures in Rhopalocera, Geometridae, and Noctuidae.

The components mentioned above represent the main structures of the male external genitalia. However, many other sclerotizations are studied in taxonomic work, for example dorsal and subventral thickenings of the anal tube (scaphium and subscaphium respectively), the various subregions of the valvae, and sclerotizations of the anellus.

Ground plan of the male genitalia in Lepidoptera

A suggested ground plan of the male genitalia in Lepidoptera has been derived (Kristensen, 1984a,c,e; see also Stekol'nikov & Kuznetsov, 1986) from detailed comparative studies of the primitive taxa *Agathiphaga* (Agathiphagidae) and Micropterigidae. A great deal of homoplasy is apparent, and the ground plan reconstruction is not definitive.

Segment A8 is probably simple because that is the condition in the trichopteran ground plan and in the early Glossata. Complexity in preglossatan moths is therefore potential misleading.

In most Lepidoptera, segment A9 is divided into a tegumen (dorsally) and a vinculum (ventrally). However, primitively the segment probably takes the form of an undivided ring, since that is the condition in the ground plan of the Trichoptera.

In the mecopteroid condition the gonopod is divided, but in Lepidoptera the valva (a gonopod derivative) is undivided except, perhaps, in Micropterigidae and Eriocraniidae where there are basal sclerotizations that may indicate a division. Thus it remains uncertain as to whether or not the gonopods of Lepidoptera are two-segmented. Kristensen suggested that the 'clasper' or 'harpe' in Ditrysia may represent the distal segment. However, since no such structure exists in non-ditrysians, the basic condition may reasonably be supposed to be undivided and the presence of a clasper or harpe in Ditrysia a reversal or an independent development.

The ejaculatory duct is enclosed by a partly sclerotized tubular aedeagus. The absence of a sclerotized aedeagus in Exoporia is assumed to represent a derived condition since the sclerotized state occurs in Trichoptera, non-glossatans and, indeed, in lower (non-exoporian) Glossata.

The 'basal plate' of Trichoptera is probably homologous with the 'juxta' of Lepidoptera. A median plate, presumably homologous with the basal plate and the juxta, is found between the bases of the gonopods in Micropterigidae, Heterobathmiidae, Eriocraniidae, and Acanthopteroctetidae.

Segment A10 is entirely separate from segment A9, a condition exhibited by some Micropterigidae, Agathiphagidae, and some Eriocraniidae.

Female

In the female genitalia (Weidner, 1934; Mutuura, 1972; Dugdale, 1974) (Figs 94-96) an oviduct runs from each of the two ovaries, and unites with its opposite number to form a common oviduct (oviductus communis). This duct continues posteriorly until it enters the invaginated genital chamber or vagina. Spermatozoa are stored in a sac-like corpus bursae and transferred either directly to the spermatheca, in which they are temporarily stored before being released into the vagina to effect fertilization, or by way of a ductus seminalis. Eggs pass through the ovipore, which is found at the end of the attenuated 'ovipositor' or is surrounded by a pair of fleshy anal papillae. (The 'ovipositor', in this sense, is not an ovipositor in the strict, morphological definition; this latter formation is constituted by ventral appendages on A8 and A9.) The ovipositor is frequently short in Lepidoptera, taking the form of a pair of broad, setose anal papillae. In various groups it is attenuated into a longer, cutting or piercing structure.

Upon this basic plan exist three major variations in the arrangement of the internal genitalia.

Internal morphology

The internal genitalia are composed of both mesodermal derivations and invaginated ectodermal structures. The three basic arrangements mentioned above involve differences in the position of the common oviduct and the number and position of the genital apertures.

Monotrysian (Fig. 94). In all non-Ditrysia other than the Exoporia, the common oviduct enters the vagina ventrally. The bursa copulatrix (corpus bursae plus ductus bursae) lies dorsal to the oviduct. This (primitive) arrangement is similar to the condition in other panorpoid insects (Dugdale, 1974). A *single* genital opening serves as both an ovipore and a copulatory pore. The ductus bursae enters the vagina directly; there is no ductus seminalis, but spermatozoa may be stored in a seminal receptacle (receptaculum seminalis) before passing to the vagina along the spermathecal duct. A cloaca, into which both anus and ovipore open, may be present.

Exoporian (Figs 95, 97, 98). Unlike the monotrysian condition, in Exoporia there are two genital apertures one for egg-laying and the other for mating. In this respect the exoporian condition resembles that of the Ditrysia. The position of the common oviduct, which opens dorsally into the genital tract and lies dorsal to the bursa copulatrix, is also similar to the arrangement occurring in Ditrysia. But there are two important differences between the exoporian and ditrysian conditions.

Firstly, no ductus seminalis exists in Exoporia, so spermatozoa released from the bursa copulatrix must presumably travel externally to the vagina, where they are temporarily stored in the spermatheca. Bourgogne's hypothesis (Bourgogne, 1946) that the spermatozoa run along a seminal groove or tract running between the copulatory pore and the ovipore is generally accepted (e.g., Dugdale, 1974). This tract is actually enclosed (at least in the primitive hepialoid genus *Fraus*, from Australia) by appression of paired transverse folds termed intergenital lobes (Figs 97, 98) (Nielsen & Kristensen, 1989). However, this enclosure does not mean that the tract is strictly internal. In some

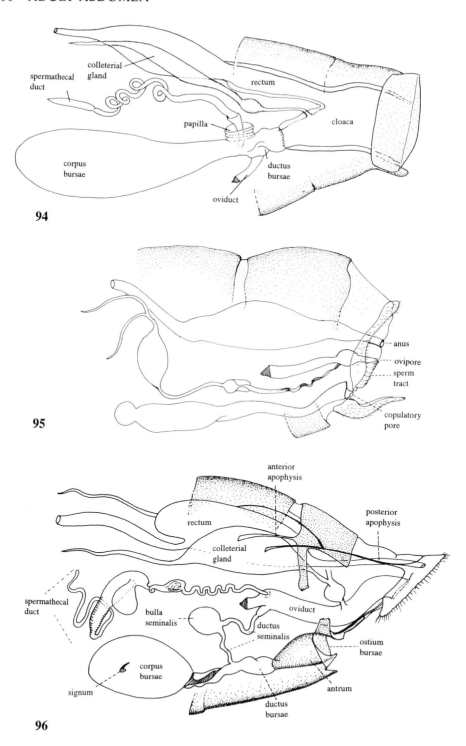

Figs 94–96. *Female genitalia (diagrammatic). 94, monotrysian; 95, exoporian; 96, ditrysian. (After Dugdale, 1979.)*

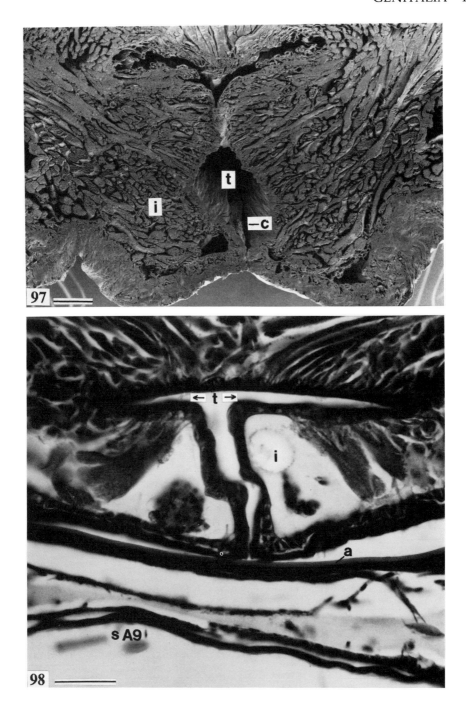

Figs 97, 98. *Female genital structure of Exoporia* (Fraus polyspila: *Hepialidae* sensu lato). *97, posterior part of intergenital lobes (i) (appressed) and surrounding area (anterior view), showing sperm tract (t) and intergenital cleft (c); 98, horizontal section through sperm tract (t) enclosed behind intergenital lobes (i) (a, floor of antrum; sA9, sternum A9). (Courtesy of Dr N.P. Kristensen and Dr E.S. Nielsen; after Nielsen & Kristensen, 1989.) Scales, 50 μm.*

cases the ovipore and genital aperture are said to be opposable, and spermatozoa are thought to be transferred directly (Dugdale, 1974).

Secondly, both the ovipore and the copulatory pore open on the same segment (A9), whereas in Ditrysia the ovipore opens terminally but the copulatory pore is situated ventromedially between segments A8 and A7, on segment A8 or on segment A7. A further difference is the universal absence of colleterial glands in Exoporia, and their usual presence in Ditrysia.

The exoporian arrangement was discovered by Oiticica (1947) who noted that certain Hepialidae had two rather than one genital apertures. At that time, Hepialidae were regarded as monotrysian Lepidoptera. Oiticica's observations were examined by Bourgogne (1946) who accepted the presence of the two apertures, but disputed Oiticica's added record of an internal ductus seminalis. Given the absence of such a structure, Bourgogne was left to account for the transfer of spermatozoa from the bursa copulatrix to the spermatheca. He hypothesized that spermatozoa travel along a furrow between the copulatory pore and the ovipore and demonstrated the presence of such a structure in those species he studied. Subsequently (Bourgogne, 1949), the genital system exhibited by the Hepialidae was called exoporian, a term now used taxonomically to include Hepialoidea and Mnesarchaeoidea, in which the female genitalia have a similar arrangement.

Ditrysian (Fig. 96). The Ditrysia include about 95 percent of all Lepidoptera. The common oviduct is situated dorsal to the bursa copulatrix, a condition similar to that found in Exoporia. Again, as with Exoporia, all members have two genital apertures, but in Ditrysia the copulatory pore opens ventrally on sternum A8, between sterna A8 and A7, or on sternum A7. Spermatozoa travel along a free ductus seminalis to the vagina. The ductus seminalis may be expanded at the same point along its length into a sac-like bulla seminalis that functions for the temporary storage of sperm. This arrangement led Dugdale (1974) to refer to the Ditrysia (*sensu* Börner, 1939) as endoporian Ditrysia, as opposed to exoporian Ditrysia (i.e., Exoporia).

Variation in internal structure

Cloaca. A cloaca is present in Zeugloptera (Dugdale, 1974), as in many Trichoptera (A. Nielsen, 1980). It is not present in Trichoptera (or related groups). The anus and gonopore are separate in Heterobathmiidae (Heterobathmiina) (Kristensen & Nielsen, 1979), but share a common chamber in Agathiphagidae (Aglossata), Eriocraniidae, and Acanthopteroctetidae. The cloaca may be present or absent in primitive (non-ditrysian) Heteroneura; it is absent in Exoporia, and present or absent in Ditrysia (Dugdale, 1974).

Spermatheca. The spermatheca (Figs 94-96) is specialized in Lepidoptera. Its duct usually bears a vesicle (Weidner, 1934) although this is rudimentary in Micropterigidae. In other Lepidoptera a constriction in the cuticular intima of the duct lumen functionally divides the latter into two compartments. One has a particularly narrow lumen, and its intima is markedly thickened; this is the efferent duct or 'fecundation canal'. In most higher Lepidoptera the fecundation canal is coiled around the wider afferent duct, and the spermathecal canal *as a whole* is usually also coiled. The spermathecal duct leads into a lagena, and a muscular utriculus. In Hepialoidea the lateral lagena is absent, although in the other exoporian superfamily, Mnesarchaeoidea, it is present. A narrow spermathecal gland leads into the utriculus.

Colleterial glands. These glands (Figs 94, 96) (termed glandulae sebaceae by Klots, 1970) open into the vagina and secrete an adhesive substance for fastening eggs to their substrate. They are absent from Exoporia, a group in which the eggs are scattered not glued.

Bursa copulatrix. The bursa copulatrix (Figs 94-96) comprises the ductus bursae and the corpus bursae. The ductus is essentially a tube of variable length, which expands into the sac-or bulb-like corpus either suddenly, or gradually without marked differentiation. The corpus sometimes gives off another smaller sac known as the appendix bursae. The corpus bursae often bears a sclerotized signum or two or more signa. Signa may take the form of bands, patches, spines or plates; plates may be plain, spined or toothed. Their form is often of taxonomic value.

The first (i.e., most posterior part of the ductus bursae may take the form of a funnel-shaped, and usually well-sclerotized, antrum. A collar-like thickening, termed a colliculum, situated just below (i.e., anterior to) the antrum may also occur. The colliculum probably functions as a valve to prevent spermatozoa from being pushed back through the ostium bursae when they are squeezed into the ductus seminalis. The terms antrum and colliculum are treated synonymously by Klots (1970), but the distinction indicated here, and followed by many lepidopterists, seems appropriate. Sometimes a ribbon-like sclerite (termed a cestum in Tortricidae) runs down the ductus for most or a smaller part of its length. The ductus bursae is occasionally spiral or looped.

External morphology

The external genitalia function for oviposition and mating.

Ostium bursae and its sclerites. The bursa copulatrix communicates with the outside via the ostium bursae. (Strictly speaking (Klots, 1970), the ostium bursae lies at that point of the ductus bursae at which the ductus seminalis arises.) This pore is usually surrounded by sclerites collectively called the sterigma or genital plate. This sclerotized complex may be divisible into a lamella antevaginalis (anterioventral of the ostium) and a lamella postvaginalis (posteriodorsal of the ostium). Although the lamellae are intersegmental sclerotizations of segments A7 and A8, the lamella antevaginalis is often fused with sternum A7 and the lamella postvaginalis with sternum or tergum A8, or both. In some cases sclerotized bands may join the anterior apophyses to the sterigma.

Modifications of segments A8-10: the ovipositor and its apophyses. The anterior edges of segments A8 and A9+10 are usually extended into a pair of internal apophyses, although they are not universally present within Lepidoptera. The apophyses function in support, particularly in the case of long narrow ovipositors, and for muscle attachment. The apophyses of tergum A8 are called the anterior apophyses and those of tergum A9 or A9+10 the posterior apophyses. In Micropterigidae they are absent. The presence of apophyses probably represent a specialized character of the Amphiesmenoptera (they are present in primitive caddisflies), therefore their absence in various Lepidoptera must represent a secondary loss (Kristensen, 1984b). Alternatively (Stekol'nikov, 1967), the absence of the apophyses in Micropterigidae might represent the primitive condition in Lepidoptera since genital and pregenital segments differ little from each other.

Whatever their derivation, lepidopteran apophyses serve as sites for the attachment of retractor muscles, which are modifications of the longitudinal muscles of the abdomen. A combination of these muscles and the enlargement of the intersegmental membrane between A8 and A9 gives the ovipositor considerable flexibility of movement allowing it to become retracted at rest and extended during oviposition. In species with long, narrow ovipositors, used for laying eggs within plant tissue or in crevices, the anterior attachment of the muscles may move cephalad as far as segment A5. This arrangement, together with a lengthening of terga, sterna, and the intersegmental membranes, permits greater mobility and a considerable capacity to retract the

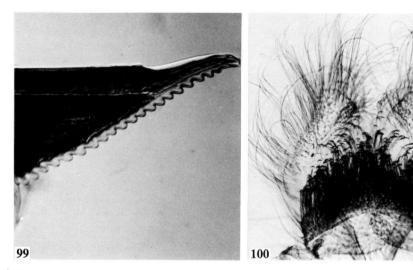

Figs 99, 100. *Ovipositors. 99,* Protaephagus capensis *(Incurvariidae) showing serrated edge of ovipositor; 100, floricomous ovipositor of* Tortricodes alternella *(Tortricidae) showing stiff spines (dark band) below ovipositor lobes. (99, after Scoble, 1980; 100, Photographic Unit, BMNH.)*

ovipositor telescopically (e.g., Stekol'nikov, 1967). In Hepialidae and some Nymphalidae, apophyses may be reduced, a feature linked to their habit of scattering eggs.

Ovipositor shape is closely related to the mode and site of oviposition. Lepidopteran eggs may be laid by randomly scattering them in the vicinity of the hostplant, by injecting them into plant tissue, by inserting them into cracks and crevices (e.g., in bark), or by sticking them onto the surface of the substrate. The first of these categories is exemplified by Hepialoidea and the ovipositor is short (Fig. 95). Incurvarioidea usually inject eggs into plant tissue, and their ovipositors are long and bear correspondingly long apophyses. In this superfamily, the tip of the ovipositor is often serrated for cutting (Fig. 99). In the many female Lepidoptera with ovipositors that are not attenuated, the segments are, nevertheless, capable of being telescoped into the abdomen and extruded for oviposition. Their length ranges from between the short hepialoid structure to the extended incurvarioid type. In some groups (notably Tortricidae), the ovipositor may be covered with stiff, broad-headed hairs, which assist in scraping dirt over eggs after they have been deposited (Fig. 100).

5

IMMATURE STAGES

Egg

At the time at which it is laid, the insect egg is composed basically of a cytoplasmic and yolky core surrounded by a membrane and a shell. The cytoplasm is divided into a boundary layer, known as the periplasm, and a reticulum within the yolk. The nucleus of the zygote is found towards the posterior end of the egg. Around the periplasm is the vitelline membrane, which in turn is bounded by the 'shell' or chorion. A wax layer is found on the inside of the chorion. One or more funnel-shaped canals known as micropyles (e.g., see Fig. 112) run through the chorion permitting the passage of spermatozoa. In terrestrial insects, oxygen uptake through the chorion is enhanced by the presence of openings called aeropyles, which communicate with air-filled spaces in the lower part of the chorion.

The chorion is produced by follicle cells while the egg is still in the ovary. In Lepidoptera, the microfibrils constituting the basis of the chorion are arranged in helicoids - a situation apparently unique to this order (Hinton, 1981). This arrangement, which is similar to the orientation of microfibrils found in arthropod cuticle, helps the eggshell to resist forces equally from any direction. Typically, the chorion consists of an endochorion and an exochorion. The endochorion bears air-filled spaces, and is supported by vertical columns or struts. This system is found in many groups of insects including the lower Ditrysia. In higher Ditrysia there are, in addition, many gas-filled layers between the outer layer of the chorion and the inner air-filled layer, although eggs of Papilionidae are of the primitive variety (Hinton, 1981).

Lepidopteran eggs (Figs 101-112; Pl. 4) may be approximately spheroids or prolate spheroids (elongated in one direction), hemispherical, flat and scale-like, spindle-shaped, ellipsoidal, flattened-spherical, or rectangular cuboid. In the Yucca moth *Tegeticula* (Prodoxidae) the egg is elongated, the dimensions being up to 20 times as long as broad. Marked elongation tends to be seen amongst those eggs inserted into plant tissue. Döring (1955) illustrated a wide range of lepidopteran eggs indicating the variation in shape. There is also variation in colour. Other works in which lepidopteran eggs have been extensively figured include those of McFarland (1973a,b) and Salkeld (1983; 1984). Hinton (1981) provides an extensive list of references to the biology of lepidopteran eggs.

Considerable variation in the shape and sculpturing of eggs exists even among supraspecific taxa (Pl. 4), but the family to which an egg belongs can often be recognized by its shape, orientation, site of deposition, and whether it is laid singly or in small or large batches. Chapman (1896) recognized two main types within the order, one termed *upright* and the other *flat*. These names are still used regularly in descriptions. In the upright variety (e.g., Figs 109-112) the micropylar axis runs perpendicular to the surface on which the egg is laid. In the flat form, the axis runs parallel to the substrate. Chapman noted that among macrolepidopterans upright eggs are found, typically, in Noctuoidea and butterflies, whereas flat eggs are more widespread. Again within macrolepidopterans, he subdivided flat eggs into 'Geometrid' types, characterised by a greater roughness, and 'Bombycid' types, which are smoother and more polished.

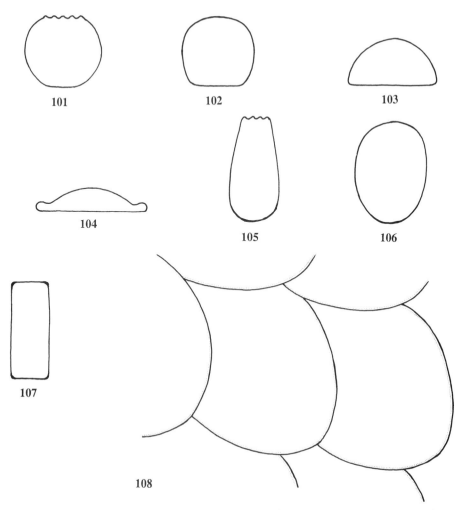

Figs 101–108. *Egg profiles. 101,* Xanthia togata *(Noctuidae); 102,* Mythimna ferrago *(Noctuidae); 103,* Cerura vinula *(Notodontidae); 104,* Craniophora ligustri *(Noctuidae); 105,* Gonepteryx rhamni *(Pieridae); 106,* Petrophora chlorosata *(Geometridae); 107,* Hemistola chrysoprasaria *(Geometridae); 108,* Acropolitis xuthobapta *(Tortricidae). (101-107, after Döring, 1955; 108, drawn from a photograph in Powell & Common, 1985.)*

Chapman considered that whereas upright eggs probably developed once in the more advanced members of the order, flat eggs developed independently twice. However, there are numerous exceptions to this generalization, and although 'upright' and 'flat' may be of use as descriptive terms they indicate no phylogenetic basis in the division of the order; both flat and upright eggs may occur within the same family (Hinton, 1981). For example, although most Geometridae have flat eggs, in some they are upright. The two kinds may occur even within the same genus, as in *Sterrha* and *Biston*. As a very broad generalization, upright eggs are characteristic of Cossoidea, Hedyloidea, Hesperioidea, Papilionoidea, and Noctuoidea. A flat egg occurs in most other Lepidoptera. Many Lepidoptera that glue their eggs together lay eggs of the flat variety, which appear upright in relation to the plant substrate. Examples are *Eriogaster lanestris* and *Malacosoma castrensis* (Lasiocampidae).

Figs 109–112. *Eggs. 109, 110, egg of* Macrosoma semiermis *(Hedylidae) showing chorion and micropyle area: 109, whole egg (scale, 130 μm); 110, micropyle area (scale, 9.0 μm). 111, 112, egg of* Adisura *sp. (Noctuidae): 111, whole egg (scale, 14.3 μm); 112, micropyle area showing micropyle (m) (scale, 1.4 μm). (109, 110, EM Unit, BMNH, specimens courtesy of Dr A. Aiello; 111, 112, courtesy of Dr M.J. Matthews/EM Unit, BMNH.)*

The chorion of lepidopteran eggs frequently bears longitudinal ribs, often connected by horizontal ridges (Figs 109, 111). Eggs may be pitted, wrinkled, or covered with a reticulate network. Sometimes they are smooth and polished. Prominent at the anterior pole of the egg is the micropylar area, which is usually surrounded by a rosette-like sculpture of the chorion (Figs 110, 112). Generally, there are four micropyles in lepidopteran eggs, but there may be more. For example, in 15 species of Notodontidae the number of micropyles were found to vary from four to twenty (Hinton, 1981).

Although rarely laid under water, lepidopteran eggs are sometimes prone to flooding. Whereas plastron respiration accounts for respiration in eggs of most other aquatic insects, in Lepidoptera the air-water interface is too small to make it effective, and, according to Hinton (1981), eggs probably survive as a result of a reduced metabolic rate. However, it was suggested (Downey & Allyn, 1981) that in eggs of Lycaenidae perforation of the chorion enables plastron respiration to occur when eggs are submerged in rain water. Eggs of Micropterigidae are typically laid in moist places.

They may be covered with brittle, rod-like structures that are able to hold water, or (as in *Neomicropterix nipponensis*, see Chapter 9) covered with gelatinous material that swells on contact with water or moisture. In their apparent need for surrounding moisture, micropterigid eggs differ from those of other members of the order (Kobayashi & Ando, 1981).

Depending on the taxon, eggs may be laid singly or in batches. While they are generally placed on or near the hostplant (or animal host in the case of carnivorous Lepidoptera), in some groups, such as Hepialidae and certain Nymphalidae, they are scattered by the female in flight.

Enormous numbers of eggs may be laid by some Hepialidae. A single female of *Abantiades magnificus* may lay more than 18000, and a similar figure has been estimated for *Xyleutes encalypti* (Cossidae) (Nielsen & Common, 1991).

Eggs are potentially a source of food for predators and a variety of means of protection have evolved (summary and references in Hinton, 1981). Many Lepidoptera lay their eggs in places concealed from many predators (e.g, in crevices in bark, or within plant tissue). Eggs of many Pyralidae and Olethreutinae (Tortricidae) are scale-like and transparent so making them difficult to distinguish from the substrate on which they are laid. Examples of eggs that mimic plant tendrils (as in *Hemiostola immaculata*: Geometridae), leaf galls (as in *Cerura vinula*: Notodontidae), and seeds (as in *Langsdorfia franckii*: Cossidae) also exist. Some eggs are disruptively coloured (as in *Gastropacha quercifolia*: Lasiocampidae). Several examples occur in which the eggs of Lepidoptera are poisonous. Those of *Danaus plexippus* (Nymphalidae) contain cardiac glycosides, and in some species of Zygaenidae hydrocyanic acid is present in the eggs. Females often cover their eggs with setae from the anal tuft. Setae from the adult may simply provide a mechanical defence (as in many Lymantriidae). But often urticating setae from the larva picked up by the anal tuft of the emerging adult from the inner lining of the cocoon are used either exclusively to cover the eggs, or mixed with the adult (non-urticating) setae. Examples of Lepidoptera making use of urticating setae in this way occur in Lymantriidae, Notodontidae (including Thaumetopoeinae), and Saturniidae (Hemileucinae).

Larva

Essentially, the larva is the feeding stage in Lepidoptera. In many adult Lepidoptera the mouthparts are functionless, and even in those adults that do feed the larva provides most of the adult resources. The larva faces two major problems - obtaining enough to eat and avoiding being eaten. Elaborate defence mechanisms, both structural and behavioural, have evolved. In this section, the basic structure of lepidopteran larvae is considered together with the relationship of structure to function. There is considerable convergence in larval structure since similar modes of life have been adopted in many groups that are not closely related. Therefore, those features of larval structure, particularly the primary setae, that are less obviously related to function are also emphasized for the value they provide in systematic studies. Comments on the natural history of the relationship between lepidopteran larvae and their foodplants are made in Chapter 7.

Head

The head capsule (Figs 113) is strongly sclerotized. Hinton (1947) interpreted the dorsal sutures and ecdysial cleavage lines of the head as indicated in the Figure. The length of the epicranial suture is related to the angle at which the head is positioned

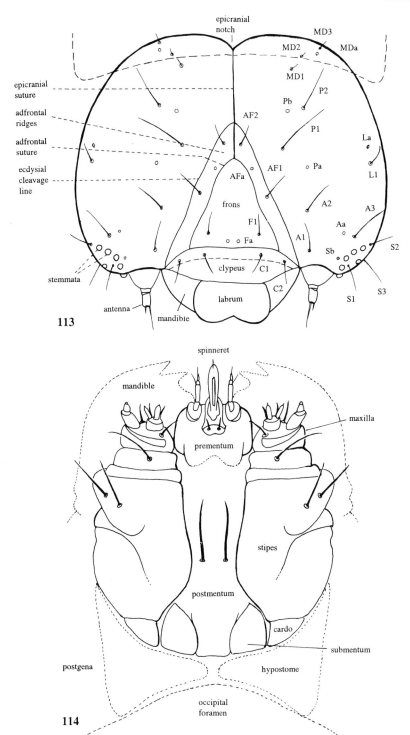

Figs 113, 114. *Larval head (diagrammatic). 113, anterior view; 114, ventral view. (113, after Peterson, 1962; 114, after Bourgogne, 1951.)*

relative to the body. In external feeders, where the head is hypognathous (see below), it is long, whereas in concealed feeders, where the head is typically semiprognathous, it is short. In miners, in which the head is prognathous, the epicranial suture is lost, and in certain mining larvae the adfrontal sutures diverge to reach the epicranial notch so that the frons loses its triangular shape and becomes trapezoidal. Laterad of the lateral adfrontal sutures are the ecdysial cleavage lines. Usually they are found only in the last instar, and represent lines of weakness along which the head capsule splits at the moult. In earlier instars, the capsule is shed intact, although there are some exceptions in which it splits (e.g., Eriocraniidae, Hepialidae, Nepticulidae). The areas between the adfrontal sutures and the ecdysial cleavage lines are termed the adfrontal areas. Adfrontal sutures with internal ridges are absent from Zeugloptera and Aglossata but present in Heterobathmiina and Glossata.

In most larvae that feed in exposed situations, the head is hypognathous and the mandibles are orientated approximately at right angles to the longitudinal axis of the body. At the other extreme, exhibited principally by leaf miners, the head is prognathous so that the mandibles lie in line with the body - an orientation associated with life in a vertically restricted habitat. Various intermediate conditions, covered by the term semiprognathous and existing between these two extremes, are characteristic of concealed feeders such as leaf-rollers and borers. In prognathous insects the articulation of the maxillary cardo to the cranium lies far behind that of the mandible. The existence of this condition in all hypognathous Lepidoptera may be explained by assuming that these insects developed from a prognathous ancestor, a reasonable assumption considering that larvae of the most primitive groups of moths are prognathous (Kristensen, 1984d). The posterior position of the articulation of the cardo is a derived lepidopteran character for although the larvae of primitive Trichoptera are prognathous, the articulations are not thus displaced (Kristensen, 1984d).

There are two components of the internal skeleton of the head. As in adults, the larval tentorium is composed of apodemes, which are invaginated rods of the head capsule, and a posterior tentorial bridge joining the tentorial apodemes. The second component of the endoskeleton of the head comprises the internal ridges along the sutures of the head-capsule. These are particularly strong in Lepidoptera compared with other insects.

Mouthparts

Lepidopteran larvae are mandibulate (Stehr, 1987). Overlying the mandibles is a well developed labrum (Fig. 113), which articulates with the anteclypeus. The shape of the labrum varies. Its anterior margin is usually notched, although in many leaf miners the notch may be lost so that the margin is straight or even convex.

The *mandibles* (Figs 113, 114) are strongly sclerotized and usually bear molar and incisor areas. With few exceptions they function for chewing. Cranial adductors and abductors are present with the adductors particularly well developed. The mandibles are opposable, and when brought together the various 'teeth' generally interlock. In sap-feeding leaf miners the mandibles are used to lacerate tissue, or they may be reduced. Mandibles articulate with the head capsule anteriorly and posteriorly. Anteriorly, a sclerotized projection of the clypeus fits a depression on the mandible, and posteriorly a condylar process from the mandible fits a corresponding depression of the head capsule.

The *maxilla* (Figs 114, 115) is composed of a proximal cardo and a distal stipes. From the stipes arise (laterally) the 2- or 3-segmented maxillary palpi, and a mesal lobe. Three palp segments are considered to be present in the lepidopteran groundplan (Kristensen, 1984b). These three segments exist in addition to the distal part of the stipes (dististipes or palpiger), a structure that resembles a palp segment. Although

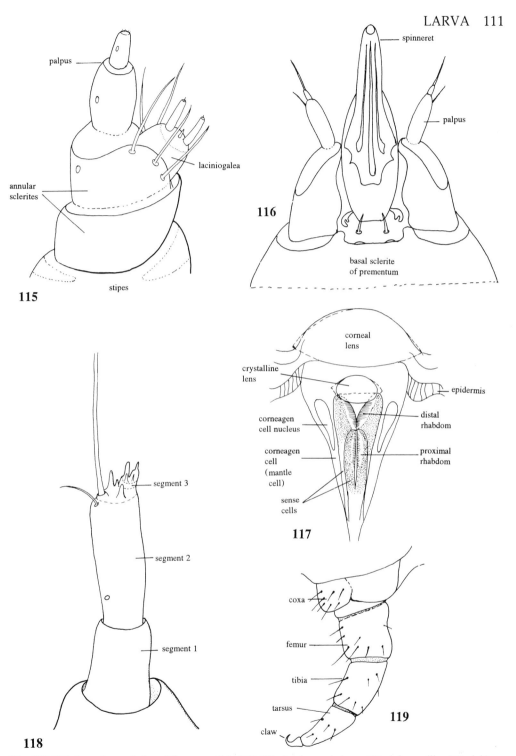

Figs 115–119. *Larval morphology (diagrammatic). 115-118, head morphology. 115, left maxilla; 116, labium, ventral view; 117, stemma of* Pyrrharctia isabella *(Arctiidae); 118, antenna; 119, thoracic leg. (115, 116, after Bourgogne, 1951; 117, after Dethier, 1963; 118, 119, after Peterson, 1962.)*

often considered to be a palp segment, the dististipes was not thought to be a segment by Hinton (1958). It is probably a composite structure composed of a basal palp segment and part of the stipes (Kristensen, 1984b).

The mesal lobe is often referred to as a 'galea', a term deriving some support from the discovery that imaginal cells within it form the adult galea (Eassa, 1953). Use of the neutral term mesal lobe seems appropriate since it remains unclear if the structure represents the galea alone or a composite structure involving both galea and lacinia (Grimes & Neunzig, 1986b). Nevertheless, Grimes & Neunzig point out that certain sensilla of the lobe are situated on a part of the lobe that is often divided from the area bearing the other sensilla. They suggest that these portions of the mesal lobe represent the lacinia and galea, and that an undivided mesal lobe may appropriately be termed a laciniogalea to indicate its derivation. A separate structure termed the lacinia is present in the most primitive Lepidoptera, but already in the Aglossata there is but a single lobe (lobarium), thought by Kristensen (1984d) to probably represent the galea.

The arrangement of the sensilla on the mesal lobe of a wide variety of Ditrysia was examined by Grimes & Neunzig (1986b). Various possible functions of the sensilla have been suggested. Since the lobes overhang the hypopharynx and their sensilla brush against the mandibles, the sensilla are ideally placed to come into contact with food. Hence, they are assumed to respond to mechanical and chemical aspects of food. There are three groups of sensilla on the mesal lobe: 3 sensilla trichodea, 2 sensilla styloconica, and 3 sensilla basiconica. The sensilla trichodea vary in shape and length. In Hesperiidae and many Pyralidae they are multibranched, and in several phycitine Pyralidae they are bifid. In larvae that bore, the sensilla are usually stout and membranous. They may function as proprioceptors to inform the central nervous system of mandibular movements, and may further provide information on the texture of food. Other than size differences there are few modifications evident in the sensilla styloconica. These sensilla are socketed and uniporous. Each includes at least four contact chemoreceptors and, in the case of the lateral member of the pair, possess an olfactory capability. Both sensilla probably respond to bending. Of the three sensilla basiconica, one or more may occasionally be lost. Their function is uncertain but it is unlikely to be restricted to mechanoreception.

The sensory equipment of the distal segment of the labial palpi in ditrysian Lepidoptera was also compared by Grimes & Neunzig (1986a). Here again the sensilla can be divided into three groups: 8 terminal sensilla basiconica; usually 2, but sometimes 3 or 4 sensilla campaniformia situated on the side of the segment; and a large, single sensillum digitiformium (usually rather less than one third of the length of the distal segment but much longer in some species) present on the dorsoposterior surface of the distal segment. The sensory functions of the distal segment so far recorded include olfaction, gustation, mechanoreception, and possibly thermo-regulation.

Some taxonomic information is indicated at the level of family according to the arrangement of the various sensilla. The authors suggested that the form of certain of the sensilla basiconica may be related to larval mode of life. Whilst in small, endophagous larvae the apical sensilla are uniporous, smooth-walled, tall, and without recognisable projections, in large, exophagous larvae they tend to be wrinkled, and probably multiporous, short, and with cuticular projections. Possibly in endophagous larvae the sensilla are mechanoreceptors and gustatory chemoreceptors, while in the exophagous larvae they also have powers of olfaction. The olfactory function may be an adaptation to the increased likelihood of depleting the food source and the need to find more. The wrinkled nature of the sensilla in exophagous larvae suggests the presence of gustatory receptors within.

The *labium* (Figs 114, 116), situated between the maxillae, is divided into a pre- and post-labium. The post-labium of most lepidopteran larvae is divided into a mentum and a submentum (postmentum). The postlabial plate of the lepidopteran groundplan was probably undivided (Kristensen, 1984b). A pair of labial palpi, which are 2-segmented or sometimes absent, are borne on the prelabium, which in turn is fused with the hypopharynx. In most Lepidoptera the composite lobe bears apically a structure of great importance - the spinneret, the organ from which silk is exuded.

The dorsal wall of the spinneret is usually shorter than the ventral wall, and the entire structure is borne on a basal sclerite termed the fusuliger. Considerable variation exists in the shape of the spinneret. It may be long and slender (as in Hepialidae), short and broad (as in some Noctuidae and some Sphingidae), or intermediate. Shape is related to habits: in larvae that form galleries in plant tissue (such as in Hepialidae and most other microlepidopterans), the spinneret tends to be long, slender and cylindrical, whereas in those that spin little or no silk it tends to be short and flat as in many miners, Noctuidae, and Sphingidae. The spinneret is a new formation characteristic of all Glossata; it is absent from Zeugloptera, Aglossata, and Heterobathmiina (Kristensen & Nielsen, 1983).

Considerable variation is also found in the hypopharynx, which forms the floor of the preoral cavity. Medially it is termed the lingua, a part that is often spinose; laterally it is frequently composed of movable paralingual lobes. Both sides of the hypopharynx are bordered by a sclerotized section of the mentum.

Stemmata

There are typically six stemmata (often, but incorrectly, referred to as ocelli with which they are not homologous) on each side of the head, arranged approximately in a semicircle and situated towards the mouth (e.g., Stehr, 1987) (Figs 113, 128). Primitively seven stemmata exist, as in Trichoptera, a number retained in *Heterobathmia* (Kristensen & Nielsen, 1983). If the dorsalmost stemma is numbered as 1, then stemmata 1 to 4 form part of an almost regular arc. Of the two ventral stemmata, one is often markedly anterior to the rest, while the other continues along the arc formed by the first four. In *Sabatinca* (Micropterigidae) 5 stemmata fuse to form a structure analogous to a compound eye. Stemmata form a fairly compact group in some taxa, for example in Hepialidae. Other variations include the loss of stemmata. In Satyrinae (Nymphalidae) stemma 3 is typically larger than the rest. In leaf miners, where the head is flattened, some stemmata are situated on the dorsal surface of the head capsule while others lie on the ventral surface. In Nepticulidae there is only a single stemma on each side of the head.

A typical stemma (Fig. 117) (Dethier, 1963; Mazokhin-Porshnyakov & Kazyakina, 1984, and references therein) is composed of light-refracting, light-isolating, and light-sensing components. The outer cornea takes the form of a biconvex lens. The corneal lens develops from three epidermal cells; sometimes the corresponding three facets are evident (summary in Dethier, 1963). Below the corneal lens is a crystalline body. Surrounding the retinular optic cells are three mantle cells or peripheral corneagen cells. In *Lymantria dispar* (Lymantriidae), the cytoplasm of these cells contains a small number of pigment granules (Mazokhin-Porshnyakov & Kazyakina, 1984). The mantle cells surround seven retinal cells with axial rhabdomes. Three of these cells are peripheral and surround four central cells. The cytoplasm of the retinal cells bears an abundance of large pigment granules. Mazokhin-Porshnyakov & Kazyakina (1984) suggested, contrary to what is often thought, that this pigment (rather than that in the mantle cells) functions as the major light-isolating factor for the rhabdom. The optic nerve runs to the protocerebrum.

The system probably produces a well-focussed image, but one that with so few cells must be poorly resolved (Dethier, 1963). However, since there are usually six stemmata on each side of the head of a lepidopteran larva a mosaic of the visual field is probably perceived. Since the stemmata are typically far apart with little or no overlap of the visual field, the mosaic will be very coarse. Side to side motion of the head, which occurs in many larvae that are moving forward, may partially offset the limited number of visual units (Dethier, 1963). The adaptive function of this probably comes into play in searching for plants, given that larvae are capable of distinguishing vertical from horizontal silhouettes.

Antennae

The antennae are generally short, 3- to 4-segmented structures situated on each side of the head between the stemmata and the mandibles (Fig. 118). Significant modifications include a length greater than that of the head in Micropterigidae, and, in miners, a reduction in the number of segments. The first two antennal segments are relatively large. Segment 2 bears various sensory structures terminally and subterminally, prominent among which is a long seta. An antennal pore is found medially and on the outside of segment 2. Segment 3 is small, and bears a cone-like sensillum and, if present, the minute segment 4. Most variation in the antennae of lepidopteran larvae is exhibited by leaf-miners and other internal tissue-feeders and involves the reduction or loss of certain segments.

Thorax

In the thorax, the prothorax (T1) differs particularly from the mesothorax (T2) and the metathorax (T3); it possesses a functional spiracle, which is actually derived from T2, and bears a more sclerotized tergal area representing the pronotum and termed the prothoracic shield (Fig. 129). A small prespiracular shield, if it occurs, is generally separated from the main sclerite. The pronotum of T2 and of T3 rarely bears a continuous shield. The pleura of lepidopteran larvae are not sclerotized as they are in adults. Sternum T1 is divided into a basisternum and a furcasternum. In T2 and T3 the sternum is less frequently divided than in T1. The spiracle on T1 is situated on the caudal margin of that segment. A reduced spiracle is present on T3 at the boundary of T2 and T3, but it does not open on the segment and lies *under* the cuticle.

The thoracic legs of lepidopteran larvae (Fig. 119) are remarkably constant in form throughout the order, but are reduced in some leaf miners and elongated in certain Notodontidae. Coxa, trochanter, femur, tibia, tarsus, and claw are present, and the first movable segment is the femur. The tarsus is 1-segmented, and bears a claw, which may be toothed. In Micropterigidae the coxa, trochanter and femur are fused to form a 3-segmented leg.

Abdomen

There are 10 abdominal segments (A1-10) (Fig. 129) which, with the exception of A9 and A10, are of a similar shape. If present at all, A11 is represented only by a pair of perianal lobes - an epiproct, situated above the anus, two paraprocts, on either side of the anus, and a subanal hypoproct, which is often thick and conical (Fig. 120). The terga of the abdominal segments are divided into several sclerites associated with the primary setae. On A10 there is usually a sclerotized shield - the anal plate (Fig. 129). This shield may be continuous with similar ones on A9 and even A8, but the presence of such shields on A8 and A9 is rare. The abdominal pleura and sterna are mostly

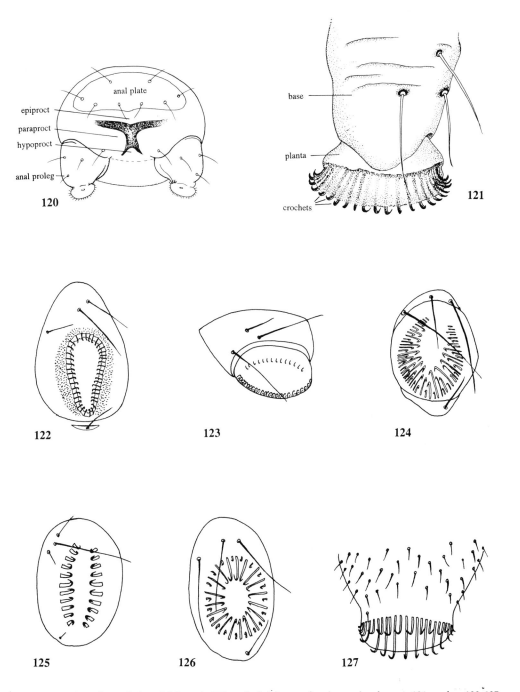

Figs 120–127. *Larval morphology (abdomen). 120, end of abdomen showing perianal areas; 121, proleg; 122-127, planta of proleg showing arrangements of crochets: 122, multiserial circle; 123, pseudocircle composed of uniordinal lateroseries and biordinal mesoseries; 124, mesal penellipse; 125, transverse bands; 126, biordinal circle; 127, biordinal mesoseries. (120, 121, after Gerasimov, 1952; 122-127, after Fracker, 1915.)*

membranous, but a series of small plates bearing primary setae are often present. There is a pair of spiracles on abdominal segments A1-8. Each spiracle is situated on the caudal margin of the segment with which it is associated. On A8 the pair is usually placed higher than in the other segments and it is also larger, whereas in A1 it is slightly lower.

Abdominal appendages called prolegs (Fig. 121, Pl. 3:1) are typically present. They occur as fleshy outgrowths of the body wall of A3-6 (ventral prolegs) and A10 (anal prolegs). A proleg comprises a base and, distally, a planta. The base may be distinguished from the planta in bearing at least primary (SV) setae (see below) and often secondary setae. The planta, by contrast, lacks setae but usually bears small hooks called crochets, structures by which larvae hold on to their substrate (Figs 121-127; Pl. 3:1). Retractor muscles attach to both the base of the proleg and to the planta, but in Micropterigidae these muscles are absent.

The main variations involve the number and size of prolegs and the number of crochets and their arrangement on the planta. Changes in numbers and size of prolegs usually involve reduction rather than increase. However, in Megalopygidae and Dalceridae there are seven pairs of prolegs (A2-7; A10) (Stehr in Stehr, 1987; Stehr & McFarland, in Stehr, 1987). Reduction is associated frequently with leaf mining larvae. In the leaf mining genus Stigmella (Nepticulidae), although prolegs occur on A2-7 they are reduced in size to ambulatory warts or cali. Prolegs are absent from Eriocraniidae and early stages of Gracillariidae, and greatly reduced in Adelidae, Incurvariidae, Tischeriidae, and Douglasiidae. In Phyllonorycter (Gracillariidae) prolegs are present in the final instars, although the larvae mine throughout their lives, a development probably relating to a significant expansion in the depth of the mine and a consequent requirement for more extensive powers of movement. The larva of Micropterix (Micropterigidae) bears a pair of abdominal appendages on segments A1-8, although the appendages of all Micropterigidae are dissimilar from those of other Lepidoptera because they lack muscles. In the Geometridae, prolegs in most species are found only on A6 and A10, and the caterpillars progress by 'looping'. In some Noctuidae known as 'semiloopers' the prolegs of A3 and A4 are lost or reduced. Looping or semilooping is typical of arboreal larvae.

The arrangement of the crochets (Figs 122-127) is related to mode of life (summary by Stehr in Stehr, 1987). Prolegs may have either a prominent base, comprising most of the structure, or a poorly developed base. In those groups where the base is insignificant, either the crochets appear to arise directly from the ventral surface of the abdominal segments, or the planta is elongate. A weakly developed proleg base is typical of the more primitive Lepidoptera, and the crochets are arranged in a circle, incomplete circle, penellipse, or in transverse bands. In those larvae in which the base of the proleg is elongated and prominent, the planta takes the form of a simple lobe. In this type the crochets are typically arranged in an anterior-posterior row along the mesal edge of the planta (mesoseries, see below). This pattern is characteristic of larvae of macrolepidopterans, which usually feed externally on plants and use their crochets to grasp the substrate (Pl. 3:1). The arrangement of crochets of microlepidopterans, which are typically concealed feeders, is less specialized and occurs as a circle, incomplete circle, or as a pair of bands stretching laterally across the planta. Primitively, crochets were probably numerous. Crochets probably evolved from numerous minute, cutaneous spicules. Reductions in number, their confinement to the planta, and increasing size probably represent specializations.

The variation in crochet arrangement (Peterson, 1962; Stehr in Stehr, 1987) is based on the number of rows in which they occur on the planta, their relative lengths, and their position. Crochets are said to be *multiserial* (Fig. 122) when there are many rows (a condition that is probably primitive) and *uniserial* when they are reduced to a single row (e.g., Fig. 126) (or a pair of bands (Fig. 125), which might reasonably be interpreted as a circle broken in two places). When the crochets are all of one length they are said to be *uniordinal* (e.g., Fig. 125). When they occur in more that one length they are termed *biordinal* (e.g., Fig. 126), *triordinal*, or *multiordinal*, whichever is appropriate. Several terms are required to cover the way in which the crochets are arranged on the planta. An *incomplete circle* is self explanatory. A *penellipse* (Fig. 124) refers to a break in an arrangement that is basically elliptical. The break may be lateral (*lateral penellipse*) or mesal (*mesal penellipse*, Fig. 124). *Transverse bands* (Fig. 125) describe a mesal and lateral interruption of a flattened circle. In some cases only one band (*single transverse band*) may be present. Externally feeding larvae typically bear a band of crochets along the mesal edge of the planta, an arrangement termed a *mesoseries* (Figs 121, 127). In addition to these variations, a row of crochets may occur in two distinct, *not intermixed* lengths, in which case the arrangement is described as *heteroideous* - as opposed to *homoideous* in which there is no sharp difference of length. A combination of these terms covers nearly all the variation found within the order.

Various other additions to the prolegs are encountered in lepidopteran larvae such as the presence of a plantar sucker in some Geometridae, and a spatulate, fleshy lobe found in most Lycaenidae, including some Riodininae.

Coverings of the body

Setae

On the head and body of lepidopteran larvae occur various setae (Figs 113, 128, 129). There is often a marked difference between the number of setae found on the first instar and those on subsequent instars. The setae of first instars are termed primary setae, and their number and position are relatively constant throughout the order. However, because there are small but significant variations in their distribution and occurrence, they can be used for systematic purposes. Primary setae are often obscured by secondary setae in later instars, although in some cases they are still identifiable. Secondary setae vary from a sparse scattering to the dense covering found in 'hairy' caterpillars. Unlike secondary setae, they cannot be homologized individually.

The regularity of position of the primary setae throughout the Lepidoptera means that they can be named in a consistent way. The nomenclature currently used most frequently follows, with minor alterations, Hinton (1946a), a system based on the principles put forward by Heinrich (1916) specifically for the head. Other terminologies were proposed by various authors, the systematic potential of the arrangement of the primaries having been recognized since the latter part of last century. The basis of the Heinrich/Hinton system is that setae are visualized as forming groups on the head and body. Each primary seta is numbered, and the number is prefixed by a letter to indicate the kind of seta and its approximate location. For example the prefix M refers to a microseta as opposed to a long, tactile seta; the letters D and SD refer respectively to the dorsal and subdorsal groups of setae on the body.

On the head occur not only primary setae but also punctures, which are relatively constant in position and number (Figs 113, 128). Pores (small sensory apertures of the cuticle) are designated by a small 'a' to distinguish them from setae. Thus the puncture associated with the stemmatal group is represented by Sa (modified by Stehr *in* Stehr (1987) from Oa, i.e., 'ocellar', of Hinton).

The setae fall into several groups, each group being distinguished by appropriate abbreviations. Stehr *in* Stehr (1987) suggested that for two reasons a more appropriate designation of the so-called 'vertical' setae, denoted by Hinton by 'V', is MD. First, the prefix M is consistent with its use to designate the proprioceptive, as opposed to tactile,

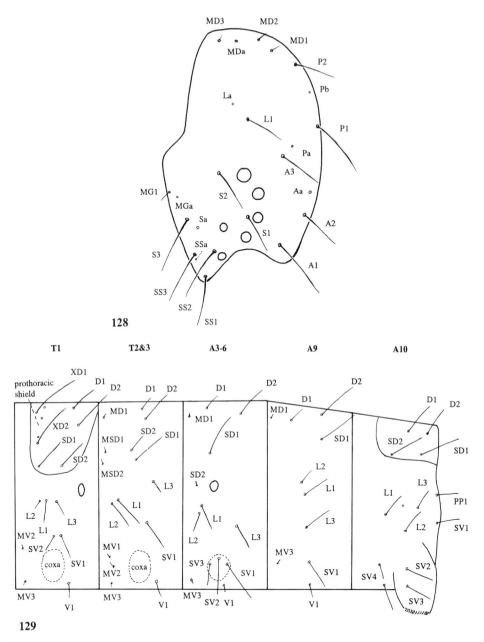

Figs 128, 129. *Larval morphology showing arrangement of primary setae (chaetotaxy). 128, head, lateral view; 129, thorax and abdomen. (128, after Hinton, 1946a, with nomenclature as modified by Stehr, 1987.)*

setae on the body because the setae in question are not tactile. Second, the term vertical refers to the vertex in this context: 'MD', meaning microdorsal, avoids any ambiguity over the meaning of the word vertical. Similarly, the genal setae are more appropriately labelled as 'MG' (microgenal).

The position of the minute setae on the head is confined to those areas retracted into the thorax. Where retraction is extensive these setae are found over a relatively large area of the head, whereas in a head that is only slightly retracted they are confined to that small area of retraction (Hinton, 1946a). The minute setae function as propriocep-tors, providing the larva with information about the relative position of the head. On the body, the proprioceptors are found near the segmental margins, positions appropriate for indicating the relative movements of adjacent segments.

The position of primary setae on the thorax and the abdomen is usually indicated on a stylized figure termed a setal map (e.g., Fig. 129). The map provides a means by which position and number of setae can be compared. On certain segments setal arrangement does not differ appreciably, but there are notable differences between the setae on T1 and the other thoracic segments, between A9 and A10 and the other abdominal segments, and between those abdominal segments with prolegs and those without.

Secondary setae are found predominantly in larvae that are exposed during feeding, and far less often among the chiefly concealed-feeding microlepidopterans. They are not regular in position although their presence gives many larvae a characteristic appearance. While very many secondary setae are filamentous, there is a great variety of forms - for example spatulate, clavate, branched, and plumose to mention but a few. Secondary setae may be sensory, defensive or may aid dispersal. Defensive hairs are typically glandular and may be irritant or otherwise toxic (see below). They are frequently arranged in tufts and brushes. Long hairs termed aerophores may facilitate wind dispersal of small larvae such as early instars of the gypsy moth.

Other elements

Cuticular structures ('sculptural elements' of Gerasimov, 1952), in the form of spinules and granules, also occur in larvae. Granules, in particular, exhibit considerable variation in shape and density. Sculptural elements may amalgamate to form other more prominent structures such as the anal plate. Hairs and setae are surrounded by cuticular pads or are sunk in depressions.

Various terms are in use to describe the different integumental outgrowths of the larval cuticle (Stehr *in* Stehr, 1987) (see Figs 130-135). However, this nomenclature does not always reflect satisfactorily the homologies between the outgrowths nor does it account entirely for their variety (e.g., see Nässig (1988) for variation in 'scoli' of Saturniidae). The simplest outgrowths are pinacula (Fig. 130) - sclerotized areas around the base of one, or sometimes more than one, seta. *Chalaza* (Fig. 131) are cone-shaped pinacula that bear a single seta, and a *scolus* (Figs 132, 133) is an enlarged chalaza with setae or branched spines. Parallel setae arise from a flat, disc-like *verricle* (Fig. 134), whilst a *verruca* (Fig. 135), which bears divergent setae, resembles a pincushion.

Variation in scoli of Saturniidae exemplifies the complexity that exists under this collective term (Nässig, 1988). Wart-like structures or their derivatives may bear mechanically piercing bristles with or without allergenic secretions, or non-piercing secretory bristles. Sometimes the bristles and the scolus may be reduced, although even if a scolus does not protrude above the epidermis its associated cupolas may contain defensive substances capable of being sprayed at enemies. Some scoli take the form of elongated horns with much reduced bristles; others, also elongated, have a hard cuticle and are extended into an apical spine-like bristle. Others have elaborate, urticating scoli.

An arrangement where three pairs of verrucae or scoli (*sensu lato*) exist on each body segment (one dorsal, one subdorsal, and one subspiracular) frequently occurs in Saturniidae (e.g., Holloway, 1987a). On A8 of many larvae of this family the dorsal pair frequently show signs of fusion. In many other Bombycoidea fusion is complete so

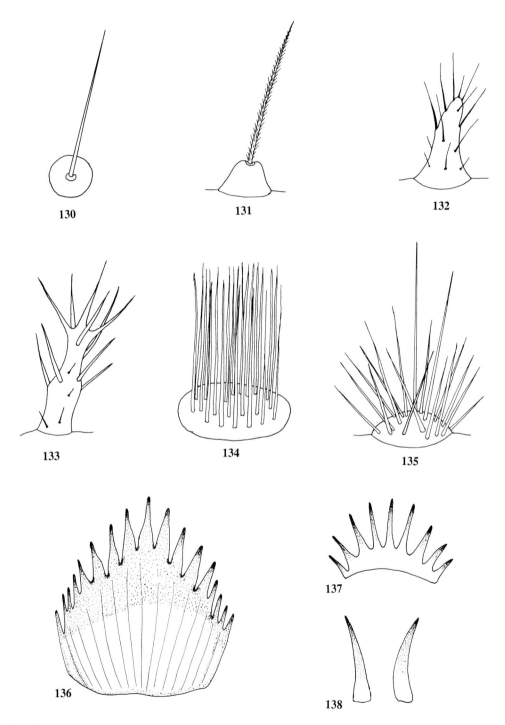

Figs 130–138. *Larval morphology. 130-135, cuticular armature: 130, pinaculum with single seta; 131, chalaza; 132, 133, scoli; 134, verricle; 135, verruca; 136-138, anal plates. (130-135, after Peterson, 1962; 136-138, after Gerasimov, 1952.)*

that a single caudal 'horn' is present alone (e.g., in Sphingidae). The 'horn' may be reduced to a simple, button-like structure in certain larvae. In many Satyrinae, Apaturinae, and Morphinae (Nymphalidae), in all Hedylidae for which larvae are known, and in some Geometridae, tergum A10 is extended into a pair of horns or anal processes. In many Nymphalidae and in Hedylidae the head capsule is extended into a pair of horns, and in certain Geometridae (typically Geometrinae) and many charaxine Nymphalidae short horns are also present. In some Notodontidae the anal prolegs are strongly modified into a pair of long filaments.

A characteristic prothoracic forked filament termed an osmeterium (Pl. 2:10, and see below) is present in the larvae of Papilionidae (swallowtail butterflies). This organ (Schulze, 1911) is tegumental and tube-like. Posteriorly, on each branch occurs an ellipsoidal, secretory gland. The osmeterium is housed within the thorax when not in use, but can be quickly everted. It is now known to release a defensive mixture of isobutyric acid and 2-methylbutyric acid (e.g., Eisner & Meinwald, 1965). However originally, the defensive value of the osmeterium was questioned by Schulze (1911), who thought that its primary function is to excrete toxic substances ingested from the foodplant.

Another eversible gland occurs ventrally between the head and the prothorax of many Lepidoptera (e.g., Yponomeutidae, Rhopalocera, Noctuoidea, Geometroidea) (e.g., Bourgogne, 1951). In *Spodoptera frugipeda* (Noctuidae), apart from an internal non-eversible sac there is an eversible, cuticular tube which when everted forms a visible external papilla (Marti & Rogers, 1988). In larvae of *Schizura concinna* (Notodontidae), the numerous cuticular folds and projections may help atomize the glandular secretion which is ejected by the insect as a fine spray (Weatherstone *et al.*, 1979). The function of the glandular secretions has usually been considered to be defensive but, at least in *Datana ministra* (Notodontidae), the substance is thought to act as a dispersal pheromone (Weatherstone *et al.*, 1986). Formic acid is well known to be sprayed by some Notodontidae when they are disturbed, so the function of the gland is almost certainly defensive in Lepidoptera as well. It is unclear as to whether the gland in the different families is homologous.

An anal comb or plate (Figs 136-138) occurs on the supraanal lobe of some or all larvae of several Lepidoptera families (Tortricidae, Gelechiidae, Oecophoridae, thyatirine Drepanidae, Hedylidae, Hesperiidae, and Pieridae. The structure enables frass to be ejected from the feeding area; as frass pellets appear at the anus they are flicked away. The anal comb probably evolved from cuticular ('sculptural') sclerotizations that became enlarged and fused at their bases. Further fusion resulted, typically, in the formation of a chitinous plate with a dentate or digitate margin (Gerasimov, 1952).

Removal of frass must help to prevent a possible build up of bacterial or fungal infestation in the case of larvae that feed in concealed sites. Frass ejection may also prevent the formation of accumulations presenting parasitoids with chemical cues (summary and references in Gauld & Bolton, 1988).

In the larvae of certain aquatic Lepidoptera (particularly the nymphuline Pyralidae), occur tracheal, or sometimes non-tracheal, gills. These are variously distributed along the body, and are discussed below (p. 128).

Defence against predators

Caterpillars have a wide range of defensive mechanisms. Essentially, these may be divided into those having no directly adverse effect on predators, for example concealment, protective resemblance, and physical protection by a case, and those that are thought to actively repel predators.

In concealment, caterpillars are either not exposed to predators or they are cryptically camouflaged. Concealed feeders such as leaf-miners (as in Pl. 3:2), borers and tunnellers (as in Pl. 3:3), leaf-rollers (as in Pl. 3:5) and tiers, and those living underground (as in Pl. 3:4) or in galls, may derive a heightened level of indirect, fortuitous protection from some predators. However, leaf miners appear to be the group of caterpillars most severely prone to attack by parasitoids, a situation apparently linked to the combined effects of their immobility and the conspicuous nature of the mine (Hawkins & Lawton, 1987). In contrast, crypsis or protective resemblance involves active defence mechanisms such that there is an increase in the noise to signal ratio (Silberglied, 1977). Although crypsis is usually thought of in terms of evading the *visual* sense of predators (e.g., as in the many twig-like looper caterpillars of Geometridae) it is not necessarily so confined. The habit, adopted by many caterpillars, of ejecting frass from the feeding site (see above) so that possibly the odour does not lead predators to the insect itself is a form of olfactory crypsis.

Camouflage is well exemplified by the caterpillar of *Thera firmata* (Geometridae) in which the presence of longitudinal stripes along the body causes it to resemble the pine needles of its foodplant (Fig. Pl. 2:1). The protective resemblance is enhanced by the shape and colour of the head, which is similar to a bud. Protective resemblance to inanimate objects also occurs in those caterpillars that look like bird droppings (as in *Acronicta alni* (Noctuidae) (Pl. 2:2), various species of *Papilio* (Papilionidae), and members of *Trilocha* (Bombycidae)). Resemblance to inanimate objects is lost or greatly reduced when a caterpillar moves, and larvae often remain motionless for much of the day (a time when they are particularly prone to predation by birds) and feed by night.

Choice of feeding time is a behavioural adaptation. Another behavioural adaptation adopted by many caterpillars is the habit of snipping away a partly eaten leaf so that the position of the larva is not betrayed.

The caterpillar of *Nemoria arizonaria* (Geometridae) occurs as two cryptically camouflaged morphs (Greene, 1989). Caterpillars arising from the spring brood resemble the catkins of *Quercus* on which they feed. Caterpillars of the summer brood feed on the leaves of the foodplant and resemble oak twigs. The polymorphism is induced by tannin content. When the tannin content is low, as it is in catkins, the caterpillars develop into catkin-like morphs. But when the tannin content is high, the situation occurring in the leaves, the caterpillars develop into twig-like morphs.

In some caterpillars, protection is derived by the disruption of body outline as in lappet moths (Lasiocampidae) where fleshy protuberances covered by long secondary setae overhang the ventral margin. Marginal hairs and protuberances may also reduce the effects of shadows cast by a larva, as may countershading of the body. Apart from disruption of the body margin, larvae are often disruptively marked and coloured.

Certain larvae decorate themselves. In some Nolinae (Noctuidae), the head-capsules of successive instars are piled on top of each other, an effect possibly providing a false target to predators. Several Synchlorini (Geometridae: Geometrinae) bear specialized hooks to which fragments of plants are attached, presumably increasing concealment.

Many caterpillars live for all or part of their lives in cases, which often are likely to provide a measure of physical protection. Such protection is probably particularly significant in those bagworms (Psychidae) that construct extremely tough cases of silk and twigs (e.g., as in Pl. 2:3), and in a species of *Amicta* from Oman in which a sizable case is made from chips of rock (G.S. Robinson, pers. comm.). Most cases, however, probably function for concealment rather than physical protection. Communal webs spun, for example, by caterpillars of Yponomeutidae, probably offer both physical protection and a means of concealment.

Secondary hairs and spines (e.g., as in Pl. 2:4) frequently play an active role in lepidopteran defence as a result of irritant or poisonous substances they carry. These defences are primarily encountered among larvae feeding in exposed conditions, particularly those living in trees. Secondary setae are usually absent in concealed feeders. Larval hairs and spines are known to have pathological effects on man, and it is from this source that the evidence from their defensive function is mainly derived. The medical term 'lepidopterism' has been coined to describe the general pathological effects of Lepidoptera on man, and 'erucism' to refer to those specifically induced by larvae (see Wirtz, 1984). 'Phanerotoxic' (allergic and toxic) effects of Lepidoptera in which venoms are injected into the skin are distinguished from 'cryptotoxic' effects where repellent volatiles are exuded by various glands (Wirtz, 1984).

The phanerotoxic structures most frequently encountered in caterpillars are 'urticating' hairs, so called because of the nettle-like irritation they typically induce. Urticating hairs are of two kinds - those with a poison gland at their base and those without. The first variety exists as undivided setae or spines occurring singly or in groups or large patches. The setae often bear detachable tips that become dislodged when brushed against the skin resulting in the release of the irritant substance from the hollow hair. Enormous numbers of urticating hairs have been reported. A last instar larva of the processionary caterpillar *Thaumetopoea processionea* (Notodontidae) bears in excess of an estimated 600 000 urticating setae, and *Euproctis chrysorrhoea* (Lymantriidae) (Pl. 2:6) over 2 000 000 (Wirtz, 1984). Those hairs that do not release a poison may, nevertheless, have strongly irritant properties through their ability to lodge in the skin and mucous membranes.

Reactions to urticating hairs or from substances transmitted via stinging spines range from mild irritation, through serious dermatitis, to shock-like symptoms, convulsions and serious haemorrhage (Wirtz, 1984). Urticating hairs of the gypsy moth (*Lymantria dispar*: Lymantriidae) caused a major outbreak of pruritic (itching) dermatitis in America in the spring of 1981, aided by the airborne dispersal of young larvae by silken threads. This is but one instance of the many examples of dermatitis caused by urticating hairs. Such episodes have occurred, particularly, due to caterpillars of Lymantriidae, Notodontidae, Bombycoidea, Arctiidae, and Megalopygidae.

Stinging spines of the larva of *Lonomia achelous* (Saturniidae) from South America delivers a potent anticoagulant causing severe internal haemorrhage in man (Arocha-Piñango & Layrisse, 1969).

Most substances transmitted by larval hairs and spines are proteinaceous. Histamine, and chemicals that simulate the release of histamine, are also present, but apparently they exist in lower proportions (Wirtz, 1984). The evidence for the defensive properties of urticating hairs is circumstantial but considerable, and it is likely that these structures provide effective deterrents to vertebrates, particularly insectivorous birds and mammals.

Cryptotoxic defence takes the form of active projection or exudation of repellent odours. For instance, larvae of Papilionidae are capable of disseminating such substances by means of the prothoracic osmeterium. Glandular cells are associated with the base of each of the tubular arms of the structure. A foul smelling, alkaline fluid is exuded from the glands of larvae of *Cossus* (Cossidae). Several Notodontidae larvae are able to emit jets of irritating fluid over 2-3 cm. The caterpillar of *Heterocampa manteo* (Notodontidae) may release a chemical when handled (probably formic acid), which causes blistering of the skin.

A further system of defence occurring in a number of caterpillars is that involving frightening the predator. Although no pathological damage is done, this form of protection is considered here as inducing an active response from the predator. Among the best known examples is the sudden display of eyespots of larvae of some Sphingidae (e.g., *Deilephila porcellus*, Pl. 2:7)), and the threat posture of *Cerura vinula* (Notodontidae) (Pl. 2:8). A striking example of a threatening form of defence is seen in caterpillars that mimic snakes (Pl. 2:9).

The final basic method of larval defence is unpalatability. This method of protection is well developed in all stages of the life history of Lepidoptera. For this reason, and because of its association with warning colour, the reader is referred to the section on colour and pattern in Chapter 3.

The evidence for the various defence strategies outlined above is largely circumstantial. Moreover, the mechanisms are by no means perfect, and are assumed to give only a measure of protection. For example, as mentioned above, leaf mining caterpillars are particularly prone to attack by parasitoid Hymenoptera, and predators have frequently evolved mechanisms to deal with noxious prey rendering their defensive properties less effective or ineffective.

One of the few experimental studies demonstrating both the effectiveness of the defence mechanism and the complexity of the evolved response was undertaken on the osmeterium-bearing caterpillar of *Eurytides marcellus* (Papilionidae), the Zebra Swallowtail butterfly (Damman, 1986). Predation is demonstratively high in this caterpillar both in spring and in summer, but predator composition differs with season. The osmeteria of the caterpillars were shown to be effective in defence against predators in spring, but not in summer. These organs deterred attacks from ants and small spiders, but failed to provide a defence against large spiders, vespid wasps, and *Trogus pennator* - a parasitoid wasp specializing in attacking *Eurytides* caterpillars. Whereas ants and small spiders are effective predators in spring, in summer they decline in importance. Correspondingly, in spring the osmeteria were shown to be important in defence while in summer other defensive strategies were used by the caterpillars. Small caterpillars wriggle, apparently preventing *Trogus* parasitoids from grasping them, and larger caterpillars move away from their foodplant when not feeding.

Diversity of food sources and feeding habits of lepidopteran larvae

Most caterpillars live on live green-plant material. Although probably most consumers of these parts of the plant occur among the macrolepidopterans, which generally feed exposed, a great many green-tissue feeders are found among the typically concealed-feeding microlepidopterans. Leaf miners (e.g., Pl. 3:2) generally feed within one or more of the chlorophyll containing layers, and leaf rollers (e.g., Pl. 3:5) and leaf tiers also usually eat green tissue. However, all parts of plants - flowers, fruit, seeds, roots, the pith of stems, and galls - are eaten by caterpillars. The main environmental impact of Lepidoptera lies in their consumption of green plants (Chapter 7), but other substrates (outlined below) are also eaten.

Mycophagy. A number of Lepidoptera are mycophagous (review by Rawlins, 1984) a term confined not only to organisms feeding exclusively on free fungi (fungivory), but one that also extends to lichenivory - a situation where the fungal hosts of algae are consumed, and detritivory, where the fungal elements are eaten incidentally. Most examples of fungivory occur in Tineoidea, a superfamily in which the larvae of several species feed in the fruiting fungal bodies. The subfamilies Nemapogoninae and Scardiinae are typically fungivorous or feed in decaying wood infested with fungi (Robinson, 1986). Primitive Psychidae are lichenivorous (Davis *in* Stehr, 1987). Many Tortricoidea feed on decaying material, but although there are facultative mycophages in this group none are yet known to be obligate. Mycophagy is widespread in Oecophoridae and likely to be high among the extensive litter-dwelling fauna of dead *Eucalyptus* leaves in Australia. Among macrolepidopterans, mycophagy is encountered in Geometroidea (Geometridae), Noctuoidea, and Papilionoidea. A few Geometridae are obligate lichenivores. Among Noctuoidea, fungivory is restricted to Noctuidae, and lichenivory is well-developed in that family and in Arctiidae. In Papilionoidea, obligate lichenivory occurs among some Lipteninae (Lycaenidae).

Plate 1

Resting posture in Lepidoptera. 1, *Yponomeuta padella* (Yponomeutidae); 2, *Autographa pulchrina* (Noctuidae); 3, *Nemoria fallax* (Geometridae); 4, *Phrygionis privignaria* (Geometridae); 5, *Smerinthus ocellatus* (Sphingidae); 6, *Laothoe populi* (Sphingidae); 7, *Emmelina monodactyla* (Pterophoridae); 8, Epiplemidae sp; 9, *Strymonidia w-album* (Lycaenidae); 10, *Thymelicus sylvestris* (Hesperiidae); 11, *Heliocheilus albipunctella* (Noctuidae), showing foveae; 12, Bogong moths (*Agrotis infusa*: Noctuidae), aestivating. (Photographs: 1, David Carter/BMNH; 2, 6–8, M.W.F. Tweedie; 3, 4, Linda Pitkin; 5, 9, P.H. & S.L. Ward, 10, P.H. Ward/Natural Science Photos; 11, Marcus Matthews; 12, Division of Entomology/CSIRO, Canberra.)

Plate 2

Larvae (caterpillars). 1, *Thera firmata* (Geometridae); 2, *Acronicta alni* (Noctuidae), early larva resembling a bird dropping; 3, Case of an African bagworm (*Eumeta* sp., Psychidae); 4, *Dirphia molippa* (Saturniidae); 5, Limacodidae; 6, *Euproctis chrysorrhoea* (Lymantriidae); 7, *Deilephila porcellus* (Sphingidae); 8, *Cerura vinula* (Notodontidae), in threat posture; 9, *Hemeroplanes* sp. (Sphingidae); 10, *Papilio thoas* (Papilionidae), with osmeteria extruded. (Photographs: 1, 6, 8, David Carter/BMNH; 2, W. Reid/BMNH; 3, 5, 7, the writer; 4, Ken Preston-Mafham/Premaphotos; 9, Phil DeVries; 10, Ken Preston-Mafham/Premaphotos Wildlife.)

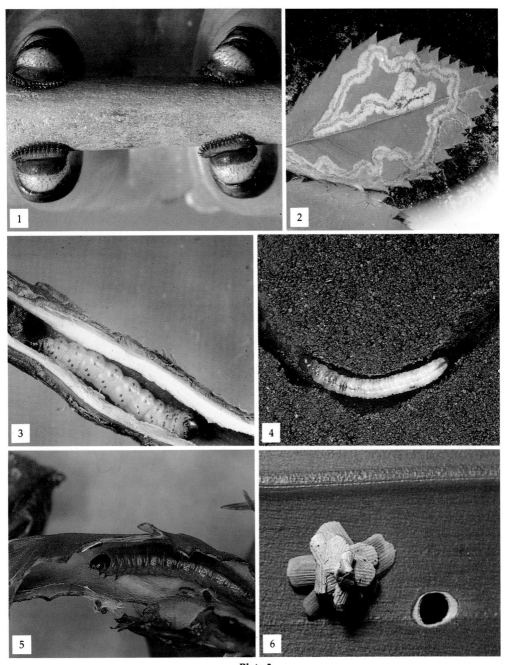

Plate 3

Larval habits. 1, *Sphinx ligustri* (Sphingidae) showing prolegs with crochets; 2, leaf mine of *Stigmella anomalella* (Nepticulidae) on wild rose; 3, *Zeuzera pyrina* (Cossidae) tunnelling in apple twig; 4, *Hepialus humuli* (Hepialidae) in tunnel in soil; 5, *Rhopobota naevana* (Tortricidae) in rolled leaf; 6, case constructed from *Strelitzia* leaf cuttings by Psychidae sp. (Photographs: 1, Photographic Unit, BMNH; 2, John Bradley; 3, Becker/BMNH; 4, Becker/Shell U.K., Ltd; 5, Martin Honey/BMNH; 6, the writer.)

Plate 4

Eggs. 1, *Pieris brassicae* (Pieridae); 2, *Attacus atlas* (Saturniidae); 3, *Clostera anastomosis* (Notodontidae); 4, *Trilocha ficicola* (Bombycidae); 5, *Cacoecimorpha pronubana* (Tortricidae); 6, *Aglais urticae* (Nymphalidae). (Photographs: 1, Becker/Shell U.K., Ltd; 2, 3, Mike Bascombe; 4, H. O'Hefferman; 5, 6, Martin Honey/BMNH.)

Mycophagy has arisen independently many times in Lepidoptera and may have evolved from detritivory, from which fungal hyphae are incidentally consumed, through an increasing intake of fungal elements resulting, finally, in obligate mycophagy involving an increased dependence on enzymes secreted by the fungi themselves (Rawlins, 1984). Such enzymes are normally used by fungi to break down their own tissue.

Animal tissue. Caterpillars are not typically associated with feeding on animal matter, but there are notable exceptions. Most instances involve scavenging, but both predation and parasitism occur. Moreover, many non-predaceous caterpillars will resort to cannibalism at times of overcrowding, a response that may be due not only to hunger, but also to opportunity and the general tendency of caterpillars to nibble objects close to them. Cannibalism is a phenomenon not infrequently encountered by those who rear larvae. Larvae often eat their egg shells and exuviae, and, with the potential for the consumption of animal material, it is understandable that obligate carnivory has occasionally arisen.

Coccidae are favoured prey of carnivorous caterpillars (summary in Hinton, 1981). Their consumption has occurred independently in several families, at least partly because of the frequency with which phytophagous larvae happen to come into contact with them. The habit of eating coccids is known in Heliodinidae, Blastobasidae, Momphidae, Cosmopterigidae, Tortricidae, Schreckensteiniidae, Pyralidae (Phycitinae), Lycaenidae, and Noctuidae. *Coccidiphaga scitula* (Momphidae) not only consumes scale insects, but also uses a scale as a portable covering holding it with modified anal prolegs. In addition, pupation takes place under the scale.

Lycaenidae provide many examples of carnivory, particularly on Homoptera and the brood of ants (Cottrell, 1984). Often, only certain instars are carnivorous. For instance in *Maculinea* the fourth instar feeds on ant brood, but the first three are phytophagous and eat buds and flowers. Predation on Homoptera or on the brood of ants permits normal development, although other forms of aphytophagy may supplement the diet (Cottrell, 1984).

Some Hawaiian members of the genus *Eupithecia* (Geometridae) actually ambush insect prey, and execute a 'strike' of considerable speed (Montgomery, 1982). The larva of *Tirathaba parasitica* (Pyralidae) feeds on caterpillars of Hepialidae.

Larvae of Epipyropidae attach themselves to Homoptera (particularly Fulgoridae) where they feed on the waxy secretions of the host (Davis *in* Stehr, 1987). Larval development takes place on the bug and pupation generally occurs away from it, although Gerasimov (1952) stated that sometimes the pupa is formed on the host. The bug may be killed or may be unharmed by the effect of the larva. Given the closeness of the relationship between larva and bug the association has often been referred to as parasitic. An association in the same mode is found in *Stemauge parasitus* (Pyralidae) in which the larvae spin threads between spines of the caterpillars of *Automeris* and sometimes *Dirphia* (Saturniidae). The pyralid larvae feed on the spinules causing fatal damage to the host.

Larvae of the subantarctic Tineidae *Pringleophaga marioni*, the adults of which are brachypterous, are omnivorous in captivity and aggressively attack and eat earthworms (French & Smith, 1983). Although this habit has not been observed in the wild, both larvae and earthworms are abundant among the macroinvertebrate fauna of Marion Island. Both are very common in the upper peat layers of many vegetation types so it seems likely that their carnivorous habit occurs, at least to some extent, in the wild.

Secretions and excretions of Homoptera or regurgitations of ants. Larvae of several Lycaenidae feed on excretions of Homoptera or regurgitations of ants (Cottrell, 1984). It appears that this source of food is inadequate, however, to support complete development. Some species actively elicit the excretion of honeydew from homopter-ans in the same way as ants, which are well known for their ability to 'milk' certain Hemiptera.

Scavenging. Scavenging has been adopted by a variety of caterpillars, and includes such peculiar habits as eating insects trapped in spiders webs and feeding on the eggs of spiders (e.g., *Batrachedra*: Coleophoridae: Batrachedrinae, and *Brachmia*: Gelechiidae), and on insect prey of pitcher plants (*Nepenthophilus tigrinus* and *Eublemma radda*: Noctuidae).

Coprophagy. Cryptoses and *Bradypodicola* (Pyralidae) feed on the faeces of sloths. When sloths descend from the trees to defecate on the ground, adult sloth moths, which live among the fur of these mammals, briefly hop onto the freshly voided faeces, deposit their eggs, and return to the sloth before it ascends the tree (Waage & Montgomery, 1976). Some larvae that feed in faeces actually consume keratin (see below) and are not, therefore, strictly coprophagous.

Keratin. Keratin feeding is rare in insects since few have the appropriate enzymes (keratinases) to break down this proteinaceous substance. In Lepidoptera the ability is confined to Tineinae (Tineidae) (Hinton, 1956; Robinson, 1980, 1988b) a group including many species that eat keratin-containing material such as horn, fur, feathers, and skin. Their capacity to digest keratinous material has caused some Tineidae (the clothes moths) to become pests of woollen products. Keratin is also found in faeces, particularly those voided by insectivores such as bats (Robinson, 1980).

Form, function and habitat

Although lepidopteran larvae are chiefly phytophagous, the range of microhabitats available to and exploited by them is considerable. The form of a caterpillar is often fairly obviously adapted to its feeding site, and the modifications have developed convergently in a number of instances.

A 'typical' (familiar) caterpillar feeds on plant tissue in an exposed position. It is cylindrical, bears hypognathous mouthparts, six stemmata, three pairs of thoracic legs, and abdominal prolegs on A3-6 and A10 with crochets arranged in a mesoseries. In exposed feeders, the base of the proleg is much larger than in concealed feeders, and the crochets are arranged in a single mesoseries of outwardly curved hooks. Exposed feeders require greater mobility than do concealed feeders for three reasons. First, they often rest in a place different from the site at which they feed. For example, caterpillars frequently feed at night and often move to positions under leaves or to sites away from leaves by day. Second, since exposed feeders are generally larger than internal feeders, they tend to eat more and are therefore more likely to need to move to a new supply. Third, while concealed feeders very often pupate within the feeding site, those that feed in an exposed position often move considerable distances to their place of pupation. (Internal feeders that pupate in the soil, for example most Nepticulidae, tend to reach the ground by means of silken threads rather than by walking.) Also, larvae that feed in exposed situations often have an impressive array of defences such as masses of urticating hairs and sharp stinging spines, or they may exhibit aposematic colours demonstrative of their unpalatability.

Different feeding patterns, niches, and body forms may be interpreted as deviations (*not* in the phylogenetic sense) from the 'typical' caterpillar.

Leaf miners. Leaf miners (Hering, 1951) offer an extreme contrast in body form and feeding site with external feeders. Most live within a vertically confined space between the epidermal layers of a leaf. In the early instars of *Phyllocnistis* (Gracillariidae), mining even occurs *within* the epidermis. Groups with larvae that are predominantly or significantly leaf mining include Eriocranioidea, Nepticuloidea, Incurvarioidea, Tischerioidea, Gracillariidae, and Elachistidae. The habit occurs sporadically in other groups, and in some (e.g., Coleophoridae) it is confined to early instars alone. The mining habit is not entirely restricted to the laminae of leaves. Thin green bark, the petioles of leaves, and the 'wings' of certain seeds (e.g., of *Acer*) are also mined. But these exceptions are associated also with living within a vertically restricted space, and are exhibited by families whose larvae are typically leaf miners. Such larvae also exhibit the characteristic adaptations associated with leaf miners. These modifications include a prognathous head, a dorsoventrally flattened body, and reduced appendages such as legs, prolegs, antennae. Often there is a reduction in the number of stemmata.

The mouthparts of miners are directed forwards either as the result of a retraction of the vertex into the prothorax, or by a lengthening of the ventral part of the head. The first of these modifications is typical of tissue feeders - those that consume parenchymatous plant cells. The second occurs in sap feeding larvae, which lacerate plant cells and imbibe the fluid released. In both tissue- and sap-miners the coronal suture of the head is absent and a frontal bridge is present. Therefore, the typical triangular appearance of the front of the head is generally lost. In a few miners in which the mine is less restricted vertically (e.g., the last instar of certain Gracillariidae, which expands the mine into what eventually becomes a pupal chamber), the triangular shape is retained. But even here the coronal suture is absent.

In most leaf miners the mandibles and hypopharynx are modified. The dorsal teeth of the mandibles tend to become better developed, while the ventral teeth are reduced. The hypopharynx, which projects beyond the labrum and is well-developed and hairy, functions in collecting sap. In tissue feeders the surfaces of the mandibles are brought together to bite and chew plant cells. In many, a considerable amount of the tissue between the epidermal layers is consumed. In sap feeders the mandibles are kept open while feeding (Jayewickreme, 1940), and the anterior part of the body is swung from side to side so that the blades lacerate the epidermal cells and release fluid. In *Phyllocnistis* (Gracillariidae) the sides of the labrum are enlarged, which may help to protect the roof of the epidermal mines from being torn by the mandibles. In some tissue feeding genera (*Lyonetia* and *Leucoptera*: Lyonetiidae) the labrum, also, may be toothed and function to tear tissue.

The maxilla and the labium are reduced to a greater or lesser extent, and the spinneret may be turned backwards. In sap feeders the spinneret is absent. Miners generally spin only in the last instar, although *Tischeria* (Tischeriidae) is an exception. The hypopharynx is often well-developed, projects forward as a tongue-like structure covered with hairs, and functions to collect sap even in tissue feeders such as *Lyonetia* and *Leucoptera*.

Borers and tunnellers. As with leaf miners, borers live within their food source. Their mouthparts are usually semiprognathous. Borers are typically pale and lack markings, and the prolegs are generally reduced or even lost. The habit of boring is widespread in Tortricidae, many of which tunnel in fruits. In wood borers, for instance Cossidae, Sesiidae, and some Hepialidae, larval development time is greatly extended, presumably as a result of the low nutrient value of wood. The larvae of numerous Pyralidae and many Noctuidae bore into a variety of plant substrates.

Several caterpillars, among them many Hepialidae and certain Noctuidae, are subterranean and feed 'externally' on underground parts of the plant. Here again, the mouthparts tend towards the prognathous condition.

Leaf-rollers and -tiers. Two notable adaptive characters occur in leaf-rollers and -tiers, although both are also found in caterpillars that have adopted other lifestyles. The crochets (p. 116) are typically arranged in a circle, or a broken circle. This arrangement, characteristic of so many microlepidopterans, is suitable for locomotion on silken webs, which are frequently spun in the rolling and tying process. Common to many caterpillars that live in rolled or tied leaves is the presence of an anal comb, a structure used to flick frass away from the feeding site (p. 121). The presence of this structure tends to be related to rolling and tying as opposed to boring. For example, in Tortricidae although both habits are widespread, the anal comb is mainly confined to the former. The mouthparts of borers and tiers are usually semiprognathous, a condition that may represent a retained primitive feature rather than a specialization for these modes of life.

Aquatic larvae. The caterpillars of a few Lepidoptera are aquatic, and face the obvious problem for an essentially terrestrial group as the Lepidoptera of obtaining oxygen. Oxygen is gained in four main ways (Bourgogne, 1951). The first involves a periodic renewal of oxygen from the surface or from bubbles released by aquatic plants. Air may simply be accumulated in the tracheae, or may form a layer, trapped by hydrofuge hairs, over the body (as in *Palustra lafoulbeni*: Arctiidae) or within roughnesses of the cuticle (as in *Nymphula nympharata*). Air stores function by helping the insect to rise to the surface and resupply its oxygen store, by enabling direct intake of air into the trachea, and by acting as physical gills in which oxygen diffuses into the oxygen-poor bubble (Wigglesworth, 1965). In caterpillars that rely on air stores, the tracheae are not greatly modified, although the spiracles may be reduced to some extent.

In the second category, oxygen is absorbed from the surrounding water through the cuticle. No air layer is developed in cuticular respiration, and the spiracles are mainly reduced and non-functional. This way of gaining oxygen is characteristic of early instars (e.g., in various species of *Nymphula* and in *Cataclysta lemnata* (Pyralidae: Nymphulinae). Later instars respire through spiracles. Similarly, in *Acentropus niveus* (Pyralidae: Schoenobiinae) the early instars absorb oxygen through the cuticle direct to the haemolymph, but later a branching mass of tracheae develop and lie against the inside of the cuticle.

The third, and most obvious, aquatic adaptation for respiration is the presence of tracheal gills. The gills may be simple or branched. In *Nymphula maculalis* (Pyralidae) they are hollow, tracheae-bearing tubes communicating directly with the haemocoel. The spiracles are reduced and generally non-functional, although in some species they are developed normally on abdominal segments A2-4 where they probably function when the larva leaves the water. In *Parapoynx stratiotata*, gills are present on all segments with the exception of T1 and A10, and are composed of filamentous tufts bearing tracheae. The gills are arranged in three rows, one supraspiracular, one subspiracular, and the other parapedal (i.e., at the base of the thoracic legs and the prolegs).

Finally, slender filaments (probably acting as blood-gills) occur in *Cataclysta*. The structures are devoid of tracheae, are unbranched, and are found along each side of the body in two rows from the mesothorax to the ninth abdominal segment. There may be as many as 18 pairs of these gills.

Use in classification

The study of lepidopteran larvae has a history as long as that of the adults, but larval stages have been relatively less extensively researched. Adults are easier to collect and store, and the much used methods of attracting nocturnal Lepidoptera to light, or netting diurnal species, have led to large accumulations of material. So, to a large extent, lepidopteran classification has developed as a result of a study of the adult. Comprehensive work on lepidopteran larvae was carried out notably by Dyar (1894),

Fracker (1915), and Hinton (1946a), and these studies were complementary to those based on adults. Both Dyar and Fracker found that, with certain significant exceptions, their work supported existing divisions.

However, larvae have been invaluable also in demonstrating weaknesses in adult classification and providing some resolution so their complementary value to adult studies should be emphasized. Sometimes, insights from larval work have had an impact on adult-based classifications or have helped to place species more accurately in existing classifications.

For example, it was very largely studies of the larva of the Neotropical genus *Heterobathmia* that led to the removal of this group from Zeugloptera (Micropterigidae) and its assignment to a suborder (Heterobathmiina) of its own (Kristensen & Nielsen, 1983). Originally, *Heterobathmia* had been placed in Zeugloptera on the basis of similarities in adult structure (Kristensen & Nielsen, 1979). Yet the larva was found not only to lack the specializations of zeuglopteran larvae, but also to share specializations with Aglossata (Agathiphagidae) and Glossata, or with Glossata alone. All these shared derived characters were found on the larval head. The similarities between adults of Zeugloptera and *Heterobathmia*, such as the occurrence of specialized armature on the epipharynx and the presence of strong spines forming a triturating basket, are apparently homoplasious (parallel or convergent) (Kristensen & Nielsen, 1983).

Another study where emphasis on larval characters led to radical hypotheses about existing classification is in the Nymphalidae (DeVries *et al.*, 1985). What began as an attempt to elucidate the relationships of *Antirrhea*, *Caerois*, and *Morpho*, developed into a work suggesting that many accepted taxa within the Nymphalidae were untenable in a phylogenetic system. Even such well known taxa as the subfamily Satyrinae and the family Nymphalidae (*sensu* Ehrlich, 1958b) were found likely to be polyphyletic. The various hypotheses of DeVries *et al.*, were derived from a numerical parsimony analysis of 92 characters of which 85 were larval and only 6 were derived from the adult.

The form of the cocoon spun by the larva suggested that what appeared to be an extraordinarily large member of the Cemiostominae (Lyonetiidae) was in fact a member of the genus (*Bucculatrix*) (Bucculatricidae) (Scoble & Scholtz, 1984). Similarities between the adults of Cemiostominae and the species in question (*Bucculatrix ingens*) were considered to be superficial, whereas the ribbed, characteristically *Bucculatrix*-like cocoon is apparently uniquely derived and indicative of phylogenetic affinity.

Pupa

Lepidoptera are holometabolous insects undergoing a complete metamorphosis with a pupal stage occurring between larva and adult. The primary function of the pupal stage in holometabolous insects is possibly to permit expansion of the wings. Lepidopteran pupae may be decticous (pupa dectica), in which functional mandibles are present (Figs 139-141), or adecticous (pupa adectica) in which functional mandibles are absent (Hinton, 1946b). Adecticous pupae may be exarate, with appendages not glued to the body, or obtect, with appendages that are cemented or fused to the body (Figs 143, 144). In Lepidoptera, all adecticous pupae are obtect (Hinton, 1946b), whereas decticous pupae are always exarate. Obtect pupae may have their appendages weakly or strongly cemented or fused to the body so that in some cases they resemble pupa exarata and are described as 'incomplete' pupae. Decticous, exarate pupae are found in the most primitive moths, whereas in the majority of Lepidoptera the pupa is adecticous and obtect. With some exceptions, the most obviously obtect pupae are found in higher Lepidoptera.

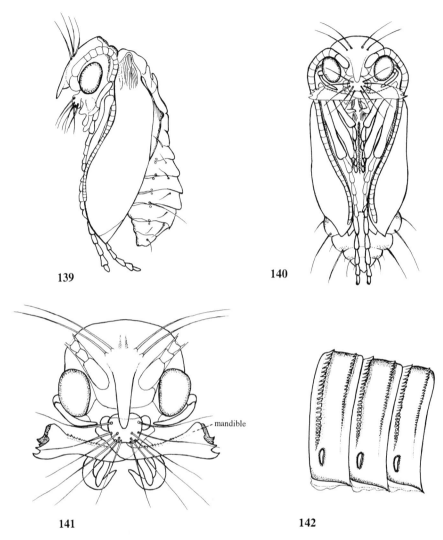

Figs 139–142. *Pupa. 139-141,* Mnemonica *(Eriocraniidae), showing decticous, exarate condition: 139, lateral view; 140, ventral view; 141, head; 142, three abdominal segments of* Cossus cossus *(Cossidae) showing dorsal spines (dorsolateral view). (139-141, after Busck & Böving, 1914; 142, drawn by Malcolm Kerley.)*

Structure

Head

On the pupal head (Mosher, 1916), the vertex, frons, and clypeus are recognizable although the sulci that typically divide these areas may be lost, particularly in higher Lepidoptera. The frons is delimited dorsally by the epicranial sulcus and ventrally by the fronto-clypeal sulcus. The clypeo-labral sulcus is rarely represented by more than a furrow. Pilifers, posterio-lateral projections of the labrum, are often well-developed.

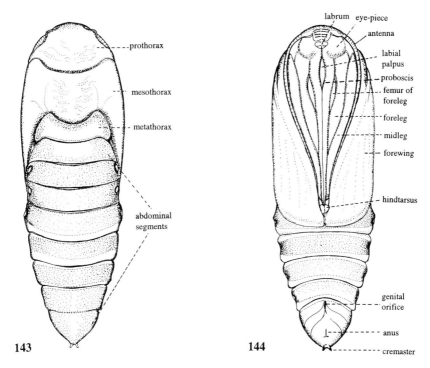

Figs 143, 144. *Pupa of* Spodoptera littoralis *(Noctuidae): 143, dorsal; 144, ventral. (Drawings by Malcolm Kerley.)*

Functional mandibles in the pupa occur in Micropterigidae, Agathiphagidae, Heterobathmiidae, Eriocraniidae, and Acanthopteroctetidae, and are suspected to occur in at least one species of Neopseustidae. Mandibular remnants are small, but visible, in many other Lepidoptera, and are situated at the lateral angles of the labrum. The maxillary palpi appear as approximately triangular or rectangular components adjacent to the eyes anteriorly and to the margin of the legs posteriorly. The galeae are situated laterally to the labial palpi, when the latter are visible, but in other instances they overlie the bases of the labial palpi. The galeae often extend to beyond the length of the wings. In some Sphingidae, the proboscis is looped anteriorly. The labial palpi are the most mesal of the various mouthparts or appendages. Although usually visible, sometimes they are virtually concealed by the galeae.

Thorax

The dorsal divisions of the thorax (Mosher, 1916) (and see Fig. 143) are usually clearly visible, although medially the prothorax may be invisible. Whereas in some pupae the mesonotum is usually longer than the pronotum or metanotum, in more generalized pupae the segments are of about equal length. Ventrally, the legs are visible, as in Fig. 144. The coxae of the forelegs, but not of the mid- or hindlegs are frequently exposed. Since the legs are folded, the femora are mostly or entirely hidden. The hindlegs are mainly concealed. The forewings virtually conceal the hindwings in Lepidoptera other than in the most primitive exarate forms although a narrow strip is visible. The mesothoracic spiracle usually lies between the prothorax and mesothorax in a dorsal position; occasionally it is situated more ventrally.

Abdomen

The abdomen (Mosher, 1916) (Figs 143, 144) is 10-segmented with the last three segments always fused and incapable of independent movement. Mobility between other segments varies, with the greatest degree of movement occurring in the primitive groups such as Micropterigidae and Eriocraniidae (Figs 139, 140) where segments A1-7 are free. A segment is said to be movable when there is mobility between its posterior margin and the next (posterior) segment. Usually, one more segment is movable in males than in females. The segments are strongly telescoped in many pupae.

Spiracles are present on segments A1-8. Those on segment A1 are rarely visible since they are covered by the wings, and those on A8 are not functional, exhibiting no distinct aperture.

Genital scars are found on segment A9 of the male. In females of Ditrysia, scars occur on A8 and A9. The scars may be rounded or slit-like. The anal scar is sited posterior of the genital scars on A10, and like them it may be round or slit-like. Segment A10 is elongated into a cremaster, a structure that may bear a number of hooked setae functioning to attach the pupa to its substrate.

The dorsal surface of many of the more primitive pupae often bears series of posteriorly directed spines (Fig. 142). As a pupa wriggles from its cocoon just prior to ecdysis, the spines prevent it from slipping back. Pupae of microlepidopterans are typically, but not universally, extruded from their cocoons prior to adult eclosion. Pupae of higher Ditrysia are usually considerably less mobile, and not usually extruded from the cocoon. However, examples of protrusible but obtected pupae exist in the main groups of Bombycoidea (Brock, 1990), an occurrence associated with abdominal spining. Sometimes ridges on the head and thorax act as a cocoon cutter. Studies of *Deilephila elpenor* (Sphingidae) demonstrated that pupal eclosion was directly related to high humidity, taking place well in advance of eclosion (Brock, 1990). Uncertainty remains as to whether pupal protrusion in Bombycoidea represents the retention of a primitive trait or an independent specialization.

Diversity, form, and function of the lepidopteran pupa

A cocoon or cell provides many pupae with varying degrees of protection from extreme temperatures, from desiccation, and from predator and parasitoid attacks (Common, 1986). However, such protection requires that the pupa develops a method of escape. The various types of pupa may be said to reflect the different ways in which the adult escapes from the cocoon or cell (Hinton, 1946b). In Lepidoptera, escape may be effected by mechanical or chemical means. Those pupae that break out of their cocoons physically do so chiefly by using articulated mandibles in decticous species, or by using a cocoon-cutter on the head in adecticous groups. Some Lepidoptera secrete a softening agent to weaken the wall of the cocoon prior to adult eclosion. The mobility of the pupa makes a contribution to the escape.

The cocoon-tearing mandibles of primitive decticous pupae are hypertrophied and moved by the powerful muscles of the adult mandibles. While the adult mandibles are sleeved within those of the pupa, the apodemes of the pupal mandibles are tightly enclosed within the mandibular apodemes of the adult (Hinton, 1946b).

In the more advanced species, the ridge or swelling (the cocoon-cutter) on the head is the functional analogue of the mandibles in decticous pupae (Hinton, 1946b). Although non-decticous pupae are obtected to varying degrees, their appendages are frequently quite mobile. The posteriorly directed tergal spines, which are often present, help the pupa to be forced forward as it wriggles to escape from the cocoon. Adult emergence from the pupal cuticle is assisted by the presence of the exuviae, against which the emerging insect may push. Without this support it was shown that the adult may be unable to escape effectively (Hinton, 1946b). To keep the pupa from falling from the cocoon after it has partially wriggled out, some Tortricidae are attached by a silken cremasteral cable.

In some lower Ditrysia and, typically, in higher Ditrysia (but see Brock, 1990, and above), the pupa is strongly obtected exhibiting increased sclerotization, a greater degree of fusion of abdominal segments and decreased mobility. Such pupae are not extruded from the cocoon at adult eclosion, and generally lack tergal spines and a cocoon-cutter. Adults usually escape from their cocoon either by way of a lid (as in Megalopygidae, Limacodidae, and some Saturniidae) or by the secretion of a softening agent (as in a few Noctuidae, Notodontidae, and Saturniidae).

In certain species pupating in subterranean cells, the pupa actually moves the insect through the soil. In some Notodontidae the pupal cremaster is used in propulsion, while in some Sphingidae there are prespiracular abdominal ridges or flanges.

The pupa may sometimes be entirely exposed, being attached to its substrate by either the cremaster and a silken girdle around the thorax or abdomen, or by the cremaster alone. An exposed pupa typical of the superfamilies Hedyloidea and Papilionoidea is found occasionally in several other groups (Common, 1986) (e.g., in Geometridae). Within Papilionoidea girdled pupae occur in Pieridae, most Papilionidae, and in Lycaenidae. The girdle is typically thoracic in Papilionidae and Lycaenidae, but abdominal in Pieridae. It is abdominal also in Hedylidae. In Nymphalidae, the pupa is suspended by its cremaster alone, and is not held in place by a girdle. Girdled pupae are found also in certain Sterrhinae (Geometridae) (*Anisodes* and *Cyclophora*). Most other exposed pupae that lack a cocoon or an earthen cell occur among microlepidopterans, particularly in Lyonetiidae, Elachistidae, Oecophoridae, and Pterophoridae (Common, 1986) although they are not necessarily entirely exposed. *Bedellia somnulentella* (Lyonetiidae) pupates in a hollow, the pupa being protected merely by a few silken strands so that it is only partially exposed, whereas in some Elachistidae the pupa is entirely exposed (Common, 1986).

Although there are exceptions, the general trend in lepidopteran pupae has been from the relatively weakly sclerotized, decticous type to increasingly sclerotized, nondecticous and obtected forms. With specialization, mobility of the pupa decreased demanding different means of escape from the cocoon.

6

HEARING, SOUND, AND SCENT

This chapter deals with communication by sound, which involves the subject of hearing, and by scent. These topics are discussed here rather than in other chapters because the various organs involved are neither confined to, nor obviously associated with, the head, the thorax, or the abdomen. Visual communication, a subject on which much important work has been carried out with butterflies, has been considered in Chapter 3 under the section dealing with the functions of wing colour patterns (p. 85). But since communication by scent, in particular, is often closely linked to communication by sight, the importance of vision, when this integration occurs, is also noted here in Chapter 6.

Hearing organs

Auditory organs are found in several families and numerous species of Lepidoptera. They vary in both position and in structural details, but most consist essentially of a tympanic membrane (tympanum) stretched across a frame which is backed by a tympanal air sac. Vibrations of the tympanum are sensed by a chordotonal organ, which is attached to the tympanum. Tympanal organs are usually found in the metathorax or the abdomen, but they also occur in the wing bases. Non-tympanal hearing organs exist in the mouthparts of adults of certain Sphingidae, and hairs sensitive to sound occur in some larvae.

Distribution and position within adult Lepidoptera

The best known lepidopteran hearing organs are the tympanal structures in the metathorax of Noctuoidea and those occurring in the base of the abdomen of Geometroidea and Pyraloidea. Tympanal organs exist also in the abdomen of Drepanoidea, Uranioidea (including Epipleminae), Dudgeoneidae (Cossoidea), and a few Tineoidea (*Harmaclona*: Tineidae). Possibly, the structures on the seventh segment of Axioidea, previously considered to be tympanal organs, are not actually hearing organs (Minet, 1983). Alar tympanal organs are found at the base of the forewing and hindwing in some Nymphalidae (particularly Satyrinae), and in the forewing of all Hedylidae. In certain Sphingidae there are ears in the mouthparts.

Precursors of the tympanal organs and their sensory basis

Tympanal organs are specialized chordotonal organs (Chapman, 1982). Chordotonal organs are stretch receptors, and usually function to record changes in the position of body segments. Their stimulation by stretch has preadapted them for use in tympanal organs where they respond to a vibrating tympanum. Chordotonal organs are composed of a number of sensory units called scolopidia. Generally subcuticular, they occur throughout the peripheral regions of the body.

134

Tympanal organs of Lepidoptera appear to have developed in conjunction with particular chordotonal organs, the homologues of which may be traceable to the most primitive members of the order (Kristensen, 1984a). Chordotonal organs in the pregenital abdomen of Micropterigidae probably represent the precursors of the abdominal tympanal organs found in higher Lepidoptera. In the abdomen of *Micropterix calthella* there is a single chordotonal organ in segment A1. The proximal end of this structure is attached to the secondary metafurcal arm, and the distal end to the lateral lobe of tergum A1. Innervation is derived from a nerve branch that runs from a ganglion supplying the metathorax and the first two abdominal segments. Two chordotonal organs, one ventral and one dorsal, occur in the pleural area of each of abdominal segments A2-7. The ventral chordotonal organs in segment A2 may be the precursor of the tympanal organs of those families in which the hearing organs lie at the base of the abdomen (Kristensen, 1984a). The proximal end of this ventralmost chordotonal organ is attached to the pleural wall, and its distal end to the upper pleural wall near the intersegmental fold of the next segment. Both the dorsal and the ventral chordotonal organs are innervated from a branch of the dorsal nerve of the anterior segmental nerve.

Tympanal organs in different families

Tineidae

The tympanal organs are abdominal and occur only in *Harmaclona* (see Davis & Heppner, 1987), a pantropical genus with 20 species of which 10 are named (Robinson & Nielsen, *in press*). The morphology of the structures has not been investigated.

Dudgeoneidae

The organs are abdominal and not dissimilar from those of Pyralidae (Minet, 1983). The unresolved taxonomic position of Dudgeoneidae is noted in Part 3.

Pyralidae (Figs 145, 146)

The tympanal organs of Pyralidae (Kennel & Eggers, 1933; Bourgogne, 1951; Minet, 1983) are situated, one on each side of the abdominal base, in a shallow cavity. The tympanal sac is enclosed by an invagination of the sternum of the first abdominal segment (*Tympanalkessel* of Kennel & Eggers, 1933; *caisse tympanique* of Bourgogne, 1951; *bulla tympani* of Maes, 1985). These structures are internal sclerites surrounded by the air sacs containing the scoloparia, and are not homologous with the tympanic cavities of Geometridae (Minet, 1983, and see below). The bulla of Pyralidae may be the homologue of the ansa of Geometridae, but the marked difference in appearance of the sclerites provides an effective distinguishing feature of the tympanal organs of the two families. The bulla is either separate, or partly or entirely united with each other in the midline. Medially, the true tympanum is divided from a whitish, opaque conjunctivum, the imaginary line separating these membranes being termed the *ligna tympani* (Maes, 1985). The difference in the angle of the one membrane to the other is diagnostic of the two main divisions of Pyralidae (e.g., Minet, 1983). In the crambiform arrangement (Fig. 145), a decided angle exists between the membranes, whereas in the pyraliform variety (Fig. 146) the membranes are in the same plane. Accessory tympana include one composed of a thin, membranous postnotum (as in Geometridae); a pair of lateral membranes on the epimera of the metathorax; and a further pair on the coxae of the hindlegs. All are in contact with the medial, tracheal air sac. The scoloparia in both Pyralidae and Geometridae are composed of four scolopidia.

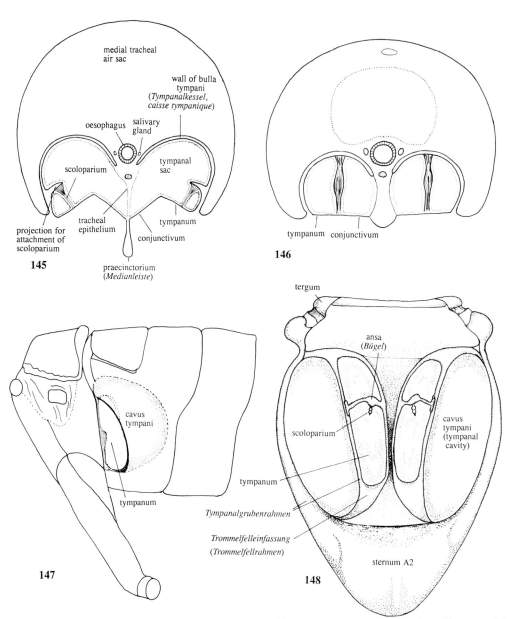

Figs 145–148. *Tympanal organs of Pyralidae and Geometridae. 145, 146, transverse sections (diagrammatic) through base of abdomen of Pyralidae: 145, crambiform type,* Crambus *(Crambinae); 146, pyraliform type,* Aphomia *(Galleriinae). 147, 148, Geometridae: 147, lateral view of metathorax and abdomen of* Epirrhoe tristata; *148, anterior end of abdomen of* Lygris prunata. *(After Kennel & Eggers, 1933.)*

An additional structure associated with, and characteristic of, the tympanal organs of the Pyralidae is the praecinctorium (*Medianleiste* of Kennel & Eggers, 1933). This component, reduced to a knob-like torulus tympani in Pyraliformes, is a median expansion of the intersegmental thoraco-abdominal membrane. It may be simple or bilobed.

Thyrididae

Species belonging to the subfamily Siculodinae have tympanal organs situated at the base of the wings (Minet, 1983; 1988a).

Geometridae (Figs 147-152)

The tympanal organs of Geometridae (Kennel & Eggers, 1933; Bourgogne, 1951; Minet, 1983; Cook & Scoble, 1992) are well-developed, paired structures found at the base of the abdomen and have their tympanal apertures opening ventro-laterally (Fig. 117). A deep, and rather flattened, tympanal cavity (cavus tympani of Cook & Scoble, 1992) extends from each organ into the second abdominal segment. The cavi tympani, which are deep hollows opening externally, are not homologous with the bullae tympani (internal structures, see above) of Pyralidae.

The oval tympanum (Fig. 151) may be observed laterally on dried specimens, but it is best seen when the abdomen is removed and viewed end-on (Figs 148, 150). The ventral anterior edge of the cavus merges with the sternum so that there is no distinct boundary between them. The free lateral margin of the cavus is a particularly strong sclerotized brace (*Grubenrahmen* of Kennel & Eggers, 1933). On the wall of the cavus another sclerotized brace (*feste Trommelfelleinfassung, Trommelfellrahmen*) partly encircles the tympanum, forming a rigid frame across which the latter is stretched.

A characteristic structure found in the tympanal organs of Geometridae is the ansa (Figs 148, 150, 151) (Minet, 1983) (*anse tympanique* of Bourgogne, 1951; Bügel of Kennel & Eggers, 1933). The ansa varies in shape, but essentially it is a curved, rod-like, invaginated sclerite of the casing of the tympanic cavity. It curves over the tympanum and is connected at its distal end to the peritympanal frame by a muscle. Contraction of this muscle is assumed to alter the tension of the tympanum.

The scolopidium is strung between the ansa and the tympanum (Figs 148-150). Each tympanal organ in Geometridae has four acoustic cells. The distal dendrites from these cells, after becoming attached to the tympanum, are folded back on themselves. Since the scolopales are therefore directed away from the tympanum the scoloparium is said to be inverse (Kennel & Eggers, 1933). The two hearing organs are served by a common air sac, which occupies part of both the metathorax and the first abdominal segment. A single accessory tympanum (Fig. 152) is found in all Geometridae except for Archiearinae. This structure lacks a scolopidium and is formed from the metathoracic postnotum which, in Geometridae, is membranous. The accessory tympanum and the two true tympana lie in contact with the common tracheal air sac. The former is regarded as an enlarged fenestra media by Minet (1933), a membrane separating the metascutellum from the crown of the first abdominal tergum. A very narrow fenestra media occurs in Archiearinae, but this would be of an insufficient size to act as an accessory tympanum.

Tympanal organs occur throughout the family with the exception of a few wingless females where they are lost secondarily.

Uraniidae (Figs 155, 156)

The tympanal organs of Uraniidae, including Epipleminae, (Kennel & Eggers, 1933; Sick, 1937; Bourgogne, 1951; Minet, 1983), occur at the base of the abdomen. They are sexually dimorphic. In females (Fig. 155), the tympanal organs open ventrally on each side of the first (visible) abdominal sternum and open ventrally. In males (Fig. 156), the organs are located at the junction of the second and third segments and open dorsally or laterally, an arrangement unlike that occurring in any other Lepidoptera.

In females, the organs are most similar in structure to those of Pyralidae. However, in Uraniidae two rather than four scolopidia are present; the tympanal sac never communicates with the central air sac; there are no accessory tympana, nor is there a praecinctorium.

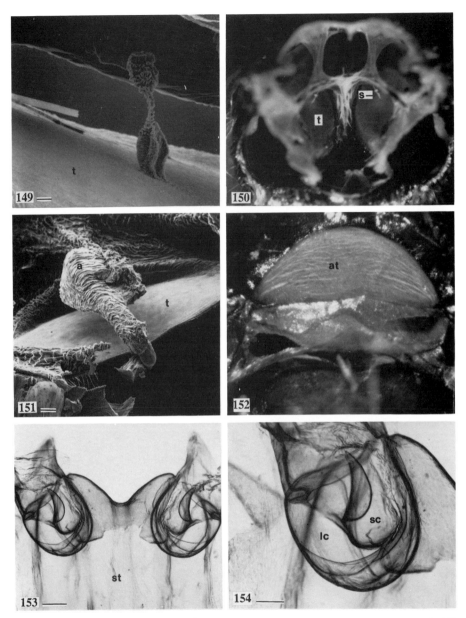

Figs 149–154. *Tympanal organs of Geometridae and Drepanidae. 149-152, Geometridae: 149, scolopidium of Geometra papilionaria (t, tympanum) (scale, 10 μm); 150, anterior end of abdomen of* Archiearis parthenias *showing each scoloparium (s) strung between ansa and tympanum (t); 151, ansa (a) curving over tympanum (t) of Pseudoterpna pruinata (scale, 22 μm); 152, metapostnotum of Sangala sacrata showing accessory tympanum (at); 153, 154, Drepanidae (Hypsidia): 153, general view of tympanal organs (st, sternum; scale 250 μm); 154, single organ enlarged (lc, large chamber; sc, small chamber; 125 μm). (149-152, after Cook & Scoble, 1992; 153, 154, after Scoble & Edwards, 1988.)*

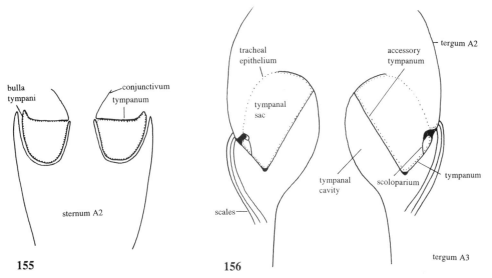

Figs 155, 156. *Tympanal organs of* Orudiza protheclaria *(Uraniidae), transverse sections through tympanal region: 155, female; 156, male. (After Sick, 1937.)*

In the male, the tympanal organ is enclosed in a large, sclerotized capsule, and is divided into an anterior tympanal sac, and a posterior tympanal cavity. The cavity communicates with the exterior via a tympanal orifice. The principal tympanum is superficial, being situated on the outside of the tympanic cavity and protected only by scales. The oblique partition is membranous and may act as a counter-tympanum.

Drepanidae (Figs 153, 154, 157, 158)

In Drepanidae (Drepaninae, Thyatirinae, Cyclidiinae, and genus *Hypsidia*) the tympanal organs (Kennel & Eggers, 1933; Gohrbandt, 1937; Bourgogne, 1951; Minet, 1983; Scoble & Edwards, 1988) are situated at the base of the abdomen. However, they are structurally distinct from those of other families. They are derived from the tergo-sternal sclerites connecting sternum A2 with tergum A1. At the base of the first visible sternum (sternum A2) occurs a pair of small oval structures (Fig. 153, 157). Each is composed of two interconnecting chambers, one small ('kleine Sternalblase' of Kennel & Eggers, 1933), and the other larger and more caudal ('grosse Sternalblase' of Kennel & Eggers). The tympanum is uniquely situated *within* the oval structures between the small and the large chambers (Fig. 158). A characteristic pleural component is also present, being composed of a prominent pleural fold supported by a conspicuous 3-armed sclerite. Anteriodorsally, this sclerite meets tergum A1 at its anteriolateral margin and divides into pre- and postspiracular components. The pleural modification is visible on dried specimens, particularly when the hair-like scales are removed or parted. The sclerites provide a framework across which is stretched an anterior counter-tympanum and posterior counter-tympanum. These membranes enclose the pleural chamber of each tympanal organ.

Nymphalidae (Figs 159, 160)

Tympanal organs occur at the base of the wings in certain Nymphalidae (Vogel, 1912; Bourgogne, 1951; Swihart, 1967). An auditory response from alar organs at the base of the hindwings of *Heliconius erato* was demonstrated experimentally (Swihart, 1967).

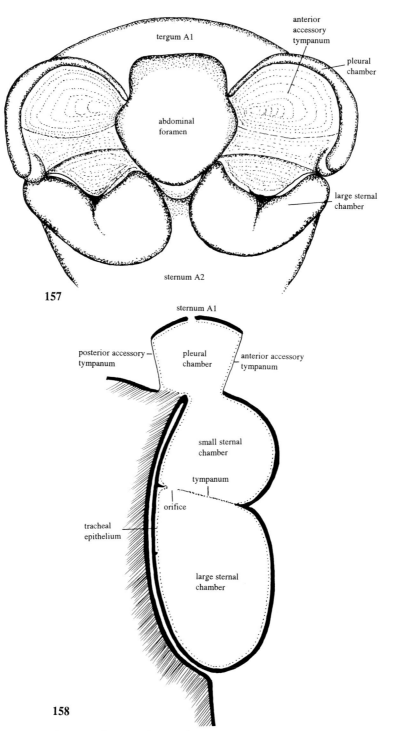

Figs 157, 158. *Tympanal organs of Drepanidae (Cilix glaucata). 157, anterior end of abdomen; 158, section (schematic) through single organ. (157, after Gohrbandt, 1937; 158, after Kennel & Eggers, 1933.)*

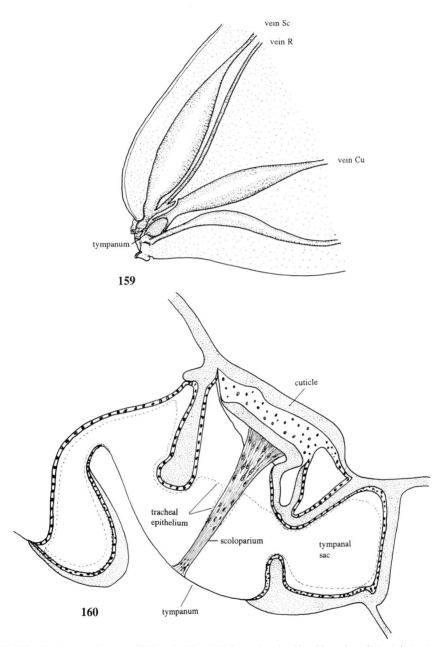

Figs 159, 160. *Alar tympanal organs of Nymphalidae. 159, base of underside of forewing of* Maniola jurtina; *160, longitudinal section of base of forewing of* Coenonympha pamphilus. *(159, after Bourgogne, 1951; 160, after Vogel, 1912.)*

Similar, but smaller, organs occur at the base of the forewing in *H. erato*. In contrast with thoracic and abdominal tympanal organs, alar organs are not sunk into a tympanal cavity. Instead, the tympanum is superficial (Fig. 159) and protected by scales. It is stretched across a scleritized frame and protected by a brush of scales. The tympanic chamber is formed from the swollen base of a wing vein.

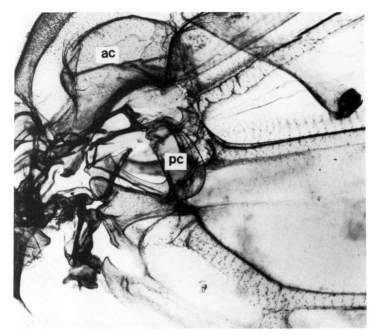

Fig. 161. *Alar tympanal organ at base of forewing of* Macrosoma hyacinthina (*Hedylidae*) (*ac, anterior chamber; pc, posterior chamber*). (*After Scoble, 1986.*)

Hedylidae (Fig. 161)

Alar tympanal organs occur in the base of the forewing in Hedylidae (Scoble, 1986; Minet, 1988a). Although their hearing capability has not been investigated, the presence of chordotonal organs has been confirmed (J. Minet, pers. comm.). Two small chambers, one within the base of vein Sc and the other within the base of the cubitus, are present. It is unclear as to whether both these chambers are involved in sound reception or just the one associated with vein Sc.

Noctuidae (Figs 162-164)

The hearing organs of Noctuidae (Eggers, 1919; Richards, 1933; Bourgogne, 1951; Spangler, 1988) are situated (Fig. 162) one on each side of the metathorax between the epimeron and the postnotum. The base of the abdomen is modified laterally to form a counter-tympanum.

The curvature of the epimeron leads to a cavity situated posteriorly in the metathorax. Within this cavity, supported by a frame, lies the tympanum - a whitish, rather lustrous membrane separated from the epimeron by a narrow, elongated nodular sclerite or epaulette. The chordotonal organ associated with the principal tympanum is composed of two scolopidia. Anterior to the nodular sclerite, the epimeron is sometimes membranous. A counter-tympanic membrane lies dorso-medial to the tympanum (i.e., considerably deeper) (Fig. 164), and is rarely visible by external examination. This membrane probably functions as an accessory tympanum.

The base of the abdomen bears a depression, the counter tympanal cavity, which faces the tympanal complex of the thorax (Fig. 163). This cavity is expanded laterally into a counter-tympanal hood. In Noctuidae the hood lies, typically, behind the first abdominal spiracle (a postspiracular hood). In a number of Noctuidae the hood is reduced or lost.

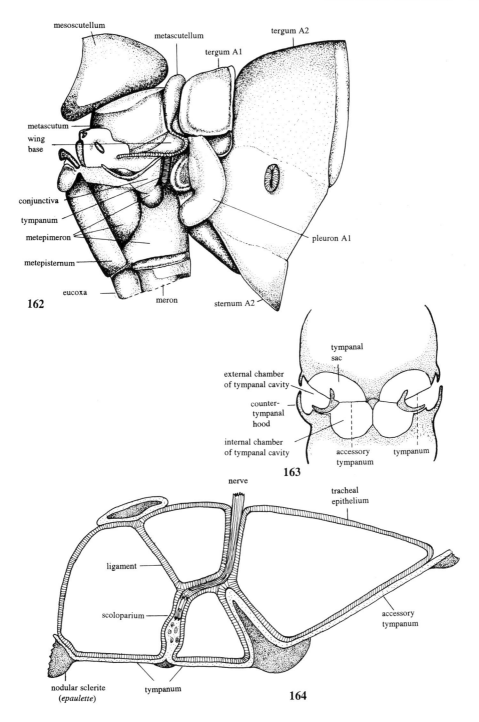

Figs 162–164. *Tympanal organs of Noctuidae. 162,* Catocala *showing lateral view of thorax and base of abdomen; 163, 164, generalised and diagrammatic sections of Noctuidae: 163, sagittal section through thorax and abdomen; 164, section through thoracic component of organ. (After Eggers, 1919.)*

Several components of Noctuidae tympanal organs vary from group to group. For example, in some Catocalinae a chitinous projection arises from the inner margin of the tympanal frame; in Acontiini (Acontiinae) the tympanal hood is reduced or lost; and in Euteliinae, Stictopterinae and Plusiinae there is additionally a pleural pouch (a term preferred to double tympanal hood, Kitching, 1987). In Agaristinae the counter-tympanum is much enlarged.

Arctiidae, Doidae, and Notodontidae

As in Herminiinae (Noctuidae), the counter-tympanal hood of Arctiidae is prespiracular, although in Lithosiinae there is no hood and no nodular sclerite. In Notodontidae (including Dioptinae and Thaumetopoeinae) and in thyretine Arctiidae the tympanum is orientated ventrally rather than posteriorly (e.g., Richards, 1933). Both the hood and the nodular sclerite are absent from Doidae and Notodontidae. At least externally, the tympanal organs are considerably more simple than in Noctuidae.

Non-tympanate hearing organs in adult Lepidoptera

Some Sphingidae (Choerocampini) exhibit evasive behaviour when exposed to ultrasound while hovering and sucking nectar. Neurophysiological investigation demonstrated hearing ability (Roeder & Treat, 1970; Roeder, 1972). Segment 2 of each labial palpus is swollen (Fig. 165), and appears to be displaced phasically by ultrasonic vibrations. The medial surface of this segment is devoid of scales, and the cavity of the segment is filled with a large sac filled with fluid (haemolymph). Each palpus lies in contact with the distal lobe of the pilifer (Fig. 165), in which is contained the sensory transducer. The medial surface of the palpus acts as an 'interface mechanism', that is a device that matches the compliance of air to that of the denser fluids of the receptor mechanism (Roeder, 1972). There is no tympanum.

The system is sensitive, but lacks the mechanical and neural refinement of a tympanal organ. The ability to distinguish between different levels of intensity is much more limited, and there is no directional response to the sound.

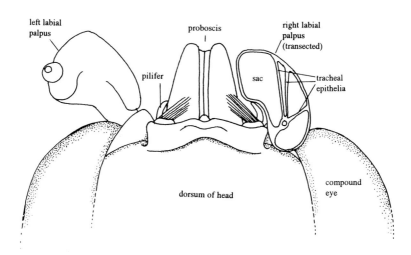

Fig. 165. *Hearing (non-tympanate) organ of* Hyles lineata (Sphingidae) *involving the labial palpi and pilifers.* (*After Roeder, 1972.*)

Auditory hairs in larvae and the responses of larvae to sound

The larvae of several species of Lepidoptera were shown (Minnich, 1936) to respond to airborne vibrations by means of sensilla. Considering the range of larvae exhibiting such a response, the phenomenon may occur generally in Lepidoptera. The responses of the larvae are almost certainly defensive (e.g., Tautz & Markl, 1978; White *et al.*, 1983).

Those sensilla sensitive to sound are situated mainly on the anterior two-thirds of the head-body (Minnich, 1925, 1936). The typical larval responses to sound takes the form either of a sudden cessation of movement, or of the contraction of certain longitudinal muscles of the body (which causes a larva to 'rear'), or both. However, *Cerura borealis* (Notodontidae) throws the two caudal tentacles over its head and then extends eversible flagella from these tentacles (White *et al.*, 1983).

In *Vanessa antiopa* (see Minnich, 1925), larvae may react to sound as a group by vigorously throwing the anterior part of the body dorsally or dorso-laterally. If the sound continues the larvae may thrash about for up to one minute before habituation occurs. If the stimulus is not continuous, but is repeated at intervals of about 5 seconds, then the larval response seems to be maintained.

In the larva of *Cerura borealis* (Notodontidae) (White *et al.*, 1983), although setae sensitive to sound are distributed sparsely over the entire body, they are concentrated on the prothorax - particularly on the two dorsal 'horns'. The larvae respond vigorously to sounds ranging from 330 to 360 Hz (E to just below F# above middle C), and less vigorously at 175 to 180 Hz (closest to F below middle C). A species of the genus *Cotersia* (Hymenoptera: Braconidae), which parasitizes *C. borealis*, makes flight sounds at 187 to 190 Hz and harmonics of 375 Hz. Also, premating 'calls' are made by the parasitoid with a major harmonic of 285 Hz. It seems, at least in part, that the behaviour of the caterpillar is a response to the sounds made by the parasitoid.

Larvae of *Barathra brassicae* (Noctuidae) with intact auditory hairs were attacked 30 percent less frequently by the wasp *Dolichovespula media* (Hymenoptera: Vespidae) than in those larvae where the hairs were removed (Tautz & Markl, 1978). Removal of the hairs led to the failure of the caterpillars to exhibit their defensive reactions - 'freezing' or writhing. Under the experimental conditions used in this work, 'freezing' was found to be a more effective defence than writhing. This situation may not pertain in nature, for writhing may cause a caterpillar to fall to the ground into concealing undergrowth. However, 'freezing' will be particularly effective in this species since the caterpillar is cryptically coloured. *D. media* is a generalist predator, and the defensive behaviour described may not be as effective against specialist caterpillar-hunting wasps. Larvae of *Barathra brassicae* seem to react to vibrations from the wing beats of flying wasps. Specialist caterpillar-hunting wasps tend to stalk their prey by walking towards them, and thus avoid providing the caterpillars with the stimulus which elicits their defensive action.

Sound production

Sounds are produced by a variety of Lepidoptera. They may be generated by specially modified structures, or produced incidentally to some other activity such as the chewing of caterpillars during feeding. Sound production is often called 'stridulation', a word derived from the Latin *stridulum* an adjective meaning creaking, hissing, rattling or buzzing. The usage is often confined to mean the noise produced by rubbing two parts of the body together, but in the present work the word is more broadly applied (closer to its derivation) to mean sounds produced by some direct action of the insect for a presumed purpose. In other words, only incidental sound is excluded from the meaning of the term.

Some confusion exists over the precise way in which sounds are actually produced in Lepidoptera. Relatively few studies have been made on living moths so much has been deduced from the examination of dead specimens. Although many organs involved in stridulation have doubtless been correctly identified, it is often difficult to be certain about their exact mode of operation.

Sounds produced by rubbing two parts of the body together or the pupal body against its cocoon

Various structures in Lepidoptera have been modified to produce sound by rubbing two surfaces of the body together. Sounds generated in this way are often amplified by a hollow organ acting as a resonator.

Wing-leg stridulation

Several Lepidoptera stridulate by rubbing parts of the body together, but many early workers misinterpreted the way in which the sounds are produced. A summary of sound production by adult Lepidoptera was provided by Hannemann (1956b) who reinterpreted earlier work. Although Hannemann deduced methods of sound production from the examination of dead specimens only, his critical observations provide convincing explanations for the exact parts of the body used.

The dorsal surface of the forewing of males of *Pemphigostola synemonistis* (Noctuidae: Agaristinae) from Madagascar is produced into a prominent blister (Hannemann, 1956b) (Fig. 166). The ventral surface of the wing bears a corresponding groove. The costa of the wing is folded over so that the subcostal vein becomes, effectively, its leading edge. This fold forms the anterior edge of the groove on the underside of the wing; the posterior edge of the groove is thickened along the base of the cubitus. This thickening is pronounced near the base of the posterior edge of the groove, where it bears a series of ribs. This ribbed ridge (the *Schrillkante*) rubs against another row of ribs on the midleg (the *Schrilleiste*) (Figs 167, 168) to produce sound. The midleg ribs run from the tibia to the distal end of the fourth tarsal segment. There are about 75 ribs on the first tarsal segment. Earlier authors considered that *Pemphigostola* stridulated by rubbing the leg against the blister on the *upper* surface of the wing, but it appears that the blister acts rather as a resonator (Hannemann, 1956b).

In *Musurgina laeta* (Noctuidae: Agaristinae), also from Madagascar, the blister on the forewing of the male is smaller than that of *Pemphigostola*. The file, found on the hindtarsus, is rubbed against the base of the first anal vein (ax_1 of Hannemann), which is strongly sclerotized and bears a sharp point against which the file rubs (Hannemann, 1956b).

In yet another agaristine (*Aegocera mahdi*) it had been suggested (Jordan *in* Hannemann, 1956b) that a series of ribs on the hindleg of the male were rubbed against the subcostal vein of the ventral surface of the hindwing during flight. However, it is very unlikely that moths with this kind of leg-wing stridulatory device stridulate during flight, and stridulation probably occurs only at rest. Sounds produced during flight are probably only incidental effects of the beating of the wings (Hannemann, 1956b).

In *Thecophora fovea* (Noctuidae) the resonating blister, over which runs a portion of the radial vein, is found on the ventral surface of the hindwing (Hannemann, 1956b). This section of the vein is heavily sclerotized in males, and at one point becomes raised

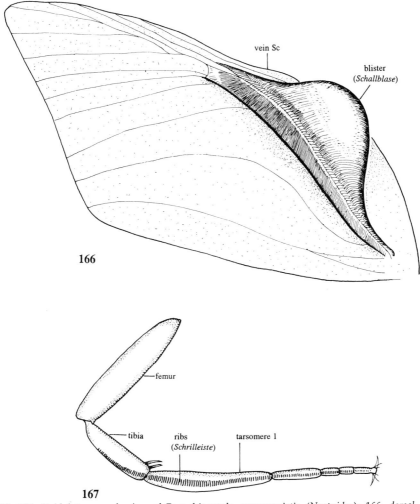

Figs 166, 167. *Stridulatory mechanism of* Pemphigostola synemonistis *(Noctuidae): 166, dorsal surface of forewing; 167, midleg, showing file. (After Hannemann, 1956b.)*

resembling the surface of a grater. It is against the grater that the file, which in this species is confined to the first tarsal segment of the hindleg, is rubbed.

A more detailed examination (Surlykke & Gogala, 1986) of the structure of the stridulatory system in *T. fovea* confirmed the view that stridulation occurs when the first tarsal segment of the midleg is rubbed against the blister of the hindwing. The file, found largely on the proximal part of the radial vein as it traverses (longitudinally) the blister and before it bifurcates, is composed of 'small, cylindrical, dome shaped, pegs' (Surlykke & Gogala, 1986) covering the base of the radius. The scraper is formed of about 100 sharp, transverse ridges. Males of *T. fovea* appear before females but do not stridulate before the females emerge. When measured by a bat detector, sounds made by the moths were found to be intense and pulsed; they continued for several minutes. The stridulatory blister seems to act as a resonator. Although *T. fovea* is palatable to bats, the moths avoid these predators by emerging in autumn just after bats had stopped flying at the study site in Yugoslavia.

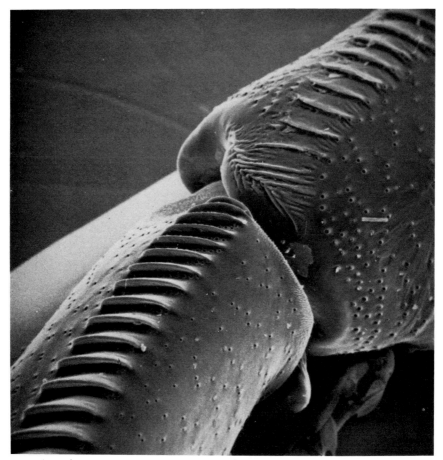

Fig. 168. Pemphigostola synemonistis (*Noctuidae*), *showing section of the stridulatory file on midleg.* (*Courtesy of Dr K. Sattler/EM Unit, BMNH.*)

In all these examples the body has become modified for stridulation in some observable way. However, the female of *Parnassius mnemosyne* (Papilionidae), produces sound simply by rubbing its hindlegs against the under surface of the hindwing. There is neither a series of ribs on the leg nor a special sclerotized ridge on the wing. Apparently stridulation in this species occurs in gravid females (Jobling, 1936) and takes place when the insects are disturbed. When stridulating, this butterfly expands its wings and uses only its forelegs and midlegs for attachment.

Frenulum on retinaculum

The retinaculum of the males of certain South American Chrysauginae (Pyralidae) is strongly bent to form an incomplete ring with its distal end thickened and pressed against the wing membrane (Fig. 169). This thickened distal end bears a file on its inside surface. It was suggested (Hannemann, 1956b) that the frenulum, which is sharp-edged, scrapes the file to produce a sound, and that the sound is amplified by means of a resonating chamber formed by a fold on the upper surface of the forewing. The distal end of the retinaculum presses against this chamber from the underside of the wing.

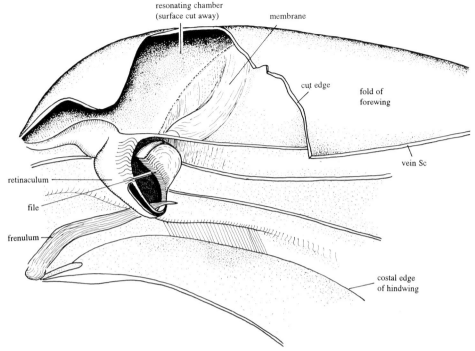

Fig. 169. *Stridulation of* Samcova *(Pyralidae), by means of the frenulum and retinaculum. (After Hannemann, 1956b.)*

Genital stridulation involving valvae and abdominal segment A8

The males of some Sphingini & Smerinthini (Sphingidae) stridulate by rubbing modified scales on the valvae of the genitalia against other modified scales, or spines, on the posterior edge of the eighth abdominal tergum (Figs 170-172) (Doesburg, 1966). In the smerinthine *Protambulyx strigilis*, the so called grater-scales are confined to a patch on the outer surface of each valva. Each scale is asymmetrically bilobed so forming two points. The more prominent of the two points acts as a plectrum, while the smaller helps to locks the scale under its neighbour. The modified scales on the eighth tergum, against which the plectral scales rub, are situated in two long rows laterally. Those that constitute the rows are elongated, and are cemented together forming a thin plate or ribbon. The plectral scales are grated against the plates by a lateral movement of the valvae.

In *Psilogramma menephron* and other Sphingini that stridulate, the eighth tergum bears, on its posterior edge, a row of small, stiff spines, rather than modified scales. However, the sound is generated in essentially the same way as it is in smerinthine Sphingidae. *Psilogramma increta* stridulates when caught (I.J. Kitching pers. comm.). A more or less continuous sound is clearly audible, and alternate anterior and posterior movements of the valvae are visible. Audiospectrograms of *Psilogramma jordani* show that the sound produced is actually composed of a series of pulses, but the structure of an individual pulse has not been precisely explained in terms of the scale-spine stridulation mechanism (Lloyd, 1974).

Genital stridulation also occurs in *Meganoton analis sumatranus* (Sphingidae) from South East Asia (Nässig & Lüttgen, 1988), and the phenomenon may be more widespread in Sphingidae than records actually show.

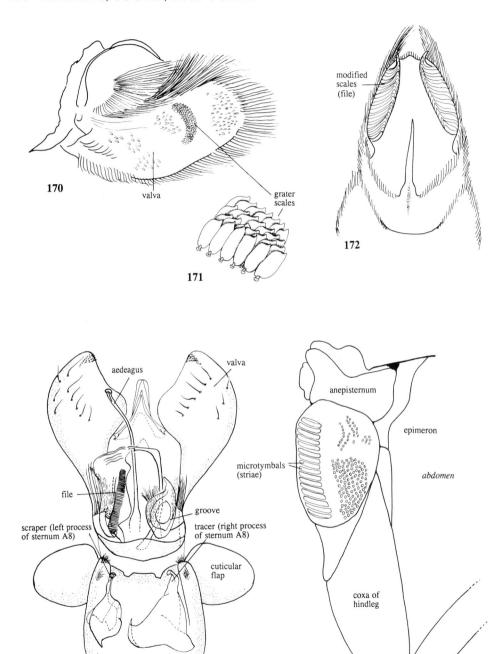

Figs 170–174. *Stridulatory organs of Sphingidae, Pyralidae, and Arctiidae. 170-172, male genitalia and sternum A8 of* Protambulyx strigilis *(Sphingidae): 170, inner view of valva; 171, specialized scales (enlarged); 172, sternum A8 showing modified scales forming file; 173, ventral view of male genitalia and sternum A8 of* Syntonarcha *(Pyralidae) showing groove and tracer system; 174, tymbal organ on metathorax in Arctiidae. (170-172, after Doesburg, 1966; 173, after Gwynne & Edwards, 1986; 174, after Blest et al., 1963.)*

Genital stridulation has also been reported (Gwynne & Edwards, 1986) in males of *Syntonarcha iriastis* (Pyralidae) (Fig. 173) from Australia where moths signal from perches on a variety of bushes. The sound producing organs are complex and include components at the base of the valvae and on the eighth abdominal sternum. A file, situated at the base of the left valva, is composed of three zones characterized by different distances between the ridges. The scraper, which terminates posteriorly in a curved tip, occurs on one of two strongly modified sclerites on the eighth sternum. At the base of the right valva is found a sclerotized area with a subcircular groove. The process of another sclerite on sternum A8 fits into this groove and probably acts as a tracer. This 'groove and tracer' system probably enables the scraper to be placed at different positions on the file and, because the file is zoned, cause differential scraping of the zones to produce sounds of different frequency. A pair of cuticular flaps of the eighth abdominal segment surround the genitalia during stridulation. They may increase intensity and resonance, or affect the directionality of the sound.

Wing against thorax

The forewing-thoracic locking system (see Chapter 3) may also be used as a stridulatory device. When the wings of a dead, but relaxed, specimen of *Xyleutes durvillei* (Cossidae), from Australia, were pulled outwards from the body, over which they are folded at rest, a sharp sound was emitted (Common, 1969a). When a living specimen was disturbed, it flexed its wings and emitted the same stridulatory sound. A similar but less distinct sound was found in a species of *Culama*, a smaller cossid. Presumably, the production of sound is a secondary development from the primary function of locking the wings and the thorax for it has not been recorded from other Lepidoptera. Furthermore, the patches of aculei on the thorax and wings are present in both sexes, whereas most stridulatory organs are confined to male Lepidoptera.

Vibratory papillae against larval epicranium

In the larvae of several Riodininae (Lycaenidae), a pair of rod-like papillae occur on the anterio-dorsal edge of the prothorax (DeVries, 1988). In *Thisbe irenea*, where these organs have been investigated in detail (DeVries, 1988), each papilla bears a series of concentric grooves and ridges. These papillae are capable of being beaten against an area of the epicranium bearing sharp granulations. When disturbed, including when initial contact is made by ants, or when the larva moves, the vibratory papillae beat much faster than when the larva is at rest. Sound is produced by the rasping of the papillae against the epicranial file. Concurrent with the vibrations of the papillae, the head capsule oscillates rapidly in and out and from side to side.

Abdominal segments of pupa against each other

In several families, sound is produced by the pupa when one or more pairs of abdominal segments are rubbed together (Hinton, 1948). Essentially, tubercles or ridges on the posterior edge of one segment are rubbed against tubercles or ridges found on the anterior edge of the succeeding segment. Such abdominal stridulation is found among Hesperiidae, Papilionidae, Lycaenidae, pyraustine Pyralidae, Noctuidae, Lymantriidae, Callidulidae, Notodontidae, and Sphingidae.

In Lycaenidae rows of tubercles on segments A5 and A6 are responsible for sound production. In most other lepidopteran pupae that stridulate the organs concerned are found on segments A4 and A5, A5 and A6, and A6 and A7. In all of these arrangements, rows of tubercles are present on the anterior part of the segment, while ridges are found on the posterior part of the preceding segment. The precise distribution

varies, tubercles may be found dorsally, laterally, or ventrally, or in certain combinations of these positions. Abdominal stridulation in lycaenid pupae is produced by
intersegmental components (Downey & Allyn, 1978; Hoegh-Guldberg, 1972). Parts of
the pupal body probably act as resonators, the entire pupa radiates sounds of a wide
frequency, and the whole body may also act as a resonator (Downey & Allyn, 1978).

Proboscis of pupa against abdomen

In *Gangara thyris* (Hesperiidae), ridges on each side of the middle of abdominal
segment A5 are rubbed against the proboscis (Hinton, 1948). The abdominal ridges are
sharp and run transversely. They scrape transverse ridges on the underside of the
proboscis (which is very long in this species) when the abdomen moves. The pupa,
situated in a cylindrical chamber in a palm leaf, stridulates when disturbed. It also
produces a sound by striking the sides of this chamber by vigorous movement.

Pupal ridges or tubercles against cocoon

The pupae of certain Sarrothripinae and Chloephorinae (Noctuidae) stridulate by
scraping tubercles or ridges on the abdomen against the wall of their cocoons (Hinton,
1948). Whereas in some species the cocoon bears no corresponding file against which
the ridges or tubercles scrape, in others, longitudinal hard ridges of silk are formed on
the inner dorsal wall of the cocoon. By scraping these ridges, pupae are said (Hinton,
1948) to produce a loud noise.

Sounds produced by the buckling of a 'tymbal' organ

Stridulation by means of the vibration of a membrane is well known in cicadas. A
similar mechanism is found in some Lepidoptera, although here the sound is produced
by the distortion of a row of microtymbals rather than a single tymbal. Tymbal organs
have been found on the metathorax of certain Arctiidae, the tegulae of some Pyralidae,
the abdomen of many Lymantriidae, and the basal abdominal sternum of Chloephorinae (Noctuidae).

Arctiidae

In many Arctiidae a row of striae is found on an air filled expansion of each
katepisternum of the metathorax (Figs 174, 175; and see Figures in Watson, 1975).
Originally, the row of striae was termed the 'striated band' (Forbes & Franclemont,
1957). Although assumed by many to function in stridulation, the mechanism was
disputed. Early authors suggested that the row of striae was scraped, but H.E. Hinton
(*in* Blest *et al.*, 1963) considered that the outer membrane of the air-filled expansion
vibrated as a result of distortion by the indirect action of certain thoracic muscles.

The mechanism was explained satisfactorily only when studies were carried out on
these structures in *Melese* (Blest *et al.*, 1963). Each stria of the band acts as a tymbal
organ in its own right. Muscle action on the striae-bearing sclerite results in serial
buckling of the striae. When the effector muscles contract, buckling starts at the most
dorsal stria and continues along the length of the band ventrally. When the muscles
relax, the striae, starting with the most ventral unit, return to their resting state. The
basic unit of sound is therefore a pulse - each pulse representing the buckling of a
single stria. Blest *et al.*, demonstrated modulation in the series of pulses of sound
emitted during both the buckling and the relaxing component of each cycle. During
buckling the frequency fell, while during relaxation it rose. Modulations were
identified between 30 and 90 Kc/sec, and each cycle of sound was repeated about 40
times per second resulting in a burst of continuous sound. The explanation of the

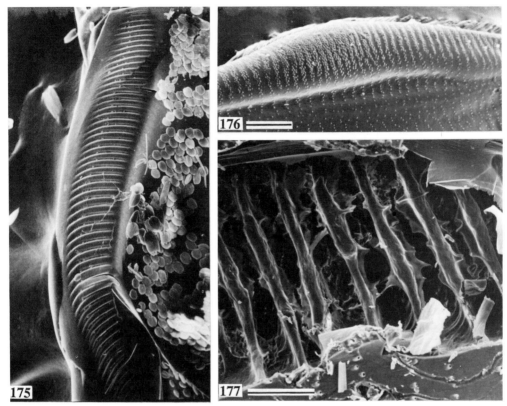

Figs 175–177. *Sound production in Arctiidae and Noctuidae. 175, metathoracic tymbal organs in* Emurena fernandez *(Arctiidae); 176, 177, fovea in* Heliocheilus stigmatia *(Noctuidae): 176, costa of forewing (scale, 42.8 μm); 177, surface of fovea removed to show lamellae (scale, 25 μm). (175, courtesy of Mr A. Watson and Mr D.T. Goodger/EM Unit, BMNH; 176, 177, courtesy of Dr M.J. Matthews/EM Unit, BMNH.)*

tymbal mechanism proposed received further support when the striated band was modelled by a steel spring with indentations hammered into it. Sounds emitted by this device were similar to a recording, which had been slowed down, of the stridulation of a moth.

The muscles responsible for buckling the katepisternum are one or more branches of the basalar muscle of the metathorax. Since this muscle presumably functions in flight, it must, for stridulation, also be capable of acting in isolation (Blest *et al.*, 1963). The tymbal organ on one side of the thorax functions independently from that of the other (there is no coupling device), but nevertheless both sound at about the same time.

Regarding control of stridulation, experiments on intact and decapitated moths by Fullard (1982b) suggest that while the clicks emitted by *Cycnia tenera* (also derived from tymbal buckling) are reflex, with an arc connecting the ear with the tymbal organ, the duration of the train of clicks is probably maintained by higher-level interneurones in the brain.

Pyralidae

Males of *Achroia grisella* (the Lesser Wax Moth) produce ultrasound when a sclerite at the base (humeral lobe) of the forewing strikes, during wing vibration, a process on the inner surface of each tegula (Spangler *et. al.*, 1984). This action results in a change of air pressure behind a scaleless, bag-like area of the anterior part of the tegula, which acts like a tymbal. The change in air pressure causes the tymbal to buckle. The frequency of the sound produced ranges from 75-130 kHz, usually with two peaks of intensity. Sound is emitted in short pulses when males fan their wings. A pulse of sound is released each time a wing is raised or lowered. The wings are coupled to the tegulae by an elongated sclerite on the tegula that arises close to the tegular process. Although they exhibit wing fanning, females lack both the forewing striker and the tegular process.

Tegular stridulation occurs also in the Greater Wax Moth (*Galleria mellonella*) (Spangler, 1985; Bennett, 1989), but the mechanism differs from that in *A. grisella*. In *G. mellonella*, a small protrusion from a basal sclerite of the forewing appears to form a link with the sclerite coupling the tegula and the wing (the tegular wing coupler). As a result, even reduced wing amplitude causes movement of the tegular wing coupler. In turn, this movement induced the tymbal on the tegula to buckle and recover. Ultrasonic pulses are emitted as a result of this mechanism. As in *A. grisella*, the tymbal appears to be the scaleless frontal part of the tegula. However, unlike the situation in *A. grisella*, the buckling of the tymbal organ in *G. mellonella* is not apparently brought about by changes of pressure in the air-filled tegula.

Lymantriidae

Structures resembling tymbal organs occur on each side of the third abdominal sternum of a wide variety of male Lymantriidae (Zerny & Beier, 1936; Dall'Asta, 1988). The striated area of each organ is located on a fold in a depression of the abdomen, and descends from a swelling, which is situated externally. The whole structure is largely obscured by scales.

Sound production by this organ has not been investigated, but assuming, as seems likely, that it occurs, a possible function is in courtship with males signalling to silent females.

Sounds produced by the expulsion of air or foam

Froth

A noxious, pungent foam or froth is expelled from the prothorax of certain adult Arctiidae. It was assumed (see Carpenter, 1938) that reports of a 'sizzling' sound were based on noise produced by the expulsion of the foam. Eltringham (1938), who investigated the morphology of two species of *Amerila*, considered that the sound was produced by the 'striated bands' now known to be tymbal organs. However, he thought that the band was scraped against an adjacent sclerite rather than subjected to distortion. The true source of the sizzling sound of Arctiidae that expel a foam from the prothorax is uncertain. An occasional audible hissing from the foam-exuding Panamanian arctiid *Belemniastis troetschi* has been reported, but this sound is not produced by the tymbal organs (Blest *et al.*, 1963).

Air through the proboscis

Sound produced by the hawkmoth *Acherontia atropos* was first noticed in the 18th century by Reamure (Dumortier, 1963). It takes the form of two short bursts - one low-pitched and pulsed, the other high-pitched (Dumortier, 1963). These sequences are repeated rapidly. As air is drawn into the dilated pharynx, the flap-like epipharynx

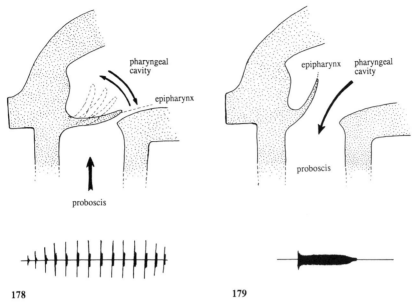

Figs 178, 179. *Base of proboscis and pharynx (diagrammatic) in* Acherontia atropos *(Sphingidae): 178, air drawn in; 179, air expelled. (After Dumortier, 1963.)*

vibrates (Fig. 178). This produces the pulsed, low-frequency sound. The high-frequency noise is produced when air is expelled through the proboscis. During this time, the epipharynx is kept raised so the air flow is not impeded (Fig. 179). *Acherontia* species are also known to emit squeaks or a lower pitched sound in the pharate adult before eclosion (Tutt, 1904; I.J. Kitching, *pers. comm.*).

Percussive sounds

Wing against wing

Several Noctuidae produce sound by the percussive action of blisters, termed fovea, on the forewings. These structures have been likened to castanets (Bailey, 1978) because they can be banged against each other as the wings vibrate. Each forewing bears a single fovea situated near the costa (e.g., as in Fig. 176). If the surface of the fovea is removed (at least in *Heliocheilus*, see Matthews, 1987), a series of lamellae cross its lower half (Fig. 177). Presumably these act to strengthen the structure.

Support for the castanet interpretation has been given for *Helicoverpa zea* (Agee, 1971), and *Hecatesia* (Bailey, 1978). *H. exultans* calls while at rest, and in this species the sound is composed of a series of rapid clicks usually grouped into bursts. *H. thyridion*, in contrast, calls during flight, a single pulse of sound being produced when the wings meet at the upstroke of each wing-beat. It appears that *H. thyridion* calls only during courtship flights, so the stridulation is under voluntary control of the moth. *Heliocheilus albipunctella* emits a buzzing sound of approximately 130 Hz while the moth clings to vegetation (Pl. 1:11) and vibrates its wings (Matthews, 1987). The wing beats are not complete and move only about 20° on each side of the vertical axis.

Although the fovea has been thought to be implicated in sound production for many years, it was not realized that the mechanism was percussive - at least in some species.

Hannemann (1956b) cited several examples of species with foveae. He suggested that sound was produced when parts of the legs were rubbed against ridges close to the fovea, and that the fovea itself acted as a resonator. These deductions were based on a study of dead specimens, and are fairly rejected only when living animals are studied. However, sounds produced by moths with fovea-bearing wings that are vibrated while the legs are in contact with vegetation are clearly not made by leg-wing interaction.

   Inachis io and *Nymphalis antiopa* (Nymphalidae) produce a sound when they rub their forewings against their hindwings. Presumably the sound is made by the abrasion of the wing veins, but there are no apparent stridulatory modifications.

Functions of hearing and sound production

Hearing in defence against bats

Hearing organs in moths probably evolved in response to bat predation, although there is growing evidence of their role in courtship (Spangler, 1988). About 85 percent of all macrolepidopterans have functional hearing organs (Fenton & Fullard, 1979), and an even greater proportion of nocturnal macrolepidopterans. Medium sized to large moths are hunted by bats, and moth hearing organs are capable of detecting ultrasonic signals emitted by these predators. Moths respond to hearing ultrasonic emissions from bats by taking evasive action, and the variation in bat hunting behaviour and in their echolocation systems may be an adaptive response to the evolution of insect ears - particularly those of Lepidoptera (Fenton & Fullard, 1981). Evasion of bats by moths has been observed most often in Noctuidae and Geometridae, but similar behaviour has been seen in Arctiidae and Pyralidae subjected to ultrasound (Spangler, 1988).

Sensory equipment of moth ears

In the ears of certain Noctuidae there are two sensory cells that respond to sound ('A cells'), and a non-acoustic cell (the 'B cell') (summary in Roeder, 1965). The function of the B cell is unknown, but it may respond to changes in pressure of the ear. The cells are found in a strand of tissue stretched across the gap between the tympanum and the skeletal frame. The two fibres of the A cells together with the single fibre of the B cell form the tympanic nerve, which leads to the thoracic ganglia of the central nervous system. By connecting the tympanic nerves with small silver hooks to an oscilloscope, it was found that moth ears respond to a wide range of frequencies used by bats (10 to over 100 kilocycles). However, being unable to differentiate one frequency from another they cannot detect any modulation in frequency. Therefore, moths are tone deaf (Roeder, 1965). Nevertheless, the organs can detect *pulsed sound* (short bursts of sound followed by silences) and the differences in the loudness of pulsed sound. Furthermore, it appears that moth ears are most sensitive to the range of frequencies mainly used by those species of bats living in the same habitat (Fenton & Fullard, 1981). Despite the simplicity of the auditory system of moths, these insects are able to perceive the general direction of origin of bat ultrasounds.

The response of moths to bat ultrasound

Bats fly faster than moths, and moths behave differently when bats are close to them than when they are far away (summary by Roeder, 1965). When played ultrasounds emitted by a loud speaker, those moths flying at a relatively great distance from the speaker turned away from it and flew quickly in the opposite direction. That this reaction was a response to the *low* intensity of the sound was demonstrated when it was observed that moths close to the speaker would fly away only when the intensity was lowered. The response can be effective only if the moths hear the bats before the bats sense their presence. The response to *high* intensity ultrasounds, which a moth experiences when close to a bat, is quite different. In the experiment, moths dived, passively dropped, or executed various turns and twists according to their position in relation to the source of the sound. 'Streak' photographs have been used to illustrate the paths taken by moths after subjection to both artificially generated ultrasounds and ultrasounds emitted by bats (e.g., Roeder, 1965).

The responses of moths to hunting bats have evolved to a high degree. For example, when *Cycnia tenera* (Arctiidae), the Dogbane Tiger Moth, is exposed to strong ultrasonic signals, it emits a series of defensive clicks (p. 152, and see below) shortly before its final line of defence against bats, which involves folding its wings and ceasing flight (Fullard, 1979). The species is most sensitive to a pulse repetition rate of 30 to 50 pulses per second (Fullard, 1984a), the very rates emitted by bats just before they close in on their prey. Repetition *rate* therefore, seems to be used as a cue by the moth to sense the proximity of a bat.

Moths on the Hawaiian island of Kaui probably respond to calls of the Hoary bat (*Lasiurus cinereus semotus*) (Fullard, 1984b). These bats appear to use a low frequency call to repel other bats and a high frequency call to hunt. The moths may also be able to use the low frequency signal, which is transmitted over a greater range than the higher frequency call, to detect the presence of a bat. Support for this hypothesis comes from the observation that pulse duration and repetition rate are similar in both the high and the low frequency calls.

A comparative study (Fullard, 1982a) has shown that in areas of high bat density and diversity, moths have ears with greater 'total sensitivity', a measure of responsiveness to all frequencies tested, than they do in areas where there is low bat density and diversity

The effect of moth hearing on bat hunting behaviour

Just as the auditory systems and associated behaviour of moths have changed in response to bat predation, so bats have evolved modifications in their hunting strategies in response to moth defence (e.g., Fenton & Fullard, 1979). Some bats probably use calls of high frequency and low intensity to reduce the chances of their being detected by moths. The resulting reduction in range of these modified calls clearly has associated limitations. Furthermore, since high intensity sounds have greater directionality the resultant loss of spread with lower intensity makes it more difficult to track a moving target.

Sound

Most lepidopteran stridulation probably functions in either defence or courtship. However, this assumption is based on relatively few detailed studies. Apart from defence or communication, Hinton (1955) suggested that certain moths might use ultrasounds to echolocate and avoid obstacles.

Defence

Against bats. Moths respond to hunting bats largely by evasion (see above), but some are thought to utilize their own ultrasonic emissions in defence. Sound produced by a moth might advertise its distastefulness, startle or confuse its attacker, or possibly interfere with the echolocation system of the predator.

One function of ultrasonic pulses produced by the tymbal organs of many Arctiidae may be to warn predators of the distastefulness of the moth (Dunning, 1968). Studies on Arctiidae demonstrate that they are generally avoided by predators. Although moths of this family are frequently distasteful and aposematically coloured, many predators (e.g., spiders, ants, bats) are probably unable to see these bright colours, particularly at night, which is when the moths are generally active. Whilst many Arctiidae advertise their distastefulness by chemical means, ultrasonic pulses produced by the tymbal organs appear to have a similar aposematic function (Dunning, 1968).

When mealworms were tossed into the air *Myotis lucifugus* bats avoided them if ultrasonic clicks of an arctiid moth species, found in the same area, were played at the same time (Dunning & Roeder, 1965). Since a flying bat probably fails to sense the odour of a moth, even one quite close by, it is likely that the sound is associated with distastefulness. Furthermore, under experimental conditions, moths that clicked were rejected more often by bats than moths that did not (Dunning, 1968).

Despite considerable differences in their 'calls' (frequency rate and repetition rate), unpalatable moths may be Müllerian mimics (Dunning, 1968) for bats tend to avoid clicking moths in the same way despite the variation. Members of the genus *Pyrrharctia*, which are more palatable, may be Batesian mimics. Another discovery (Dunning, 1968; Blest, 1964) was that the more palatable the species, the easier it was to elicit the clicking response.

A difficulty with the aposematic hypothesis is that when bats were presented with (palatable) mealworms, they failed to associate clicks with distastefulness, even though they had several chances to learn.

Possibly, ultrasonic pulses emitted by moths may in some way interfere with the bats' echolocation system (Dunning & Roeder, 1965). This suggestion was investigated by Fullard et. al. (1979) who found that ultrasounds of *Cycnia tenera* (Arctiidae) are similar (in terms of power and frequency-time structure) to those emitted by the bat *Eptesicus fuscus* as it closes on its prey. Bats are known to suffer fatal collisions with solid objects, and may interpret the sounds emitted by the moths as echoes from such objects. The clicks may disrupt a bat's processing system and perhaps startle it into taking avoiding action. Horseshoe bats (*Rhinolophus*) emit sounds of constant frequency, unlike the frequency modulated emissions of many other bats, and may even suffer jamming of their sonar.

The importance of ultrasonics in the defences of moths against bats was underlined by a study (Fullard & Barclay, 1980) of the differential predation pressure on spring and summer Arctiidae at a site in Ontario, Canada. Although moths emerging in the spring had ears sensitive to bat ultrasonics, the insects did not produce clicks. In contrast, in species where the moths emerge later in the season, sounds are produced by the tymbal organs. The emission of sound in those species flying in the summer may be a response to increased predation pressure from lactating female bats and their young. Thus while spring-moths respond to hunting bats at a distance by flying away, summer-moths appear to require the short-range defence provided by stridulation. This interpretation (Fullard & Barclay, 1980) is further supported by the observation that in spring, bats tend to forage mainly over water, where they feed predominantly on chironomid midges and mayflies, before returning to their roosts at dawn. In summer, however, foraging is undertaken by lactating females, which tend to hunt close to the roost to which they return during the night to nurse the young. When the young are able to hunt, they also fly initially near the roost putting further pressure on the moths.

Against other predators. Defence against predators other than bats has been less well investigated. A study of the protective displays of a wide range of New World Arctiidae demonstrated that although most of the species examined were capable of producing sound, few of them did so when handled (Blest, 1964). Furthermore, sound production in Arctiidae appears to be of minor importance in defence against birds and captive primates, which is to be expected given that these predators are visually-orientating animals. Only a limited correlation between protective coloration and the threshold for sound production was observed when the moths were handled. The only correlation was that in the more palatable species of the Phaegopterini the threshold for sound production was lower than in more distasteful species.

Sounds produced by *Parnassius* and *Vanessa* (Nymphalidae) may be defensive, but no experimental work has been carried out to test this view.

The purpose of sound production in pupae of Lycaenidae is uncertain. Four possibilities were considered by Hoegh-Guldberg (1972). The suggestion that lycaenid pupae stridulate to instigate congregation may be dismissed on the grounds that most lycaenids are not gregarious. The view that female pupae of this family stridulate to keep emerged males in the vicinity before emergence, met with the objection that some pupae make sounds months before they are mature. Furthermore, male pupae also stridulate, and also, as far as we know, lycaenid adults lack hearing organs. The idea that myrmecophilous pupae stridulate to attract ants is weakened by the observation that some lycaenid pupae stridulate without having an ant association.

So, largely by a process of elimination, Hoegh-Guldberg (1972) concluded that the function of stridulation in pupae was mostly defensive, a view also reached by others (Hinton, 1948; Downey & Allyn, 1978). Pupae that rattle against dead leaves may alarm parasites (although perhaps more because of their motion than the sound produced), but the activity may well attract birds. Although hardly convincing, support for the defensive function of sound follows from the observation that pupae capable of sound production *do* in fact make a sound when disturbed. However, there is no convincing direct evidence that pupal sounds have any function.

Sound production is very rare in lepidopteran larvae. However, the larvae of certain Lycaenidae, particularly Riodininae, use sound probably to attract ants (DeVries, 1988, and p. 151), and gain protection from the ant-association. Larvae of species capable of producing sound seem to be myrmecophilous, while those that do not stridulate are amyrmecophilous.

Sound produced by the larva of *Diurnea fagella* (Oecophoridae), which scrapes its metathoracic, terminally hooked, legs back and forth against a leaf of its shelter, appear to provide a measure of defence against other arthropods (Hunter, 1987). Spiders, for example, were found to elicit sound production and to be deterred from entering the larval shelter. Conspecific larvae also caused sound to be produced, and it appears that this reaction together with other defensive behaviour (wrestling, biting, and regurgitation) helps the defender to retain its shelter.

Courtship

Males of *Psilogramma* (Sphingidae), with abdominal hairpencils displayed, were observed to stridulate while flying around females (Mell, 1922) suggesting that the primary function of sound production in this species is for courtship. The use of acoustic signalling in lepidopteran courtship has been established by studies on certain Pyralidae (Dahm *et al.*, 1971; Zagatti, 1981; Spangler, *et al.*, 1984; Bennett, 1989), and in a member of the Arctiidae (Sanderford & Conner, 1990).

In *Achroia grisella* (Pyralidae) (Spangler *et al.*, 1984), although pheromones emitted by a male may attract females, ultrasounds produced by tymbal organs on the tegulae (see above) induce the female to orientate itself towards, and guide it to, the male. In a dense cluster of males found in the vicinity of honeycombs (i.e., near the egglaying substrate and larval food), the sources of emitted pheromones are unlikely to be easily located. By contrast, an ultrasonic signal is strongly directional and may effectively guide females to individual males. Observations show that females are indeed attracted by this stimulus.

Sounds produced by *A. grisella* are of a high frequency. However, since high-frequency sounds become attenuated rapidly, the distance at which they are effective is limited. Hence, females are attracted at short distances. An advantage of short distance signals is that predatory bats have little chance of being attracted. Furthermore, the high-frequency sounds of the communicating moths would be less subject to interference from low frequency sounds in the environment.

Ultrasounds produced by genital stridulation in *Syntonarcha iriastis* (Pyralidae) (see above) also function in courtship (Gwynne & Edwards, 1986). As in *Achroia grisella* the male attracts the female, a situation uncommon in lepidopteran courtship. However, while in *A. grisella* the sounds are emitted by moths close to the larval food source, in *S. iriastis* the males stridulate from perches on bushes of plants other than the larval hosts. The ultrasounds of *iriastis* are of a much lower frequency than those of *grisella*, making them less rapidly attenuated; signals from the moths were detected by a bat detector from a distance of 20 metres. Since the signals from *iriastis* are effective over a comparatively great distance, they serve to orientate females to males from the outset. By contrast, in *grisella* the moths initially depend on pheromonal attraction. A disadvantage of the long-distance sound-signalling is that it increases the chance that bats will be attracted to the moths.

Tymbal organs in Arctiidae produce sounds associated with defence (see above). However, courtship calling occurs in the Polka-Dot Wasp Moth (*Syntomeida epilais*) (Sanderford & Conner, 1990), an example in which acoustic signals have been converted from defensive signals to sexual communication. Males and females engage in antiphonal calling, which may continue for long periods prior to mating. The sounds differ between males and females, possibly functioning for gender recognition.

Scent

The nature of pheromones

Hormones are endocrine secretions that are released from one organ and travel to another within the same individual where they have a measurable physiological effect on the target organ. By contrast, pheromones are ectohormones - exocrine secretions of one *organism*, which release a behavioural or developmental response in another organism or other organisms. In Lepidoptera, pheromones are associated particularly with mating behaviour - as attractants and as 'aphrodisiacs'.

Pheromones are involved in intraspecific interactions. As such, they belong to that more general class of substances termed *semiochemicals*, which includes any chemical involved in communication between organisms. By contrast, *allelochemicals* are semiochemicals affecting any aspect of the biology of *other* species.

Glands

Many pheromone-producing glands are modifications of the epidermal cells of the integument, and they exhibit considerable histological uniformity. Whereas in females the glandular epithelium is typically found between the posterior abdominal segments, in males scent glands are more variable in position and size, occurring in a variety of areas of the body and ranging from single cells to groups of cells.

The histology of pheromone gland cells in Lepidoptera (see McColl (1969) for a summary of the studies of H. Barth) has been examined in relatively few species. A basal zone, a medial secretory zone, and an apical zone are generally present in glandular cells of both males and females. A glandular cell is largely composed of a medial secretory zone, and in its secretory phase such a cell is vacuolate. The apical zone includes a rhabdorium composed of parallel fibres. Sometimes the basal zone is absent.

Scent dissemination

The conventional view of the role of pheromones in reproductive behaviour in Lepidoptera is that females use scent to attract males, typically over long distances, whereas males release scent at close quarters to prepare the female for mating. There are exceptions to this general picture but nevertheless, scent dispersal is best considered separately in males and females.

Females

Female Lepidoptera typically remain stationary and allow their scent to waft, carried by air currents, from glands situated usually between abdominal segments A7 and A8 or A8 and A9 (summary in McColl, 1969). Frequently, there are no special distributing organs; scent molecules apparently diffuse through the cuticle, and pores are not usually present. Telescoping of the terminal abdominal segments helps protect the glandular areas and, particularly, prevents scent from being released at inappropriate times. When 'calling', many females extend and curve their abdomen so encouraging the release of scent.

There are, however, sophisticated aids to scent dispersal even in females. For example, the area for evaporation can be increased by having scales associated with the glandular parts, as in *Euploea core* (Nymphalidae: Danainae) (see McColl, 1969). More complex means of dispersal are found in some other species where a brush of scales, with a retractor muscle to permit its withdrawal, is used for scent dispersal.

Males

In males, unlike females, a great diversity of scent disseminating structures is located on a variety of body parts (Boppré, 1984b). Various attempts (summarized by McColl, 1969) have been made to classify this diversity. All, according to McColl, are unsatisfactory because since they are based largely or partly on the bodily location of the organs, they do not permit ready comparison of the more fundamental factors of histology and morphology.

A scheme (see McColl, 1969) based on structure and function rather than by location on the body is adopted here. Male scent organs fall into six basic types. In type 1, a single, isolated scent scale arises from a hypodermal gland cell. Such organs are found only on the wings. Type 2 organs are composed of aggregations of glandular cells, each member of which bears a scale or hair. Although sometimes found on the legs and on the abdomen, they usually occur on the wings. Aggregations of glandular cells with scales and hairs are also found in type 3 organs, but they differ from those of type 2 in being concealed by folds in the cuticle (e.g., on wings or intersegmental membranes of the abdomen). Sometimes the glands may be located on eversible tubes of cuticle from the abdomen. Type 4 includes 'mixed organs' so termed because scales or hairs

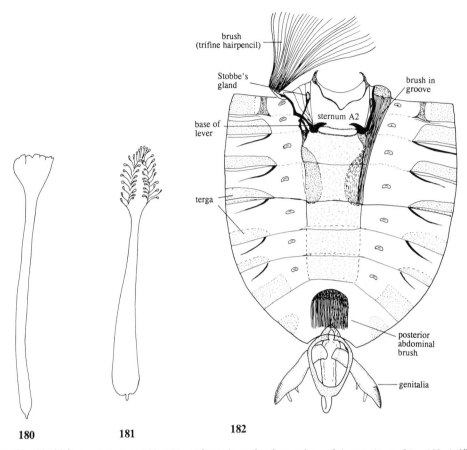

Figs 180–182. *Male scent organs. 180, 181, androconia scales from wings of* Argynnis paphia; *182, 'trifine' hairpencil of* Phlogophora meticulosa *(Noctuidae) and Stobbe's gland. (180, 181, after Sellier, 1973; 182, after Birch, 1970a.)*

associated with glandular cells are interspersed with scales or hairs not so associated. This arrangement effectively increases the surface area for pheromone evaporation per unit area of glandular epithelium. As with type 3 organs, type 4 organs are concealed. A refinement of type 4 organs is seen in type 5 where the distributing hairs or scales are located in an area separate from the glandular area. The distributing hairs may be adjacent to the glandular area or remote from it. Scales or hairs may be present on the glandular areas, reduced in number, or entirely lost. There are no 'non-glandular' distributing scales or hairs mixed with those directly associated with the cells. In type 5 organs the glandular area is covered by the remote distributing pencil even when scent is not being actively disseminated. By contrast, in type 6 the distributing pencil, which is also remote from the glandular area, comes into contact with the glandular area just before or during scent secretion. When scent is not being secreted, the pencil is remote ('binate' or 'dual organs' of Boppré & Vane-Wright, 1989). The pencil itself may also be glandular.

The scales of type 1 organs (e.g., as in Figs 180, 181) are often elongated, and terminally frayed, the divisions presumably assisting scent dissemination. Frequently, these scales are found in rows, streaks, or even spots, but they differ from type 2 organs in being found under normal covering scales rather than in forming discrete and isolated patches. The scales have many shapes (e.g., flattened, hair-like, fan-like, spoon-shaped). Evaporation from scent scales on wings is so effective that these scales are restricted to butterflies with a fairly slow wingbeat.

Type 2 organs were recorded from many Papilionoidea, and some Hesperiidae and Arctiidae, but discrete patches of scent scales are present in other families (e.g., Nepticulidae, Geometridae) (McColl, 1969).

The concealed scent organs of type 3 are better protected than unconcealed ones, and evaporation is more effectively reduced when dissemination is not required. The most simple concealment occurs in Lepidoptera in which a glandular area on one wing is overlapped by the other wing (e.g., the forewing overlap of glandular areas on the upper surface of the hindwing in a number of butterflies and some Arctiidae). More complex arrangements occur where the glandular area lies within a fold of the wing margin, a situation seen widely in the Lepidoptera.

In group 3, McColl included what Janse (1932) called 'coremata', long, eversible membranous tubes with glandular cells from which the hair-like scales arise (Fig. 183). Typically, these coremata are paired structures situated on the intersegmental membranes of various abdominal segments. Some arise from the valvae (e.g., in many Geometridae); occasionally, the coremata are branched. The hairs are either found along the length of the coremata or are bunched at their distal ends. The large surface area of coremata promotes the evaporation of chemical signals, and the fact that they are concealed when not in use means that pheromones are not released in an uncontrolled fashion (Boppré & Schneider, 1989). In *Creatonotos* (Arctiidae), surface area is increased both by the number of hairs and the ultrastructure of each hair (Boppré and Schneider, 1989). Based on hair dimensions and number, the corematal hair surface area of *Creatonotos* was estimated to be as much as 4.5 cm^2, a figure that is increased if the lattice-like structure of each hair is taken into account.

Type 4 organs are structurally intermediate between types 3 and 5; unlike type 3, they bear odour-disseminating hairs not directly associated with the gland cells. But, in contrast with those of type 5, these hairs are intermixed with the glandular hairs and not located separately from them. Type 4 organs are far less common in Lepidoptera than are the other arrangements. Some species of *Spodoptera* (Noctuidae) have type 4 organs on the valvae of the genitalia, which are extruded, by haemolymph pressure, with the genitalia prior to mating.

There are more examples of type 5 (dual) organs in Lepidoptera than of any other category. Whereas in type 5 the distributing pencil covers the glandular area when scent is not being dispersed, in type 6 the pencil is brought into contact with the glandular area only during scent dispersal. In some species with type 5 organs, the glandular areas are not concealed by folds or pockets, but are protected only by the distributing pencil (e.g., *Gnophodes parmeno*: Nymphalidae). Type 5 organs are found on many parts of the body (e.g., legs, wings, abdomen).

The complex hair pencils associated with abdominal (Stobbe's) glands are included in this category (Fig. 182). In *Phlogophora meticulosa* (Noctuidae), the structures are large and well developed (Birch, 1970a); such organs are widespread in trifine Noctuidae, and their structure is probably unique (Birch, 1972). (McColl viewed them as similar to those found in many Sphingidae, but more complex.) A trifine hair pencil occurs on each side of the abdomen. The brush is attached to an arm, which, with its associated muscles, acts as an everting lever. The other end of the arm joins the posterio-lateral corner of sternum A2. At rest the brush lies in a well-developed groove of the abdomen, in which are found scales associated with gland cells. The secretory cells of the Stobbe's glands are situated in an invagination of sternum A2 and release scent, via hairs, onto the brush organ. The glands are active immediately after the pharate adult emerges from the pupa, and only then, so the secretion is stored for later

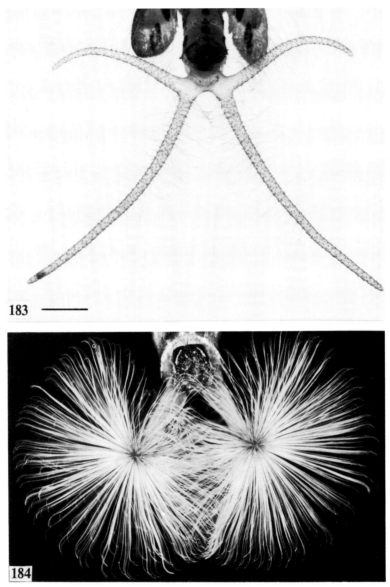

Figs 183, 184. *Male scent organs. 183, coremata of* Creatonotus gangis *(Arctiidae) (scale, 5 mm.); 184, abdominal hairpencils (expanded) of* Tirumala petiverana *(Nymphalidae: Danainae). (184, courtesy of Dr M. Boppré; after Ackery & Vane-Wright, 1984.)*

use either on the brush or on the scales found within the pocket. The secretion from Stobbe's glands is a precursor of the actual scent (Birch, 1970a).

The most highly developed organs, type 6 (binate or dual organs, are mainly found in Danainae (Nymphalidae) where their particular form is unique (Ackery & Vane-Wright, 1984). In this category, the distributing abdominal hair pencil comes into contact with glandular alar areas only at times of scent dispersal. Eversion of integumental sheaths between abdominal segments A8 and A9 causes the hairpencils or brushes to be displayed (Fig. 184). Extrusion occurs by haemolymph pressure, and the organs are retracted by muscles.

It seems that many Danainae transfer scent by way of cuticular dust formed in the alar organs or the abdominal brushes. Scent scales on the wings are found throughout the subfamily, and are located in patches or pockets on the hindwings. Contact between the hairpencils and alar organs does indeed occur in a number of Danainae, but this contact is by no means universal within the family (a weakness of the concept of category 6) even where both organs exist (Ackery & Vane-Wright, 1984; Boppré & Vane-Wright, 1989). The contact may allow different components required for pheromone production, isolated on either the pencils or the alar organs, to be brought into association.

Function and mode of operation of lepidopteran pheromones

In Lepidoptera, pheromones seem principally to be sex pheromones. They may act either as long range attractants or, in courtship, at close range. Typically, females secrete long range pheromones (at least in the case of moths) while in males scent is generally released close to the female during courtship. However, there are exceptions to these tendencies. The entire subject of the role of pheromones, particularly in female butterflies, is uncertain for not only are experiments on females difficult to carry out, but other factors, particularly visual cues, are also involved (Silberglied, 1984; Boppré, 1984b).

Female pheromones

The first attractant to be chemically identified was from the female of *Bombyx mori* (Bombycidae) (the Silkworm Moth). Isolation of the compound began in Germany in the 1930s by Adolf Butenandt, but its identity was not established until the 1950s (see Birch & Haynes, 1982; Schneider, 1984a). The substance, termed bombykol, was found to be an alcohol (E-10,Z-12-hexadecadien-1-ol), and since so little of it could be extracted from any one gland, extracts were taken from around half a million. The term pheromone was actually coined at the time of the chemical identification of bombykol.

It was originally assumed that each species of Lepidoptera released one pheromone. But as analytical methods became more sophisticated, attractants of individual species were shown to comprise two or three chemical components. In *B. mori* a second compound, an aldehyde called 'bombykal', was discovered subsequent to the identification of bombykol. Extracellular recordings demonstrated that male sensilla on the antenna responded to both bombykol and bombykal. In *B. mori*, the aldehyde is apparently a behaviour inhibitor, but in many cases different chemical components act synergistically.

The release of attractants and the various components involved in their composition may give rise to a very complex system, a complexity demonstrated effectively by two closely related species of Lymantriidae (*Lymantria dispar*, the Gypsy Moth, and *L. monacha*, the Black Arches or Nun Moth). Studies on the pheromones and behaviour of these species (summarized by Schneider, 1984a) showed that females of both secrete the pheromone 'disparlure'. Disparlure exists as two optical isomers, (+)-disparlure and (−)-disparlure. Females of *L. dispar* release only (+)-disparlure, while in *L. monacha* 90 percent of the pheromone released is (−)-disparlure and only 10 percent occurs as (+)-disparlure. In the males of *dispar*, receptors, situated on the antennae, for both isomers occur in about equal numbers whereas in *monacha* only (+)-disparlure receptors are present. Apparently, the male of *dispar* is repelled by the (−)-disparlure isomer of the pheromone released by the female of *monacha*, whereas the male of *monacha* responds only to the (+)-disparlure isomer. It is unclear why *monacha* males are not also attracted to females of *dispar* given that the diel rhythms of the two species overlap.

To be effective, studies on the biology of sex attractants in nature need to account for a variety of factors such as the blend of constituents (including optically active forms), environmental conditions such as wind, times of calling and emergence, and the physiological state of the moths. Such complexity makes the planning of experiments and the interpretation of results extremely difficult.

Male moths respond to the scent of females by flying upwind towards the source of the odour. It has been suggested (summary in Birch & Haynes, 1982) that they do this either by chemotaxis, along a concentration gradient, or by anemotaxis, that is by wind steering. There is little evidence to support the first of these suggested mechanisms. Probably, male moths fly chiefly within the plume of scent released by the female and check that they are progressing by comparison of their position with stationary objects. The characteristic zig-zag path of their flight does not support the chemotaxis hypothesis, for if moths were to fly along a concentration gradient their path would be more likely to be direct to the source.

Although flight upwind is the most obvious overall feature of the response of male moths, sex attractants actually release a succession of responses. The series of reactions to a calling female may progress as follows: raising of antennae, wing fluttering, zig-zag flight upwind, landing, wing fluttering while walking, approach to female, stimulation of female by 'hair-pencilling' (e.g., Birch & Haynes, 1982). A given response may be elicited by a single chemical component or by all of them. Thus within a series of responses a single pheromone may be required at a particular stage if the next response is to be elicited. However, the complexity of male behaviour is a response not only to chemical (pheromonal) stimuli but also to visual and mechanical cues.

The role of pheromones in female butterflies (e.g., Boppré, 1984b) differs, at least somewhat, from that in female moths in having short-range rather than long-range effects. Certainly evidence strongly suggests that male butterflies respond initially to females on the basis of visual stimuli rather than by scent (e.g., Silberglied, 1984). However, female scent is almost certain to play various roles in later stages of courtship, even well before the final stages. Possible functions, which may also interact with visual stimuli, include directing the male towards the female, and discrimination by the male between sexes, species, and female receptivity. Female butterflies may release odours that repel males, but the source of these odours may be from the male as members of that sex are known to transfer scent to females during mating.

Pheromones are likely to be involved in the attraction of adult males to female pupae of butterflies belonging to at least one species of Papilionidae, certain Heliconiinae, and one species of Lycaenidae (summary in Boppré, 1984b). But here again, attraction to pupae appears to depend on a complex interaction of visual and pheromonal cues. For example, males of *Heliconius hewitsonii* have been observed searching all vertical objects that resemble a pupa (Gilbert, 1984), so scent would appear to provide close-range cues in such situations.

Male pheromones

Male Lepidoptera typically utilize pheromones at short range in courtship. For example, in *Eumenis semele* (Nymphalidae) (the Grayling butterfly) the male manoeuvres his wings so that scent scales are brought into contact with the antennae of the female (Tinbergen 1951). 'Hair-pencilling', in which extrusible male brush-like organs are wafted in front of the female, is well known in Danainae (Nymphalidae) (e.g., in *Danaus gilippus berenice*; the Queen butterfly, see Brower, *et al.*, 1965). The importance of hairpencils in males of *Phlogophora meticulosa* (Noctuidae) is demonstrated by the fact that their removal prevents mating (Birch, 1970b).

Although male scent is undoubtedly important in courtship, the precise way in which the pheromones work is uncertain. Some male pheromones ('aphrodisiacs') may stimulate the female to mate. But the initial function of scent organs of Danainae and Noctuidae seems to be for releasing pheromones that inhibit flight of the female. This action induces those females that are courted first in the air (e.g., butterflies) to land, and then prevents females that attract males from a 'resting' position (e.g., nocturnal Lepidoptera) from flying away (Birch, 1974). Male Queen butterflies release, from their hairpencils, a dust (cuticular transfer-particles; pheromone transfer-particles of Schneider, 1984b). This pheromone-bearing dust contains the flight inhibitor (a ketone) and a glue (a terpenoid alcohol) to stick the particles to the antennae of the female (Pliske & Eisner, 1969). Hair-pencilling by the male may continue after the female has alighted. In *Phlogophora meticulosa* (Noctuidae), there is some experimental evidence to suggest that benzaldehyde, a substance extracted from the scent organs, makes female moths less active. In this species the male flies to a stationary female, a response typical of nocturnal moths. Females of Noctuidae do not apparently adopt an acceptance posture. But in some species of other families (e.g., *Plodia interpunctella*: Pyralidae) the female raises the tip of her abdomen in response. The pheromone responsible for this response might, in this instance, reasonably be termed an aphrodisiac because it induces mating. However, in most species, it is not known whether pheromones act as stimulants or as flight-inhibitors; both may have an initial function in courtship. Therefore great caution is needed when using the term aphrodisiac to refer to pheromones released by males (e.g., Boppré, 1984b).

In some moths, females appear to be attracted by scent to stationary males, the reverse of the situation typical of Lepidoptera. In *Eustigmene acrea* (Arctiidae), males form aggregations or leks within which many individuals display long abdominal coremata (Willis & Birch, 1982). Conspecific males and females are attracted to the leks, probably by scent (although to date no pheromone has been isolated), and females flying into the lek are mated. Although no chemical secretion was identified from the coremata, in structure and mode of inflation the organs resembled those of *Utetheisa* (see below), another species of Arctiidae, from which a pyrrolizidine alkaloid (PA) has been identified. PAs are known to be essential for successful courtship (short distance effect) by males of certain Lepidoptera (see below), but confirmation is still required that a long range chemical attractant is actually released from the coremata of *E. acrea*. Willis & Birch considered the possibility that the coremata might complement the bright white and yellow males enhancing a possible visual attractant. However, they pointed out that since at least one (the second) peak of male display occurs well after dark, visual attraction seems unlikely, or at least severely limited.

Coremata are also displayed by males of *Creatonotos* (Arctiidae) (Schneider, 1984b; Boppré & Schneider, 1989). Corematal odour reaching isolated females induces them to fly upwind to the males. In *Creatonotos*, the coremata contain the same pyrrolizine (hydroxydanaidal) as *Utetheisa* (Schneider, 1984b).

The pheromone found in the coremata of *Utetheisa* (Culvenor & Edgar, 1972; Conner *et al.*, 1981) and *Creatonotos* (Schneider *et al.*, 1982), is synthesized from PAs taken up by the larvae from PA-containing foodplants. Corematal size, at least in *Creatonotos*, depends on the quantity of PAs consumed (Schneider *et al.*, 1982). However, the exact function of the corematal pheromone is uncertain because males without coremata or with reduced coremata are still able to mate. But since PAs are also used in Lepidoptera as defensive substances (p. 81), corematal size may indicate to females the level of protection they would gain from the PA-containing spermato-phore of a particular male (Eisner, 1980). Many *adult* Arctiidae also visit plants containing PAs (Pliske, 1975a). In most species, visitation was strongly male-biased, but in some it was female-biased, and in others neutral. But although PAs taken up by these moths *may* be used in the production of pheromones (as in Ithomiinae and Danainae), direct evidence for this view is lacking (Pliske, 1975a).

In Danainae and Ithomiinae (Nymphalidae), the production of male courtship pheromones (hydroxydanaidal and, less frequently, danaidone and danaidal) depends on PA precursors (summary by Ackery & Vane-Wright, 1984). In these Lepidoptera, PAs are derived by the adult, not the larva, from withered PA-producing plants. PA-derived pheromones have been identified in several Danainae and, given the many species of Danainae and Ithomiinae in which adults feed on PA-producing plants (e.g., Pliske, 1975b), the presence of these substances is probably widespread or even universal within these groups.

Other functions for male pheromones have been suggested (e.g., Birch, 1970c; Schneider, 1984b). Some may be species-specific signals. Others may induce the final stages of egg formation in the female (e.g., yolk deposition). Since the male often exposes his hairpencils as he flies upwind towards the female, his scent is released into the plume of scent from the female, which may inhibit other (competing) males. Potentially, pheromones may act as allomones (interspecific semiochemicals - allelochemics - used as chemical messengers for the benefit of the releasing organism) to deter predators during courtship and mating - a time at which moths are often exposed for long periods. Although such allomones have been identified in other insect orders, they have not been definitely recorded from Lepidoptera.

Pheromones of male Lepidoptera are typically small molecules which, appropriately for substances acting over short distances, are both volatile and unstable. In contrast, pheromones of females tend to be large molecules such as unsaturated alcohols (e.g., Birch, 1970c).

Pheromones and speciation

If attractants released by females are a blend of components, then small changes in the components, or in their ratios, might be effective 'isolating mechanisms'. Chemical differences have been suggested as a major means of effecting isolation. In pheromones of some closely related Tortricidae, differences in the ratio of chemical components are known to occur. Furthermore, it has been established that females of many closely related species of moths 'call' at different times of day or night, an observation that has led to the suggestion that the system evolved as an isolating mechanism. For example, the plume moths *Platyptilia carduidactyla* and *P. williamsii* (Pterophoridae) are attracted to the same synthetic sex attractant, but while *P. carduidactyla* calls during the first half of the night, *P. williamsii* calls during the second half (Haynes & Birch, 1986). Artificial removal of temporal differences led to cross-attraction of males, resulting even in attempted copulation. In nature, temporal differences between the two species was considered to be the 'primary factor preventing interspecific cross-attraction and courtship' (Haynes & Birch, 1986). Several other examples involving the same phenomenon in Lepidoptera were noted. Yet Haynes & Birch doubted that male pheromones are of great importance as isolating mechanisms since, because they act so late in courtship, their use would still require the expenditure of much energy in flight to the female. Moreover, it appears that at least in Noctuidae, no species specificity has been found in the courtship pheromones of related species (Birch, 1974). Nevertheless, while in males of the butterfly species *Colias philodice* and *C. eurytheme* (Pieridae) visual cues are used to locate and recognize females, females of *P. philodice* rely entirely on male pheromones for recognition while in *P. eurytheme* both visual and pheromonal cues are used. In this example, then, male pheromones do play a significant role in mate recognition or, depending on ones view, isolation.

An alternative view to the isolation concept of speciation involves a concept of the incidental breakdown of specific-mate recognition systems of species in allopatry (Paterson, 1985). Perhaps lepidopteran attractants should be regarded as recognition signals. The composition of blends in (geographically) isolated populations might certainly change in allopatry without the need to involve the concept of character displacement when species come together again. Different times of calling could be viewed in the same way.

PART II

7

ENVIRONMENTAL AND ECOLOGICAL IMPORTANCE OF LEPIDOPTERA

The ecological and environmental importance of Lepidoptera stems largely from the fact that their larvae eat plants, and because all stages of these insects are consumed by insectivorous predators or parasitoids. Caterpillars do not feed exclusively on plants, so mention is made of non-phytophagous habits in Lepidoptera. The main ecological importance of *adult* feeding is pollination, a subject briefly considered below.

Larvae, and even to a very limited extent adults, play a part in decomposition either by breaking down live plant (or occasionally animal) tissue or by continuing the process of decay by detritivory.

The potential for environmental impact

Ultimately, it is the number of individuals that are relevant, but given the enormous number of lepidopteran species it would be surprising if the group did not have a significant environmental impact. Estimates of the number of lepidopteran species vary considerably. Laithwaite *et al.* (1975) gave an approximate figure of 165 000 for *described* (taxonomic) species, a figure based on the comprehensive index to Lepidoptera at The Natural History Museum, London, with allowances made for synonyms. The *actual* number of insect species may have been underestimated (Stork, 1988), although for a strongly moderating view see Hodkinson & Casson (1991). This underestimate, if valid, may primarily concern parasitic Hymenoptera and weevils, so the increase in numbers of species of Lepidoptera may not be proportional to the increase in those groups. Nevertheless, there are numerous undescribed species housed in museum collections, and unquestionably the number of extant lepidopteran species is greater than the number described. An upper limit of 500 000 species has been suggested by Gaston (1991), well over twice the number of validly described species.

Despite the great number and diversity of seed plants that have evolved since the angiosperm radiation in the Cretaceous, relatively few insect *orders* have exploited what is a considerable resource. But although only 9 out of 29 extant orders of insects are almost wholly or partly phytophagous, nearly half of all insect *species* are phytophagous (Southwood, 1973; Strong *et al.*, 1984). It seems that although plant feeding requires that a number of formidable barriers are overcome, when these barriers are surmounted available niches are plentiful. Lepidoptera about equal Coleoptera in number of phytophagous species, and include a little under twice the number of phytophagous species of Hemiptera, the order with the next greatest number of plant feeding species (Strong *et al*, 1984: table 1.2). These observations are at least suggestive of the impact of Lepidoptera as primary consumers.

Natural systems are generally held in dynamic equilibrium and species diversity and numbers of individuals of Lepidoptera are constrained by plant diversity, plant defences, other barriers to plant feeding, predation and parasitization, and competition for resources. Lepidoptera become of considerable concern to mankind when species develop pest status usually as a result of imposed environmental changes such as crop monoculture. In environmentally balanced ecosystems, Lepidoptera, like other insects, are more appropriately viewed as the 'glue' and 'building blocks' of the system because, for example, they both consume large quantities of plants and are themselves consumed by other animals in large numbers (Janzen, 1987).

The diversity of plants consumed

Two issues regarding foodplant diversity are considered here - the taxonomic range of plants, and the variety of substrates available (e.g., leaves, roots, galls, seeds).

Although it is with *flowering* plants (Angiospermae) that Lepidoptera share their diversification, their larvae also feed on a variety of substrates other than angiosperms. For example, one major group of Lycaenidae (Lipteninae) eat lichens. Fungi are eaten by a number of caterpillars either facultatively or obligatively. Several Lepidoptera feed on non-vascular plants (bryophytes, which include mosses and liverworts), particularly mosses. Among moss feeders are included representatives of a range of families including, for example, Psychidae, Oecophoridae, Gelechiidae, Tortricidae, and Pyralidae. Certain Nymphulinae (Pyralidae) feed on algae on rocks in aquatic habitats. Caterpillars of Lithosiinae (Arctiidae) feed on epiphytic algae of trees and rocks or on the algal component of lichens (Habeck *in* Stehr, 1987). Included in the spore-producing vascular plants are lycopods (clubmosses), a group whose members are mostly extinct. Nevertheless, Lepidoptera have been recorded from the group (e.g., the caterpillars of *Catoptria furcatellus*: Pyralidae, and *Euptychia*: Nymphalidae). Several species feed on ferns, but the only family that appears to be predominantly, or perhaps exclusively, fern-feeding is the Callidulidae (J. Weintraub, *pers. comm.*), and there appears to be a major association with ferns of the tribe Lithinini (Geometridae: Ennominae) (Rindge, 1986; Holloway, 1987b). The 21 species of microlepidopterans recorded from ferns (Powell, 1980) were spread across five families and four superfamilies. Among macrolepidopterans, fern-feeding seems to be widespread among the Lithinini (Geometridae) (Rindge, 1986; Holloway, 1987b).

All higher plants are vascular seed-producers, and are divided into Gymnospermae (including, for example, cycads and conifers), and Angiospermae (flowering plants). There are many species of Lepidoptera that feed on gymnosperms (particularly conifers), but the great diversification of the group appears to be associated with the rise of the angiosperms, which radiated in the Cretaceous. The caterpillars of the two most primitive groups of Lepidoptera (Zeugloptera: Micropterigidae, and Aglossata: Agathiphagidae) seem to be associated with primitive plants, although in the case of Zeugloptera the association is not as clear as with Aglossata. Although Micropterigidae caterpillars have been collected from leaf litter it is not known whether they feed on general detritus, fungal hyphae, mosses, or a combination of these three substrates. Therefore, it is unclear if the association with fungi or moss is primary. However, at least some Micropterigidae consume *green* plant material (Lorenz, 1961; Carter & Dugdale, 1982), an association that presumably must be secondary since Micropterigidae probably originated before the angiosperms.

The relationship of *Agathiphaga* (Agathiphagidae) with *Agathis*, the Kauri Pine is quite possibly primary. The Lepidoptera probably originated in the Jurassic (Whalley, 1986), a time at which *Agathis* was known to occur extensively in the Australasian region (White, 1986). Caterpillars of the next lineage of Lepidoptera, Heterobathmiina, mine the leaves of *Nothofagus* the Southern Beech (Fagaceae), which are dicotyledonous angiosperms with a long fossil record.

With the possible exception of Agathiphagidae on *Agathis* and Heterobathmiidae on the Southern Beech, speculation about the historical associations of the Lepidoptera with early plants is largely fruitless; the uncertainty over the primary association of Micropterigidae caterpillars exemplifies the point. But the wide range of plants accepted secondarily by lepidopteran larvae shows how effective the exploitation of this food source has been. The close historical association of Lepidoptera and plants is underlined further by the wide variety of modes of feeding adopted (e.g., mining, boring, external feeding), and the whole range of plant structures accepted (e.g., root, stem, leaf, seed, fruit) (see Chapter 5).

Why do caterpillars eat plants?

The most obvious answer to this question is that plants offer an enormous amount of food, which can be converted into caterpillar- and, subsequently, adult protein, and can provide the necessary calories to support lepidopteran activity. But to obtain nutrients, Lepidoptera have to overcome certain barriers or hurdles (Southwood, 1973). The three main hurdles, as discussed by Strong *et al.* (1984), involve nutrition, attachment, and desiccation.

Concerning nutrition, the protein content of plants may be suboptimal, the balance of the nutritional components may be inadequate, and the composition of the plant may change under different conditions. Compared with that of carnivorous insects, the efficiency of phytophagous insects in converting their food into growth is relatively low. However, in Lepidoptera it is high compared with the rate of conversion in other seed-plant feeders. The mean E.C.I. (an index of efficiency of converting food ingested into growth) for various orders of phytophagous and predatory insects was summarized by Southwood (1973). Figures show, for example, in Phasmida (dry weight) an E.C.I. of 8.0, compared with 17.8 for Heteroptera (Hemiptera), and 20.9 for Lepidoptera (dry weight).

A further major hurdle to successful feeding on plants is the problem of remaining attached. One solution has been the development of mechanisms for holding, and most caterpillars use crochets on the planta of the abdominal prolegs (Chapter 5) for attachment. The most highly developed arrangement is found in macrolepidopterans, the predominant external feeders in Lepidoptera. The need for special solutions for attachment is avoided in those Lepidoptera living within plant tissue, such as in the many microlepidopterans the caterpillars of which are leaf miners, or borers of stems, roots, galls, and seeds.

The third hurdle to plant feeding is the threat of desiccation. Largely avoided by internal feeders, and reduced in concealed feeders such as leaf rollers, it is a serious problem faced by those insects that live externally. Even on plants with a high water content, air movement may pose problems of desiccation. Water loss is reduced in insects by the presence of a waterproof cuticle, and when water is lost it can be replenished by drinking from water droplets. To these general solutions may be added the behaviour of resting on the underside of leaves, or elsewhere, away from more exposed sites, and the tendency to feed at night, a time when relative humidity is higher.

In addition to these three obstacles, Lepidoptera also face the problem of finding the correct food source (Southwood, 1973). In Lepidoptera the choice is usually made by the adult at oviposition, but the greater the degree of foodplant-specificity the greater has become the recognition of, and dependence on, secondary plant chemicals as cues. Paradoxically, these very chemicals probably evolved in plants as protective agents - particularly against herbivorous insects.

Plant defences and their effects on Lepidoptera-plant relationships

Caterpillars are able to eat plant tissue if they surmount the obstacles summarized above. Once attacked, plants respond by evolving chemical or physical defences against phytophages. Examples of physical defences include the presence of a thick cuticle and of hairs and spines on the epidermis. Cuticular teeth on holly leaves seem to inhibit feeding of the oligophagous caterpillar of *Lasiocampa quercus* (Lasiocampidae). Resins of conifers may provide a physical defence by drowning insects (e.g., caterpillars of the Pine Shoot Moth, *Rhyacionia buoliana* (Tortricidae), were found to be drowned in resins of conifers as they invaded the buds of young trees, see Southwood, 1973).

In chemical defence, evidence suggests that the great range of secondary plant substances have evolved in response to insect attack (Southwood, 1973). Examples of these compounds, which cut across phyto-taxonomic boundaries, are alkaloids, betaines, glycosides, tannins, flavenoids, essential oils, caponons, and organic acids. They often directly inhibit feeding. In other cases the deficiency of some amino acids may result in a plant becoming inadequate as food. For example, the protein content of oak leaves falls while the tannin content rises. Proteins and tannins also combine to form relatively indigestible complexes.

In turn, the evolutionary response of the plants is met by development of resistance by the insects to the plant defences. Such a response is seen particularly in those trees with which an insect species has been associated for a long time. The ability to switch to a new hostplant will be affected by the similarity of the old and the new host. Similarity between hosts reflects not only taxonomic similarity, but also similarity in secondary plant substances. Other factors such as water content, leaf structure and architecture, and overall chemistry are also important. Since secondary plant substances may be found in plants that are not related taxonomically, hostplant switching often causes a considerable taxonomic incongruence in hostplant range. Eventually the stage is reached where some secondary plant substances even become cues for insect feeding; insects may get 'hooked' on the chemicals (Southwood, 1973). Furthermore, and notably in Lepidoptera, secondary plant substances may be used by insects in their own defences (see Chapter 5).

On what plants caterpillars feed, and why

Insect-plant relationships have attracted much attention from ecologists and chemists, although the subject is still an 'infant science' (Strong et al., 1984). Systematists have been interested to know to what degree there is congruence between phylogenies of plants and phylogenies of insects. This is not the place to summarize existing reviews of insect-plant associations. Instead, I consider three questions that seem to be related to the general picture of the natural history of Lepidoptera and their hostplants. First, to what degree is there congruence at low taxonomic levels between Lepidoptera and their hostplants? Second, if there *is* an association, does it extend to higher taxonomic levels? Third, if there is little or no congruence, why? These questions lead ineluctably to a consideration of the much debated, but ambiguous, concept of coevolution.

The numerous records of caterpillars and their hostplants demonstrate little in the way of association of lineages of Lepidoptera with lineages of plants. The only review of a major component of the Lepidoptera and their hostplant associations was conducted by Powell (1980) who examined records of microlepidopterans from the literature and from unpublished observations of his own and of others. Despite the acknowledged inadequacy of sampling from the tropics, the very area in which most microlepidopteran speciation has taken place, Powell stated: 'It appears that funda- mental evolutionary radiation in Monotrysia [*sensu* Common, 1970] and Ditrysia occurred through specialized larval feeding in niches or horizons within communities (i.e. detritophagy, root and stem-borers, leaf miners, or external foliage- and seed- or flower-feeders) rather than along botanical evolutionary lines.' Although some lepi- dopteran families are strongly associated with particular groups of plants (e.g., 90 percent of Eriocraniidae feed on Fagales, 80 percent of Ethmiinae (Oecophoridae) feed on the Polemoniales-Lamiales branch of Asteridae, Powell, 1980), Lepidoptera have not evolved in parallel with plants. High levels of specificity between particular lepidopteran and plant lineages resulted rather from secondary colonisation. There is little, if any, evidence for such 'parallel cladogenesis' between Lepidoptera and plants (see below). Powell concluded that 'Microlepidoptera are believed to have radiated in several independent lines, but no superfamily is dependent exclusively or even primarily on one angiosperm branch. . . . More significantly, no major microlepidop- teran family is diversified (i.e. more than 12% of hostplant records) in association with a primitive angiosperm subclass.'

A somewhat different pattern was found for macrolepidopterans in a comparison of hostplant selection by Canadian and British species (Holloway & Hebert, 1979). The likelihood of a new plant being accepted as a host seemed to depend primarily on taxonomic criteria (i.e., the plant is more likely to be fed upon if it is related to others of the group fed upon by the moth), and secondarily on ecological factors (e.g., plants sharing the same habitat). A general *tendency* was found for related moths to feed on related plants. A further finding was that conifer feeders are less host-specific than moths feeding on angiosperms, presumably because angiosperms have evolved a greater variety of secondary plant substances, which has led to specialization by the moths. Two possible explanations may account for the ecological association of the macrolepidopterans and their hostplants (Holloway & Hebert, 1979). First there may have been persistent exposure, but of low frequency, to plants that were originally not the usual hosts. Such exposure would be related to 'mistakes' in egg laying or in the accidental transfer of caterpillars to a plant growing near the normal host. Second, caterpillars are, in a sense, adapted to other environmental factors in the habitat of the usual host (e.g., moist or dry soil, rate of predation or parasitism).

The study of Holloway & Hebert is exceptional for moths because it dealt with groups for which relatively good data were available, namely the larger moths of trees and shrubs in Britain and Canada. Foodplant records for butterflies are much more numerous permitting substantial reviews (e.g., Ehrlich & Raven, 1965; Ackery, 1988). Several instances of taxonomic association between groups of butterflies and of plants occur. For example Papilionidae often feed on Aristolochiaceae, Pierinae on Cappari- daceae and Brassicaceae, Ithomiinae on Solanaceae, and Danainae on Apocynaceae and Asclepiadaceae (Ehrlich & Raven, 1965). But these taxonomic associations are more likely to have arisen as the result of similarity between plant chemistries rather than by a history of congruent evolution between plants and insects at the species level (i.e., parallel cladogenesis, see below).

Coevolution

Coevolution has been defined in both ecological and phylogenetic ways. The word was originally phrased in ecological terms in a study of butterflies and their foodplants (Ehrlich & Raven, 1965) to account for the reciprocal evolutionary relationships between two groups of organisms 'with a close and evident ecological relationship, such as plants and herbivores.' Within this definition, the term coevolution should be confined to *particular* reciprocal associations of two populations rather than a *general* reciprocal evolutionary response of a plant to a whole array of selective pressures placed on it by phylogenies (diffuse coevolution) (Janzen, 1980). Coevolution has been described as '*reciprocal* evolutionary change in interacting species (Strong *et al.*, 1984). Species A evolves in response to selection imposed by species B; species B then evolves in response to the change in A.' The coevolutionary model was outlined (Strong *et al*, 1984) as follows: plants bearing mildly toxic phytochemicals having, say, a physiological function are fed upon by insects. Plants respond to the insect attack by producing more noxious phytochemicals - substances that can appear independently in plants that are only distantly related. The effect is to reduce insect feeding leading to possible evolutionary radiation of the plants. Insects that evolve tolerance of these chemicals might even use them as attractants, and benefit from a reduced competition from other phytophages.

Coevolution (ecological sense) may be rare, occurring only as 'coevolutionary vortices in an evolutionary stream', with 'some insect-plant interactions [being] sucked into intense, escalating whirlpools of reciprocal adaptations'. But most may be the product of diffuse coevolution (Strong *et al.*, 1984).

But coevolution has also been defined in phylogenetic terms, for example under Fahrenholz's rule (Hennig, 1966) where total congruence would mean the corresponding speciation of a parasite in parallel with the speciation of its host. The concept applies equally to Lepidoptera and their hostplants, the former being equivalent to parasites and the latter to their hosts. Host-parasite (or phytophage-foodplant) coevolution may be resolved into two components (Brooks, 1979): *co-accommodation*, referring to mutual adaptation of parasite and host, and *co-speciation*, referring to the cladogenesis of a parasite tracking the cladogenesis of its host. In terms of Lepidoptera and their hostplants, the question is whether associations between these insects and the plants on which they feed evolved by parallel cladogenesis or, rather, by colonization by the insect after plant speciation (Miller, 1987a). Effectively, if parallel cladogenesis cannot be demonstrated then neither, strictly speaking, can coevolution. Following earlier authors, Miller accepted the view that coevolution means stepwise coevolution - 'that if lineages of organisms have affected each others' evolution in a stepwise manner, then there will be congruence between the phylogenies of the two groups. The key problem, then, is to distinguish between cases of association by descent (= stepwise coevolution) and colonization (= sequential evolution . . .).'

No convincing evidence for evolution by parallel cladogenesis has been discovered (Miller, 1987a). Even those few studies claiming a high level of stepwise coevolution were often based on work in which either the phylogeny of the insects or that of the plants, or of both, was inadequately known or based on *a priori* acceptance of the coevolutionary theory. The study of coevolution of Papilionidae butterflies, particularly at the species level, and their hostplants was also constrained by the lack of plant and, indeed, papilionid cladograms. But what limited deductions were made suggested that patterns of hostplant associations in Papilionidae are the result more of repeated colonization of plants belonging to a fairly small number of families (Miller, 1987a).

In an examination of the primitive moth genus *Heterobathmia* (Heterobathmiidae) and its *Nothofagus* hostplants (Humphries *et al.*, 1986), no evidence was found to support 'association by descent' (i.e., parallel cladogenesis). Rather, the analysis of these authors pointed to colonization.

Two studies involving Lepidoptera and their foodplants were used by Strong *et al.* (1984) to illustrate coevolution (ecological sense). But neither demonstrate convincingly coevolution in the phylogenetic sense. In the genus *Papilio* and its Apiaceae hostplants (e.g., Berenbaum, 1983), it was claimed that the degree of plant speciation is related to the level of complexity of coumarins, which are the secondary plant substances involved. Plant genera with angular coumarins are more speciose than genera with linear coumarins, and genera with linear coumarins are more speciose than those with hydroxycoumarins. Although plants with angular coumarins are better protected than those with linear- or hydroxycoumarins, there are more species of *Papilio* associated with plants with angular coumarins than with those with linear or hydroxycoumarins. It appears that the ability of *Papilio* species to feed successfully on plants with complex coumarins frees them from most competition.

This example does not demonstrate coevolution in the phylogenetic sense (Miller, 1987a) for the phylogeny of *Papilio* is uncertain, and no phylogeny for the Apiaceae was provided in the study. Hence, until properly generated phylogenies (cladograms) are compared, it remains questionable as to whether it is strictly an example of coevolution in the (phylogenetic) sense here adopted.

The second example of a possible coevolutionary interaction also involves a butterfly genus, *Heliconius* (Nymphalidae). *Heliconius* caterpillars feed only on young foliage of various passion vines (*Passiflora*), and the female searches specifically for such foliage. Passion vines may be severely damaged or destroyed by *Heliconius* caterpillars, and the plants appear to have evolved defences, such as attracting predators and parasitoids of the caterpillars. For example, structures that mimic the eggs of the butterfly occur on the stipule of the leaves, the leaf lamina, and the tendril tips, probably giving the *Passiflora* protection because the butterflies tend to avoid plants on which eggs have been already laid. The butterflies obtain their nitrogenous food from the flowers of cucurbitaceous vines, and are attracted by these vines, which in turn depend on the insects for pollination. The excessive production of male flowers over female flowers may be a coevolved response to the need to attract the butterflies.

However, as in the *Passiflora/Papilio* example, there is no convincing evidence to suggest the occurrence of parallel cladogenesis of *Heliconius* and its hostplants. The incongruence between the existing phylogenies of both the plants and the butterflies suggest that most, if not all, heliconiine evolution occurred after the radiation of the Passifloraceae (Mitter & Brooks, 1983). A doubt about accepting either the coevolution-ary or the colonization explanation exists in this example because neither the phylogenies of the butterflies not of the plants have been strictly based on the techniques of phylogenetic systematics.

Nevertheless although parallel cladogenesis remains unrecorded, evolutionary associations between Lepidoptera and their foodplants obviously do occur. Methods now exist for the analysis of the degree of congruence between cladograms of 'hosts' and 'parasites' enabling us to better understand how certain biological systems came into being.

The consequences of plant consumption

It is tempting to discuss the economic aspects of lepidopteran natural history in terms of their direct economic significance to man. After all, Lepidoptera are important competitors for food when they eat crops: Tortricidae, Noctuidae, and Pyralidae include members that are very serious agricultural pests. But pest status is nearly always caused by an upset of natural balance through monoculture, the killing of natural predators, or both. The present work is concerned with the economic importance of Lepidoptera in nature: what these insects eat, what eats them, their requirements for survival, and the subtleties of their natural role.

The most obvious natural role of Lepidoptera stems from their association with plants. Having overcome barriers generally too great for most other insects to surmount, the consumption of green plants by lepidopteran larvae has developed alongside the evolution of plants, and has probably been significant in the radiation of the latter. The pressures from Lepidoptera on plants is such that if their natural controls are removed, they become serious pests. There are numerous examples, but two that cause much damage are the Noctuidae genera *Spodoptera* (the armyworms), and *Helicoverpa* and its relatives (the corn earworm complex).

Pollination

A plant-associated, but largely unexplored, role of Lepidoptera is in the pollination of flowers. Although Lepidoptera are probably of little commercial importance in the pollination of crops (Free, 1970), certain groups (particularly Sphingidae, Hesperiidae and Papilionoidea, but also many Noctuidae and Geometridae) undoubtedly play an important, although largely unassessed, role in pollination in nature. Barth (1985) stated that Lepidoptera are fourth in order of importance regarding pollination, being exceeded by Hymenoptera, Diptera, and Coleoptera. Lepidoptera are incidental pollinators, for with few exceptions they have no organs for collecting pollen. The main exception is seen in female *Yucca*-moths (Incurvarioidea: Prodoxidae), which make pollen balls using the so-called maxillary tentacle (see Chapter 2). Also, pollen is gathered by *Heliconius* (Nymphalidae) using the proboscis (see Chapter 2).

Flowers considered as typically adapted to pollination by Lepidoptera are either those with a long corolla, or those that are zygomorphic with a substantial keel acting as a perch. These flowers are considered to be associated respectively with hoverers and perchers. Hoverers include Sphingidae, moths attracted, typically, by long-tubed, white flowers, with strong nocturnally released fragrances, and a copious production of nectar. However, Sphingidae visit many other kinds of flowers, at least in Costa Rican dry forest, including those usually pollinated by other organisms such as bees, bats, and humming birds (Haber & Frankie, 1989). Furthermore, although flowers with long-tongued corollas restrict access to their nectar to long-tongued moths, long-tongued Sphingidae do not limit their visit to long-tubed flowers.

The importance of Sphingidae as pollinators (summary by Haber & Frankie, 1989) lies in their capacity to carry large pollen loads, to cover long distances, and to move readily between plants. Their longevity, and their habit, in both sexes, of cruising around large areas leads to outcrossing of plants that are widely separated. In contrast, and diminishing their efficiency as outcrossers, is the opportunistic nature of their feeding habits; pollen loads on individual moths are regularly found to be derived from different species of plants. In Costa Rican dry forest, Sphingidae are estimated to be the primary pollinators of about 10 percent of the trees. Also, they pollinate various species of shrubs, herbs, lianas, and epiphytes (Haber & Frankie, 1989).

An exception to the hovering strategy typical of Sphingidae is that adopted by those that pollinate flowers of the epiphytic cactus *Hylocereus costaricensis* from northwestern Costa Rica. When feeding on nectar from these flowers, Sphingidae (particularly *Manduca sexta*, *M. rustica*, and *M. ochus*) hover only briefly at the mouth of the corolla before plunging right into the flower and crawling as far as possible into its base, an action that causes the entire body to get covered by pollen (Haber *in* Janzen, 1983).

Lepidoptera other than Sphingidae usually perch on flowers. Butterflies frequent Asteraceae and other groups with inflorescences composed of small tubular flowers (e.g., *Buddleia*: Buddleiaceae). These flowers provide a suitable perch for the insect while it probes individual flowers for their nectar. Other flowers associated with butterfly-feeding are the so-called 'butterfly flowers' (e.g., see Barth, 1985). Such flowers are zygomorphic with platforms on which Lepidoptera perch. For example, in the Meadow Pea (*Lathyrus pratensis*), two petals form the 'wings' of the flower and cover the keel within which the stamens and pistil are found. The 'wings' provide the landing stage for butterflies.

Most flowers pollinated by Sphingidae and other moths open by night or during the evening or produce scent or nectar, or both, only at dusk or at night, although a number of pollinating Lepidoptera (besides butterflies) feed by day. Whereas Sphingidae tend to pollinate white or pale flowers, butterflies are usually attracted to those with bright colours (often red) or white. The actual visual cues in the attraction of Lepidoptera (and other insects) to flowers are often complex. Flowers exhibit various patterns to which these insects respond (e.g., Barth, 1985). Such visual signposts take the form, for example, of bright colours in the centre of a flower contrasting with a duller colour at the periphery, or of spots or lines, which guide potential pollinators to the nectar. Frequently, these patterns are visible only in the ultraviolet range of the spectrum. Often, for example, the centre of a flower may be UV absorbing while the periphery is UV reflecting. *Colias* butterflies (Pieridae) seem to visit most frequently flowers with this pattern, although they respond also to 'visible' colours. It appears that this flower pattern represents a common target for these butterflies, and is indicative of a particular nectar quality (Watt *et al.*, 1974).

Very little quantitative information has been gathered to give an idea of just how important Lepidoptera are as pollinators. The same limitation applies to the question of whether or not, and if so the extent to which, lepidopteran pollinators (and other pollinators) and their hostplants have co-evolved (Feinsinger, 1983). What can be said (Proctor & Yeo, 1973) is that the tropics are the main centres of zoophilous pollinators. Since exposed, cold and damp conditions are unfavourable to pollinating insects, the relative scarcity of Lepidoptera in temperate regions may have a significant effect on the composition of plant communities.

The importance of Lepidoptera in pollination in tropical seasonal forest in Costa Rica was outlined by Janzen *in* Janzen (1983) based partly on the observations of W.A. Haber. Sphingidae are known to transport pollen distances of several kilometres, an observation that underlies their significance in outcrossing of plants. Since Sphingidae visit flowers in a regular pattern they are, at least theoretically, extremely significant pollinators. Also, medium to large sized butterflies tend to follow a similar pattern of flower visiting in Costa Rica. By contrast, small moths are likely to act as short-distance pollinators since they are frequent visitors to sprays of small flowers present throughout the year. With other small insects, they seem to be able to move pollen over several hundred metres, perhaps, as Janzen suggested, wind-aided.

Lepidoptera as environmental indicators

Lepidoptera, in particular macrolepidopteran moths and butterflies, may be used to indicate environmental quality (e.g., level of habitat degradation), and to categorize habitats into smaller components (e.g., Holloway, 1977, 1980, 1984, 1985b; Kudrna, 1986; Erhardt & Thomas, 1991).

Holloway's work deals with macrolepidopteran moths of rainforests of the Far East (particularly those of Borneo, New Caledonia, Norfolk Island, Sulawesi etc.). Although organisms have been more extensively used as indicators of aquatic habitats, their use as indicators of terrestrial habitats has been far less extensive, although interest is growing. The subject is potentially of value to conservation since it deals with such questions as the environmental quality of habitats under regeneration, the size of areas selected as national parks, and the early monitoring of changes in patterns of land management.

Using organisms to monitor natural or man-induced environmental changes is much cheaper than measuring directly, say, primary production. To be effective as indicators, organisms need to be easily assessed and sensitive to early changes in the environment. So understanding the conditions normally prevailing is a prerequisite for using organisms as indicators. For an organism to be appropriate as an indicator, a number of criteria need to be fulfilled. First, the group in question should be, as far as possible, widely distributed; second, it should be sensitive to environmental change at the level of species; third, it should be capable of easy sampling in the field and of identification in the laboratory; and fourth, confidence in the results from a particular group of organisms will be increased if the group is diverse. Potentially, insects meet many of these requirements, but problems of identification in many groups pose a major problem. Holloway considered that Lepidoptera (and certain Homoptera) meet most of the criteria. Although the state of lepidopteran taxonomy is very far from complete, identification of many species of macrolepidopterans is possible on external characters alone. Many others can be identified by examination of the genitalia. Furthermore, Lepidoptera are easily sampled by the use of light traps, although problems of sampling occur because the composition of samples is affected by factors other than environmental quality.

Essentially, the methods used by Holloway for moths involved quantitative comparison of samples taken from different habitats in a given area in terms of number of species shared. In the Gunong Mulu National Park in Sarawak, Holloway (1985b) sampled 28 localities in six different forest habitats in a study aimed at assessing the value of Lepidoptera in categorizing rainforest. The light trap samples were scored for five different groups of moths. Similarities between samples were assessed by cluster analysis, and the results were presented visually as dendrograms and linkage diagrams. The study demonstrated the differential value of particular groups of moths in the categorization. For example, Sphingidae and most Noctuidae tend to be poor discriminators of different habitats since they are often ecologically widespread and have a high level of mobility. Arctiidae discriminated well between upper montane and lower montane areas and, to some extent, lower *montane* samples segregated from other *general* lower altitude samples.

In a survey of the larger moths of rainforests in New Caledonia and Mount Kinabalu in northern Borneo, Holloway (1980) found that despite the view that Lepidoptera are commonly fairly host specific, the correlation between macrolepidopterans and floristic diversity does not always hold. Macrolepidopterans may sometimes be more effective indicators of vegetation type (structural diversity). Structural diversity includes spatial and architectural components. The spatial component of rainforest includes the number of foliage layers, an indication of the openness of the vegetation. Architectural factors include the proportions of leaves, flowers, buds, shoots, branches, trunks, and roots.

Unquestionably, certain groups of Lepidoptera tend to be associated with particular habitats. However, the use of these organisms as effective indicators of environmental quality requires further ground work, chiefly to decide precisely which groups are most appropriate and to what exactly they respond. It is very likely that, among moths, Geometridae will be of particular value as indicators given their relatively restricted mobility. Sampling methods and differences in temperature and rainfall have profound effects on light trap catches so, clearly, such factors have to be taken into account for monitoring to be effective. A further limit is set by the level of taxonomic resolution, an observation demonstrating the importance of taxonomy to environmental science.

Diurnal Lepidoptera, in particular butterflies, are useful habitat indicators in temperate (e.g., Kudrna, 1986; Erhardt & Thomas, 1991) and tropical (Brown, 1991) regions. The special value of butterflies lies in their conspicuousness, relative ease of identification, and the absence of a need for light trapping. Moreover, much ecological and distributional information is available for these insects, particularly in Europe, and it has been argued (Kudrna, 1986) that, as a group, butterflies have sufficiently broad ecological requirements that changes in their diversity are likely to indicate changes in a wide variety of invertebrates. For all European butterfly species, Kudrna (1986) compiled indexes of their potential in resisting habitat change and of their actual vulnerability. Since butterfly diversity, at least in grassland habitats, seems to decline more quickly than plant diversity when these habitats cease to be managed traditionally (Erhardt & Thomas, 1991), even a rough idea of the resistance to habitat change of any butterfly species may be of considerable value.

In more specific terms, Erhardt & Thomas (1991) have compared the effect on different groups of (mainly) diurnal Lepidoptera on the effect of changing-patterns of management of temperate grasslands. For example, mowing grasslands can cause a decrease in diversity of Lepidoptera both by removing flowers on which adult butterflies depend for nectar, and by killing larvae. Considerable variation occurs in changes in species diversity depending on taxonomic group. For example, in Lycaenidae, the number of species closely parallels that of vascular plants during the process of succession of grassland to woodland climax vegetation. In contrast, Geometridae show an opposite trend with the number of stenotopic species (i.e, those restricted to a few narrow niches) increasing with plant succession. The maximum number occurs at the climax vegetation type - woodland.

Lepidoptera as prey

Lepidoptera are predated or parasitized at all stages in their life history, so their ecological importance is by no means restricted to their position as primary consumers. The importance of Lepidoptera as prey is suggested directly, by studies on their predation, and indirectly by the diversity of their defence mechanisms (e.g., Blest, 1963; Carter et al., 1988).

The most important predators of Lepidoptera are birds, bats, parasitoids, and, probably, other small mammals, although predation by other animals (e.g., earwigs, ants, various reptiles, and humans) also occurs. The dragonfly Orthetrum austenti (Libellulidae) appears to have specialized in preying on butterflies, at least at a site in Lagos, Nigeria, where many observations of this behaviour were made (Larsen, 1981). Ants are voracious predators of lepidopteran larvae. In parts of mainland Europe, the predatory habits of Formica rufa and F. lugubris (Formicidae) are utilised for protection of trees, particularly from the ravages of defoliating Lepidoptera (Pavan, 1981). Programmes have been developed in which colonies of ants are transported to regions where they are needed.

Analysis of life table data for 14 species of Lepidoptera (Dempster, 1983) identified predation as one of the two main features in determining the fluctuation in abundance from one generation to the next. (The other was the failure of females to lay their full complement of eggs.) Also, arthropods are probably more important predators of eggs and early larvae, while vertebrates, particularly birds, are more important predators of older larvae and pupae. But important as Lepidoptera are likely to be as a food source, it is questionable as to whether natural enemies frequently regulate their population levels in a density-dependent way. Instead, populations may be limited by the availability of resources (Dempster, 1983).

Birds

Although case histories of Lepidoptera as prey are not numerous, those that do exist indicate the importance of these insects. In a study of food consumed by Elegant Trogon (Trogon elegans) nestlings in Costa Rica (Janzen, 1987), large Hawkmoth caterpillars constituted at least 74 percent of the items brought to the nestlings, and over 90 percent of the weight of items brought during the main nesting period. It is likely that many other insectivorous, medium-sized birds depend on Hawkmoth caterpillars as food for nestlings. The importance of these caterpillars is emphasized by the fact that the nesting period of the Trogons may be linked to the time at which large Hawkmoth caterpillars are present.

All stages of Lepidoptera (but particularly larvae) form a very large portion of the nestling diet of woodland dwelling Blue Tits (*Parus caeruleus*) and Great Tits (*P. major*) in Britain (summary and references in Cowie & Hinsley, 1988). In the case of Blue Tits, Lepidoptera formed 73 percent of the diet in woodland and 40 percent of the prey selected by birds breeding in suburban gardens, where artificial food represented a large component of the diet.

The examples of the Trogons in a tropical habitat and Titmice in temperate habitats demonstrate the importance of Lepidoptera, and particularly caterpillars, as prey of nestlings of insectivorous birds. It seems very likely that the findings are more generally applicable.

Several studies have been carried out on predation of adult Lepidoptera by birds, but it is unlikely that they form anything like such a regular source of food as do larvae. *Most* attacks by birds on adult Lepidoptera are probably opportunistic. However, exceptions exist in the case of large, seasonal aggregations (e.g., as occurs in the Monarch butterfly, see Chapter 3). Also, *Galbula*, a genus of jacamar (order Piciformes) have adult Lepidoptera consistently in their diet (Sherry *in* Janzen, 1983).

Avian predation on adult Lepidoptera is not restricted to palatable species. For example, unpalatable ithomiine butterflies (Nymphalidae) were preyed upon by the tanager *Pipraeidea melanonota melanonota* in southeastern Brazil during the winter (Brown & Neto, 1976). Entire insects were not eaten; instead, the gut contents were squeezed out and consumed by the birds. Probably, the unpalatable substances are concentrated in the external parts of the body so that, by feeding in this way, they are not ingested by the tanagers. Similarly, certain birds have adopted a way of reducing their intake of cardenolides in Monarch butterflies by breaking off the wings of these insects where the highest concentrations of the chemicals occur (see also Chapter 3).

Further avian predation on Lepidoptera was observed on moths in northern Venezuela (Collins & Watson, 1983). Over a period of several months, 30 species of birds were recorded feeding on moths remaining, early in the morning, on the walls of a research station and its surrounding vegetation. Of 908 bird attacks, 764 moths were eaten.

Bats

The other major group of vertebrate predators on Lepidoptera are insectivorous bats. Bats are crepuscular, nocturnal, or both, and they prey on moths, most of which are active at these times. Evidence for bat predation of Lepidoptera has come from direct observation, experimentation, and the presence of remains of moths gathered beneath bat roosts. Also, there is evidence (see Chapter 6) that bats and moths are involved in a coevolutionary response, with moths developing physiological and behavioural means of evading predatory bats, and bats 'responding' to overcome moth defences.

An insectivorous bat may eat one-quarter to one-half of its body weight each night. If this figure is a fair estimate, a 20 g bat would consume at least 1.8 - 3.6 kg per night, and a colony of 100000 individuals perhaps 180000 to 300000 kg per year (Hill & Smith, 1984). The lepidopteran component in these figures is probably high.

Other small mammals

The effect of predation by small mammals is very likely to have been underestimated since less work has been carried out on them compared with birds. However, indicative of just how important small mammals may be are results from a study of predation by *Sorex araneus* (the Common Shrew) on *Operophtera brumata* (Geometridae) (Buckner, 1969). This shrew was found to prey extensively on cocoons of the moth, leading Buckner to suggest that small mammals were the most important predators of *O. brumata* cocoons, at least in Whytham Wood in Oxford.

Humans

Insects in general, including numerous species of Lepidoptera, are eaten by people in many parts of the world (e.g., Bodenheimer, 1951; DeFoliart, 1989). In certain areas, a considerable percentage of the total animal protein consumed may be insectan (summary and references in DeFoliart, 1989). Extensive use of Lepidoptera is made in Mexico, Africa, and S.E. Asia with caterpillars of the larger moths (particularly Saturniidae) being frequently recorded. Apart from their value in silk production, the Silkworm (*Bombyx mori*: Bombycidae) provides a source of food in parts of Asia where the pupae are eaten (DeFoliart, 1989).

The distribution of caterpillar consumption by peoples in central and southern Africa is widespread, and the variety of species consumed extensive (Silow, 1976). In Zaire, an advanced classification of caterpillars, based largely on edible kinds, has been developed by certain ethnic groups. In terms of both number of species and individuals, most of the caterpillars eaten by the Mbunda people are Saturniidae. Other families represented include Sphingidae, Notodontidae, Noctuidae, Lasiocampidae, and Limacodidae.

Caterpillars, as well as other insects, are perhaps used more often as a relish than a source of vital nutrients. However, the considerable numbers gathered for food indicates their importance as human prey. At least in Zaire, caterpillar-eating has declined among those people exposed to European habits and values (Silow, 1976). Given the high food-conversion efficiency, compared with those animals conventionally used as meat in the world as a whole, insects (not least Lepidoptera) offer a considerable potential as a source of food for man (DeFoliart, 1989).

Parasitoids

Parasitoids, particularly hymenopteran, attack Lepidoptera extensively. Ichneumonoidea (the largest group of parasitoids) attack lepidopteran hosts disproportionately highly. For example, although only 12 percent of British insect species belong to the Lepidoptera, 46 percent of Ichneumonoidea use them as hosts (Gauld, 1988).

The evolutionary association of parasitoids and their lepidopteran hosts is close and complex (Gauld, 1988). Two major strategies have evolved in hymenopteran parasitoids. Idiobionts immobilize or kill their hosts while koniobionts allow the host to continue living for a time after the egg has been attached to the host or implanted within it. Within both these modes of life both ecto- and endoparasitoidism are represented. But since the egg of a parasitoid laid on or near the host is prone to damage or to being eaten, endoparasitoidism (if it can be achieved) provides better protection than ectoparasitoidism. So it is understandable that ectoparasitoids predominate on hosts that live concealed lives. Such ectoparasitoids occur on, for example, endophytophagous Lepidoptera and leaf rollers and tiers, or on pupae protected by dense cocoons. It has been suggested (Gauld, 1988) that ectoparasitoids on cocooned pupae may use silk as a cue for egg laying.

Ectoparasitic koinobionts of Lepidoptera usually attack those species in which the larva leaves its concealed feeding site to pupate elsewhere. Koinobiont endoparasitoids are able to exploit hosts that live fully exposed. In Britain, most of them attack Lepidoptera.

Silk production

The production of silk is widespread among lepidopteran larvae. Produced by specialized salivary glands, it is extruded via the spinneret. The most obvious function of silk in the lives of Lepidoptera is for the production of the cocoon, but most published information on the composition of silk has been generated as a result of the commercial exploitation of the cocoons of the Silk Moth *Bombyx mori* (Bombycidae). Unless otherwise stated, the following details refer to the larvae of this moth.

The labial glands of all lepidopteran larvae are homologous with the true salivary glands of other insects. However, in lepidopteran larvae the glands producing silk and saliva are derived from the mandibular glands. The labial glands, particularly long in Bombycidae and Saturniidae, are four times the length of the body in *Bombyx mori*. They are arranged in a series of folds. Anteriorly, the ducts unite and enter the spinneret. Immediately posterior to the spinneret they diverge into a pair of silk ducts, each broadening somewhat into a reservoir forming a double, U-shaped loop, and then into a coiled and twisted secreting gland. Just before the ducts unite anteriorly, a pair of accessory glands (glands of Lyonet or Filippi) open into them. The function of these glands is uncertain but it has been suggested that either their secretions lubricate the common duct and facilitate the movement of the silk, or that a substance is produced that causes the threads of silk in each gland to adhere. The secretory cells of the glands, which surround their cavities, are particularly large and have branched nuclei. In the silk ducts these cells are more flattened. A peritoneal membrane bounds both ducts and glands externally, while internally they are lined by cuticle.

Silk is composed of two main components - fibroin and sericin (e.g., see Handschin, 1946). Fibroin, the tough inner protein, is surrounded by sericin a water soluble gelatinous protein. The precursor of fibroin is fibrinogen, which is secreted by the posterior part of the silk gland. Fibrinogen, a water soluble compound, is denatured on contact with air to insoluble fibroin. The chief amino acids of fibroin are glycine, alanine, and tyrosine. Sericin is secreted in the middle section of the gland; its main constituent amino acids are serine, alanine, and leucine. The levels of those amino acids involved in silk production are high in the mature larva but low in the adult moth. Sericin forms a relatively brittle coating to the fibroin after the silk has been spun, and this component is removed in industrial silk production. The substance is readily soluble in warm, soapy water.

Bombyx mori has been reared for the production of silk since at least 2600 BC. The species is thought to have been native to the Himalayas, but today it is unknown in the wild. The first silk farmers (sericulturists) were the Chinese, who manufactured and exported silk goods to Europe via India. Silk goods were much in demand in the Roman Empire. The export value of silk products to the Chinese was evidently considerable, for they went to extreme lengths to prevent stock leaving the country: export of breeding material was illegal, and was a capital crime. In the fourth century, however, some eggs were smuggled out of China to Persia and finally to Asia and Europe. The success of rearing silkworms in northerly climes has always been limited, and commercial silk production, restricted by the conditions and cost of production, has never flourished in these areas for long. However, several attempts have been made including production of silk at Lullingstone Castle in Kent earlier this century (Hart-Dyke, 1949). Silk from Lullingstone was used in the Coronation robes and train of the wedding dress of Queen Elizabeth II.

Although silkworms other than *Bombyx mori* have been exploited in the production of silk, large scale rearing has been successful only with the former. The chief reasons for the suitability of this silkworm are the high fecundity of the moth, the high quality of the silk produced by the larvae, and the relative ease of growing the foodplant (mulberry) in adequate quantities. Furthermore, the larvae of *B. mori* are not prone to wander, so they are easy to keep in breeding trays. Many wild silkworms have more active larvae making them inappropriate for commercial rearing indoors. In ancient times the silk of the European *Pachypasa otus* was used in the dress of the Romans, but this species is not exploited today.

Silk from Tusseh silkmoths (*Antheraea mylitta*) continues to be produced, particularly in India. These insects live wild in forests and their silk is harvested (see Feltwell, 1990). A detailed review of the silk industry and the uses of silk are provided by Feltwell (1990).

PART III: MAJOR TAXA

8

INTRODUCTION TO GROUPS

The purpose of Part III is to provide a guide (although not an identification guide) to the diversity of the Lepidoptera. Some groups are better known or more thoroughly studied than others and inevitably the text reflects this imbalance. The sections on Micropterigidae, Agathiphagidae, and Heterobathmiidae are lengthy relative to their comparatively small number of species because as primitive Lepidoptera their structure and biology has helped to establish a picture of the lepidopteran groundplan.

The choice of classification is a compromise, for lepidopterists adopt various different systems. Moreover classifications are constantly being modified as phylogenetic relationships between taxa are better understood. But despite variations in the system a general consensus exists partly because the classification of the order continues to rest on many long established families. To that extent existing classifications exhibit considerable stability. The guiding principle adopted in Part III is that the taxa accepted should be, as far as possible, monophyletic entities, the term monophyletic being used in the sense accepted by phylogenetic systematists or cladists. Shortcomings or problems with the classification are pointed out wherever possible.

The organization of Part III

In Part III, the Lepidoptera are treated under the headings of superfamilies, families, and sometimes subfamilies. But this section is also divided into four chapters reflecting the way in which lepidopterists refer informally to large groupings of families, but groupings that are definitely not, or possibly not, monophyletic. Chapter 9 deals with the primitive, homoneurous moth families the members of which are, with the exception of Exoporia, monotrysian and small. Chapter 10 includes the early Heteroneura, those moths in which the female genitalia are monotrysian, but where the venation differs between forewing and hindwing. Chapter 11 is concerned with the lower Ditrysia, an assemblage of families of small to medium small ditrysian, heteroneurous moths the larvae of which are mostly concealed feeders. The moths covered in Chapters 9-11 are widely referred to by lepidopterists as 'Microlepidoptera', although sometimes Pyralidae are excluded. Chapter 12 deals with macrolepidopterans, which, as in the lower Ditrysia, are ditrysian and heteroneurous. However, macrolepidopterans are usually larger than other Lepidoptera (mean wingspan, taking into account all the species of macrolepidopterans and microlepidopterans, is definitely greater), and the larvae of most species feed in exposed situations (although some are borers) and have crochets arranged in a mesoseries on the prolegs.

The families included in each superfamily are introduced firstly by considering factors such as general diversity and distribution, and the basis, collectively, of their monophyly. Each family is then treated by some brief introductory remarks, followed by a summary of adult structure, immature stages, general biology, classification, and phylogenetic relationships. Since many superfamilies are monotypic, a separate discussion of the superfamily and family concerned would be superfluous, so details presented for the superfamily apply also to those of the family.

Space does not permit the illustration of large numbers of specimens, so each family is illustrated by one, or occasionally more than one, specimen. The specimens were selected, as far as possible, to show a typical member of almost every family. But it has not been possible to illustrate the diversity of appearance, which is often considerable, within each family. The Lepidoptera are particularly difficult to illustrate with a limited number of examples because of this intrafamilial variation in colour pattern, and sometimes wing shape. Conversely there exists much convergence in pattern and wing shape between different families.

Classification of the Lepidoptera

Current classification

Recent classifications of the Lepidoptera at the family level and above have been presented by Munroe (1982), Minet (1986), Nielsen (1989), and Nielsen & Common (1991). The divisions above those of superfamily were considered by Kristensen (1984b). The classification of superfamilies and families adopted in the present work is given in Table 1.

Despite a broad consensus in recent classifications, the placement of certain families is uncertain and changes are constantly occurring. Moreover, many superfamilies are monotypic, that is they include only one family (and sometimes one genus), so the classification contains redundant names (names that refer to one and the same taxon). For example, the definition of Heterobathmioidea, Heterobathmiidae, and *Heterobathmia* are equivalent. The increase in the number of superfamily taxa in recent classifications compared with the relatively modest number in those of earlier works has not resulted entirely through an increase in number of described species. With the growing emphasis on ensuring that classifications include only taxa that are strictly monophyletic, there is now a tendency to remove those families, or taxa of lower rank, that compromise the monophyly of any given superfamily. Since these taxa are often not able to be assigned to some other existing superfamily, their rank is usually elevated to that of a new superfamily. For example, Schreckensteiniidae were removed from the Yponomeutoidea and are now placed in the monotypic superfamily Schreckensteinioidea. Sometimes the removal of one or more families from a superfamily results in that superfamily being left with a single family. Such action has also increased the number of monotypic superfamilies. For example, Minet (1983) argued that the Geometroidea, which had included up to 12 families (e.g., Fletcher, 1979), lacked a specialised character common to all of them and probably did not represent a monophyletic group. Consequently he removed all except Geometridae from Geometroidea.

This increase in the number of superfamilies in the Lepidoptera is concerning not only for the rise in the number of redundant names, but also because the existence of further names makes it virtually impossible, particularly for biologists other than lepidopteran systematists, to remember, and therefore use, them. But if the development of a natural classification is a goal, it may be argued that pragmatic considerations are less important. Of course, with further study, many monotypic superfamilies may be united; indeed re-amalgamation of certain macrolepidopteran superfamilies has already been proposed (Minet, 1991).

A possible alternative to this scenario would be to leave certain small disassociated groups *incertae sedis* within the order or within one of the taxa between the order and the superfamily. The objection to this proposal is the arbitrary nature of deciding

Table 1 List of superfamilies and families of Lepidoptera

MICROPTERIGOIDEA
  Micropterigidae
AGATHIPHAGOIDEA
  Agathiphagidae
HETEROBATHMIOIDEA
  Heterobathmiidae
ERIOCRANIOIDEA
  Eriocraniidae
  Acanthopteroctetidae
  Catapterigidae
LOPHOCORONOIDEA
  Lophocoronidae
NEOPSEUSTOIDEA
  Neopseustidae
MNESARCHAEOIDEA
  Mnesarchaeidae
HEPIALOIDEA
  Hepialidae s.l.
INCURVARIOIDEA
  Heliozelidae
  Adelidae
  Crinopterigidae
  Incurvariidae
  Cecidosidae
  Prodoxidae
NEPTICULOIDEA
  Nepticulidae
  Opostegidae
PALAEPHATOIDEA
  Palaephatidae
TISCHERIOIDEA
  Tischeriidae
TINEOIDEA
  Tineidae
  Eriocottidae
  Acrolophidae
  Psychidae
  Galacticidae
GRACILLARIOIDEA
  Gracillariidae
  Bucculatricidae
  Roeslerstammiidae
  Douglasiidae
YPONOMEUTOIDEA
  Yponomeutidae
  Ypsolophidae
  Heliodinidae
  Lyonetiidae
  Glyphipterigidae

GELECHIOIDEA
  Oecophoridae
  Lecithoceridae
  Elachistidae
  Pterolonchidae
  Coleophoridae
  Agonoxenidae
  Blastobasidae
  Momphidae
  Scythrididae
  Cosmopterigidae
  Gelechiidae
COSSOIDEA
  Cossidae
  Dudgeoneidae
  Ratardidae
TORTRICOIDEA
  Tortricidae
CASTNIOIDEA
  Castniidae
SESIOIDEA
  Sesiidae
  Brachodidae
  Choreutidae
ZYGAENOIDEA
  Zygaenidae
  Heterogynidae
  Megalopygidae
  Limacodidae
  Dalceridae
  Epipyropidae
  Cyclotornidae
  Chrysopolomidae
IMMOIDEA
  Immidae
COPROMORPHOIDEA
  Copromorphidae
  Carposinidae
SCHRECKENSTEINIOIDEA
  Schreckensteiniidae
URODOIDEA
  Urodidae
EPERMENIOIDEA
  Epermeniidae
ALUCITOIDEA
  Alucitidae
  Tineodidae
PTEROPHOROIDEA
  Pterophoridae

HYBLAEOIDEA
  Hyblaeidae
THYRIDOIDEA
  Thyrididae
PYRALOIDEA
  Pyralidae
GEOMETROIDEA
  Geometridae
URANIOIDEA
  Uraniidae
  Sematuridae
  Epicopeiidae
DREPANOIDEA
  Drepanidae
AXIOIDEA
  Axiidae
CALLIDULOIDEA
  Callidulidae
  Pterothysanidae
HEDYLOIDEA
  Hedylidae
HESPERIOIDEA
  Hesperiidae
PAPILIONOIDEA
  Papilionidae
  Pieridae
  Nymphalidae
  Lycaenidae
BOMBYCOIDEA
  Lasiocampidae
  Anthelidae
  Lemoniidae
  Eupterotidae
  Apatelodidae
  Carthaeidae
  Cercophanidae
  Oxytenidae
  Saturniidae
  Brahmaeidae
  Endromidae
  Bombycidae
  Sphingidae
  Mimallonidae
NOCTUOIDEA
  Oenosandridae
  Doidae
  Notodontidae
  Lymantriidae
  Arctiidae
  Noctuidae

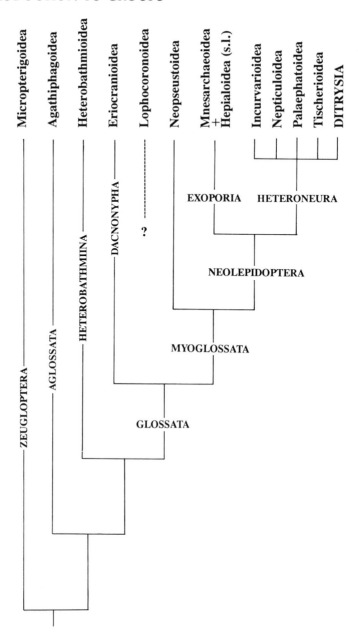

Fig. 185. *Cladogram of main lineages of Lepidoptera.*

which group would be sufficiently small to leave *incertae sedis*. A further possibility would be to leave such a group in its current superfamily until it can be reassigned to another and simply note its anomalous position.

Finally, a more radical solution might be to drop the use of superfamily names in lepidopteran classification leaving a list of families. Monophyletic assemblages could then be indicated as 'family groups', or possibly treated as subfamilies of the oldest name.

In the present work I generally follow (rather reluctantly) the current broad consensus, for at least it has the advantage of striving for a more natural classification of the order.

The phylogenetic relationships and names adopted above the level of superfamily (Kristensen, 1984b) are presented in Fig. 185. It is easier to understand the classification of the Lepidoptera above the level of superfamily from a branching diagram than from an indented list, partly because not all the monophyletic taxa have received names but also because several names are coincident in what they include (e.g., Zeugloptera and Micropterigoidea; Aglossata and Agathiphagoidea), their definitions being equivalent. Reference is made to various of these names in Chapters 9-12.

Historical perspective

Most of the literature quoted in the systematic treatments in Part III is fairly recent. This emphasis is more apparent than real, and implies no disregard of the work of earlier lepidopterists. Instead, it reflects the purpose of this book, which is to provide a summary of lepidopteran classification rather than a taxonomic history of each family. Nevertheless, current treatments are but modifications and extensions of past work, and our legacy to the past is substantial. In particular, the various systems adopted in recent years were built, and therefore rest, on those proposed in the 18th and 19th centuries (e.g., see Emmet, 1991; Scoble, 1991).

Linnaeus. The most prominent landmark in lepidopteran classification is the 10th edition of Linnaeus' *Systema Naturae* (Linnaeus, 1758), which represents the beginning of zoological nomenclature as accepted by the *International Code of Zoological Nomenclature*. Nevertheless, writings on lepidopteran natural history date back to Aristotle (384-322 BC), and significant contributions were made by many other, but later, pre-Linnaean authors - Albertus Magnus, Marcel Malpighius, Jan Swammerdam, Reaumur, and De Geer to mention but a few.

But it was Linnaeus who provided a classification, comprehensive for the time, in which we see the germ of subsequent systems. Linnaeus (1758) recognized three main divisions of the Lepidoptera: *Papilio, Sphinx,* and *Phalaena; Phalaena* was divided into seven subgroups. Today, these names form the basis for nine superfamily names. Following the Linnaean sequence these superfamilies are: Papilionoidea, Sphingoidea, Bombycoidea, Noctuoidea, Geometroidea, Tortricoidea, Pyraloidea, Tineoidea, and Alucitoidea.

Linnaeus based his classification on features of the antennae and the wings, particularly the position of the wings at rest, and on whether the insects flew by day or by night. He also took account of the mouthparts. Thus, in his descriptions, Linnaeus noted that the antennae of butterflies were clubbed, and that some members of his 'Bombyces' lacked a tongue. In fact, a considerable variety of features, referring to both larvae and adults, is to be found in his species descriptions.

After Linnaeus. Lepidopteran classification developed from Linnaeus' foundation in several ways. Numerous additional species were described, Linnaeus' divisions and subdivisions were further divided and formal categorical ranks were established for them, and many supraspecific taxa were added to those introduced by Linnaeus.

Just under 20 years after the publication of the 10th edition of *Systema Naturae*, Denis & Schiffermüller (1775) produced a work on the Lepidoptera of the Vienna area. Their system was based on Linnaeus' classification, but they added further observations on structure and examined many additional species. These authors subdivided the Linnaean groups and based the name of many of the subdivisions on both larva (e.g., Larvae Punctatae) and adult (e.g., Ph[alaenae] Geometrae Unicolores). The Danish entomologist Fabricius also described many new species and named some of the Linnaean subgroups (Fabricius, 1775), while Latreille (1796) added further supraspecific groups, most of which represent genera in the modern sense.

By adding a taxonomic level between the Linnaean 'genus' and the species, Schrank (1802) appears to have provided a category representing the modern lepidopteran genus (Emmet, 1991). Many modern genera were established by Jacob Hübner, a designer and artist, who described and illustrated many of the genera (his *Stirpes*) recognised as such today (Hemming, 1937). The lepidopteran genus was firmly established by Ochsenheimer (1807-1835), in a series of works completed by Treitsche, in his synthesis of the lepidopteran fauna of Europe.

Probably the most influential of all nineteenth century works on the higher classification of the Lepidoptera, was the six volume study of G.A.W. Herrich-Schäffer published at irregular intervals between 1843 and 1856 (Herrich-Schäffer, 1843-1856). Although, as for earlier works, it was based on the European fauna, many of the higher lepidopteran taxa recognized today were established therein. To a significant extent, these taxa were established on wing venation. Herrich-Schäffer provided numerous morphological illustrations, which are remarkable in their detail and execution.

A further dimension to the classification was added by the incorporation of information on Lepidoptera from parts of the world beyond Europe. Apart from describing numerous new genera and species, Edward Meyrick (particularly for microlepidopterans) and Sir George Hampson (particularly for macrolepidopterans) also contributed to the higher classification. Meyrick (1895) proposed a classification of the order as a whole in which he divided the Lepidoptera into ten 'phyla'. Meyrick, like Herrich-Schäffer, relied to a significant extent on wing venation because he considered the pattern to be of low adaptive ('physiological') value and so likely to be altered little by external factors. Hampson's contribution to lepidopteran higher classification focused mainly on Noctuoidea (Kitching, 1984), Pyraloidea, and Thyridoidea.

Since the mid nineteenth century many other systems have caused the earlier classifications to be modified and expanded. Some writers emphasized other character sets (e.g., from larvae (Fracker, 1915), or pupae (Mosher, 1916)). Probably the major contribution to the higher classification of the order concerns its division above the superfamily. Landmark works are those of Börner (1925, 1939) who proposed a fundamental division of the Lepidoptera into Monotrysia and Ditrysia on the basis of the structure of the female genitalia. The Monotrysia (*sensu* Börner), which included only 5 percent of the Lepidoptera, are not monophyletic. But Börner's recognition of the Ditrysia as a natural group was an important step in understanding the phylogenetic structure of the order, as was his appreciation of the systematic value of many morphological characters in the classification and diagnosis of lepidopteran families and superfamilies.

Our understanding about the way in which monotrysian moth families are related to each other has been advanced by the studies of N.P. Kristensen and E.S. Nielsen (summary in Kristensen, 1984b). Their work is concentrated mainly on adult morphology, although significant larval finds have also been made (e.g., Kristensen & Nielsen, 1983).

Primary divisions of the Lepidoptera. Several attempts were made to divide the Lepidoptera into fundamental groups, that is groups above the level of superfamily. These divisions include Nocturni (moths) and Diurni (butterflies), Heterocera (moths) and Rhopalocera (butterflies), Microlepidoptera and Macrolepidoptera, Homoneura and Heteroneura, Jugatae and Frenatae, Stemmatoncopoda and Harmoncopoda, and Monotrysia and Ditrysia. These divisions are unnatural dichotomies because the first name in each of these pairs does not represent a monophyletic group (see, for example, Hennig, 1966). The second name in each pair probably does represent a monophyletic group, although there is very considerable doubt as to the monophyly of the

Macrolepidoptera. So although it is reasonable to refer, collectively, to those Lepidoptera with similar venation in forewing and hindwing as homoneurous, the noun Homoneura should be avoided because it does not refer to a monophyletic group. The Rhopalocera (a term introduced by Duméril, 1823) include the butterflies; butterflies were called Diurni by the very early writers. Provided the Hedylidae (see Chapter 12) are now included here (although they do not have clubbed antennae, nor are they primarily diurnal), the Rhopalocera are monophyletic; but the Heterocera are not.

The Frenatae (Comstock, 1893) are equivalent to the Heteroneura (Tillyard, 1918), and the Jugatae to the Homoneura. Frenatae include those groups in which the jugal lobe of the forewing is not extended into a jugum but where coupling occurs by means of a frenulum and retinaculum (see Chapter 3) (a system often secondarily reduced or lost), and those groups where the venation is heteroneurous. Frenatae and Heteroneura probably represent monophyletic groups. However, Jugatae and Homoneura are not monophyletic. A further problem with the Jugatae-Frenatae division is that the presence or absence of a functional jugum is sometimes difficult to assess, and a jugal lobe certainly occurs outside the Jugatae (*sensu* Comstock).

Harmoncopoda is approximately equivalent to Macrolepidoptera, and perhaps also represents a monophyletic group. Harmoncopoda include those Lepidoptera in which the crochets of the larval prolegs are arranged in a mesoseries rather than in a circle. Those Lepidoptera with crochets occurring in a circle were termed Stemmatoncopoda by Karsch (1898), but the grouping is not monophyletic.

The name Ditrysia was introduced by Börner (1939), the word referring to the presence of two separate genital apertures in the female genitalia, one for mating and one for oviposition (see Chapter 4). The division is well established in lepidopteran classification, the group including over 95 percent of the order. Although actually ditrysian, in Exoporia the two apertures occur on the same segment and there is no internal ductus seminalis so the arrangement is not homologous with that of the Ditrysia. The remaining moths were termed Monotrysia by Börner (1939), but that grouping is not monophyletic, and the term has fallen into disuse or (Davis, 1986) has been restricted.

9

PRIMITIVE LEPIDOPTERA

Although primitive Lepidoptera comprise under 1 percent of all species of the order, among them are included those groups with the greatest range of morphological variation. For example, among the groups considered in this chapter are those with functional mandibles in the adult and those with a sucking proboscis. Of those with a sucking proboscis, one group lacks intrinsic muscles in that organ, while such muscles occur in the proboscis of members of the other. The female genitalia may be either monotrysian or exoporian in primitive Lepidoptera. However, in all families of primitive Lepidoptera, the venation is homoneurous, and the wing coupling jugal.

Detailed studies carried out on primitive Lepidoptera over the last two decades (Kristensen, 1984b and references therein) has led to a much improved understanding of the lepidopteran ground plan. It has also enabled a proposal for the basic branching pattern of the lepidopteran cladogram (Fig. 185) to be constructed, although this arrangement is by no means definitive (e.g., Kristensen, 1990b). From this diagram, it can be seen that the primitive Lepidoptera do not form a monophyletic group and they are not treated as a taxonomic entity. However, they share many primitive features.

With the exception of Hepialidae, primitive moths are small or very small. Their larvae are correspondingly small, and usually live within plant tissue, particularly as leaf miners - a habit adopted at a very early stage in lepidopteran diversification.

MICROPTERIGOIDEA

Micropterigidae

A single family, Micropterigidae, is included in the superfamily. Micropterigidae are probably the most primitive group of Lepidoptera although Agathiphagidae (Aglossata) may occupy that position (N.P. Kristensen, pers. comm.).

One hundred and eighteen species of these very small moths (the wingspan is about 10 mm) have now been described, but many others represented in collections remain unnamed (Kristensen, 1984b). Micropterigidae have been collected from all major zoogeographical regions. The genus *Micropterix*, which occurs in the Palaearctic and Oriental regions, contains the most species. The other genera have been informally treated as the *Sabatinca*-group by Kristensen & Nielsen (1979), and they are found variously in Asia, Australia, New Zealand, southern Africa and South America.

Adult (Fig. 186)

The head is rough-scaled. Several sulci are present including an epistomal sulcus joining the anterior tentorial pits. External ocelli are prominent (except in the fossil *Parasabatinca*), and chaetosemata are present. The intercalary sclerite is well developed in some species (Kristensen, 1984b). The antennae range from moniliform to filiform; sensilla auricillica are absent. Functional mandibles occur each with a double condylar articulation with the head capsule. Paraglossae, premental lobes of the labium, are present (absent from other Lepidoptera). The maxillary palpi are flexed between segments 1 and 2 and between segments 3 and 4, an arrangement representing a ground plan character of the order (Kristensen, 1984b). The organ of vom Rath is present.

The epiphysis is present or absent, and the tibial spur formula is 0-0-4. The wing venation is homoneurous and more complete than in most other Lepidoptera (Fig. 58). In the forewing, veins Sc and R_1 are sometimes forked. Basally, the anal veins of the forewing are usually forked to form a double loop. Vein M_4 is absent as a free vein. The wing coupling is jugate. A hindwing-body lock occurs (Kristensen, 1984a) (see Chapter 3), in addition to the forewing-metathoracic wing locking device.

Abdominal sternum A2 is present but small and of variable shape (Kristensen & Nielsen, 1980; Kristensen, 1984a). Paired, unsclerotized fenestrae are present on sternum A1. The ventral dilator muscle of spiracle A1 is present, a situation considered primitive with respect to the derived condition in Trichoptera where the muscle is absent (lost). A pair of glands is present on sternum A5; in many species the opening of each gland is found at the end of a structure protruding from the sternum. Considerable variation exists in the structure of the male genitalia. Sternum A8 is more or less desclerotized, an arrangement probably representing a derived character for the family as a whole. Sometimes segment A9 (i.e., the tegumen dorsally and the vinculum ventrally) comprises a completely fused ring. Anterior and posterior apophyses are absent from the female genitalia, and the ovipositor is not retractable (Kristensen & Nielsen, 1983).

Immature stages

The eggs of *Micropterix* range in size from 0.38 × 0.24 to 0.53 × 0.43 mm depending on species (Heath, 1962). They are ovoid, and covered with numerous clubbed rods that are formed within one hour after oviposition (Chauvin & Chauvin, 1980).

The larva (Lorenz, 1961; Yasuda, 1962; Davis *in* Stehr, 1987) is small and recognizable from its rugose cuticle and the presence of eight rows of scale-like setae, which are paired on four longitudinal ridges on each segment. The head is retractable and prognathous. There are five ocelli and a well-developed antenna comprising three segments and a terminal seta. There is no spinneret. Both the thoracic legs and the abdominal prolegs end in claws. The coxa, trochanter, and femur of the thoracic legs are fused. A unique medial seta occurs on the head, but most setae are probably homologous with those in other Lepidoptera (Davis *in* Stehr, 1987). Unique to the larvae of Micropterigidae is the honeycomb-like structure of the dorso and dorsolateral trunk exocuticle (Kristensen, 1990a). A liquid-filled space separates the exocuticle from the pro/endocuticle.

The pupa (Mosher, 1916) is decticous, but although the mandibles are functional, and large compared with most Lepidoptera, they are not hypertrophied. The first seven abdominal segments are movable while the remainder bear no distinct sutures and are not independently movable. The appendages are free, not fused to the body. Pupation occurs in a tough, silken cocoon.

Biology

In *Micropterix* and *Epimartyria*, eggs may be laid singly or in groups of between 2 and 45 (Davis *in* Stehr, 1987; Tuskes & Smith, 1984). In *Neomicropterix nipponensis* they are covered with gelatinous material, which swells when in contact with water, and presumably functions in the retention of water (Kobayashi & Ando, 1981). This material, which may occur in the eggs of all species, resembles a similar layer in the eggs of many Trichoptera.

The food substrate of the larvae varies between species. Heath (1976) suggested that most species feed on minute particles of leaf-litter or possibly fungal hyphae, a view based on the observation that larvae are found in the soil. However, *Micropterix aruncella* and *M. calthella* required, at least in the laboratory, fresh, photosynthetic angiosperm tissue for their development (Carter & Dugdale, 1982). Larvae of an unidentified species of *Micropterix* were collected from tussocks of Cocksfoot grass (*Dactylis glomerata*) (Luff, 1964) where they were thought to feed on dead leaves. *M. aureatella* may feed on mycorrhizal fungi among the lower litter of oak, beech, and bilberry communities (Carter & Dugdale, 1982). Larvae of *Neomicropterix* from Japan, and of *Sabatinca* and its relatives from New Zealand and New Caledonia, appear to feed on bryophyte swards (Yasuda, 1962; Dugdale, 1975). Those of the nearctic species *Epimartyria pardella* have actually been observed eating the underside of liverwort thalli (Tuskes & Smith, 1984).

Adults are typically diurnal and visit flowers to feed on pollen or, particularly in *Sabatinca*, the spores of ferns (Gibbs *in* Kristensen, 1984b). However, females of at least some species come to light at night. In Australia, New Caledonia, and New Zealand Micropterigidae are usually associated with rainforest, but in South Africa *Agrionympha capensis* has been swept in the early morning from the low Macchia ('Fynbos') vegetation (Acocks, 1975) of the Cape even at the hottest times of year (G.W. Gibbs pers. comm.). The South American species *Hypomartyria micropteroides* and *Squamicornia aequatoriella* are found in forests (Kristensen & Nielsen, 1982). In Britain, *Micropterix* species are found on a variety of flowers: oak, sycamore, hawthorn, sedges, and many herbaceous plants have been recorded. They generally occur in damp areas, but *M. aruncella* is found in drier habitats such as hillsides and downland (Heath, 1976).

A species of *Sabatinca* from New Caledonia pollinates the flowers of *Zygogynum*, a member of the primitive, vessel-less family Winteraceae (Magnoliales) (Thien *et al.*, 1985). The moths seem to be attracted to the plant by scent. The flower-sites are used by the moths as locations for mating, as well as for feeding. The pollen tetrads of the male plant are embedded in an oily, lipid matrix, which helps them to stick to the insects. Both *Sabatinca* and *Zygogynum* belong to groups with fossil records that extend back at least as far as the early Cretaceous, suggesting that the association of these moths and the plants is ancient.

Fossil history

The earliest fossil Micropterigidae, *Parasabatinca aftimacrai*, was described from three specimens from Lebanese amber dating from at least 100 Myr BP (Lower Cretaceous) (Whalley, 1977; 1978; 1986). The moths are mandibulate, lack a proboscis, and have maxillary and labial palpi similar to those of *Sabatinca* and *Micropterix*. Epiphyses are present. However, unlike other Micropterigidae ocelli are not visible and a branch of vein Sc of the forewing is apparently absent.

Classification

The classification of the family is under review (J.S. Dugdale, G.W. Gibbs, & N.P. Kristensen). Currently, there are 10 genera (Kristensen & Nielsen, 1979). In the present classification (summary by Kristensen, 1984b), the genus *Sabatinca* is generally accepted as being unnatural. The closely related genera *Paramartyria*, *Palaeomicroides* and *Neomicropteryx* are found in eastern Asia, and the nearctic genus *Epimartyria* may also belong to this complex. Three species of *Agrionympha* are found in South Africa (Whalley, 1978), and two species from South America were described and placed in the *Sabatinca* group (Kristensen & Nielsen, 1982). Preliminary support for the monophyly of the *Sabatinca* group was given by Kristensen & Nielsen (1982). The best character appears to be the membranous condition of sternum A8 in males. The presence of strongly moniliform, rather than less distinctly moniliform, antennae suggests that *Parasabatinca* belongs to the *Sabatinca* group rather than to *Micropterix* (Kristensen & Nielsen, 1979). *Undopterix sukatshevae* from the Lower Cretaceous was originally assigned to the Micropterigidae (see Kristensen & Nielsen, 1979; Whalley, 1986), but Kristensen & Nielsen pointed out that no derived micropterigid characters are apparent. Nevertheless, vein M_4 is absent from the wings (Whalley, 1986) as in all Lepidoptera except the primitive aglossatan genus *Agathiphaga* (Agathiphagidae). This absence, together with the general pattern of venation in *Undopterix*, is at least not incompatible with its assignment to Micropterigidae.

Some doubt exists over the monophyly of Micropterigidae (Sonnenschein & Häuser, 1990). In most Lepidoptera both apyrene (anucleate) and eupyrene (nucleate) spermatozoa are produced, a situation that is possibly specialized for the order. However, although both apyrene and eupyrene spermatozoa have been observed in *Epimartyria*, only eupyrene spermatozoa have been found in *Micropterix*. Therefore, either the Micropterigidae are not monophyletic, or dichotomous spermatogenesis has arisen twice in Lepidoptera, or it has been lost secondarily in *Micropterix*.

Phylogenetic relationships

Micropterigidae were earlier treated as a separate order of insects on the grounds that they were thought to share specialized characters with Trichoptera (Chapman, 1917; Hinton, 1946a). However, 26 probable specializations shared between this family and all other Lepidoptera were listed by Kristensen (1984b) so the evidence that Micropterigidae are true Lepidoptera seems overwhelming.

Heterobathmia, originally assigned to Micropterigidae (Kristensen & Nielsen, 1979), was later removed to Heterobathmioidea (Kristensen & Nielsen, 1983).

AGATHIPHAGOIDEA

Agathiphagidae

A single family, Agathiphagidae, with two species in one genus is included in the superfamily. Of the two species (Dumbleton, 1952) one, *Agathiphaga vitiensis* was described from Fiji, and the other, *A. queenslandensis* from Queensland, Australia. *A. vitiensis* has since been recorded also from Vanuatu, the Solomon Islands, and New Caledonia (summary in Kristensen, 1984d). The moths are small to medium-sized microlepidopterans and superficially resemble caddis-flies. They have greyish brown forewings, and at rest the wings form a steep roof over the body. Both species were reared from larvae from seeds (cones) of Kauri pines (*Agathis*: Araucariaceae).

The primitive nature of *Agathiphaga* is apparent from the homoneurous venation and the presence of mandibles with a double articulation in the adult, and the decticous condition of the pupa, which has hypertrophied mandibles (Dumbleton, 1952). *Agathiphaga* was originally assigned to Micropterigidae, but similarities between the two taxa are actually primitive features thus giving no support for their phylogenetic association. A study of the larva led Hinton (1958) to suggest that *Agathiphaga* belonged to the Dacnonypha (see under Eriocranioidea, below) because it had none of the distinguishing features of the Micropterigidae larva. *Agathiphaga* was raised to family rank by Kristensen (1967), who retained the genus in Dacnonypha. Speidel (1977) erected the suborder Aglossata for *Agathiphaga* distinguishing it from both Zeugloptera (Micropterigidae) and Glossata (all other Lepidoptera, including Dacnonypha). Doubt remains as to whether the Aglossata are the sister-group of all other Lepidoptera or of the Heterobathmiina + Glossata (Kristensen, 1984b, c, d).

Adult (Fig. 187)

External ocelli and chaetosemata are absent. Lobe-like mandibles are present, each mandible having a double articulation with the head-capsule. Apart from a rudimentary tooth, the mandibular surfaces are relatively smooth, and the structures, although large, are probably not functional in the adult. As in Micropterigidae, Agathiphagidae have a lacinia on each maxilla, and although galeae are present they do not form a haustellum. The maxillary palpi are 5-segmented and folded. The labrum is trilobed and not reduced, but there is no complex of spines on the epipharynx although scattered setae do occur. However, as in Micropterigidae and Heterobathmiidae, a triturating basket is present on the hypopharynx. The labial palpi are 3-segmented but have a palp-like prementum (Kristensen, 1967). Paraglossae are absent.

An epiphysis and a single short spur are present on the foreleg. Both midleg and hindleg bear two pairs of spurs on the tibia. The venation is homoneurous. In the forewing, vein Sc is forked to form Sc_1 and Sc_2. Vein M_4 is present (Common, 1970, 1973, Kristensen in Hennig, 1981) (absent from other Lepidoptera but present in Trichoptera). Three anal veins are represented; although all are separate at the base of the wing, they unite before reaching the margin. In the hindwing Sc is also forked. Wing coupling is jugate. A forewing-metathoracic locking-device is present. Striae project beyond the apical margin of each wing scale, and transverse ribs are absent (Common, 1973).

In the abdomen, the gland opening on sternum A5 is particularly long and twisted. Its form is probably unique to Agathiphagidae (Kristensen, 1984a). In the male genitalia (Kristensen, 1984c), segment A9 forms a complete ring and is not divided into a dorsal tegumen and a ventral vinculum. The valvae are long and curved. Two paramedian (superior) lobes comprise tergum A10. Between these lobes occurs a median lobe possibly forming part of tergum A11. The genital chamber is complex and distinctive with a spiny plate, a pair of presocii, and a pair of socii (terminology of Kristensen, 1984c) occurring in its roof. There is no sclerotized median plate. In the female, the ovipositor is extensible but not piercing. There are two pairs of anterior apophyses, one ventral the other dorsal, besides the single pair of posterior apophyses. (Dumbleton, 1952; Common, 1973; Kristensen, 1984b, c, d.)

Immature stages

The larva (Dumbleton, 1952; Kristensen, 1984d) is a yellowish, grub-like insect with the body much wider than the head. The head is prognathous, but appears hypognathous since the anterior segments of the body are longer dorsally than ventrally. No corneal lenses are present where stemmata are expected but two stemmatal vestiges occur (Kristensen, 1984d). There are neither adfrontal ridges nor distinct ecdysial lines. All primary head setae present in the lepidopteran groundplan are present; however, they vary from one side to the other in position, length and even presence or absence. A complete hypostomal bridge is present, found elsewhere in Lepidoptera only in Micropterigidae. The spinneret is absent. Thoracic legs and abdominal prolegs are absent.

The pupa is decticous and exarate, the mandibles being hypertrophied and bent.

Biology

The ovipositor is extensible rather than piercing, so the eggs are probably laid on the surface of the cones of the foodplant *Agathis*, possibly in crevices, and not injected into plant tissue. The larva hollows out seeds of *Agathis* during feeding, and pupates within the chamber so formed. A cell is formed in which the chamber is coated with a mixture of frass and a resinous material, pupation occurring within the cone. Emergence is

assisted by the action of the hypertrophied pupal mandibles which, although probably inhibited from moving to their full extent, chisel a hole in the cell. Emergence holes in the seeds of *Agathis robusta* are about 2.5 mm in diameter. Most of the pupa emerges from the cell before the pupal skin splits at eclosion, an action probably assisted by the action of tarsal claws. Apparently the pupa may remain in the cone for up to three years. Since infestation levels can be high, the larva sometimes causes considerable damage to kauri pines, at least in Queensland.
(Dumbleton, 1952; Common, 1973; Kristensen, 1984d; Robinson & Tuck, unpublished.)

Phylogenetic relationships

The relationships between Agathiphagidae (Aglossata), Heterobathmiidae (Heterobathmiina), and the Glossata are uncertain due to conflicting sets of characters. The preferred view is that Heterobathmiina is the sister group of Glossata, but a sister group relationship between Aglossata and Heterobathmiina cannot be ruled out (Kristensen & Nielsen, 1983).

Larvae of *Agathiphaga* differ from those of Glossata in lacking adfrontal sutures, ecdysial lines, and a spinneret, and by the presence of a complete hypostomal bridge (Kristensen, 1984d).

HETEROBATHMIOIDEA

Heterobathmiidae

A single small genus, *Heterobathmia*, in a single family, Heterobathmiidae, from temperate South America is included in the superfamily. Initially, two species were described in a subfamily of their own within the Micropterigidae (Kristensen & Nielsen, 1979). Later (Kristensen & Nielsen, 1983), after the larva was discovered, *Heterobathmia* was elevated to the rank of suborder (Heterobathmiina). About 10 species (mostly undescribed) have been discovered. The moths are small (wingspan about 10 mm), and the ground colour of the forewings is a metallic lead-grey with purplish, reddish, and bronzy reflections. Many white scales are present. Some species resemble Eriocraniidae (Kristensen, 1984b).

The larva of *Heterobathmia* mines the leaves of *Nothofagus* the Southern Beech, and differs fundamentally from the larva of Micropterigidae in appearance, structure, and habits.

The revised position of *Heterobathmia* means that many characters originally considered to be shared derived features of that genus and Zeugloptera are now treated as homoplasious. The shared derived characters of *Heterobathmia* and Glossata outnumber those of *Heterobathmia* and Zeugloptera.

Adult (Fig. 188)

The compound eyes are relatively small and bear a few interfacetal hairs. External ocelli are present. Above and slightly medial of each ocellus occurs a chaetosema. A further pair of elongate chaetosemata are situated on the dorsum of the head. Postinterocellar and epicranial sulci are prominent. Scales are sparse on the head, and not evenly distributed. Each antennal base lies very low on the head between the compound eyes. The flagellum is filiform, and the intercalary sclerite in the membrane between the scape and the pedicel is weak. The mouthparts are basically similar to those of Micropterigidae. A complex epipharyngeal armature exists, and there is a broad but shallow depression on the apical part of the epipharynx around which are found microtrichia and a pair of prominent, asymmetrical sclerites. From the left hand spoon-shaped sclerite and the adjacent area arises a brush of long, backwardly directed microtrichia. The base of this brush is effectively situated in the middle of the epipharynx. The hypopharynx bears a triturating basket of spines in a depression. Mandibles are well developed and bear both incisor-like, and molariform teeth. The maxillary palpi are 5-segmented, and small laciniae and galeae are present. The labial palpi are 3-segmented.

An epiphysis is present on the foreleg, and the tibial spur formula is 0-0-4, but there are various spines on all legs. The wings are narrow, elongate, and covered with a single layer of scales. The scales are lamellar and of the primitive type (imperforate and lacking a lumen). On the obverse surface of the scales on the upperside of the forewing, the areas between the longitudinal ridges bear circular to elliptical plates. Several cross-veins are present, but Sc-R is absent. Vein Sc is not forked nor is R_1. Vein CuP is present in both wings. In the forewing the base of vein 1A is forked, a character apparently unique within Amphiesmenoptera. The jugal lobe is folded under the forewing. Wing coupling is jugate, and the forewing-metathoracic locking device is present.

Abdominal tergum A1 is sclerotized only at the margins, and sternum A1 is considerably reduced. Sternum A2 bears a pair of unsclerotized fenestrae or windows. Weakly sclerotized, but smooth plates (probably 'tuberculate plates') occur in the pleural membranes. Glands on sternum A5 are present as conical, scaly structures with a subapical orifice and an associated glandular reservoir. In the male genitalia the tegumen and vinculum (i.e., the tergum and sternum of segment A9) are not fused, so a complete ring does not occur. Each corner of the tegumen is produced into a pair of characteristic processes, one long and anteriolateral, the other arising from the same point and curved medially. Anterior and posterior apophyses of the female genitalia are absent. The ovipositor cannot be retracted into the abdomen, and the genital tract opens separately from the rectum on segment A9.
(Kristensen & Nielsen, 1979.)

Immature stages

The egg is laid singly, in a jelly-like secretion on the underside of a leaf of *Nothofagus*.

The larva is a leaf miner on *Nothofagus*. It bears three pairs of thoracic legs. There are no abdominal prolegs with crochets, although thickenings occur dorsally on segments A1-8 and ventrally on segments A2-8. There are seven stemmata, and the trochanter is subdivided (both traits as in Trichoptera).

The pupa is decticous, with much enlarged mandibles each of which is bent distinctly near its base.
(Kristensen & Nielsen, 1983.)

Biology

Heterobathmiidae occur in temperate South America in the northern part of the Patagonian region of Argentina and in southern Central Chile. The adults fly during late winter and early spring. As in Micropterigidae, they are thought to feed on pollen since they are mandibulate and have been observed visiting the flowers of the larval foodplant *Nothofagus*. Larvae mine the leaves of *Nothofagus* during the spring and spend the rest of the year in an oval, silken cocoon in the soil. There are probably three instars. The mine commences as a gallery and soon widens into a blotch. The hypertrophied mandibles are used to free the pharate adult from the strong silken cocoon, which is buried in the soil from 8-15 cm below the surface.
(Kristensen & Nielsen, 1983.)

Phylogenetic relationships

The phylogenetic position of Heterobathmiidae remains unresolved, but the family may represent the sister group of the Glossata (Kristensen & Nielsen, 1983). The attributes shared with Micropterigidae, and which originally led to the placement of *Heterobathmia* within that family, involve mostly losses or reductions. Examples are the absence of apophyses in the female genitalia, the shortening of the labial palpi, and the absence of mesotibial spurs. The presence of a similar triturating basket on the hypopharynx of both Micropterigidae and Heterobathmiidae may be a retained primitive character. *Heterobathmia* shares as many putative specialized characters with Aglossata as with Glossata, but these traits are very possibly regressive (Kristensen & Nielsen, 1983). These authors listed six characters considered to be derived at least in the groundplan of Heterobathmiidae plus Glossata. Sensilla auricillica are present on the antennae (these somewhat leaf-like sensilla occur in *Eriocrania* (Eriocraniidae) (Davis, 1978)); crossvein Sc-R is lost; the pupal mandibles bear apical teeth; a Y-shaped adfrontal ridge occurs on the front of the head of the larva; the hypostomal bridge on the head of the larva is incomplete, the hypostoma (that part of the subgena lying behind the mandible) being unsclerotized medially (in Zeugloptera and Aglossata there is a complete sclerotized connection); and the hypostoma bears a pair of prominent ridges.

ERIOCRANIOIDEA

The suborder Dacnonypha originally included Eriocraniidae and Mnesarchaeidae (Hinton, 1946a), but the association of these two families was based on shared primitive characters. Moreover, Mnesarchaeidae have exoporian female genitalia (Dugdale, 1974), and Mnesarchaeoidea and Hepialoidea are now associated in Exoporia. Subsequently, Hinton added *Agathiphaga* to the Dacnonypha (see above); later, Acanthopteroctetidae, Lophocoronidae, and Neopseustidae were included. Lophocoronidae and Neopseustidae were later removed from Dacnonypha, and Acanthopteroctetidae are only dubiously associated. Recently, another family, Catapterigidae, has been described (Zagulajev & Sinev, 1988) from some specimens collected in the southeastern Crimea. The family is probably most closely related to Acanthopterocteti-dae (Zagulajev & Sinev, 1988; Kristensen, 1990b).

Eriocraniidae

Eriocraniidae are small moths confined to the Holarctic region. Within the Palaearctic they extend across the region into northern Japan within an area approximately matching the distribution of one of their main hostplants, *Betula pendula* (Betulaceae) (Davis, 1978). There are fewer than 30 species.

The moths are typically diurnal and often moderately iridescent with a wingspan of around 6 to 16 mm. The larvae are leaf-miners, mostly on Fagales, and pupation occurs in a cocoon in the soil. Eriocraniidae are the first moths to have developed a functional proboscis. They are the most primitive Glossata.

Adult (Fig. 189)

The compound eyes bear a few interfacetal microtrichia. External ocelli are prominent or absent, and chaetosemata are present on the vertex, and possibly elsewhere on the front of the head. The tentorium bears a pair of dorsal arms (structures present in primitive Trichoptera, Neopseustidae and Mnesarchaeidae and absent from Micropterigidae and Agathiphagidae (Kristensen, 1968b)). An intercalary sclerite occurs between the scape and the pedicel of the antenna. Scattered chemosensory sensilla auricillica are found on the antennae. The labrum is fairly large, and approximately pentagonal, but the epipharynx lacks the characteristic armature of Micropterigidae and Heterobathmiidae although some papillae are present (Kristensen, 1968b). The mandibles are distinct, but small compared with Micropterigidae and Heterobathmiidae. They are sclerotized proximally, but membranous distally. Mandibular 'teeth' are not well developed, nor are there discrete or prominent articulations with the head capsule although weak condyles occur in some species (Davis, 1978). Mandibles are not used for feeding but muscles occur in the pharate adult, which function to move the hypertrophied mandibles for escape of the adult from its cocoon (Kristensen, 1968b). Maxillary galeae form a short, but coilable, proboscis, and laciniae are membranous vestiges or absent. The food canal is formed by the union of the two grooved galeae, which unite to form a channel. These grooves are composed of a series of overlapping plates. Dorsal and ventral linking mechanisms unite the galeae, but this union is relatively weak for the proboscis was seen to split apart when one moth pressed the structure against the glass of a petri dish while imbibing water (Kristensen, 1968b). Each galea is operated by two sets of extrinsic muscles, one a galeal levator and the other a flexor (Kristensen, 1968b). Unlike the condition in more advanced moths (Myoglossata) there are no *intrinsic* galeal muscles in Eriocraniidae. The maxillary palpi are 5-segmented and folded, although the flexor muscle between segments 2 and 3 is lost so that each palpus appears almost 4-segmented (Kristensen, 1968b).

The epiphysis is present or absent, and the tibial spur formula is 0-1-4. The wing venation is homoneurous. Vein M is present within the cell of the forewing, and CuP is preserved in both wings. Wing coupling is jugate, and the forewing-metathoracic locking device is present. The broad scales on the dorsal surface of the forewing are not perforated. The dorsal and ventral laminae of these scales are fused to a considerable degree and lack a large lumen, and their terminal edge is finely serrated.

At the base of the abdomen, the first sternum is distinct but much reduced (Davis, 1978) and resembles that of *Micropterix* (Kristensen & Nielsen, 1980). The fourth sternum of most female Eriocraniidae bears a pair of membranous windows demarcating the presence of internal reservoirs. Reservoirs are present in males, but there are no corresponding windows. Each gland opens via a duct by way of a tubercle on sternum A5. In the male genitalia (Birket-Smith & Kristensen, 1974) gonopods are represented by proximal and distal sclerotizations, and the valvae are very small and generally not movable. The aedeagus is supported ventrally by a median plate thought to be composed of elements homologous with those forming that structure in Micropterigidae and Heterobathmiidae. The aedeagus is divided into dorsal and ventral components, the ejaculatory duct being found in the dorsal part. In the female genitalia, two pairs of elongate apophyses are present, and the ovipositor is retractile and long. The posterior apophyses are fused posteriorly to form a piercing organ with small teeth laterally. An extra pair of apophyses arise from near the apex of the posterior apophyses. A cloaca is present, and is situated ventrally on segment A10. The vagina is encircled by a distinctive vaginal sclerite, and colleterial glands are present. (Davis, 1978.)

Immature stages

The egg is inserted into leaf-tissue, but egg morphology does not appear to have been studied.

The larva has an approximately cylindrical head and body, unlike the typically flattened condition occurring in most leaf-miners. The head is prognathous and partially retracted into the body. Cranial setae are greatly reduced. The vertex is deeply divided and the adfrontal sclerite broad. A spinneret is present. Only a single stemma is present; although rudimentary it is innervated. The antenna is 3-segmented. Thoracic legs are absent but paired ventral swellings (calli) occur. Prolegs (and crochets) are absent from the abdomen.

The pupa is decticous and bears hypertrophied mandibles similar to those in Agathiphagidae. Its cuticle is thin with only the mandibles well sclerotized. An elongate frontal ridge extends as far as the labrum. All the appendages are free and the abdominal segments are capable of independent movement with the exception of segments A8-10, which are fused. Pupation occurs in a silken cocoon in the soil. (Busck & Böving, 1914; Davis, 1978.)

Biology

Typically, eggs are inserted into leaf buds by the long cutting ovipositor of the female (Davis, 1978). Prior to egg deposition the female makes an incision into the leaf tissue, a process resulting in the formation of a leaf scar. At one side of this scar a swelling (termed an egg pouch) develops around the egg.

The larvae (Davis, 1978; Davis & Faeth, 1986) are leaf miners forming blotch or linear-blotch mines usually on members of Betulaceae, but also Fagaceae. At least in the larva of *Eriocrania sangi*, hairs on the basal segment of the maxillary palpus may function to push food that may not have been ingested into the mouth. Other hairs on the hypopharynx may function to mop up sap (Jayewickreme, 1940).

The moths are typically diurnal and univoltine, emerging in late winter to early spring. They fly in sunshine and sometimes swarm (Heath, 1976). They have been observed to suck the sap of injured leaf buds (Kristensen, 1984b). Adults of the European species *Eriocrania semipurpurella* were shown to imbibe water (Common, 1973).

Acanthopteroctetidae

The family includes a single genus, *Acanthopteroctetes*, with about five species. It is restricted to western North America (Davis, 1978). Although in both Eriocraniidae and Acanthopteroctetidae the proboscis lacks intrinsic muscles, this absence is primitive (unless secondary loss has occurred). Hence it does not support a sister-group relationship between these two taxa. Among the differences between the families is the presence of 'hollow' rather than 'solid' wing scales in Acanthopteroctetidae. Acanthopteroctetidae and Eriocraniidae are currently associated mainly on the grounds of general similarity.

Adult (Fig. 190)

Acanthopteroctetidae are fairly small and slender-winged moths with a wingspan of about 11-16 mm. External ocelli are absent, mandibles vestigial, maxillary palpi 5-segmented, and the labial palpi 2-segmented and lacking vom Rath's organ.

The epiphysis is absent. As in Eriocraniidae, a single spur is present on each midleg rather than a pair of spurs. The wing scales are hollow. The forewing-metathoracic locking device is present, and wing coupling is jugate. Vein M_1 is stalked with $R_4 + R_5$. On the third abdominal pleuron of *A. unifasciata* a pair of filamentous, pedunculate glands occurs. Sterna A4 and A5 are unmodified unlike the condition in Eriocraniidae. In the female, the distinctive vaginal sclerite occurring in Eriocraniidae is absent, and the walls of the vagina and the ductus bursae are thickened and folded. The genitalia

are modified for piercing, and eggs are inserted into leaf tissue in *A. unifasciata* (Davis & Frack *in* Stehr, 1987).
(Davis, 1978.)

Immature stages

Larval information is restricted to *A. unifasciata* (Davis and Frack *in* Stehr, 1987). The body is subcylindrical. Six stemmata are present on the head. The ecdysial cleavage lines are close to the adfrontal suture, unlike their position in Eriocraniidae where they curve away markedly. Thoracic legs are present, and on the dorsum and the coxal membrane of each thoracic segment occurs a pair of what are probably ambulatory calli. Ventral abdominal prolegs are absent but an anal pair is present. Crochets are completely absent. A pair of sucker-like discs are also present on both dorsal and ventral sides of abdominal segments A1-7.

The pupa, which has hypertrophied mandibles, resembles that of Eriocraniidae.

Biology

The larva of *A. unifasciata* forms a full-depth mine on leaves of *Ceanothus* (Rhamnaceae). Pupation occurs in a cocoon in debris beneath the hostplant.

The sucker-like structures found on both dorsal and ventral sides of the larva may help the insect to maintain its position between the upper and lower walls of the mine (Davis & Frack *in* Stehr, 1987).

Phylogenetic relationships

There are no convincing specialized characters to unite Acanthopteroctetidae and Eriocraniidae. Certain morphological differences between the families exist, notably the presence of hollow wing scales in Acanthopteroctetidae and 'solid' scales in Eriocraniidae, and the phylogenetic position of Acanthopteroctetidae remains uncertain (Kristensen, 1990b). The closest relatives of Acanthopteroctetidae are probably the **Catapterigidae** (Zagulajev & Sinev, 1988; and see above).

LOPHOCORONOIDEA

Lophocoronidae

The family (Common, 1973) includes 3 species in a single genus *Lophocorona* from southern Australia. Originally assigned to Dacnonypha (Common, 1973), Lophocoronidae have now been removed (see Common, 1990) to Lophocoronina, a taxon of a rank equivalent to Dacnonypha. The moths are very small, pale, and variously patterned with fuscous markings. The immature stages are unknown.

Adult (Fig. 191)

The head is covered with long hairscales. Sulci are absent, external ocelli and chaetosemata are absent. The dorsal arm of the tentorium is present. Mandibles are present, but not functional. The maxillary palpi are 5-segmented, and the proboscis is about half the length of a palpus. Intrinsic muscles are not developed in the proboscis. A small invagination is present on the terminal segment of the 3-segmented labial palpus.

The epiphysis is lacking, but a small spine occurs posteriorly just below the midpoint. The tibial spur formula is 0-2-4. Spines are also present postmedially and distally on the midleg and hindleg. The venation is homoneurous. A vestige of the humeral vein is present, and vein Sc is forked. The jugum is small. Short frenular bristles are present on the forewing, and the wing scales are hollow (Kristensen, 1986). The forewing-metathoracic locking device is present.

The abdomen bears paired ventral glands opening on protuberances on sternum A5. The female genitalia are of the piercing variety (Kristensen, 1986) with a serrated ovipositor.
(Common, 1973; 1990.)

Immature stages

The immature stages are unknown, but since the adults lack mandibular muscles the pupa must be adecticous (Kristensen, 1990b).

Biology

The proboscis is thought to be functional and used to imbibe water as an adaptation to the relatively arid environments in which lophocoronids have been discovered (Common, 1973). Although larvae, as for other immature stages, are unknown, they are probably internal feeders in plant tissue (e.g., leaf miners and borers) given the piercing condition of the ovipositor.

Phylogenetic relationships

Although Lophocoronidae were tentatively placed in Dacnonypha, they lack shared derived characters with either Eriocraniidae or Acanthopteroctetidae. Their hollow wing-scales contrast with those of Eriocraniidae, which are solid. The family may represent the sister group of Exoporia + Heteroneura, of Exoporia alone, or even of Neopseustina + (Exoporia + Heteroneura) (Kristensen, 1990b).

NEOPSEUSTOIDEA

Neopseustidae

A single family, Neopseustidae, with 10 species is included. The moths are usually broad-winged and weakly scaled with a wingspan of between 14 to 27 mm (Davis, 1975c). The distribution of the family is disjunct with species occurring in the Assam region of northeastern India, in Burma, Taiwan and southwest China, and in the temperate forested regions of Argentina and Chile (Davis, 1975c; Davis & Nielsen, 1980; 1985).

Adult (Fig. 192)

On the head occur many sulci, well-developed subgenal processes, and a pair of large compound eyes (Figs 1-3). External ocelli are absent. Chaetosemata are present, one pair on the front of the head, the other pair on the vertex. An intercalary sclerite is generally present in the membrane between the scape and the pedicel (absent from *Archepiolus*). Dorsal arms of the tentorium are well-developed. Mandibles are quite large and roughly triangular in shape, but they are not functional; however, adductor and abductor muscles, which presumably operate the pupal mandibles, are present (Kristensen, 1984b). There are no mandibular 'teeth'. The galeae are united to form a short proboscis bearing a unique double food canal (see Chapter 2 and Fig. 16). The maxillary palpi are 5-segmented. There is no deep depression at the end of the apical segment of the 3-segmented labial palpus; the sensilla occurring within only a shallow depression. The proboscis of *Neopseustis meyricki* is short (about 3 mm) and bears a narrow food groove. The galeae are linked dorsally by broad, scale-like projections situated just above the food groove, and ventrally by sclerotized projections 'zip-scales') found below (ventral to) the food groove. Intrinsic proboscis muscles are present but relatively simple compared with those in higher Lepidoptera. They are composed of longitudinal fibres running along the length of each galeal lumen.

The prothoracic arm is free (i.e., separate from the pleuron), and arises from a bridge between the sternum and the pleuron. The furcal apophyses of the mesothorax are fused, a condition unlike many other families of primitive Lepidoptera (Davis, 1975c). The epiphysis is present but much reduced in *Apoplania* and *Synempora*. The tibial spur formula is 0-2-4. The wing scales of Neopseustidae are perforated. The upper surface of the forewing at least, seems to lack 'primitive type' scales. Microtrichia cover the surfaces of the wings densely. The wing venation is homoneurous with a relatively long discal cell and several cross veins. A jugum is present, although not elongated. An 'anal pocket', best developed in *Neopseustis*, is formed in the forewing by the folding of the base of vein 1A over the base of CuP. It may be used to help hold the wings together when they are folded in the resting position, but it does not appear to be a mechanism for coupling the wings in flight (Davis, 1975c).

In the abdomen of the female there usually occurs a pair of membranous windows, but there are no glandular openings on sternum A5. The genitalia of both males and females (Davis, 1975c) exhibit several unusual features. In the male of *Neopseustis* there is no sclerotized aedeagus, but such a structure occurs in other neopseustid genera. In *Neopseustis* the end of the ejaculatory duct is simply surrounded by the anellus, and there is no eversible vesica. A pair of digitate processes ('parameres' of Davis, 1975c) arises from the opening of the ejaculatory duct. These structures are absent from Neopseustidae with a sclerotized aedeagus, so they are probably involved in sperm transfer. Usually, prominent rather valva-like lobes arise from the tegumen, but the true valvae are very small. In the female, there is a pouch-like invagination of the intersegmental membrane of sterna A7 and A8. The apophyses are short and heavily sclerotized, and are fused apically to form a blunt ovipositor. Terminally, the ovipositor is serrated, and presumably functions as a rasping organ. The colleterial gland is undivided, unlike the situation in other Lepidoptera where these glands (where present) take the form of a pair of sacs. The oviduct, anus, spermatheca, and copulatory pore open into a common chamber (cloaca), although since the anus is divided from the spermathecal opening by a vestigial sternum A9, this cloaca is incomplete.

(Kristensen, 1968a; Davis, 1975c.)

Immature stages

Unknown. The presence of large mandibular lobes, together with their associated muscles, suggests that the pupa is decticous (Kristensen, 1984b).

Biology

The moths are primarily crepuscular or nocturnal, although *Synempera andesae* from Argentina is active by day and resembles day flying Adelidae (Davis & Nielsen, 1980). The ovipositor is probably rasping, so the eggs are likely to be inserted into plant tissue and the larva is presumably a borer or miner (Davis, 1975c). In *Synempora*, the wings are steeply roofed over the body at rest. In *Apoplania valdiviana* the wings are flattened at rest and the antennae are bent back and lie *under* the wings.

Phylogenetic relationships

Many primitive features are exhibited by Neopseustidae such as homoneurous venation, the absence of a frenulo-retinacular mechanism to couple the wings, a free profurcal arm, and long dorsolateral tentorial arms. But more specialized characters, particularly the presence of intrinsic proboscis muscles and perforated wing scales, are also present. Currently Neopseustidae are placed in Neopseustina, a taxon of rank equivalent to Dacnonypha and Lophocoronina.

EXOPORIA: MNESARCHAEOIDEA + HEPIALOIDEA

Most Exoporia belong to the superfamily Hepialoidea, the ghost and swift moths - a family of worldwide distribution. However, the New Zealand family Mnesarchaeidae currently is given separate superfamily rank (Mnesarchaeoidea), a status that seems questionable.

The name 'Exoporia' is based on the arrangement of the female genitalia. Although Exoporia may reasonably be described as ditrysian because there are two genital apertures (one for mating and the other for oviposition), the arrangement differs from that occurring in true Ditrysia by the absence of a free ductus seminalis (see Chapter 4). Instead, in Exoporia there is a seminal gutter between the ostium and the ovipore (Figs 97, 98) or a system by which the two pores are opposable within a genital pouch (Dugdale, 1974). Unlike the condition in monotrysians, where the common oviduct enters the copulatory chamber ventrally, in Exoporia (and Ditrysia) it enters dorsally. Furthermore, there are no colleterial glands in Exoporia. Besides the specialized condition of the female genitalia, the presence of three other derived characters of the adult lends additional support to the monophyly of the Exoporia (Kristensen 1978a; 1984b): the antennal scape articulates with the head by means of two condyles (there is an additional dorsal as well as the usual ventral condylous connection); many of the wing scales bear specialized, secondary and second-order ridges; and the male genitalia have a membranous rather than a sclerotized aedeagus lacking extrinsic protractor/retractor muscles (Nielsen & Kristensen, 1989). In addition, there are shared specializations in the developing embryo (Kobayashi & Gibbs, 1990).

The membranous condition of the aedeagus is presumably linked to the presence of a sclerotization, known as the trulleum (mesosome), in the diaphragma of the male genitalia. The presence of the trulleum is a distinctive exoporian feature, the details of which vary at all taxonomic levels in the group. In the primitive condition, the lateral components of the trulleum are free from the arms of the vinculum.

MNESARCHAEOIDEA

Mnesarchaeidae

A single family, Mnesarchaeidae, with one genus, *Mnesarchaea*, is included in Mnesarchaeoidea. The family is restricted to New Zealand. Originally assigned to Micropterigidae, and later to Eriocraniidae, the hepialid affinities of *Mnesarchaea* were first noted by Mutuura (1972) and confirmed, on the basis of the exoporian female genitalia, by Dugdale (1974). Six species have been described, but 14 species are now known (N.P. Kristensen pers. comm). On the basis of colour pattern and male genitalia they fall into three groups. The moths are small, with a wingspan of between 8 mm and 12 mm (Gibbs, 1979).

Adult (Fig. 193)

The frons and vertex are covered with narrow, upright scales, and the head capsule is devoid of sulci apart from a weak transverse, interantennal sulcus. The compound eyes are relatively large, and external ocelli and chaetosemata are absent. The antennae are filiform. Small mandibles are present in the form of unarticulated lobes. The maxillary palpi are 3-segmented, and the galeae are short but unite to form a functional proboscis. The labial palpi are 3-segmented, and the organ of vom Rath is absent. Wing coupling is jugate, and the forewing-metathoracic locking device is present. In the female genitalia (Dugdale, 1974) tergum A8 and sternum A8 enclose a chamber in which the anus and ovipore are situated in the roof and the ostium in the floor. It was suggested (Dugdale, 1974) that the opposable nature of the ostium and ovipore provides the means by which spermatozoa are transferred in the absence of an internal ductus seminalis.
(Kristensen, 1968a.)

Immature stages

The eggs are oval, and the chorion is smooth and leathery.
   The larva is elongate, with thoracic segments narrower than those of the abdomen. Puncture Fa lies lateral to F_1, and seta AFa is absent, features also characteristic of Hepialoidea. The spinneret is very long, and the final instar spins a cocoon covered with debris. There are four instars.
   The pupa is adecticous with the mandibles being small and not articulated. Backwardly pointed dorsal spines are arranged in two rows on A3-7, and a single hook occurs on each side of the anterior part of the wing. On A8 there is a large hook pointing anteriorly and assumed to function to hold the posterior part of the pupa in the cocoon prior to, and during, adult emergence. Appendages are mainly fused to each other but free from the abdomen, and there is considerable mobility of the abdominal segments.
(Gibbs, 1979.)

Biology

The moths are diurnal and most active on bright, humid days. Males appear to fly more frequently than females. At rest, the wings are folded roof-like over the body, and a conspicuous fringe along the posterior part of the moth is displayed. The antennae are held erect and motionless. When mnesarchaeid moths settle, the wings are rapidly and briefly separated then returned to their normal resting position. This cycle is carried out several times. Mnesarchaeidae are found on or near the floor of moist forests, or on damp roadside banks.
   Eggs are placed on, rather than cemented to, the substrate by the female; they are not scattered. Larvae occur among mosses and liverworts found on soil in moist places, rotting logs, or the trunks of trees. They feed on a range of substrates in the so-called periphyton layer. In captivity, silken galleries were spun by larvae.
   Pupation occurs in a broad, spindle-shaped cocoon.
(Gibbs, 1979.)

HEPIALOIDEA

Hepialoidea are composed of the ghost and swift moths, family Hepialidae (*sensu stricto*), plus 11 small genera (Nielsen & Scoble, 1986; Nielsen & Kristensen, 1989). The group is represented in all zoogeographical regions. Several of the 11 smaller genera have been given family rank (e.g., Prototheoridae, Neotheoridae etc.) while several have not. In the present work they are all treated as Hepialidae (*sensu lato*).

Hepialidae *(sensu lato)*

Anomoses ('Anomosetidae') is Australian, and *Neotheora* ('Neotheoridae') occurs in Brazil. *Palaeoses* from Australia, *Genustes* from Assam, Burma, and Thailand, *Ogygioses* from Taiwan, and *Osrhoes* from Colombia are associated in 'Palaeosetidae' even though it is highly dubious that the 'family' is monophyletic (Kristensen, 1978a). The South African genera *Prototheora* and *Metatheora* constitute the 'Prototheoridae'. A further genus, *Afrotheora*, from south and central Africa has also been described. *Fraus*, from Australia, *Antihepialus*, from southern Africa, and *Gazoryctra*, from the Holarctic region, represent the remaining genera falling outside Hepialidae *sensu stricto*.

The monophyly of the Hepialoidea is supported by two characters (Kristensen, 1978a). First, whereas in all non-hepialoid homoneurans at least the first three of the four branches of the radial sector vein (R_s) terminate distally at the costal edge of the wings rather than at the termen, in Hepialoidea vein R_4 ends at the termen. Second, the maxillary palpi and galeae are reduced, vestigial or even absent. Even where the proboscis is about as long as the head, it does not appear to be coilable. The sperm transport system of the female genitalia may be unique to Hepialoidea (Dugdale, 1974) because instead of taking the form of a pouch in which the two genital apertures may be opposed, it is composed of a sperm canal either enclosed by muscle tension or by fusion of the medial edges of the intergenital lobes.

The male genitalia are complex structures, and the homologies of the various components still in doubt. A reinterpretation of traditional views was presented by Nielsen & Kristensen (1989). The structure referred to as the tegumen, or teguminal plates, appears to be a derivative of A10 and perhaps also A11, but not tergum A9 as has been generally thought. The intermediate plate (Nielsen & Scoble, 1986) is a descriptive term but may not be of a different morphological origin from the pseudoteguminal plates. The structure appears not to be a derivative of the base of the valva (gonopod) as suggested previously. The trulleum, a sclerite hinged to the distal (anteriodorsal) margin of the juxta, is fused to the pseudoteguminal plates in Hepialidae *sensu stricto* but it is free in Hepialidae *sensu lato*. A long tubular, membranous phallus is present in hepialoids. But although membranous, it is not simply an eversible portion of the ejaculatory duct, a structure which itself runs to the apex of the phallus. Correlated with its membranous nature, the phallus lacks extrinsic retractor and protractor muscles.

Anomoses (Anomosetidae of Kristensen, 1978b)

During the course of its taxonomic history, *Anomoses* has been placed in a family of its own (Anomosetidae) or assigned to Micropterigidae, Eriocraniidae or Prototheoridae. Although the family is now known to be hepialoid its precise relationships are uncertain (Kristensen, 1978a,b).

Only one species, *A. hylecoetes*, a small, dull brown moth, is known. It occurs in rain forests of New South Wales and Queensland. The life history is unknown.

Adult

On the head, sulci are indistinct. The dorsal arms of the tentorium are well developed, and small, distinct, but non-functional mandibles exist. A very short proboscis is present, and each maxillary palpus is reduced to one minute segment. Although sensilla are present on the apex of each labial palpus, the third segment is not invaginated. An intercalary segment is present between the scape and the pedicel, and extends, within a pouch, into the lumen of the scape. There is a dorsal and a ventral, cranial condyle in the antennal socket.

The prothoracic furca forms a bridge between the sternal discrimen and the pleural suture. From this bridge, and near to the pleural suture, arises the free furcal arm. There is no epiphysis. Tibial spurs are present on midlegs and hindlegs - the formula is 0-2-4. Vein Sc is forked in the forewing but not in the hindwing. Vein CuP is present in both wings, and cross vein M_2-M_3 links M_2 and M_3. Aculei are present only at the base of the wings; they are not scattered over the whole surface. A jugum is present as are frenular bristles. There is an upper layer of normal-type scales and a lower layer of primitive-type scales on the wings. Secondary ridges are present on both kinds. A pouch (assumed to be a scent pouch), which is formed from an upfold in the wing, is present on the hindwing. The pouch houses a hair-pencil, composed of hollow scales, arising from the hind margin of the wing.

Strong lateral ridges are present on abdominal tergum A1. The ridges are weaker on tergum A3. Three tuberculate plates are present on the pleural membranes of segments A2-6. A trulleum and juxta are present in the male genitalia. In the female, the genital aperture leads into a chamber that continues laterally to form a pair of lateral chambers. Between the ovipore and the copulatory aperture occurs a medial groove presumably used for sperm transport. A spermathecal papilla, absent from other Hepialoidea, is present from which the spermathecal duct appears to arise. (Kristensen, 1978b.)

Neotheora (Neotheoridae of Kristensen, 1978a)

Neotheora is known from a single species (*Neotheora chiloides*) from southwestern Brazil. The species, of which only one specimen has been collected, is phylogenetically isolated within the Hepialoidea. The moth is of medium size, the length of the forewing being 18 mm. The forewings are brown and the hindwings darker greyish brown.

Adult

The head is densely covered with hair-like scales as are the relatively long, porrect labial palpi. External ocelli are absent and, unlike the condition in *Anomoses*, there are no special processes on the vertex. The dorsal arms of the tentorium are well-developed. The mandibles are very small, and the proboscis, although short, is about the length of the head capsule but not coiled spirally.

Although a metascutal-forewing locking device is present, aculei are not generally scattered over the wings. The jugal lobe is well-developed in the forewing, and long frenular hairs are present in the hindwing. There are no strong frenular bristles. As in *Anomoses* the wings are covered with an upper layer of 'normal-type' scales and a lower layer of 'primitive type' scales. The legs of the only known specimen are incomplete, but the epiphysis is present and the hindleg appears to bear two pairs of tibial spurs. Vein Sc of the forewing is forked. The cross vein linking M_2 and M_3 is clearly present in the forewing, and indicated in the hindwing. Vein CuP is present in both wings, but is weak and short.

Lateral ridges are present on abdominal tergum A2, but not tergum A3. There is a single tuberculate plate in the pleural membrane of segments A2-6. In the female genitalia there is a longitudinal, folded area of ventrum A9 and A10. The folds may represent a seminal gutter running from the copulatory pore to the ovipore. A spermathecal papilla has not been observed. (Kristensen, 1978a.)

Phylogenetic relationships

Although definitely hepialoid, *Neotheora* is difficult to place within the superfamily (Kristensen, 1978a). The presence of a cross vein linking M_2 and M_3, and the occurrence of a modified intercalary sclerite between the scape and the pedicel, suggests that Neotheoridae belong within a group formed by Hepialidae, 'Prototheoridae' and 'Anomosetidae', (i.e., excluding 'Palaeosetidae'). (Kristensen, 1978a.)

Palaeoses, Genustes, Ogygioses, and *Osrhoes* ('Palaeosetidae')

It is unlikely that these four genera will remain united in 'Palaeosetidae' for there is no evidence to support their monophyly (e.g., see Kristensen, 1978a; Common, 1990). *Palaeoses* occurs in Australia where there are two species (one unnamed) both of which have been collected in rainforest. In the male, abdominal segment A6 bears a pair of prominent, lateral protuberances (Common, 1990).

   Genustes and *Ogygioses* occur in the Oriental region, the former (with a single species) has been recorded from Assam and India, and the latter (with two species) from Taiwan (Issiki & Stringer, 1932). A hairpencil, concealed in a fold, occurs in the hindtibia of males of both genera.

   Ogygioses caliginosa adults have been observed resting singly or in small groups on leaves of grasses, ferns, mosses, rock surfaces, and even hanging from spiders threads. Males have been observed swarming in groups of 10-15 in the afternoons (Kuroko, 1990), and mating pairs have been found on leaves of grasses. Light intensity appears to be an important cue for swarming, the range lying between 2000 and 10 000 lux with the optimum being 3000-4500 lux. A temperature of 15°-17.5°C seems also to be a prerequisite for swarming.

   Osrhoes occurs in South America.

Prototheora-group (Prototheoridae of Meyrick, 1920)

The group includes a few species of small hepialoid moths found exclusively in South Africa. The Australian genus *Anomoses*, which was included later, has subsequently been shown not to belong to the *Prototheora*-group. Seven species, five of which were assigned to *Prototheora* and the other two to *Metatheora* were recognized by Janse (1942). Further species have been found by Davis (in preparation).

   The moths are fairly drab with greyish brown or fuscous forewings and greyish hindwings. The labial palpi are hairy and porrect, and the wings are folded roof-like while the moth is resting. The wings are narrower than in Hepialidae *sensu stricto*.

   Specimens occasionally come readily to light. Most have been caught in, or in the vicinity of, pockets of forest in Natal, and particularly in or near the coastal forests of the southern part of the Cape Province. The juvenile stages are unknown.

Adult

Compound eyes are large and external ocelli are absent. Mandibles are vestigial. The proboscis is short but distinct, and extends to about the distal end of the first segment of the labial palpus. The maxillary palpi are minute, and the labial palpi relatively long. The dorsal arms of the tentorium are long. A modified intercalary sclerite is present between the scape and pedicel of the antenna. The flagellar segments of the antenna are submoniliform.

Small lateral processes occur on the prothorax. The epiphysis is present, and the tibial spur formula is 0-2-4. Aculei are scattered over the surface of both forewings and hindwings, and the wings are covered with only 'normal type' scales; 'primitive type' scales appear to be absent. The cross vein linking M_2 and M_3 is present in both forewing and hindwing, and a rather weak 'anal loop' occurs in the forewing. Wing coupling is jugate.

The sclerotized aedeagus is lost, a trulleum in the male genitalia is present, and the female genitalia are exoporian. Davis (in preparation) has observed some genital features unique to Prototheoridae, the male bearing a 'conjugal pouch' and the female a 'conjugal process'.

(Meyrick, 1920; Janse, 1942; Kristensen, 1978a,b; 1984b.)

Phylogenetic relationships

Although these moths belong to Hepialoidea, their position within the superfamily is uncertain (Kristensen, 1978a).

Fraus, Gazoryctra, Antihepialus, and *Afrotheora*

The composition of the Hepialidae is not straightforward. Hepialidae used to include the Hepialidae *sensu stricto* (Nielsen & Scoble, 1986) plus the genera *Fraus, Gazoryctra, Antihepialus,* and *Afrotheora*. The grouping was defined by the complete absence of tibial spurs. However, hindtibial spurs are in fact present in *Gazoryctra* (Viette, 1949) and in some species of *Bipectilus* (Nielsen, 1988) so this character is not actually a shared derivation. The spurs are absent from *Fraus* and *Antihepialus*, but in these genera the genitalia and scales differ from Hepialidae *sensu stricto* (see below). Furthermore, in *Fraus* some species have a proboscis while in Hepialidae *sensu stricto* the proboscis is absent. At present, therefore, Hepialidae *sensu stricto* exclude *Fraus, Gazoryctra, Antihepialus,* and *Afrotheora*. The removal of the four genera from Hepialidae *sensu stricto* makes the latter monophyletic.

The four genera do not form a monophyletic group, but collectively they lack the two specialized characters of Hepialidae *sensu stricto* (see below) from which they are excluded. First, unlike the condition in Hepialidae *sensu stricto*, the trulleum of the male genitalia is never fused or hinged with the pseudoteguminal lobes, Second, the pupa, where known, fails to conform to the condition present in Hepialidae *sensu stricto* where there are two rows of teeth on segments A3-7 dorsally, and a ventral row on segment A7 (Nielsen & Kristensen, 1989).

Fraus, an Australian genus of 25 species, has been studied by Nielsen & Kristensen (1989). The genus is probably monophyletic. The moths are fairly small with a wingspan of 19-45 mm, have 2-segmented labial palpi, and a head capsule without tufts of scales below the antennal sockets. Some species are small, slender-bodied moths with narrow wings, while others are medium sized, stout-bodied, and fairly broad-winged. Three characters suggest their monophyly. First, the scales on the upper surface of the forewings are of a characteristic shape, often bearing a 'tooth' at each corner; second the antennae are typically bipectinate, an arrangement assumed to be independently derived; and third there is, uniquely, a patch of hair-scales at the base of the hindwing in females. *Fraus* occurs in various habitats (forests, woodland, and mallee) in Australia. Eggs are dropped singly or in groups of up to four. In two species they are catapulted for up to a few centimetres. The larva of *Fraus simulans*, from Tasmania, feeds on grasses and forms tunnels in pastures. The tunnels are silk-lined and increase in length during development extending eventually to about 15 cm. Pupation occurs in the tunnel, and the pupa is capable of considerable movement within.

Afrotheora includes eight species from Central and southern Africa (Nielsen & Scoble, 1986). Its members are characterized by an antennal scape with long, bristle-like setae, and by the trulleum being composed of two sclerotized rods separated by a membrane. The moths are small to medium sized hepialoids. The immature stages are unknown. *Afrotheora* may be related to another Afrotropical genus - *Antihepialus*.

Hepialidae (*sensu stricto*)

The Hepialidae is a cosmopolitan family with about 500 species. Although particularly numerous in the Indo-Australian area, they are well represented in the Afrotropics and the Neotropics. The moths are generally medium sized. However, some are small, and others very large. The wingspan ranges from 20 mm to 200 mm within the family, and wing shape is also quite variable, ranging from rather narrow to broadly triangular. The larvae frequently live underground and feed on roots or tubers of plants. The monophyly of the group is supported by two possible specializations, one in the male genitalia, and the other concerning the pattern of spining of the pupa.

Adult (Fig. 194)

The compound eyes are fairly large, external ocelli are absent, and there is a well-developed dorsal tentorial arm. The antennae are short, and usually fairly complex with a flagellum often bipectinate, tripectinate or dentate. The extreme reduction of the maxillary palpi and the galeae (there is no functional proboscis) is characteristic, although these features do not strictly define the group.

An epiphysis, sometimes reduced, is present on the foreleg. There are no tibial spurs, a condition shared by *Fraus* and *Afrotheora*, but not by other Exoporia. The wings are often a drab fuscous or brown, but sometimes green, variegated, or strikingly marked with silver patches. In both forewing and hindwing there is a humeral vein, and vein M is forked from near the base of the cell. In the forewing, vein Sc may be simple or forked, and from CuP a cross vein joins CuA and 1A. In the hindwing Sc is simple. There is a pronounced jugum on the forewing, but no frenulum or frenular bristles.

In the male genitalia the trulleum is fused laterally with the pseudoteguminal lobes (Nielsen & Scoble, 1986; Nielsen & Kristensen, 1989), unlike the condition in the non-hepialid (*sensu stricto*) hepialoids where it is free.

Immature stages

The egg is nearly spherical and without ornamentation.

The larva (Wagner *in* Stehr, 1987) is cylindrical, and almost prognathous. Stemmata form two vertical rows. On the head, puncture Ga is absent, and Fa is unusually lateral of seta F_1 (Hinton, 1946a). The prothoracic shield is enlarged such that the L-group setae all occur upon it, although L_3 lies sometimes just off the shield. Prolegs are present on A3-6 and on A10. In first instars the crochets are uniserial and uniordinal and arranged in a circle, while in other instars they are multiserial, uniordinal, and arranged in an ellipse.

In the pupa (Mosher, 1916), the mandibles are rudimentary and maxillary palpi absent. The arrangement of the dorsal teeth or spines appear to be characteristic. Segments A3-7 each bear two transverse rows dorsally, and on A7 occurs also a ventral row often connected to the anterior dorsal row (Nielsen & Kristensen, 1989). Considerable mobility exists in the abdomen, but the appendages are fixed to the body and to each other. There is no cremaster.

Biology

Feeding habits. Hepialid larvae are phytophagous, phytophagous with an early mycophagous stage, or possibly, but rarely, exclusively mycophagous (Grehan, 1989). The adults do not feed. Our knowledge of larval feeding habits (summarized by Wagner *in* Stehr, 1987; Grehan, 1989) is limited mainly because the concealed, often subterranean, habits of the larvae make them difficult to study.

Tunnels may be formed within stems or roots, on which the larvae feed, or in the soil. Leaf-feeding also occurs, with larvae emerging from tunnels in the soil to eat material at ground level. Leaf-feeders may consume grass, herbs, and moss. Certain predominantly moss-feeding species from New Zealand live in bogs, and their tunnels may reach below water level (summary in Grehan, 1989).

The polyphagous habits of larvae of most species has led to various species becoming pests of lawns, pasture, fruit, vegetables, and decorative flowering plants. Foodplant range is often particularly broad in species feeding on roots or within tree trunks (Grehan, 1989). But not all hepialid larvae are polyphagous. *Leto venus*, a species restricted to the forests of the southern Cape of South Africa, appears to bore exclusively in trunks of *Virgilia capensis* (Fabaceae) (Janse, 1945).

Early instars of several species are mycophagous (review by Grehan, 1989) in that they feed on dead wood and sometimes on fungal fruiting bodies. In *Aenetus virescens* the change from feeding on decaying to non-decaying wood is preceded by a fusion and darkening of the larval pinacula before the first moult (Grehan, 1989). The extent of mycophagy in Hepialidae may be underestimated.

Courtship. Males of several species fly in swarms. At dusk females of *Hepialus humuli* are attracted to shimmering swarms of white, hovering males, probably by both visual and olfactory stimuli (Mallet, 1984). Male swarming in this species is entirely independent of ambient temperature, humidity, and wind. The cue responsible for inducing swarming is a fall, to a critical level, in light intensity (Mikkola, 1974). After approaching the female, the male follows her down to the vegetation where mating takes place. However, males of *Oncopera* in New South Wales appear to pursue females by hawking flights rather than by attracting them (Barton Browne *et al.*, 1969). In *O. alboguttata*, hawking males fly after females as they rise from the ground at around sunset. They collide with the females and the pair falls to the ground where mating occurs. In *O. rufobrunnea* and *O. tindalei*, females fly into clouds of hawking males from which one or more males pursue them, although in these two species the female is not knocked out of the air.

The general view that males pursue females in Hepialidae was questioned by Mallet (1984) who suggested that emphasis should rather be placed on the orientation of females to males. Whereas the various scent organs on male Lepidoptera are usually considered to release pheromones onto the female at close range (but see Chapter 6), it is those Hepialidae bearing rosette organs, thought to have a long range action, on the hindtibia that apparently swarm. In *Hepialus humuli* the tibial brushes are fully everted during swarming. In species lacking tibial scent organs, males seem to be attracted to females. Further support for the view that females are attracted to males comes from the 'calling' habits of the North American species *Sthenopis auratus*. Male moths hang from the tip of fern fronds and fan their wings so pushing air over metatibial scent brushes (McCabe & Wagner, 1989).

10

EARLY HETERONEURA

The four superfamilies treated here do not, collectively, form a monophyletic group. All include small or very small moths with heteroneurous wing venation and monotrysian female genitalia. Wing coupling is effected by frenular-retinacular coupling. The larvae of most species are leaf miners.

INCURVARIOIDEA

This superfamily comprises a large group of monotrysian Lepidoptera with representatives worldwide. Incurvarioidea include small to very small moths. Many species are fairly drab, but several are metallic. Larval habits are varied, but all species construct portable cases at least in the last larval instar. Endophagy is widespread within the group, but many species are exophagous and feed on detritus. The Yucca moths (Prodoxidae) are symbiotic with *Yucca* (Agavaceae) (see Chapter 2).

The adults (Nielsen & Common, 1991) lack external ocelli and chaetosemata. The groundplan condition is as follows. The maxillary palpi are 5-segmented, and the haustellum, which is fully functional, is approximately 3/4 the length of the maxillary palpi. The labial palpi are short and drooping with lateral bristles on segment 2. The venation is generalized heteroneurous, and microtrichia occur extensively on the wings. On the hindwing pseudofrenular bristles are usually present in addition to a frenulum. The forewing-metathoracic locking device is present. The epiphysis is usually present. Three specialized characters suggest that the superfamily is monophyletic (Nielsen & Davis, 1985). These are 1) the presence of a sagittate juxta, 2) the occurrence of strong muscle 'tendons' ('guy-wires') attaching to the vagina or vaginal sclerite (Dugdale, 1974), and 3) the presence of a portable case - at least in the final larval instar.

The superfamily is composed of six families (Nielsen & Davis, 1985).

Heliozelidae

Heliozelid moths (Emmet, 1976; Nielsen, 1985a; Nielsen & Davis, 1985; Common, 1990) are very small and frequently dark in ground colour with silver spots or fasciae on the forewings. Heliozelidae are found worldwide, but there are no representatives in New Zealand. According to Nielsen (1985a) there are 13 genera, of which several are undescribed. The adults are diurnal and fly in sunshine while the larvae of most species are leaf-miners. The final larval instar constructs an oval case from the epidermis of the mined leaf, the shape of which gives the family its colloquial name of 'shield bearers'.

Adult (Fig. 195)

Two of the three specialized characters for the family proposed by Nielsen & Davis (1985) are cephalic. The anterior tentorial arms are curved dorsally, and the mandibles are particularly small; the scales on the vertex of the head are appressed to the

213

cranium, although exceptionally the head is rough-scaled; and the maxillary palpi are reduced, rarely 5-segmented (Nielsen, 1985a). The antennae are fairly short. The wings are lanceolate, the forewing broad and the hindwing narrower, with the venation strongly reduced. In the hindwing, crossvein M-CuA is absent, the third of the specialized characters of the family (Nielsen & Davis, 1985). The tibial spur formula is 0-2-4. A pectinifer is present on each valva. The ovipositor of the female is of the piercing kind, a condition typical of Incurvarioidea.

Immature stages

The larvae (Davis *in* Stehr, 1987) mine leaves and, sometimes, petioles or twigs. Thoracic legs are usually absent; when present, they are well developed or represented by ambulatory calli. Abdominal prolegs are typically absent or rudimentary; sometimes crochets, arranged in multiserial rows, are present. When reduced, prolegs are represented by calli. In some species of *Antispila* and *Coptodisca*, the calli of most segments are fused. A callus, or calli, also occurs dorsally on A8.

The pupa (Mosher, 1916) has short antennae and a long labrum projecting down over the labial palpi. Appendages are free, and most abdominal segments mobile. The spiracles of *Antispila pfeifferella* are projected from the sides of the abdomen on stalks, a feature possibly characteristic of the entire family (Emmet, 1976).

Biology

The egg, which is laid singly, is often inserted beneath the bark of a twig of the hostplant. From this position, the larva mines from the twig into the petiole of the leaf and then into the lamina. The egg may also be inserted directly into a leaf or a petiole. Sometimes the larva mines for a time in the midrib. At the end of its mining life the larva constructs an oval case from two pieces of epidermis cut at the end of the mine, which leaves a characteristic oval hole in the leaf. A wide range of plants is mined. In South America one species feeds on *Nothofagus* (Fagaceae), and species in both South America and Australia mine the leaves of members of the Myrtaceae. Several South African species mine the leaves of Myrtaceae, including the grapevine. In North America, *Coptodisca splendoriferella* is occasionally a major pest of apple (see Davis *in* Stehr, 1987). Pupation occurs in the case, usually on the ground. Sometimes the pupal cases are attached to twigs.

Adelidae

Fairy moths are small, and recognizable by their particularly long antennae (Fig. 196). In the male these are often two to three times the length of the forewing, and in the female, although generally somewhat shorter, they usually extend beyond the tip of the forewing. The moths fall essentially into the diurnal, metallic *Adela* group (best represented in the eastern Palaearctic) and the crepuscular, drab, or yellowish *Ceromita* group (best represented in southern Africa and South America). Larvae seem to feed on dead leaves or on low plants, unlike the typically leaf-mining families Incurvariidae and Heliozelidae.

Adult (Fig. 196)

The relative length of the male antenna exceeds that of other Lepidoptera. A well-sclerotized intercalary sclerite is present. Pointed pegs (possibly derived from chaetiform sensilla, E.S. Nielsen, 1980) occur on certain proximal segments of the flagellum of males of *Nemophora* and *Adela*. In *Adela* these are strongly curved, and in some species of *Nemophora* several pegs may occur on a single segment. The proboscis is longer than the head capsule and extends beyond the maxillary palpi. Unlike other Incurvarioidea, excepting a single species of Prodoxidae, the proboscis is scaled. In most Adelidae the maxillary palpi are much reduced both in length and number of segments. Males of some species have greatly enlarged eyes. The labial palpi are 3-segmented. The sensory invagination (organ of vom Rath) on the apical segment of the labial palpus may be deep or shallow.

The tibial spur formula is 0-2-4. The wing venation is hardly reduced. The wings are coupled by a single (composite) spined frenulum in males, and 3-5 frenular setae in females. In Adelidae these setae are better developed than in other Incurvarioidea (Davis, 1986). The frenular setae arise from separate sockets. Distal to the frenulum in both sexes are some pseudofrenular setae. There is generally a marked reduction of aculei over the upper surfaces of the wing membranes.

Pectinifers (comb teeth) are absent from the valvae of many Adelidae, but present in others. In the Palaearctic genus *Nematopogon*, they may be borne on stalks or the stalks may be absent (Nielsen, 1985b). In the female genitalia the tip of the ovipositor is flattened laterally, a condition considered independently specialized for the family (Nielsen & Davis, 1985). The cloaca is slender and tube-like (E.S. Nielsen, 1980). The number of ovarioles is high (10-12), contrasting with the usual number of four polytrophic ovarioles in most Lepidoptera (E.S. Nielsen, 1980).
(E.S. Nielsen, 1980; Nielsen & Davis, 1985; Nielsen, 1985b.)

Immature stages

Eggs are inserted singly into plants.

The larva (Davis *in* Stehr, 1987) is typically prognathous with 6 stemmata. The absence of seta AF2 is a specialization (E.S. Nielsen, 1980; 1985b). Thoracic legs are well developed but prolegs (present on A3-6 and A10) are much reduced. Crochets, arranged in multiserial rows, are restricted to segments A3-6.

The pupa has appendages weakly adhering to the body.

Biology

In many of the metallic species the moths fly in sunshine and may swarm, particularly around groups of flowers, and also trees and bushes. Swarming is associated with two structural adaptations found only in males (E.S. Nielsen, 1980) - pegs on the antennae, and specialized, dorsally enlarged (holoptic) eyes that meet each other in the midline. In some species of *Nemophora* the eyes are divided into a swollen dorsal component and an unmodified ventral component. The facets of the upper part of the eye are much larger than those of the lower. In species with secondarily reduced eyes, swarming occurs infrequently or not at all. Oord (1981) reported swarming of about 25 males of *Nemophora degeerella*. A resting female apparently elicited hardly any response to the swarm, but later a female was observed being carried to a leaf where mating took place. It is unclear if and how antennal pegs play a functional role in swarming.

In the life history of a typical adelid species (Davis *in* Stehr, 1987), the first instar larva usually mines leaves or bores into plant ovaries. Subsequently, probably from the second instar onwards, it lives in a case constructed from plant material, and feeds on leaf litter, or lives among low growing vegetation. Pupation occurs in the larval case.

Crinopterigidae

Crinopterigidae were treated as a separate family by Nielsen & Davis (1985), who were unable to place the Mediterranean species *Crinopteryx familiella* in any other incurvarioid family. Petersen (1978) recognized the incurvarioid (his incurvariid) association of the species.

Adult (Fig. 197)

Maxillary and labial palpi are well developed, and the proboscis extends to only about half the length of the labial palpi. There is no pectinifer on the valva of the male genitalia but a strong spine occurs. The juxta does not taper to form the typical sagittate shape found in other members of the superfamily. The ovipositor is long. (Petersen, 1978; Nielsen & Davis, 1985.)

Immature stages

The larva feeds on *Cistus* (Cistaceae), and lives in a case formed from the epidermis of the leaf. The pupa bears numerous dorsal spines.

Incurvariidae

The family is widespread, but absent from New Zealand. There are about 13 genera worldwide, and the group is particularly diverse in Australia where there are probably many more than the 55 species already known (Common, 1990). Only one species has been collected from the Afrotropical region (Scoble, 1980). The moths are often fairly drab, but several are metallic. Currently, membership of the family depends on the absence of well-defined pectinifers from the valvae of the male, and the presence of flattened, scale-shaped spines on these structures (Nielsen & Davis, 1985). Typically, the larvae are leaf miners in their first instars after which they construct cases from the leaf. Subsequently they usually become skeletonizers but may continue as leaf miners.

The groundplan condition is illustrated best by the primitive members of the family, which are found in the southern hemisphere.

Adult (Fig. 198)

The head may be entirely rough scaled, or the scales of the frons appressed to the head. The antennae are not elongated, maxillary palpi are well developed, and 4- or 5-segmented, and labial palpi are 3-segmented. The proboscis is short. An epiphysis is usually present, but is absent from the Australian genus *Perthida*, and the tibial spur formula is 0-2-4. The wing venation may be slightly reduced. Although many flattened, scale-like spines on the valvae of the male genitalia may occur (a character probably specialized for Incurvariidae, Nielsen & Davis, 1985), pectinifers are absent.

Immature stages

The larvae are typically of a similar appearance and habits to those of Adelidae - prognathous and often with reduced thoracic legs and prolegs. Among the diagnostic characters given by Davis *in* Stehr (1987) is the presence of a single, transverse band of crochets on A3-6 rather than several bands. In *Paraclemensia* and *Vespina*, crochets are present on A10.

The pupa is relatively mobile and the appendages are free.

Biology

Typically, the larvae are leaf miners in the first instar and subsequently case bearers. However, there are some exceptions including a gall dweller, a mode of life rare for Incurvarioidea other than in Cecidosidae. The generalized habit is exemplified by the holarctic genus *Alloclemensia* (Nielsen, 1981). *A. mesospilella*, for example, cuts an oval case from its blotch mine and, after attaching the case to the same leaf, becomes a leaf skeletonizer. Pupation occurs in the case, which falls to the ground when the larva is fully fed. A similar life history is found in *Paraclemensia*. *P. acerifoliella*, known as the maple case bearer or maple leaf-cutter, is a pest of sugar maple (*Acer saccharum*: Aceraceae) in North America (Nielsen, 1982b). This species also feeds on members of a variety of other plant families. Larvae of many Incurvariidae cut additional pieces of leaf as they mature, and use them to enlarge the early case (e.g., *Paraclemensia caerulea*, and *Vespina quercivora* (Davis, 1972)).

The foodplant families of Incurvariidae are predominantly Myrtaceae, Proteaceae, and Fabaceae (Nielsen, 1985a). The most primitive Incurvariidae are found in Australia and feed on *Eucalyptus* (Myrtaceae), but many Australian species feed on Proteaceae. Larvae of the South American genus *Simacauda* probably feed on Myrtaceae (Nielsen & Davis, 1981), while those of the South African species *Protaephagus capensis*, the only true incurvariid known from the Afrotropical region, mine the leaves of various Proteaceae (Scoble, 1980). The larva mines until mature and only then makes a case by cutting an oval piece of epidermis from both lower and upper surfaces of the leaf and joining them. The mined leaf, and the case formed, resembles that typical of Heliozelidae. Similar larval habits are found in the Australian genus *Perthida*, which includes the jarrah leaf-miner, a pest of jarrah (*Eucalyptus marginata*) (Common, 1969b). Several larvae of this species may mine a single leaf causing a reduction in tree girth (Mazanec, 1974). Dense swarms of moths have been reported (Wallace, 1970).

The essential oils of eucalypts are presumably defensive, and larvae of *P. glyphopa* avoid the oil glands of *Eucalyptus*, which secrete essential oils, during the first instar. In subsequent instars, although they chew through the glands, the larvae do not ingest the oil, but wipe the substance against frass pellets or walls of the mine (Mazanec, 1983).

Cecidosidae

The family occurs in southern South America, and the larvae live in galls on Anacardiaceae. The gall association of the Cecidosidae is rarely encountered elsewhere in Incurvarioidea.

Adult (Fig. 199)

Maxillary and labial palpi in the adult moth are rudimentary, and the proboscis is absent. The epicranial suture is lacking, an absence considered to be specialized for the family. Other specialized features are found in the thorax: of particular note is the lack of the typical frenular-retinacular wing coupling system in males. The tibial spur formula is 0-2-4. In the metafurcasternum, the posterior dorsal tendons are lost (Davis, 1986). The first abdominal segment lacks a tergosternal process, which is assumed to be a secondary reduction. The male genitalia are typically incurvarioid, for example the valva bears a pectinifer, as are the female genitalia with their long, cutting ovipositor. (Becker, 1977; Nielsen & Davis, 1985.)

Immature stages

As in Prodoxidae, the larvae are endophagus. Legs on the thorax and prolegs on the abdomen are reduced or lost. Stemmata are also entirely lost. Numerous secondary setae are found on the larva.

The pupa of *Eucecidoses minutans* is adecticous, obtect, and incomplete. Dorsal spines are present (Becker, 1977).

Biology

The larvae live in galls on Anacardiaceae. Galls induced by a South American cecidosid on *Rhus* drop to the ground where larval activity causes them to flip or twitch.

Prodoxidae

The family includes the North American *Yucca* moths (e.g., see Davis, 1967) and other Incurvarioidea with a pair of stellate signa on the corpus bursae of the female genitalia, and a rounded posterior edge to sternum A7 on the abdomen of that sex (Nielsen, 1982a). The family is distributed across the holarctic region widely (Nielsen, 1982a), and the group also occurs in southern Argentina and Chile (Nielsen & Davis, 1985). Prodoxidae larvae are internal feeders and, as in Cecidosidae, they do not construct cases.

Adult (Fig. 200)

The proboscis is long except in *Lampronia*. In *Tridentaforma* the proboscis is scaled, a character otherwise found only in Incurvarioidea among Adelidae. The forewing is usually not metallic. The frenulum-retinacular wing coupling system is reduced in males of *Parategeticula*. Pectinifers are present on the valvae of males of many species. In the female, sternum A7 is strongly rounded, and the bursa copulatrix bears a pair of stellate signa.

Immature stages

The egg of *Prodoxus phylloryctus* is elongate reniform and slightly broader at one end (Wagner & Powell, 1988).

In the larva, thoracic legs are present or are replaced by ambulatory calli. Prolegs are absent or vestigial. Crochets are generally absent, but occasionally occur on each of A3-6 where they take the form of a transverse row.

Biology and classification

The family is divided into Lamproniinae and Prodoxinae mainly on the basis of larval habits (Frack - see Davis *in* Stehr, 1987). In Lamproniinae young larvae feed in seeds or the receptacles of fruit chiefly of Rosaceae and Saxifragaceae. Subsequently, they overwinter in a hibernaculum in the soil or at the base of the plant before emerging and boring into new buds. Exceptionally, as in *Lampronia fuscatella*, they feed within galls the formation of which they induce. Pupation occurs either in feeding galleries or in a cocoon outside the plant.

Larvae of Prodoxinae typically bore into seeds or stems of Agavaceae. In *Tegeticula* and *Parategeticula*, development of seeds used for food is ensured by the adult female, which pollinates the plant. However, *Greya* does not feed in Agavaceae and only the early instars feed within the fruits, and in *Prodoxus phylloryctus* the larva mines the semi-succulent *leaves* of *Yucca baccata* (Wagner & Powell, 1988). Pupation occurs either in a cocoon in the soil, or within larval galleries at the feeding site.

There are 6 genera of Prodoxinae (Wagner & Powell, 1988) with fewer than two dozen species. In structure (Davis, 1967), the head is rough-scaled in these medium to small moths. The proboscis is reduced, and is not coiled when the moth is at rest, and the food canal is poorly developed. The folded maxillary palpi are long and usually 5-segmented. In females of *Tegeticula* and *Parategeticula* a structure referred to as a 'maxillary tentacle' occurs, which functions to gather pollen from Yucca flowers (see Chapter 2). Females with improperly formed tentacles do not gather pollen. The tentacles are extensible, and covered with short hairs. In the male of *Parategeticula* the frenulum is absent.

NEPTICULOIDEA

Nepticuloidea include two families, Nepticulidae and Opostegidae, although it would be reasonable to combine them in a single family. Both are widespread, but the former has around six times the number of described species. The larvae are predominantly leaf-miners; adults are small to minute moths with a characteristic expanded antennal scape ('eye-cap').

The monophyly of Nepticuloidea is suggested by eight shared derived characters (Davis, 1986). The enlarged antennal scape is most easily observed with the occurrence of an expanded scape in other families presumed to be secondary. Also on the antennae, sensilla coeloconica are reduced or lost, whereas in other monotrysians they are well developed. The pseudofrenular setae of the hindwing are better developed in Nepticuloidea than in other Lepidoptera. The basisternum is poorly developed (Minet, 1984). The hindtibiae are densely covered with spinose setae (Nielsen, 1982c) unlike the condition occurring in other monotrysians. There are three characteristic features of nepticuloid larvae: stemmata are reduced in number to one on each side of the head; crochets are lost from the prolegs; and the hypostoma is extended forward (Heinrich, 1918). Crochets are sometimes secondarily lost in the more endophagous larvae of Tischerioidea and Incurvarioidea (Davis, 1986), but this loss is assumed to be independent of that in Nepticuloidea.

Nepticulidae

Nepticulidae, the most speciose of the monotrysian families with around 600 described species and probably many undescribed (Nieukerken, 1986), include the smallest Lepidoptera. On average, the wingspan is about 5-6 mm, but in some species it is under 4 mm. The larvae mine leaves, occasionally leaf-petioles and bark of angiosperm plants, and sometimes the parenchyma of winged seeds of *Acer* (Aceraceae). The adults are either rather drab, fuscous and ochrous moths, or metallic typically with one or more shining silver fasciae and an iridescent background of gold or purple.

The evidence for the monophyly of the Nepticulidae comes mainly from characters of the male genitalia (Scoble, 1983), but a unique type of sensillum is found on the antenna (Nieukerken, 1986; Nieukerken & Dop, 1987) (see Chapter 2).

Adult (Fig. 201)

The head is rough-scaled. External ocelli are absent, and chaetosemata are present on the vertex. Mandibles are generally reduced, and the maxillary and labial palpi are 5-

and 3- (rarely 2-) segmented respectively. The galeae form a short, but functional (Downes, 1968), proboscis not usually extending beyond the labial palpi. Rarely, the structure is about 2.5 times the length of the head capsule. The antennae are fairly short. A sensillum vesiculocladum (Fig. 38) is found on many of the flagellomeres (Nieukerken, 1986; Nieukerken & Dop, 1987) (see Chapter 2). Other sensilla on the antenna of Nepticulidae include sensilla trichodea, of which there are few, sensilla chaetica, of which there are two kinds, and sensilla coeloconica.

The prosternal bridge is only weakly sclerotized in Nepticulidae (Minet, 1984), but still probably represents the derived condition (secondarily reduced) characteristic of the Heteroneura (Nieukerken, 1986; Davis, 1986). The epiphysis is absent and the tibial spur formula 0-2-4. The wing venation is strongly reduced in both wings, although somewhat less so than in Opostegidae, with many veins being coalesced or lost. The so-called 'anal-loop', formed by the free proximal lengths of the first and second anal veins in the forewing, which then unite, is present in several species from southern Africa (Scoble, 1983), but absent from most members of the family. The venation of the hindwing is even further reduced than in the forewing. The forewing and the hindwing of the male are coupled by a frenulum and a retinaculum. The retinaculum is subcostal and unusual in comprising a series of looped scales instead of a more discrete hook. Sometimes, costal (pseudofrenular) setae are also present. Pseudofrenular setae alone occur on the hindwing of the female. The forewing-metathoracic wing locking device is present. Specialized scales, presumably scent scales, are present on the surfaces of the wings in males of many species. Hairpencils also occur in males of several species on either forewing or hindwing.

In the male genitalia a pectinifer occurs on the valvae of Pectinivalvinae (Scoble, 1982).
(Scoble, 1983; Nieukerken, 1986.)

Immature stages

The flattened, oval egg is laid singly on the plant substrate (usually a leaf), and appears as a minute shiny black object at the beginning of a mine.

In the larva, the head is strongly flattened and prognathous, there is a single stemma on each side, and the antenna (1-segmented) is strongly reduced. The head capsule is retracted deeply into the prothorax. Unlike the condition in Opostegidae, the epicranial notch is deep. The body is slightly flattened. Thoracic legs are absent, but paired ambulatory calli are found on T2-3, and are prominent in the large and cosmopolitan genus *Stigmella*. Such calli are usually present on abdominal segments A1-7. Prolegs and crochets are absent.

In the pupa, the coxae are broadly exposed. Appendages are free, and spines occur on the dorsum.
(Davis *in* Stehr, 1987.)

Biology

Most Nepticulidae spend their entire larval growth period within a mine. The cocoon is silken, elliptical, and typically spun by the larva among leaf litter or in the soil. In many species, overwintering occurs as a cocooned larva. While most larvae mine one leaf, and make only one mine, species belonging to the European *Trifurcula* subgenus *Fedalmia* mine two or three leaves moving from one leaf to another via the petioles (e.g., Emmet, 1976).

Several fossil leaf mines of Nepticulidae are known from the Lower Cretaceous (summary in Skalski, 1990).

Opostegidae

The family, cosmopolitan in distribution, has relatively few species compared with the Nepticulidae. Davis (1989) listed 106 species in 6 genera, but there are probably many undescribed species. The moths are typically small and white, but on average they are larger than Nepticulidae.

Adult (Fig. 202)

Scales are usually appressed to the head, but a tuft of hair-like scales arises from between the antennal bases. External ocelli are absent. The antennal scape is greatly enlarged. On nearly all the antennal flagellomeres occur 3 typically palmate, branched sensilla ascoidea (Fig. 37). Sensilla vesiculoclada and coeloconica are lacking as are microtrichia. The proboscis is reduced, and the food canal poorly developed. The wing venation is very strongly reduced - in some species to merely 4 or 5 veins in the forewing. Vein CuP is absent, the anal fold often being mistaken for CuP. Only scattered aculei are found on the surfaces of the wings. The frenulum is lost in both males and females. Modified pseudofrenular scales arise in closely set rows ranging from 4 to nearly 20. They curl, hook-like, over a ventral expansion of vein M in the forewing (Nielsen, 1985c). The forewing-metathoracic locking device is present. In the male genitalia, the valva bears a pedunculate pectinifer. In some species the aedeagus is not sclerotized. In the female, the ovipositor is short and almost truncate. (Davis, 1989.)

Immature stages

The egg is elongate-oval with a smooth, transparent chorion.

The larva is prognathous and has a flattened head with reduced mouthparts. Each mandible bears a large spinose seta. The setae of the head are much reduced, with but a single stemma on each side of the head. The body of the larva is strongly elongated, slender, and cylindrical. On both thoracic and abdominal segments are found a pair of conspicuous, elongate pronotal sclerites similar to those of Nepticulidae. Paired ambulatory callosities are present on the meso- and metathorax, but the larva is essentially apodal. Opostegidae larvae differ from those of Nepticulidae by the shallower epicranial notch. Also, the hypostoma of the strongly depressed head is lengthened anteriorly much more than in Nepticulidae.

The pupa is similar to that of Nepticulidae. (Heinrich, 1918; Davis, 1989.)

Biology

The life history was reviewed by Davis (1989). Foodplant families include Rutaceae, Saxifragaceae, Ranunculaceae, and Fagaceae. Larvae are miners in the cambial layer under bark, in leaves, petioles, or in the stalks of flowers. The larva of the neotropical species *Notiopostega atrata*, which mines the petiole of *Nothofagus dombeyi* (Fagaceae), moves into the cambium and then down the trunk, and may travel up to 7 meters (Carey *et al.*, 1978). Some Hawaiian species are leaf miners (Zimmerman, 1978). The large spinose seta on each mandible possibly functions to sweep small particles of food into the mouth or to clean the dorsal surfaces of the lower mouthparts (Heinrich, 1918). As far as is known, pupation occurs in a cocoon outside the mine.

Adults may be diurnal, crepuscular, or nocturnal (Davis, 1989). Most are white but *Notiopostega atrata*, which flies only by day, is almost black. The moths tend to burrow into small crevices, a habit supported by the expanded antennal scape (which may protect the eye), the somewhat depressed body, and the spinose legs, which may help the insects to grip the substratum as they move into crevices (Davis 1989).

PALAEPHATOIDEA

This superfamily comprises a single family, Palaephatidae, and is represented by 28 species in five genera from southern Argentina and Chile (Davis, 1986), and a single genus in eastern Australia with four described species (Nielsen, 1987). Further species have now been discovered in Australia, and one is known from South Africa (E.S. Nielsen, pers. comm.). In South America the moths are associated with moist temperate forests (Davis, 1986), and in Australia they have been collected near rainforest or in wet sclerophyll forest (Nielsen, 1987).

Palaephatidae

The adults are very small to moderately large for microlepidopterans, the wingspan ranging from about 8 mm to 16 mm. The forewings are generally dull and vary from dark fuscous to brownish, or brownish yellow, to pale buff or cream. Occasionally there is some bronzy iridescence, and one species has bright yellow wings. White or cream markings are frequently encountered.

Adult (Fig. 203)

External ocelli are usually present (absent from one genus). Chaetosemata are absent. The compound eyes vary in size; in one diurnal group they are reduced. Mandibles are functionless, but prominent in *Azaleodes*, and the proboscis is fairly short. The maxillary palpi are 5-segmented in all species except for those in the genus *Apophatus* where the number of segments is reduced to 4 and in which the length of the structures is also reduced. The labial palpi are 3-segmented and bear a small sensory pit subapically. The antennae usually extend to one half or just over one half the length of the forewing. A pecten is often present on the scape, and an intercalary sclerite occurs in the membrane between the scape and the pedicel. There are no specialized sensilla of the kinds found in the Nepticuloidea on the antennae of Palaephatidae.

The epiphysis is present, and the tibial spur formula is 0-2-4. The wing venation is little reduced. The wings are coupled by a frenulum-retinacular system. In the male the retinaculum is a simple fold of the costal margin of the forewing; in the female the retinaculum is represented merely by a few relatively stiff scales on the subcostal vein and sometimes a weak ridge. The frenulum of males is composed of a stiff, composite bristle, but pseudofrenular bristles also occur. In females only 2 to 4 frenular setae are present, but they are much enlarged. These setae arise from closely associated sockets, an arrangement similar to the general ditrysian condition. Aculei are generally evenly scattered over the surfaces of the wings, but they may be absent from the basal parts. The wing-thorax locking system is present. Various androconial organs, mostly concealed in pockets, occur on the wings. An elongated pocket extends along the hindwing in three of the four species of *Palaephatus* subgenus *Palaephatus*. This pocket, formed by folds of the wing membrane and closed by a tongue and groove system along its length, houses a hairpencil. The ultrastructure of the hairs shows them to be spongy. Other sex scales are also found in the pocket.

Although distinct, the male genitalia resemble those of other monotrysian Heteroneura - particularly Nepticulidae. The juxta is absent or reduced. The valva lacks a pectinifer. The female genitalia display three characteristics (Davis, 1986) tentatively supporting the monophyly of Palaephatidae. The ovipositor bears a medial sensory ridge (possibly homologous with a raised medial area on segment A10 of female Tischeriidae); minute spicules are present on the utriculus of the spermathecal canal of the females of most species; and the ductus bursae is characteristic, bearing numerous spicules and a V-shaped sclerite - modifications which are sometimes secondarily lost. (Davis, 1986; Nielsen, 1987.)

Immature stages and biology

The only larval record is that of a specimen of *Sesommata holocapna* which was found spinning together twigs of *Diostea juncea* (Verbenaceae) (Karsholt *in* Davis, 1986). Pupation occurred in a loose shelter between the spun twigs.

The pupa, which was derived from this larva, is adecticous and obtect. No cocoon cutter is present on the frons. Proleg scars occur on abdominal segments A3-6. Short spines are present dorsally on segments A2 to A9 + 10, implying that the pupa extrudes from its shelter shortly before adult emergence.

Phylogenetic relationships

The Palaephatidae may be the sister group of the taxon Nepticuloidea + Tischerioidea (Davis, 1986), or Palaephatoidea + Tischerioidea may represent the sister group of the Ditrysia (Nielsen, 1987). The first suggestion is based on the sharing of a non-piercing ovipositor, the absence of a cloaca, and the reduction or loss of the juxta. The second is supported by the reduction in number of the frenular setae, the absence (loss) of a pectinifer on the male valva, and the medial sensory ridge on the ovipositor of both Palaephatidae and Tischeriidae (assuming that that structure is homologous in the two groups).

TISCHERIOIDEA

Tischeriidae

There is one family, Tischeriidae, with a single genus, *Tischeria*. The moths are small with a wingspan of around 5-11 mm. The colour of the forewing varies - pale whitish or yellow, yellow, bronzy, dark grey or blackish. According to Braun (1972) there are about 65 species, undoubtedly a conservative estimate. The family is primarily Holarctic with over half the species described occurring in the United States of America (Davis *in* Stehr, 1987). Some species have been described from the Neotropics, the Afrotropics, and Indo-Malaya, but none occur in Australasia. The larvae are predominantly leaf-miners.

Adult (Fig. 204)

Only the North American species have been studied in detail (Braun, 1972). The head narrows ventrally, the vertex is smooth scaled or rough scaled, and the front of the head is smooth scaled. Maxillary palpi are minute, the proboscis is scaled near its base, and the labial palpi are short to very short. The antenna is about as long as the forewing. Although the scape bears characteristic projecting scales, it is not expanded into an eye-cap. Sensilla trichodea on the antenna are filamentous and recurved, a situation not yet observed in any other Lepidoptera (Davis, 1986).

Aculei are present on parts of both forewing and hindwing. The forewing is lanceolate, and the hindwing, which may be narrower or sometimes broader than the forewing, tapers sharply towards the apex. The forewing venation is reduced, mainly through fusion, although not to the extent found in Nepticuloidea. The hindwing venation is considerably reduced and appears similar to that of many Nepticulidae.

The wings are coupled by a frenulum and retinaculum system. In the female, the frenular bristles arise from closely associated sockets, a condition similar to that found in primitive Ditrysia and in Palaephatidae (Davis, 1986). The forewing-metathoracic wing locking device is present. The metafurcal apophyses are fused to the secondary arms of the furca sternum, a specialized character of the family (Davis, 1986).

Characteristic of the male genitalia is the forked aedeagus, with the phallotreme (the external opening of the aedeagus duct) situated at the base of the fork. In the female genitalia the ovipositor lobes bear short, peg-like setae. Anteriolateral to these lobes is a second pair of lobes bearing both short thick setae, and long slender setae. A structure similar to the medial sensory ridge of Palaephatidae occurs on segment A10 (Davis, 1986). The ventral anterior margin of segment A8 is extended into an additional pair of apophyses. Each additional apophysis articulates with the anterior apophysis on the same side. The posterior apophyses are often strongly swollen at the base.
(Braun, 1972.)

Immature stages

The egg is elliptical. It is cemented to the leaf surface and covered by an iridescent adhesive that spreads in a circle (Braun, 1972).

In the larva (Davis *in* Stehr, 1987), the body is depressed, and the segments often markedly rounded laterally. The head is strongly depressed, and has a deep epicranial notch. Typically, the 4-6 stemmata form a horizontal line. Thoracic legs are much reduced or absent; when present they are composed of two minute segments or they take the form of small pads. A pair of ambulatory calli are found on the dorsum of T1-3. There are three L-group setae on T1. Abdominal prolegs are present on A3-6 and A10; although only weakly developed they bear crochets. Crochets on A3-6 are multiserial and arranged in two bands or in a broken ellipse; those on A10 bear two or three rows.

The pupa is incomplete with all appendages free (Braun, 1972). Abdominal segments A3-7 in the male and A3-6 in the female are free, and the dorsum of the pupa bears distinct spines (Mosher, 1916; Braun, 1972).

Biology

Eggs are typically laid on the upper surface of leaves. Tischeriid larvae are leaf miners. The mine may expand rapidly from the initial short linear section into a 'trumpet mine', or may become enlarged into a blotch. Frass is ejected through a hole cut by the larva in the surface of the leaf at the beginning of the mine. The main function of the silk spun by larvae throughout their lives is probably to provide them with a surface on which they may more easily walk. In some species a circular 'nidus' of silk is constructed to which the larva retreats when disturbed. Foodplant families include Fagaceae, Rosaceae, Rhamnaceae, Malvaceae, Ericaceae, Asteraceae, Tiliaceae, Combretaceae, and Sterculiaceae.
(Braun, 1972.)

Phylogenetic relationships

Davis (1986) considered Tischeriidae to represent the sister group of the Nepticuloidea, while Nielsen (1989) suggested that, together with Palaephatidae, they might be the sister group of the Ditrysia. The sister group relationship between Tischeriidae and Palaephatidae was suggested mainly because both share a medial ridge on the ovipositor. However, the problem of the phylogeny of the monotrysian Heteroneura and their relationship with the Ditrysia remains unresolved.

11

LOWER DITRYSIA

All Lepidoptera other than those already considered have ditrysian female genitalia, and are grouped together in the monophyletic taxon Ditrysia (Börner, 1939). Those groups referred to here as Lower Ditrysia do not form a monophyletic group. However, typically they are smaller than most higher Ditrysia (macrolepidopterans), and the larvae are more often concealed- rather than external-feeders.

Relationships between ditrysian superfamilies are poorly resolved, with minimal taxonomic structure evident (Nielsen, 1989). All ditrysian superfamilies excepting Tineoidea, Gracillarioidea, Yponomeutoidea, and Gelechioidea, have a tortricid-type abdominal sternum 2 and were termed, collectively, Apoditrysia (Minet, 1986). Within the Apoditrysia, that group of superfamilies in which the pupa has lost the mobility of the first four abdominal segments was called Obtectomera (Minet, 1986). The Obtectomera include all those superfamilies from Epermenioidea onwards. However, the pupae of certain Bombycoidea are not obtect and it not certain that this condition represents a reversal (Brock, 1990). If these incomplete bombycoid pupae exhibit the primitive condition then the Obtectomera are not monophyletic. So these proposed divisions of the Ditrysia are by no means established.

TINEOIDEA

The Tineoidea used to include a diverse assemblage of families, but gradually many of those assigned to it have been removed to other superfamilies. Analysis of the superfamily by Robinson (1988c) suggests that only three families besides Tineidae (Eriocottidae (including Compsoctenidae), Acrolophidae, and Psychidae) should strictly be included in Tineoidea. This revised grouping excludes Gracillariidae, Bucculatricidae, and Roeslerstammiidae, families frequently included in Tineoidea. The two resultant assemblages were termed the 'tineoid lineage' and the 'gracillarioid lineage' by Robinson (1988c), but no specialized character has been found to suggest that these taxa are sister groups and here they are treated as Tineoidea (*sensu stricto*) and Gracillarioidea (see below).

The monophyly of the Tineoidea is supported by six characters (Robinson, 1988c). Three of these have undergone reversal within the group to the presumed primitive state, and one is convergent with a character in the gracillarioid lineage (Gracillarioidea). The characters include: erect scales on the frons; loss of the leaf mining habit in the larva; a short proboscis with disassociated galeae; lateral bristles on the labial palpi; an elongated, telescoped ovipositor; and a pair of ventral ovipositor rods - structures that are additional to the apophyses. These rods are rarely lost, but where they are lost the ovipositor is usually reduced. The shortened condition of the proboscis is considered to be convergent, since it also appears in the Bucculatricidae (Gracillarioidea). The last three characters undergo reversal to the presumed primitive state within the lineage.

The wing venation in Tineoidea is sometimes reduced, usually related to a narrowing of the wings. The wings are coupled by a frenulum and retinaculum, and the forewing-metathoracic locking device is present.

Tineidae

Although widely recognized as a cohesive taxon, the Tineidae are heterogeneous (Nielsen, 1978), and their monophyly is uncertain (Robinson, 1988c). The only character supporting the monophyly (the presence of a single row (as opposed to a frequently disordered double row) of overlapping scales on the antennal flagellomeres) is convergent, being found in Gracillariidae (Robinson, 1988c). Its distribution within the Tineidae requires further assessment.

Tineidae are generally small moths, mostly drab but occasionally colourful. The larvae feed on a variety of substrates, but they are typically detritivorous, fungivorous, lichenivorous, or keratinophagous. Unlike most Lepidoptera they do not consume live plant material. There are about 3000 named species (Davis *in* Stehr, 1987).

Twelve subfamilies are recognised by Robinson and Nielsen (*in press*), but the monophyly of two (Meessiinae and Myrmecozelinae) is not established.

Adult (Fig. 205)

The head is rough-scaled, often with a yellowish tuft between the antennae. External ocelli and chaetosemata are absent. The proboscis is reduced, and the maxillary palpi are typically 5-segmented, long and folded but may be reduced. The labial palpi vary from drooping, through porrect to ascending and typically bear bristles on the second segment. The epiphysis is present on the foreleg, and the tibial spur formula is 0-2-4; the hindtibia is rough-scaled. The venation is typically almost complete for Heteroneura; vein M and the chorda are often present in the cell of the forewing, and M is sometimes forked. In some species some venational reduction occurs. Occasionally a pterostigma occurs in the forewing. The ovipositor is generally long and extensible. (Nielsen, 1978; Common, 1990.)

Immature stages

Generally oval and slightly flattened, the eggs are usually laid singly in crevices (Davis *in* Stehr, 1987).

In the larva (Hinton, 1956; Davis *in* Stehr, 1987), the head is hypognathous and bears a variable number of stemmata - usually 1 or 6 on each side, but sometimes 2 or 5. Occasionally stemmata are absent. On the thorax there are usually 3 L-group setae on T1, but in Scardiinae there are 2. Thoracic legs and prolegs are well developed. The crochets of the ventral prolegs are typically arranged in a uniserial ellipse or circle.

The pupa bears dorsal abdominal spines and protrudes from its case when the adult emerges. There is extensive mobility between the abdominal segments.

Biology

The larval habits are reviewed briefly by Davis *in* Stehr (1987). Many species are detritivores feeding on dried plant material (e.g., as in Setomorphinae and Hieroxestinae). Scardiinae, or fungus moths, with 111 species (Robinson, 1986), and Nemapogoninae are the only Tineidae that feed exclusively on fungi. Scardiine larvae feed in the fruiting bodies of bracket fungi or in decaying wood infested with the hyphae of bracket fungi. Most form loose webs with frass over the substrate. They feed at or just below the surface of the substrate and pupate just below it. Hieroxestinae (Oinophilidae) mostly consume fungi or plant debris, but some species of *Opogona* infest sugarcane or stored tubers. Some live in caves and probably feed on bat guano. Certain detritus-feeding Tineidae live in ants nests. The clothes moths, the *Tinea pellionella* complex (Tineinae), or case-making clothes moths, comprise a group of 11 species

(Robinson, 1979). The larvae, which feed on wool and other keratinous material, are pests of industrial and household significance. Food substrates of tineid larvae feeding on items other than those associated with man include 'animal corpses, bird's nests, bird-pellets, mammal burrows and weathered carnivore faeces' (Robinson, 1979). The keratinophagous habit is very rare in insects (Hinton, 1956).

Phoretic behaviour occurs in the adults of some species of *Amydria* and *Ptilopsaltis* (Davis, *et al.*, 1986). The moths were found riding on the backs of two species of spiny pocket mice from Costa Rica. They were capable of remaining on the mouse even when the mammal was very active. Only females moths were collected from mice. The elongate pretarsal claws of this sex may be related to their holding ability.

The adult of *Choropleca terpsichorella* (Dryadaulinae) 'dances'. The moth (known as the dancing moth (Zimmerman, 1978) runs in tight circles and exhibits a crab-like sideways gait (Robinson, 1988a). The function of the dance is unknown: it is performed solo so appears not to be used in courtship.

Eriocottidae

The family includes Eriocottinae and Compsocteninae (Nielsen, 1978), taxa united by the presence of an additional pair of dorsal apophyses, and the membranous rather than sclerotized condition of sternum A8 of the female (Robinson, 1988c). There are about 70 species currently assigned to the family, although Robinson (1988c) stated that many others, at present in Psychidae, should be placed in *Compsoctena*. The moths are small to medium in size, not brightly coloured, and generally lack distinctive markings. Eriocottinae range through the southern Palaearctic, and may possibly occur in Taiwan and the Afrotropical region (Nielsen, 1978). The subfamily is represented by a single species in Australia (Nielsen *in* Common, 1990). Compsocteninae occur in the tropical parts of the Afrotropical region and in Asia (Nielsen, 1978), and have also been recorded from South America (Davis, 1990).

The early stages are unknown.

Adult (Fig. 206)

The head is rough scaled, and external ocelli are present or absent. The labial palpi are 3-segmented, and the maxillary palpi 5-segmented or reduced. The wing venation is almost complete for the heteroneurous condition. The chorda (the base of $R_4 + R_5$) may be present in the cell of the forewing, as may that of vein M. Vein M is present in the hindwing. The ovipositor is long, and telescoped into the final segments of the abdomen. It has a pair of ventral rods additional to the dorsally positioned anterior apophyses. In most Eriocottidae a pair of X-shaped rods occur in segment A8, structures that may be unique to the family.
(Nielsen, 1978.)

Biology

Virtually nothing is known of the general biology of the family. The long ovipositor led Nielsen (1978) to suggest that the larvae may be endophagous, in the expectation of eggs being inserted into crevices in parts of plants or internally into plant tissue. The moths have been collected by night but they also fly by day.

Classification

Prior to the study of Nielsen (1978) who observed the ditrysian condition of the female genitalia, *Eriocottis* was misplaced in the monotrysian family Incurvariidae. Two genera (*Eriocottis* and *Deuterotinea*) are included in Eriocottinae. The Compsocteninae (reviewed by Dierl, 1970) comprise a single genus *Compsoctena* to which about 50 species are currently assigned.

Acrolophidae

The family, with about 250 species (Robinson, 1988c), is restricted to the New World. Possible specializations were summarized by Robinson (1988c). The postgenae of the larval head are fused medially along a broad front (Hinton, 1956); the furcal apophyses of the adult metathorax are reduced and stumpy with the posterior tendons arising from the secondary furcal arms, not from the apophyses (Davis, 1986); and the ovipositor is short and stubby (a secondary reduction) with a concomitant loss of ventral rods. The arrangement of the furcal apophyses may indicate a reduction suggesting that the family should be subsumed within Psychidae (see below).

Acrolophidae moths are either large, robust and noctuid-like, or small, more delicate and tortricid-like. The wingspan ranges from about 10 to 40 mm (Hasbrouck, 1964), with males generally being smaller than females. The moths are predominantly brown.

Adult (Fig. 207)

The head and palps are rough-scaled. External ocelli are absent. The compound eyes may have long or short, delicate interfacetal setae, or these setae may be lacking; in several species the eyes are 'lashed', with hair-like scales arising around their margins. The antenna of the male varies from simple moniliform to strongly bipectinate, while that of the female is always simple. The proboscis is not developed and the maxillary palpi are minute and 2-segmented. The labial palpi of males are large, 3-segmented and upcurved, or are sometimes recurved over the head and thorax. In the female of any particular species the labial palpi are shorter than in the male; they are never recurved over the head and thorax. The wing venation is complete heteroneurous. (Hasbrouck, 1964.)

Immature stages

The egg is oval with prominent ridges (Hasbrouck, 1964).

In the larva (Davis *in* Stehr, 1987), the cuticular spines around the planta are slightly enlarged compared with those of Tineidae; otherwise the larvae are similar to those of Tineidae.

The pupa is heavily sclerotized (Forbes, 1923).

Biology

The larvae form underground tunnels opening at the surface and extending up to two feet or more. Typically, they feed on roots of grasses and have been observed to attack young maize plants (Hasbrouck, 1964). Others are coprophagous (Davis *in* Stehr, 1987). Larvae of *Acrolophus pholeter* feed on faecal pellets of the gopher tortoise and on decaying plant debris within the burrow of this reptile (Davis & Milstrey, 1988); certain other species are found in rodent burrows.

Psychidae (including Pseudarbelidae and Arrhenophanidae)

Psychidae are known as bagworm moths because their larvae construct prominent larval cases (bags). Females are frequently, but not exclusively, larviform, spending their entire life within the bag. Cases are made from a tube of silk to which material, such as twigs, leaf debris, sand grains, and small particles, are attached. Not all females are larviform; many are winged and some are brachypterous. The larviform state is the most specialised condition.

There are about 600 species of which around 500 are found in the Old World (Davis *in* Stehr, 1987). The moths are small to medium sized, typically fairly drab, but occasionally brightly coloured. Sometimes extensive scaleless areas occur on the wings.

The monophyly of the family is possibly supported by the presence of a corethrogyne (a dense tuft of hair at the tip of the abdomen in females), and the loss of the gnathos in the male genitalia (a convergent character) (Robinson, 1988c). Possible larval specializations include reduction of crochets on the prolegs, the presence of four pairs of ventral setae on the labrum, the occurrence of all three lateral setae on the prespiracular pinaculum, and the fusion of the spiracle and the prothoracic shield (Davis *in* Stehr, 1987).

Adult (Fig. 208)

The head is rough-scaled. External ocelli may be present or, usually, absent, and chaetosemata are absent. The proboscis and maxillary palpi are often vestigial or absent, and the labial palpi are short, in primitive genera, or vestigial. The antennae are often bipectinate. The epiphysis is present or absent; when present it may be long. The tibial spur formula is 0-2-4 primitively but in specialized species spurs are reduced in number or are absent. Females are frequently legless, generally brachypterous and may lack wings entirely. In winged individuals (i.e., all males and some females) vein M is often present in the cell of the forewing and often forked, and the chorda is sometimes present. Venation is generalized heteroneurous. The abdomen of males is often highly extensible so that it can be inserted into the bag of the female for mating. In the female the abdomen bears, terminally, a tuft of long hair-like, deciduous, backwardly pointing scales, the corethrogyne, which are mixed with the eggs at oviposition.
(Davis, 1964; 1975a; Nielsen & Common, 1991.)

Immature stages

The larva (Davis *in* Stehr, 1987) is hypognathous, and the head is variously, and darkly, pigmented. Six stemmata are present. Thoracic legs are well developed. Prolegs are present on A3-6 and on A10 and the crochets are arranged in a uniordinal penellipse.

The pupa (Mosher, 1916) superficially resembles that of Hepialoidea. The appendages are fused to each other and to the body and considerable mobility occurs between the abdominal segments of the abdomen. Spines, toothed ridges, and setae occur on the dorsal surface of the abdomen. In wingless females, pupal wing cases are not developed, whereas in apterous females of other families they are present.

Biology

'Micropsychina' (Psycheodinae) typically form bags from silk and minute particles of detritus or sand, while 'Macropsychina' (Psychinae) usually cover their bags with large fragments. The form of the bags may be of some taxonomic importance, but this is by no means always so (Davis, 1964). Larvae often live on stones or rocks in fairly open habitats. When found on forest trees, they select those on the edges of forests or in clearings. Since females of most species never leave the bag, they are mated within the bag. Eggs are laid inside the pupal exuviae, which itself remains in the bag. In temperate regions at least, larval development usually takes more than one season, and may last two to three years. Lichenivory is widespread; larvae of other species feed on trees and shrubs. Several species are important pests. Tea, coffee, and citrus plants are

often attacked as are fruit trees, and vegetables and conifers are also damaged. Pupation occurs in the case.
(Kozhanschikov, 1956.)

Classification

Psychidae have been divided into the 'tineiform' Psycheodinae (Micropsychina) and the 'bombyciform' Psychinae (Macropsychina) (Kozhanschikov, 1956). Other works have split the family into several subfamilies, but no satisfactory phylogenetic analysis has been established, and for practical purposes Kozhanchikov's division is used here. In Psycheodinae veins 2A and 3A of the forewing are fused for almost their entire length, and in the female the legs are usually well-developed and the ovipositor long. Females are sometimes alate. In Psychinae veins 2A and 3A are fused for only a short distance, the females are always apterous and without legs, and have a short ovipositor.

   Arrhenophanidae, included in Psychidae by Robinson (1988c), are a small group of moths from Central and South America (Bradley, 1951). The vesica of the aedeagus in the male genitalia is characteristic, being permanently everted and extremely long. A membranous sac within the abdomen (termed an apotheca by Bradley, 1951) houses the coiled vesica. The bursa copulatrix of the female is thread-like, and correspondingly long. *Harmaclona*, a genus considered by Bradley (1953) to belong to the *Arrhenophana*-group, is a typical tineid despite the presence of a tympanal organ (Robinson & Nielsen, *in press*). The Pseudarbelidae are regarded as Psychidae by Robinson (1988c), who doubted even that they constitute a monophyletic group.

Phylogenetic relationships

Psychidae are probably most closely related to Acrolophidae, a suggestion based on three shared specializations. These are: the absence of an antennal pecten; a rudimentary or lost proboscis; and the extension of the metafurcal apophyses to meet, and fuse with, the secondary furcal arms. In Acrolophidae the furcal apophyses do not meet the secondary furcal arms. If this lack of fusion indicates a reduction, Acrolophidae should be subsumed within Psychidae.
(Robinson, 1988c.)

Galacticidae

The *Homodaula* group of Kyrki (1984) was raised to family rank as Galacticidae by Kyrki (1990), who removed the assemblage from the Yponomeutoidea. The family includes about a dozen species, and is widely distributed in the Old World. One species has been introduced into North America. The moths are small, and have been placed provisionally in Tineoidea by Minet (1986) and Common (1990).

Adult (Fig. 209)

The head is smooth-scaled. External ocelli and chaetosemata are absent. The maxillary palpi are short, 2-segmented structures, and the labial palpi are short and weakly curved upwards. Terminally, the abdomen of the female is extensible.
(Common, 1990.)

Immature stages and biology

The body of the larva, at least in Australian species, is pale with dark markings. On the prothorax, group L is trisetose. The crochets on the ventral prolegs are arranged in a circle. The larvae live in webs.

The pupal appendages are fused, and the dorsum bears spines. The exuviae are protruded from the cocoon prior to adult emergence. (Common, 1990.)

Phylogenetic relationships

The pupa, which is spined dorsally and protruded from the cocoon prior to adult emergence, is unlike this stage in Yponomeutoidea. Therefore, Galacticidae are assigned, at least provisionally, to the Tineoidea, which they appear to resemble most closely.

GRACILLARIOIDEA

Four families are included - Gracillariidae, Bucculatricidae, and Roeslerstammiidae (Robinson, 1988c), and Douglasiidae (G.S. Robinson, pers. comm.). In the first three families possible specialized characters include the reduction of the maxillary palpi, the compressed lateral condition of the ovipositor, the presence of an additional inner line of crochets on the larval prolegs, and the occurrence of larval hypermetamorphosis.

Reduction in the wing venation may occur, the wings are coupled by a frenulum-retinacular system, and the forewing-metathoracic locking device is present.

Gracillariidae (Lithocolletidae)

The moths are very small and often colourful, frequently with shining white markings, long antennae, and narrow wings with prominent fringes. At rest, the front of the moth is typically raised while the back touches the substrate. However, in some the head is kept down while the abdomen is raised. The larvae are mostly leaf miners, at least in the early stages. Worldwide in distribution, Gracillariidae include over 1600 species (Davis *in* Stehr, 1987). The family includes *Phyllocnistis*, a genus sometimes, but unnecessarily, assigned to a separate family.

Adult (Fig. 210)

The head is nearly always smooth-scaled. External ocelli and chaetosemata are absent. The proboscis is well developed, and the maxillary palpi are 4-segmented and porrect, or sometimes reduced. The labial palpi are 3-segmented and usually ascending. The epiphysis is present, and the tibial spur formula is 0-2-4. The wings are lanceolate or narrow-elongate, the hindwings being even more narrow than the forewings. The venation may be almost full or much reduced; the cell of the forewing is long. The ovipositor is not extensible. (Vári, 1961.)

Immature stages

The eggs are laid singly, and glued to the surface of the plant tissue to be mined.

The larvae (Davis *in* Stehr, 1987) are sap-feeders in the early instars; later instars are tissue-feeders. Hypermetamorphosis occurs when the mode of life changes. The number of larval instars varies from 4 to 11. Early (sap-feeding) instars are prognathous with a dorsoventrally flattened head. The body is also flattened. The labrum is enlarged and the edges sometimes serrated. Mandibles are scissor-like for piercing the cells of plant tissue. The hypopharynx is usually broad and hairy, and the labium is reduced with the spinneret absent or vestigial. Thoracic setae are much reduced or absent, and legs, and prolegs are absent or vestigial. In tissue-feeding instars the head and body are cylindrical and the mandible bears distinct chewing cusps. Thoracic legs and abdominal prolegs are usually present. Crochets are uniordinal and usually biserial, and are arranged in a lateral penellipse, in bands, or in a circle.

In the pupa, the antennae and proboscis usually extend beyond the tips of the wings. Minute spines are present on the dorsum, and pupal exuviae protrude from the cocoon when the adult emerges.

Biology

The predominant biological feature of the family is the mining habit of the larvae. This usually occurs in leaves, but young bark is also mined. The tissue-feeding stages of some of the more primitive species feed outside the mine, although the larvae are concealed in folded or rolled leaves. Mines are typically blotches. Pupation may occur outside the mine in a chamber formed by a rolling-over of the leaf, or within the leaf mine. Chambers in mines are sometimes expanded by the contraction of drying silk (Hering, 1951). The genus *Phyllocnistis* forms a long, epidermal mine terminating in a small blotch. The larva is a sap-feeder throughout its development apart from the final instar, which does not feed and in which the legs and mouthparts, with the exception of the spinneret, are lost.

A wide variety of foodplants is accepted, and the larvae may cause damage to fruit trees and ornamental plants.
(Davis *in* Stehr, 1987.)

Phylogenetic relationships

It is uncertain as to whether the closest relatives of Gracillariidae are Roeslerstammiidae or Bucculatricidae + Roeslerstammiidae (Robinson, 1988c). With Roeslerstammiidae, the Gracillariidae share elongated antennae, an ostium bursae positioned anterior to sternum A8, and one or more dagger-shaped signa (modified as single and cruciform in Roeslerstammiidae). With Bucculatricidae + Roeslerstammiidae, male Gracillariidae share a retinaculum arising between the costa and vein Sc (rather than from vein Sc itself), a bisetose L-group on the meso- and metathorax of the larva, and a unisetose SV group on A1-5 of the larva. The position of the retinaculum is convergent with the situation in non-tineid Tineoidea (*sensu stricto*). The hypermetamorphosis of the larva was considered homologous to that found in certain Bucculatricidae (Robinson, 1988c), but nothing is known about the occurrence of this feature in Roeslerstammiidae.

Bucculatricidae

The family is excluded from Yponomeutoidea since pleural lobes are lacking from A8 of the male abdomen (Kyrki, 1984). A single widespread genus, *Bucculatrix*, is included. The larva spins a distinctive ribbed cocoon, and the first flagellar segment of the antenna in males is notched in many species. The family includes mostly small, drab moths, with larvae that are typically leaf miners in their early instars and external feeders in later instars.

Adult (Fig. 211)

The head is distinctively subtriangular when viewed anteriorly, and usually has a tufted vertex; rarely it is smooth-scaled. External ocelli are absent. The proboscis is

short, the maxillary palpi rudimentary, and the labial palpi minute. At the base of the antenna, the scape is expanded somewhat, although only rarely does it reach the size of the characteristic eye-cap of certain other groups. In males, the first flagellar segment is often deeply notched; in some species the notch is slight, and in others it is absent. The epiphysis is present, rarely absent, and the tibial spur formula is 0-2-4. The wings are lanceolate, and their venation somewhat reduced. There are no pleural expansions of A8 in males. The genitalia of many females bear a characteristic signum in which a series of spined ribs form a ring, or incomplete ring, encircling the corpus bursae near its posterior end.
(Braun, 1963.)

Immature stages

The eggs (Braun, 1963) are usually flattened and ovoid. Sculpturing is variable.

In the larva (Braun, 1963) the body is typically flattened in early instars but more rounded in later stages. On the abdomen setae L_1 and L_2 are widely separated on A1-8. Prolegs are elongate and slender. The crochets of the ventral prolegs are arranged in a single transverse row, those of the anal prolegs in two transverse rows.

Pupation takes place in a characteristic ribbed cocoon (Jäckh, 1955; Braun, 1963). The appendages of the pupa are free, and there is considerable mobility of the abdominal segments. Minute abdominal spines are present on the dorsum, and the pupal exuviae are exposed at adult emergence.

Biology

The eggs are glued to the leaf of the foodplant, although one North American species is reported to have an elongated egg standing perpendicular to the leaf. First instar larvae eat their way through the egg shell and enter the leaf directly. Typically, the larvae are leaf-miners in their early instars and then become external leaf-skeletonizers. Sometimes the larvae mine throughout their lives. Occasionally, late instars eat complete holes in the leaf. *Bucculatrix* mines are very narrow, linear, full-depth channels, which expand into a blotch in species with larvae that mine throughout their lives. As a miner, the larva is flattened and apodal. In larvae that do not mine throughout this stage of the life, the third instar leaves the mine and moults into a cylindrical external feeder with thoracic legs and long slender prolegs. These habits correspond to what Braun (1963) regarded as one of the two main divisions of *Bucculatrix*. In the other division, known from North America, larvae live either in stem-galls, which they presumably induce, throughout their lives or mine leaves in their early stages and bore into stems in later instars. In at least one species there is a non-feeding instar between the last instar and the pupa. This stage represents a hypermetamorphosis with the non-feeding larva lacking a spinneret and prolegs. The situation resembles that occurring in a South African gall-associated species in which a relatively immobile penultimate instar without prolegs is transformed into a mobile final instar with well developed prolegs (Scoble & Scholtz, 1984).

The cocoon and its construction is distinctive (Lyonet, 1832; Jäckh, 1955; Emmet, 1985). The spindle-shaped, ribbed structure is characteristic of the group, and its construction is remarkable for being spun from the outside. Before construction, two 'palisades' of vertical silk strands are often spun, each forming a semicircle around either end of what will become the cocoon. It has been suggested that these strands guide the larva in its subsequent spinning. The outer ribbed cocoon is lined with an inner tube.
(Braun, 1963; Davis *in* Stehr, 1987.)

Phylogenetic relationships

The position of *Bucculatrix* has been disputed for most of its taxonomic history, the most recent debate being concerned with the question of whether the genus should be treated as a subfamily (Bucculatricinae) of Lyonetiidae or placed in a family of its own within Tineoidea. Kyrki (1984) argued that Lyonetiidae belonged to Yponomeutoidea, a superfamily defined on the presence of pleural lobes on segment A8 of males (see under that superfamily below). Since *Bucculatrix* lacks these lobes, Kyrki proposed that the group should be placed as a family of its own in Tineoidea. Robinson (1988c) tentatively suggested that *Bucculatrix* is related to Gracillariidae and Roeslerstammii-dae. *Bucculatrix* was divided into two subgenera by Scoble & Scholtz (1984) to account for the discovery of a species from southern Africa with glossy white wings and large eye-caps resembling a cemiostomine Lyonetiidae, with a larva that spun a typical *Bucculatrix*-like cocoon. This species (*B. (Leucoedemia) ingens*) probably represents the sister group of all other members of the genus. If Lyonetiidae and Bucculatricidae are correctly assigned to different superfamilies, and *ingens* to *Bucculatrix*, then the external features of *ingens* must be convergent with those of typical Cemiostominae.

Roeslerstammiidae

Roeslerstammiidae include *Roeslerstammia* and those genera previously placed in Amphitheridae (Kyrki, 1983b). The family occurs in the Palaearctic region, in S.E. Asia and the Orient, and in Australia. Three specialized characters support the monophyly of the group (Robinson, 1988c): a brush of long hair scales on the outer surface of the valva of the male near the base; a single cruciform signum in the corpus bursae of the female; and stalked or completely fused veins M_3 and CuA_1, in the hindwing.

Adult (Fig. 212)

The vertex of the head is rough-scaled, and the frons smooth-scaled. External ocelli and chaetosemata are absent. The antennae usually extend beyond the wings. The maxillary palpi are reduced and the labial palpi are long and recurved. The epiphysis is present and the tibial spur formula is 0-2-4. Many species have metallic forewings. The wing venation is fairly complete for Heteroneura but in the forewing cell the base of vein M and of the chorda is vestigial or absent. In the male genitalia, the valvae bear a small pad of long hair scales near the base. The ovipositor is short, and a single cruciform signum occurs in the corpus bursae.
(Kyrki, 1983b.)

Immature stages

In the larva the L-group is bisetose on the meso- and metathorax, and the SV group is unisetose on the abdominal segments. The prolegs bear crochets arranged in a bilateral penellipse.

In the pupa the prothoracic femur is exposed, and the abdomen is spined dorsally. (Moriuti, 1978; Kyrki, 1983b.)

Biology

Larvae of *Roeslerstammia erxlebella* (see Kyrki, 1983b) mine the leaves mainly of *Tilia* (Tiliaceae) and *Betula* (Betulaceae) in their first and second instars. Subsequently, they live in a weakly developed web and feed externally. A cocoon is formed under a fold of the edge of the leaf, within which pupation takes place. In Australia, the larvae of

Thereutis, Macarangela, and *Nematobola* are also leaf-miners initially and external feeders later (Common, 1990). In *Agriothera elaeocarpophaga,* which occurs in Japan and India, the larva binds several leaves of the foodplant (*Elaeocarpus*: Elaeocarpaceae) together and feeds from within the shelter (Moriuti, 1978).

Phylogenetic relationships

Kyrki (1983b) noted that the male genitalia are similar to those of Eriocottidae, but that as in Gracillariidae the larva is a leaf miner with crochets on its prolegs, and the apophyses in the female genitalia are short.

Douglasiidae

This family of very small moths is widely distributed, but includes only about 20 to 25 species (Agassiz, 1985; Heppner *in* Stehr, 1987). Most occur in the Palaearctic, but the family is also represented in North America and in Australia. As in most Gracillariidae, the adults rest with the front of the body raised. Few species have been reared from the larval stages but, where known, Douglasiidae larvae mine leaves, tunnel in flowers, or bore into stems of members of Boraginaceae, Roseaceae, and Lamiaceae (= Labiatae) (Gaedike, 1974; Agassiz, 1985; Heppner *in* Stehr, 1987).

Adult (Fig. 213)

The head is smooth-scaled, external ocelli are present, and chaetosemata are absent. Maxillary palpi are vestigial and the labial palpi short and drooping. Epiphyses are present and the tibial spur formula is 0-2-4. The wings are lanceolate and the venation slightly reduced. A short ovipositor is present.
(Gaedike, 1974; Common, 1990.)

Immature stages

The larva is hypognathous, fusiform, and has reduced prolegs lacking crochets. Some of the setae are very long.
The pupa bears abdominal spines on the abdomen.
(Heppner *in* Stehr, 1987.)

Biology

The larvae mine leaves, tunnel in the petioles of flowers, or bore into stems (Heppner *in* Stehr, 1987). The European species *Tinagma balteolella* spins together flowers of *Echium vulgare.*
Pupation takes place within a cocoon in the larval site.

Phylogenetic relationships

The group was placed in Gelechioidea by Forbes (1923) and Yponomeutoidea by Heppner *in* Stehr (1987). Kyrki (1984) allocated the family to Tineoidea (*sensu lato,* i.e., including Gracillarioidea) based partly on the absence of pleural lobes on segment A8 of the male, structures typical of Yponomeutoidea, but the superfamilial assignment of Douglasiidae remains uncertain according to Robinson (1988c). The reduction in length of the palpi and the raised position of the body of the moth when at rest suggest a possible, but tentative, placement in Gracillarioidea.

YPONOMEUTOIDEA

The composition of the superfamily was clarified by Kyrki (1984), who suggested that the monophyly of the newly constituted group (based on *Yponomeuta*) was supported by the presence of pleural lobes on abdominal segment A8 of the male. The lobes vary in size. Sometimes fused to the tergum, they often take the form of prominent flaps covering the valvae of the genitalia and may resemble a pair of second valvae. The superfamily includes a diversity of very small to medium sized moths.

Adult

External ocelli are present or absent; chaetosemata are generally absent. The proboscis, usually well developed, is not scaled at its base. The maxillary palpus is 1 to 4 segmented; it may be minute to moderately developed in size. The antennae are simple, and a flap of scales or a pecten may be present or absent on the scape.

The epiphysis is present and the tibial spur formula is typically 0-2-4. The wing venation is usually fairly complete, but is sometimes reduced significantly. A pterostigma (a thickening of the forewing between veins Sc and R_1, which sometimes extends towards the base of the wing along the costal margin) is often present. Wing coupling is achieved by a frenulum-retinacular system, and in the female there are usually only two frenular bristles. The forewing-metathoracic wing coupling device is present. Vein R_5 reaches the termen.

At the base of the abdomen, sternum A2 is of the 'tineoid type', with apodemes and sternal rods but without extended anteriolateral corners. In many Yponomeutoidea there is a transverse rim near the anterior margin of sternum A2 (Kyrki, 1983a), a structure that may be a derived character of the superfamily (Kyrki, 1984). If this interpretation is correct, the absence of the rim from certain Yponomeutoidea would imply a secondary loss. Apart from variation in the ratio of length to width, the form of the sternum is fairly uniform (Kyrki, 1983a). Paired lobes usually occur on the pleuron of segment A8, but occasionally they are present on tergum A8. These lobes vary in size, and are secondarily lost on occasion or fused to the tergum. Between the pleural lobes and the genitalia, a pair of ventrolateral coremata generally occur. In the male genitalia, there is usually a pair of teguminal processes rather than a well developed, medial uncus. In the female, the anterior apophyses usually have both a dorsal and a ventral branch.
(Kyrki, 1984.)

Immature stages

Cranial setae P_1 and P_2 of the larva form a line approximately horizontal in frontal view, rather than the more usual vertical arrangement. On the prothorax, 2 or 3 lateral setae (L) are present and on the abdominal segments L_1 and L_2 are far apart, rarely close together. The pupa is not protruded from the cocoon when the adult emerges and, related to this, there are no rows of tergal spines. The appendages are firmly fixed to the body.
(Kyrki, 1984.)

Classification

The exact composition of individual yponomeutoid families, and their relationship one to another is uncertain. Different workers have adopted different systems. Fifteen generic groups of Yponomeutoidea were recognized by Kyrki (1984) with a further 12 genera unassigned to any group. Kyrki (1983a) drew up a tentative list of the component families to include Yponomeutidae (*sensu lato*), Ypsolophidae (his Ochsen-heimeriidae), Heliodinidae, Lyonetiidae, and Glyphipterigidae. This infrasuperfamilial arrangement disrupts the existing family concepts as little as possible. It was adopted by Karsholt *et al.* (1985) and Minet (1986), following Kyrki's implicit recommendation. A further attempt to clarify the classification has been presented (Kyrki, 1990) but many uncertainties still exist particularly when species from areas other than Europe are taken into account.

Yponomeutidae

Kyrki (1983a; 1984) included a wide range of genera (slightly modified by Kyrki & Itämies, 1986) in what he termed Yponomeutidae *sensu lato*. This arrangement is adopted here although the composition of the family will undoubtedly change with further study, particularly of those genera from outside Europe. In what follows, the definition of the family is broader than is often conceived for Yponomeutidae; for instance, the *Argyresthia* and *Plutella* groups frequently have each been given family status and Kyrki (1990) treats the *Plutella* group and the *Acrolepia* group as Plutellidae. The Yponomeutidae, in the wider sense, may be monophyletic, but the group has not yet been shown to be so. However, with some exceptions the gap between the costa of the hindwing and vein Sc + R suddenly narrows from about the distal third of the wing, a character not found in other Yponomeutoidea (Minet, 1986).

The family is cosmopolitan, best developed in the tropics and includes around 800 species of very small to small-medium moths. Sometimes drab, Yponomeutidae may also be conspicuous in colour and pattern. Some (e.g., *Argyresthia* and *Zelleria*) rest with the head down and the body raised (e.g., Common, 1970: fig. 36.26).

Adult (Fig. 214)

The head may be smooth-scaled or bear long hair-like scales. External ocelli may be present or absent, and chaetosemata are sometimes present. The maxillary palpi vary markedly in length and in number of segments, and the labial palpi are ascending, or may be short and drooping. Sometimes the tibial spurs are reduced in number. The wings vary from narrow and lanceolate to relatively broad. The forewing usually has a pterostigma. The venation is usually slightly reduced. In the hindwing, veins Rs and M_1 are usually separate rather than stalked. Tergal spines are sometimes present on the abdomen. The ovipositor is short with hairy lobes or, as in the *Argyresthia* and *Plutella* groups may be long and narrow.
(Kyrki, 1984, 1990; Munroe, 1982; Common 1990.)

Immature stages

The head is hypognathous. Prolegs, ranging from long and slender to short, bear crochets that are usually uniordinal, sometimes tending to biordinal, and are arranged in uniserial to multiserial circles or broken circles or, rarely, transverse bands.
(Heppner *in* Stehr (1987): under families Acrolepiidae, Argyrestiidae, Yponomeutidae, and Plutellidae.)

Biology

In many species (e.g., the *Yponomeuta* and *Plutella* groups), the larvae spin webs over leaves or flowers, are gregarious, and typically skeletonize leaves. Some genera (e.g., *Argyresthia*) mine needles of conifers (Freeman, 1972). Several important pests are found in the group including the cosmopolitan Diamond-back moth (*Plutella xylos-tella*), the larva of which damages vegetables. In the *Prays* group, *P. citri* is a pest of citrus, and last instars of *P. oleae* attack olives. Included in the *Argyresthia* group is the Australian species *Ogmograptus scribula* (the scribbly gum moth), the larva of which is responsible for long scribble-like mines in the bark of some species of *Eucalyptus* (Common, 1970). Members of *Argyresthia* and its relatives are typically miners or borers, particularly on conifers where they attack cones and needles. Larvae of *Acrolepia* and its allies are chiefly leaf miners, but also bore into buds, stems, seeds, and bulbs. The leek moth (*Acrolepiopsis assectella*) is a pest of leeks and onions.

Classification

Six subfamilies were proposed by Kyrki (1990), but the phylogenetic relationships between them is uncertain.

Ypsolophidae (= Ochsenheimeriidae)

The family is equivalent to the *Ypsolopha* group of Kyrki (1984), including the genus *Ochsenheimeria*. It occurs in the Palaearctic as far as North Africa and eastern Russia, and is represented in northwestern India and, as a result of accidental introduction, in the U.S.A. (Davis, 1975b). The moths are small with slender to stout bodies, drab in appearance, and diurnal. In *Ypsolopha* the forewings are fairly narrow and elongated with the apex often produced or falcate. Larvae mine the leaves of grasses and other monocotyledons, and bore into their stems.

Structure and biology

In the adult (Fig. 215), external ocelli are present or absent. The maxillary palpi are 4-segmented and moderate to long, but 2-segmented in *Ochsenheimeria*. Segment 2 of the labial palpus may or may not bear a scale brush. A pterostigma occurs in at least some species. In the forewing the base of veins 1A + 2A is forked prominently. Abdominal segment A8 is reduced, and the aedeagus usually has a caecum. The apophyses of the female genitalia are long, the ductus seminalis joins the ductus bursae near the ostium, and the signum comprises an elongated plate with two transverse ridges. In the larva, crochets are uniserial. The pupa lacks a cremaster and bears movable segments.

In *Ochsenheimeria* (Davis, 1975b; Davis *in* Stehr (1987); Karsholt & Nielsen, 1984) external ocelli are well developed, and compound eyes very small, features that are associated with diurnal habits in moths. In the female the ovipositor is extensible and elongated, and the ostium is asymmetrical. Larvae, of which few are known, usually mine leaves of their foodplants (primarily grasses - Poaceae - but also other monocotyledons of the families Cyperaceae and Juncaceae). Later, they become stem borers and are able to transfer from one stem to another.
(Kyrki, 1984, 1990; Moriuti, 1977.)

Heliodinidae

The family includes a group of small, metallic-coloured, diurnal moths with about 55 described species (Heppner *in* Stehr, 1987). The group is worldwide in distribution. Many species previously described in Heliodinidae are stathmopodine Oecophoridae, assigned to the former because the habit of raising the hindlegs at rest is found in both groups. Heliodinidae may be distinguished from Stathmopodini by the absence of scales at the base of the proboscis. *Schreckensteinia*, also sometimes placed in Heliodinidae, was excluded and assigned to its own superfamily by Minet (1983).

Adult (Fig. 216)

The head is smooth-scaled, external ocelli are present, and chaetosemata are absent. Antennae are typically thickened in the male. The proboscis is developed, maxillary palpi are very small, and the labial palpi are 3-segmented, drooping or porrect. The forewing is narrow lanceolate or narrow triangular, and the hindwing slightly narrower than the forewing. Vein CuP is only weakly developed. There is no great elongation of the ovipositor.
(Munroe, 1982.)

Immature stages and biology

The larva bears well developed thoracic legs, and small, slender, relatively short, prolegs, which are sometimes reduced. Crochets are arranged in a uniordinal circle. The larvae mine or skeletonize leaves, and bore into fruit, and sometimes live communally in a web.
  The pupa is often flattened and bears lateral ridges; but dorsal spines are absent.
(Heppner *in* Stehr, 1987.)

Lyonetiidae

Lyonetiidae include about 200 named species of small moths with representation worldwide. Three generic groups (often given subfamily status) remain in what was once a more inclusive family (Kyrki, 1984). These are *Bedellia* (Bedelliinae), the *Lyonetia* group (Lyonetiinae), and the *Leucoptera* group (Cemiostominae). However, the monophyly of this more restricted group remains uncertain although the presence of an immobile pupa is possibly an independently evolved specialization (Minet, 1986). The Bucculatricidae were excluded from Lyonetiidae by Kyrki (1984) and transferred to Tineoidea (*sensu lato*) (see under Tineoidea, above).
  The moths are small with the forewings narrowly lanceolate (*Bedellia* and the *Lyonetia* group) or broadly lanceolate (*Leucoptera* group).

Adult (Fig. 217)

The front of the head is smooth-scaled and the vertex rough-scaled. External ocelli and chaetosemata are absent. The proboscis is short and the maxillary palpi rudimentary. In *Bedellia* the forewings are ochreous irrorated with fuscous, whereas in the *Lyonetia* and *Leucoptera* groups they are mostly shining white with striking black and yellow-orange markings, which take the form of bars and an apical spot. In some species the ground colour is grey. In the forewing, vein R_5 terminates on the costa. The antennal scape is often expanded to form an 'eye-cap'. In the *Leucoptera* group, the abdominal terga are finely spined. The ovipositor is specialized for piercing in the *Lyonetia* group, but not in the other groups.
(Kuroko, 1964.)

Immature stages

The egg is typically flat and oval.

The head of the larva is prognathous or semiprognathous and the body is cylindrical in later instars but more flattened in early instars. There are usually 6 stemmata. Thoracic legs are usually fully developed, but in the first two instars they are frequently absent. The L-group is usually trisetose on abdominal segments A1-8, but in *Bedellia* it is bisetose. Although typically absent from the first two instars, in final instars prolegs are present on A3-6 and A10. Crochets are usually arranged in a uniordinal, uniserial circle, but the circle is not always complete.

The pupal appendages are fused to the body, and the abdominal segments are immobile.

(Davis *in* Stehr, 1987.)

Biology

Eggs are either laid on the surface of the leaf of the foodplant (in *Cemiostoma* and *Bedellia*) or inserted into the leaf (in the *Lyonetia* group). Most species are leaf miners, many mining one leaf throughout their larval life, others moving to different leaves. The cocoon is usually white and spindle-shaped. *Bedellia* is a leaf-miner on Convolvulaceae including sweet potato (*Ipomoea batatas*).

Phylogenetic relationships

Bedellia was treated as a family by Kyrki (1990) in his tentative classification of Yponomeutoidea. In that account it was suggested that relationships between Heliodinidae, Bedelliidae, and Lyonetiidae (restricted sense) were unresolved.

Glyphipterigidae

The composition and phylogenetic position of this family, the sedge moths, has been clarified by Heppner (1982a, b). The group is cosmopolitan. The moths are small, often bear metallic markings, and are typically diurnal.

Adult structure and biology

In the adult (Fig. 218), the head is usually smooth-scaled, external ocelli are present (e.g., as in Fig. 5), and chaetosemata nearly always absent. The antennal scape lacks a pecten. The maxillary palpi are small, and the labial palpi rather weakly upturned. Both pairs of wings are fairly broad. In the forewing, vein R_5 reaches the apex or the termen, and CuP is represented near the margin. Vein CuP is often reduced in the hindwing. The pleural lobes on A8 of the male are fused to the tergum so that this sclerite appears exceptionally wide and almost surrounds the end of the abdomen (Kyrki, 1984). In *Orthotelia* (Orthoteliinae), the lobes are well developed (Kyrki & Itämies, 1986).

Immature stages

Eggs are deposited singly.

The larva is hypognathous and characterized by the presence of enlarged thoracic spiracles on cone-like protuberances, which are particularly prominent on T1. Spiracles take a similar form on the abdomen, and those on A8 are borne on a particularly large cone. Prolegs are vestigial on A3-6, and crochets are either sparse, uniordinal and arranged in a lateral penellipse, or are absent.

The pupa is not protruded when the adult emerges and dorsal spines are absent. The thoracic spiracles are elevated.

(Heppner *in* Stehr, 1987.)

Biology

The biology of relatively few species is known. Larvae of most species bore in seeds or stems, others into leaf axils, and a very few mine leaves. Pupation occurs in the channel formed by the larva. Most hostplants are monocotyledons belonging to Juncaceae and Cyperaceae, but several other families of plants are also recorded. Specimens of *Diploschiza kimballi*, a species belonging to a genus of New World Glyphipteriginae, were reared from a sample of aquatic weeds. Members of the European genus *Orthotelia* (Orthoteliinae) live at pools on shore meadows, and their hostplants (e.g., *Iris* and *Scirpus*) often grow in shallow water. The larva mines leaves and stems in spring and early summer, and pupates within the stem (Kyrki & Itämies, 1986).

Classification

The family is divided into two subfamilies, Glyphipteriginae and Orthoteliinae (Kyrki & Itämies, 1986). The widespread European genus *Orthotelia* was added to Glyphipterigidae on the basis of several shared larval specializations between *Orthotelia* and other glyphipterigids. The lateral group on the prothorax is bisetose, adfrontal punctures are absent from the head, the spiracle on A8 is enlarged and sited posterior to the normal position, A8 and A9 have sclerotized dorsal shields, and the foodplants are monocotyledonous. Six shared derived characters for the Glyphipteriginae were suggested (Kyrki & Itämies, 1986): their diurnal habits, the smooth-scaled head, a reduction of the base of vein M, the reduction of vein CuP in the forewing, the approximated or stalked condition of veins M_3 and CuA in the hindwing, and the fusion of pleural lobes to tergum A8. The presence of what Heppner (1981) called 'secondary valvae' in a species of the New World genus *Diploschizia* (*D. minimella*) suggests that the pleural lobes may not be fused to tergum A8 universally in glyphipterigines.

Phylogenetic relationships

Glyphipterigidae were shown to belong to Yponomeutoidea by Kyrki (1983a, 1984). The thorax-abdominal articulation is 'tineoid'- not 'tortricoid'-like, and although there are no free pleural lobes on A8 of the male, the form of tergum A8 suggests that lobes are present but secondarily fused with the segment.

GELECHIOIDEA

The superfamily is a large grouping of very small to medium- and occasionally large-sized moths with concealed larvae. There are about 12,000 species of which almost one third are found in Australia. While the superfamily is almost certainly monophyletic, several of its constituent taxa are not. Three well known diagnostic characters of Gelechioidea may represent specializations for the superfamily (Minet, 1986). These are: the imbricate scale-covering of the base of the proboscis; the clasping of the base of the proboscis by the maxillary palpi; and the strongly recurved condition of the labial palpi. The scaling at the base of the proboscis is not unique to Gelechioidea (it is present notably in Pyralidae); the maxillary palpi are reduced in certain Gelechioidea so therefore they are not folded over the base of the proboscis; and the labial palpi are not always sickle-like, they may be only weakly ascending or porrect, and they are recurved in certain non-Gelechioidea.

Adult

The head is smooth scaled. External ocelli are present or absent, and chaetosemata are absent. The antennae are generally either simple or ciliate, but rarely they may be pectinate or bipectinate. The proboscis is present and anteriorly its base is covered by a dense mass of imbricated scales. The maxillary palpi, usually 4-segmented, are short and folded over the base of the proboscis, or they are reduced (Coleophoridae, Agonoxenidae, Elachistidae, and some Oecophoridae). The labial palpi are usually recurved or upturned, but in some instances they are porrect.

The epiphysis is present and the tibial spur formula is 0-2-4. The forewing is usually narrow, but may be relatively broad. A pterostigma is often present. The venation, almost complete in the groundplan, is significantly reduced in some narrow winged groups. Vein M is rarely present in the discal cell, and vein 1A may be present or absent. The hindwing ranges from very narrow to broad. Vein CuP is sometimes absent. Occasionally the wings are reduced. Wing-coupling is effected by a frenulum-retinacular system, and the forewing-metathoracic locking device is present.

The abdominal terga are frequently spined, often extensively. Sometimes the spines are arranged in transverse bands. The male genitalia may be symmetrical or asymmetrical, and sometimes the valvae are reduced with their function transferred to modified pregenital segments. In the female the ovipositor is often extensible. (Hodges, 1978; Common, 1990.)

Immature stages and biology

Eggs are usually laid singly, and are sometimes inserted into flowers or crevices (Hodges, 1978).

The larvae have concealed lives and are endophagous or exophagous. Some live, for example, in mines, galls, and rolled leaves, others feed externally but from within portable cases. The L-group of the prothorax is trisetose, rarely (Momphidae) bisetose; on the abdomen setae L_1 and L_2 are close together; and the crochets of the prolegs are arranged in a circle, an ellipse, or rarely in two transverse bands (Nielsen & Common, 1991).

The pupa (Nielsen & Common, 1991) lacks spines on the dorsal surface of the abdomen, and is not protruded from the larval case at ecdysis. Considerable mobility exists between the abdominal segments.

Classification

The actual ranks given to taxa at the family and subfamily levels in Gelechioidea vary considerably between different classifications, and no consensus exists. The system adopted here is essentially that of Hodges (1978). Contrasting approaches were adopted by Zimmerman (1978), who incorporated most families into Gelechiidae, Minet (1986) and Nielsen & Common (1991), who increased the number of families, and Minet (1990b) who proposed a significantly modified classification of the superfamily based to a considerable extent on characters from immature stages.

Oecophoridae

The grouping is probably not monophyletic. It has been based on primitive characters - the presence of vein CuP in the forewing, and the well separated condition of veins Rs and M_1 basally. Since the family is considered to be unnatural a description is inappropriate. Instead, the various subgroups, which are mainly monophyletic, are discussed. Within Hodges' conception of Oecophoridae fall broad-winged and narrow-winged moths. There are over 4000 described species of which more than half occur in Australia.

Depressariinae form a large, cosmopolitan group particularly well represented in temperate areas (Hodges, 1974). The group is sometimes given family rank (e.g., Common, 1990) and Minet (1990b) proposed that it should be treated as a subfamily of Elachistidae. The moths (e.g., Fig. 219) are broad-winged gelechioids and characteristically flattened, a feature implied by their name. The abdominal terga usually lack spines. The male genitalia are symmetrical and the gnathos is composed of a single spinose lobe or sometimes two separate knobs. The ovipositor is not extensible. The larvae are leaf rollers or tiers, feed on seeds and flowers or, rarely, mine leaves. Unlike Oecophorinae larvae, which feed on detritus, larvae of Depressariinae consume live plant tissue.

The depressed shape of the moths permits them to squeeze into narrow spaces, and they often overwinter in thatch and haystacks (Holloway et al., 1987).

Ethmiinae, often given family rank (e.g., Sattler, 1967; Powell, 1973), have about 250 described species (Powell, 1973), mainly in the genus *Ethmia*. The group is cosmopolitan with its greatest diversity in the northern Neotropical region. Minet (1990b) proposed that the group should be treated as a subfamily of Elachistidae.

The moths (e.g., Fig. 220) are small to large gelechioids, nocturnal or diurnal depending on species, and with forewings commonly bearing many black spots on a pale background. Their principal distinguishing features (Sattler, 1967) are the presence of a 'segmented' costal lobe on the valva of the male, and the occurrence of a pair of ventrocaudal processes on the pupa (structures also present in Agonoxenidae - Zimmerman, 1978). In the hindwing, vein M_2 is usually closer to M_3 than to M_1. The abdominal terga are not spined. The apex of the gnathos is spinose.

The biology of 15 species of *Ethmia* was studied by Powell (1971). Eggs are laid singly, and each is approximately rectangular in outline. When disturbed most larvae wriggle energetically backwards while others feign death. There are probably five instars in most species. Larvae tend to seek soft woody material in which to pupate. The cremaster of the pupa is vestigial, and this stage of the life cycle is probably anchored by a pair of anteriorly directed processes. Most species feed externally on Boraginaceae or Hydrophyllaceae, although some other foodplants are also known (Powell, 1973).

Peleopodinae was described by Hodges (1974) for a small group of New World species. The moths (e.g., Fig. 221) are broad-winged, and differ from other Oecophoridae in having veins Rs and M_1 stalked in the hindwing. Little is known of the larvae, but one species lives under a web spun on a leaf. Pupation occurs in this shelter.

Minet (1986) gave the group family rank and added three other genera from the Old World.

Autostichinae (e.g., Fig. 222), a small group, are Indo-Australian and include genera that had been placed mainly in Gelechiidae or Xyloryctinae (Hodges, 1978). The presence of spines arranged in transverse rows at the posterior edge of the abdominal terga of both autostichines and xyloryctines led Minet (1986) to treat Autostichinae as Xyloryctinae.

Xyloryctinae are fairly widely distributed but occur mostly in Indo-Australia, where there are 250 described species (Common, 1990). The subfamily (Common, 1990) includes small to large moths (e.g., Fig. 223). External ocelli are absent, the antennae of males are ciliated or pectinate, and the scape lacks a pecten. Vein CuP is present in both wings, and Sc + R₁ diverges from Rs well before the end of the cell in the hindwing. Tergal spines are typically present in transverse bands on the posterior region of each abdominal tergum. The crochets on the prolegs of the larva are biordinal and arranged in a circle or an ellipse.

The larvae live concealed lives, but may emerge from a hidden position to feed on lichens. Galleries may be made of silk and debris, or they may be formed of silk and soil. Some species tie leaves or bore into bark or the wood of branches. Whilst some species actually feed on bark, others drag leaves into their burrows. Xyloryctinae feed on several families of plants, and a few are lichenivorous.

Stenomatinae (see Hodges, 1978) include small to medium-sized, broad-winged, chiefly neotropical moths (e.g., Fig. 224). Minet (1990b) proposed that they should be treated as a subfamily of his broadly conceived family Elachistidae. Most species are assigned to either *Stenoma* or *Antaeotricha*. The hindwing is usually considerably broader than the forewing. Veins Rs and M₁ of the hindwing are usually stalked. The abdominal terga of the adults lack spines and the apex of the gnathos is spinose. The larvae of many species are leaf tiers, some are borers, and others are leaf miners. Pupation takes place within larval shelters.

Oecophorinae were divided by Hodges (1974) into two tribes (Oecophorini and Pleurotini) to which he later (Hodges, 1978) added a third, the Stathmopodini. The most prominent of the three tribes is the Oecophorini, which are particularly well represented in Australia. Pleurotini are confined to the Old World apart from a single species in North America. They differ from Oecophorini in details of the labial palpi and female genitalia (Hodges, 1978).

Oecophorini (Oecophorinae of Nielsen & Common, 1991) include numerous species. The subfamily (Hodges, 1978; Nielsen & Common, 1991) includes mainly broad-winged moths although there are a few species with apterous females. The maxillary palpi are not reduced, and the labial palpi are commonly strongly recurved. In the fore-wing, veins R₄ and R₅ are stalked, and in the hindwing Rs and M₁ are separate and run parallel. The abdominal segments are densely spinose, and the male genitalia are symmetrical. Crochets on the larval prolegs are arranged in a biordinal, or rarely triordinal, ellipse.

The wings are folded steeply over the body, and the adults walk with a slow waddling gait (Common, 1980). Oecophorine larvae exhibit a variety of habits (Nielsen & Common, 1991). Some live in portable cases, others tunnel into wood, flowers, or the soil, and some are leaf tiers. A large number in Australia live amongst, and feed upon, leaf debris on the ground in *Eucalyptus* forests. Larvae of *Myrascia* store distasteful oils from their myrtaceous hostplants in a diverticulum of the foregut which they eject if prodded (Common & Bellas, 1977). Pupation occurs in the larval shelter, or the pupa may be exposed and attached by its end so that it stands upright as in *Hypertropha*, from Australia, (Common, 1980).

Although few oecophorines are pests, *Hofmannophila pseudospretella* is an exception. This species (Carter, 1984) is widespread, and occurs in stored products including, for example, grain, dried fruit, seeds and furs. It is a common domestic pest. Although the larvae feed on the stored substrate it appears that they are often detritivorous scavenging on general organic debris.

Stathmopodini were considered by Hodges (1978) to be very closely related to Oecophorini because of similarities in the head, genitalia, and larval feeding habits. The moths are small, narrow-winged gelechioids with distinctive hind legs bearing whorls of stiff bristles. At rest the hind legs are raised in most species. They are found mostly in the Afrotropical and Indo-Australian regions. A number are colourful and diurnal. The larvae (Common, 1990) who treats the group as Stathmopodinae) are borers or miners; some are predatory either on scale insects or the eggs of spiders.

Lecithoceridae

The family is predominantly tropical or subtropical and occurs mainly in the Oriental and Australasian regions. There are, nevertheless, several representatives in the Palaearctic. The moths are fairly small, broad-winged, and mainly fuscous, yellow, cream, or grey.

Adult (Fig. 226)

The head is smooth-scaled, external ocelli are absent, and the antennae are often thickened in males and long. The maxillary palpi are 4-segmented and folded over the base of the proboscis, and the labial palpi are recurved. The venation may be reduced. (Gozmany, 1978; Common, 1990.)

Immature stages and biology

The egg is ribbed and, at least in one Australian species, upright.

The larva bears dense, branched secondary setae on verrucae. The crochets are arranged in a uniordinal mesal penellipse or in a pair of transverse rows.

At rest, the head of the moth is directed downwards and the antennae are extended anteriorly.
(Common, 1990.)

Elachistidae

The family comprises a group of small moths many of which are white or grey variously marked with fuscous, others of which are predominantly dark and marked with white. Represented worldwide, the group is best developed in the northern hemisphere, particularly North America and the European and Mediterranean parts of the Palaearctic. The moths are mostly diurnal, and the larvae are typically miners in monocotyledons. The Fennoscandian and Danish fauna was detailed by Traugott-Olsen & Nielsen (1977), the North American species by Braun (1948). Minet (1990b) suggested that the family should be much more inclusive.

Adult (Fig. 227)

The head is smooth scaled, and external ocelli are generally absent. The maxillary palpi are very short, and the labial palpi usually long, porrect to recurved, or secondarily short and drooping. The scape of the antenna is large, and often forms an 'eye-cap'. Whereas the forewings range from relatively broad to narrow, the hindwings are always narrow. Brachyptery is rare. The venation is fairly complete even in the hindwings of more primitive species. In some species the hindwing venation is significantly reduced. In the hindwing, $Sc + R_1$ is remote from Rs, and Rs nearly always runs along the axis of the wing, a character that is diagnostic. The apex of the gnathos is spinose. The ovipositor is short.

Immature stages and biology

The larvae (Wagner *in* Stehr, 1987) exhibit considerable diversity in body shape. Some are typical of generalized gelechioids while those mining grasses are highly modified (e.g., they have strongly flattened heads). The head is prognathous, well sclerotized, and capable of being retracted into the thorax. Those that feed on grasses and sedges bear flattened clavate setae on each side of the claw of the thoracic legs. Crochets on the abdominal ventral prolegs are uniordinal (or rarely biordinal) and are arranged in a circle or as transverse bands. Most larvae are miners but some are stem borers. Whereas some species feed on dicotyledons, most mine the leaves of monocotyledons.

The pupal appendages are firmly attached to the body. Pupation typically occurs outside the mine in a cocoon or a flimsy covering of silk. The pupae of some species are fully exposed and attached to the substratum by a girdle.

Classification

Minet (1990b) suggested that Elachistidae should be broadened to include Agonoxeninae, Elachistinae, Stenomatinae, Cryptolechiinae, Hypertrophinae, Ethmiinae, and Depressariinae. This radical proposal was based largely on the presence of a modification of the usually mobile abdominal segments of the pupa, which restricts their lateral movement. The anterior margin of the posterior segments is produced into a point where it articulates directly with the posterior edge of the previous segment.

Pterolonchidae

The group is small with 10 species in *Pterolonche* (9 from the Mediterranean region and 1 from South Africa), and 1 species, from South Africa, in *Anathyrsa*. In the adult (Fig. 228), the wings are lanceolate and the body and legs are slender. Hodges (1978) stated that they showed certain similarities to Pleurotini (Oecophoridae), but differ in having a reduced proboscis, fused R_4 and R_5 in the forewing, and in genital structure.

Coleophoridae

The family includes Coleophorinae and Batrachedrinae. Batrachedrinae were transferred to Coleophoridae from Momphidae by Hodges (1978).

Adult (Fig. 229)

External ocelli are absent. The maxillary palpi are small or minute, and the labial palpi of Batrachedrinae are drooping. In the forewing one or two veins are usually absent, and at least one vein is absent from the hindwing. Spines are present on the abdominal terga; in Batrachedrinae they are covered by scales, in Coleophorinae they are not. The apex of the gnathos is usually spinose.

(Common, 1990 - as Batrachedridae and Coleophoridae.)

Immature stages, biology, and classification

Coleophora (Coleophorinae) is a homogeneous genus with about 1000 species (Sattler & Tremewan, 1978) of small, narrow-winged moths. The genus is cosmopolitan but most species occur in the Holarctic. The maxillary palpi are minute and not folded over the base of the proboscis, and abdominal terga A1-7 each bear two patches of spines arranged longitudinally rather than transversely. The epiphysis is sometimes lost or reduced. The adults rest with their antennae porrect.

In the larva (Stehr *in* Stehr, 1987), the head is semiprognathous. Ventral abdominal prolegs are reduced and bear transverse bands of uniordinal crochets. Sometimes crochets are absent from A6. The anal prolegs are usually short. The first instars are leaf miners. Second instars construct cases, sometimes characteristic of the species, from fragments of leaves, frass and silk. Subsequent instars continue to live as case bearers and pupation occurs within the case. Case bearing stages are often termed 'external' leaf miners because the larva extends into the leaf as a typical miner while the case remains attached to the surface of the leaf. Some species of *Coleophora* are leaf skeletonizers, feed on flowers or live in galls or leaf rolls.

In *Batrachedra* (Batrachedrinae) (Hodges, 1966) the moths are small, drab and narrow-winged, and are represented in all the main zoogeographical regions. Hodges (1978) removed the genus from Momphidae because of differences in genitalia and venation. He assigned them to the Coleophoridae mainly because vein CuA_2 usually arises at almost a right angle from the cell, and because the larvae of some species make similar cases to some species of *Coleophora*. Larvae feed on a variety of substrates for example on flower heads and seed heads, as miners of pine needles, as inquilines of galls, and as borers in leaves of *Agave* (Agavaceae). Those of *Batrachedra stegodyphobius* live as commensals in nests of the communal spider *Stegodyphus* from southern Africa (Pocock, 1903), and those of *B. arenosella*, from Australia, on scale insects. (Common, 1990.)

Agonoxenidae

The family is small, and its relationships within the Gelechioidea uncertain. Two subfamilies, Agonoxeninae and Blastodacninae, are included (Hodges, 1978). The moths are narrow-winged and small. Agonoxeninae are found from Australia through the islands of the western Pacific to Hawaii (Bradley, 1966). Blastodacninae occur mainly in the Holarctic region.

Adult (Fig. 230)

External ocelli are absent. The maxillary palpi are reduced, 1-segmented structures, and the labial palpi are recurved with their terminal segments flattened. The venation is reduced slightly in the forewing and considerably in the hindwing. The abdomen lacks dorsal spines. (Bradley, 1966; Common, 1990.)

Immature stages

In the larva of Agonoxeninae (Bradley, 1966) the crochets of the ventral prolegs are irregularly uniordinal and biserial and arranged in a circle. In the larva of Blastodacninae (Stehr *in* Stehr, 1987) the crochets are uniordinal and arranged in a mesal penellipse to a mesoseries. There are more secondary setae than in Agonoxeninae.

Biology

Agonoxena larvae feed on coconut and other palms (Arecaceae). That of the coconut flat moth *A. argaula*, which is of economic importance as a result of the damage it does to the leaves of young palms, lives between folded leaflets of its host.

Blastodacninae larvae are endophagous in the leaves, stems, and fruit of various members of the Rosaceae; in later instars they may roll leaves.

Blastobasidae

The family (Adamski & Brown, 1989) includes, at a very conservative estimate, 300 species of small to medium sized moths with narrow wings. The group is represented worldwide, but their greatest diversity is in the tropics and subtropics.

Adult (Fig. 231)

The head is usually smooth scaled. External ocelli are absent. The antennal scape is long and bears a pecten. In some species the first flagellar segment of males is dilated at the base and deeply notched. Although usually conspicuous and upcurved, labial palpi are sometimes strongly reduced. A pterostigma is present between veins Sc and R1. The venation of the forewing is hardly reduced, but there is some fusion of veins in the hindwing. The base of CuA_2 in the forewing makes nearly a right angle to the Cu trunk. In females, a subcubital retinaculum is present. Spiniform setae occur on abdominal terga A1-8 in males and A1-6 or A1-7 in females. In the male genitalia, a long internal aedeagal sclerite is present. In the female, sternum A8 is plate-like, and the ovipositor long and extensible.
(Hodges, 1978; Adamski & Brown, 1989.)

Immature stages

The larva (Stehr *in* Stehr, 1987) is hypognathous to semiprognathous. Six stemmata occur on each side of the head. A pit on the submentum is present. Crochets are arranged in a uniordinal or biordinal circle on the short ventral prolegs and in a transverse band on the anal prolegs.

The pupa (Common, 1990) is fairly broad and lacks a cremaster.

Biology

The larvae are usually scavengers, but some feed in acorns and others are phytophagous (summary in Adamski & Brown, 1989). Many of the detritivores live in nests of externally feeding caterpillars, flowers or seed heads, or in the burrows of boring insects; some are entomophagous (e.g., on scale insects), and others bore into seeds or fruits (Stehr *in* Stehr, 1987). Phytophagous habits are probably underestimated (Adamski & Brown, 1989).

Classification

A cladistic study (based on the North American species) by Adamski & Brown (1989) confirmed the monophyly of the Blastobasidae and established their division into monophyletic subfamilies, Blastobasinae and Holcocerinae.

Momphidae

Only the genus *Mompha* is included (Hodges, 1978) although two others genera, *Chrysoclista* and *Synallagma* were considered to be associated by Stehr *in* Stehr (1987). *Mompha* includes over 100 species of narrow-winged, small to medium-small moths. Most species occur in North America but the genus is represented in most other areas. They are apparently absent from Asia, the Orient, and Australia.

Structure, immature stages and biology

The adult (Fig. 232) is similar in general morphology to Blastobasidae, but the forewing lacks a pterostigma, the gnathos is absent from the male genitalia, a pair of hook-shaped signa are present in the corpus bursae of the female genitalia, and the larvae feed on living plant tissue (Hodges, 1978). Spines are present on the abdominal terga.

In the larva (Stehr *in* Stehr, 1987), the L-group on segment T1 is bisetose (atypical of Gelechioidea). Crochets are uniordinal and arranged in a circle or a penellipse on the

ventral prolegs. Larvae are associated with Onagraceae and have been collected from stems and flower buds.
(Hodges, 1962; 1978.)

Scythrididae

These small and frequently darkish, but sometimes paler greyish, moths have often been associated with Yponomeutoidea but the scaled base of the proboscis is characteristically gelechioid. Although more than 370 species have been described, this total represents a small fraction of the diversity that actually exists (Landry, 1991). Distribution is worldwide. The wings are narrow and the venation is reduced; certain species are flightless. The moths are drop-shaped when the wings are folded over the body at rest.

The monophyly of the family is based on several specialized characters (Landry, 1991). These are the narrow ductus bursae, the broad ductus seminalis, the absence of a signum on the corpus bursae, the shape of the metathoracic furca, the presence of only 2 subapical spurs on tarsomeres 1-4, the immobilisation of the aedeagus through fusion with other parts (ankylosis), and the stalking of veins R_4 and R_5 with R_4 extending to the costa and R_5 to the termen.

Adult (Fig. 233)

External ocelli are present or absent. The maxillary palpi are very small, and the labial palpi, which are ascending, diverge from each other. Only 2 subapical spurs occur on tarsomeres 1-4. In the forewing, veins R_4 and R_5 are stalked, and vein M_3 usually coalesces with CuA_1. Vein CuP is present. In the hindwing, veins M_2 and M_3 may be separate, stalked, or coalesced. The abdomen is broad, particularly in the female. The abdominal terga lacks spines. The male genitalia are extremely variable, and often asymmetrical, and the aedeagus is ankylosed. In the female the ovipositor is long.
(Bengtsson, 1984; Landry, 1991.)

Immature stages

The egg (Bengtsson, 1984) is ellipsoidal, of the upright variety, and with a flattened top and base.

The larva (Bengtsson, 1984; Stehr *in* Stehr, 1987) bears conspicuous secondary setae, particularly at the bases of the proleg. The crochets are triordinal (sometimes quadri-, uni-, or biordinal) and arranged in a circle (sometimes incomplete).

The pupa (Mosher, 1916; Bengtsson, 1984) has only the last three segments movable.

Biology

The adults are mainly diurnal, but tropical and subtropical representatives are often nocturnal. When disturbed, the moths usually jump from the vegetation on which they rest and remain motionless. Larvae usually live in weak webs and feed externally on buds or leaves although a leaf miner of grasses, and a species where the larva lives in a leaf fold, are known.
(Bengtsson, 1984; Stehr *in* Stehr, 1987.)

Cosmopterigidae

The family is represented worldwide and includes over 1200 species, with many occurring in Hawaii (Hodges, 1978). The moths are typically small or very small, and are sometimes brightly coloured. The wings are typically narrow, or very narrow, but occasionally they are broader.

Three subfamilies, Cosmopteriginae, Chrysopeleinae, and Antiquerinae were recognized by Hodges (1978).

Adult (Fig. 234)

External ocelli are present or absent. The maxillary palpi are 4-segmented and are folded over the base of the proboscis. The forewing is broadly lanceolate and lacks a pterostigma. The venation ranges from almost completely to significantly reduced. In the forewing veins R_4 and R_5 are stalked, with the latter extending to the costa. Vein CuP is vestigial or absent. The hindwing is narrower than the forewing. In the male the uncus is absent, the gnathos usually takes the form of two asymmetrical lobes lacking apical setae, the vinculum is reduced, and the aedeagus is fused (ankylosed) with the anellus. In more primitive species the valvae are simple, the aedeagus free, the vinculum well developed, and the gnathos lobes symmetrical.
(Hodges, 1978; Common, 1990.)

Immature stages and biology

The larvae (Hodges, 1978; Stehr *in* Stehr, 1987) are difficult to distinguish from those of Gelechiidae and Oecophoridae. Most species of *Cosmopterix* and its relatives mine the leaves of monocotyledons, but Urticaceae and Moraceae have also been recorded as foodplants. Others bore into stems, feed in seeds, scavenge, or prey on scale insects. Feeding habits of Chrysopeleinae include mining in pine needles and in stems, leaf rolling, fruit boring, gall and flower dwelling.

Gelechiidae

The family is a large and diverse group of small to medium-sized moths represented worldwide. There are more than 4000 species described and many more await description (Hodges, 1986).

Adult (Fig. 235)

External ocelli are present or absent. The maxillary palpi, usually 4-segmented, are distinctly folded over the base of the proboscis, and the labial palpi are 3-segmented and upturned or recurved with segment 3 long and narrow. The pterostigma is often present. The wing venation is not generally reduced significantly. In the forewing R_4 and R_5 are usually stalked and the latter extends to the costa; vein CuP is absent. The hindwing is approximately trapezoidal, and the termen is frequently sinuate. Veins M_4 and CuA_1 are rarely separate, and CuP is generally absent. The abdominal terga are rarely spined; infrequently the male genitalia are asymmetrical.
(Munroe, 1982.)

Immature stages and biology

The larva (Stehr *in* Stehr, 1987) has biordinal crochets arranged in an ellipse or in two transverse rows. Rarely, the prolegs are absent. Gelechiid larvae vary greatly in their modes of life being, for example, leaf tiers or rollers, seed-feeders, leaf miners, stem, fruit, or tuber borers. Pupation occurs in a cocoon within the larval shelter or among detritus on the ground; the exuviae are not produced on adult emergence.

The larvae of many species are serious pests. For example *Pectinophora gossypiella* damages balls of cotton, *Phthorimaea operculella* bores in stems of Solanaceae (e.g., potato, tobacco, tomato), *Sitotroga cerealella* infests stored grain (Stehr *in* Stehr, 1987), *Scrobipalpa heliopa* (which damages tobacco and egg plant) has spread, probably from Australia, to other parts of the Old World (Common, 1990).

Classification

Three subfamilies were recognised by Hodges (1986) mainly on the basis of the structure of the base of the abdomen. Gelechiinae have well developed sternal apodemes (free anterior structures) and venulae on segment A2; Pexicopiinae have well developed apodemes but the venulae are absent or weakly indicated; and in Dichomeridinae apodemes are absent and the venulae well developed. Included here are Symmocinae (excluded from Blastobasidae by Adamski & Brown, 1989). Their larvae are scavengers and lichen feeders (Munroe, 1982).

COSSOIDEA

The superfamily includes the Cossidae and, provisionally, Ratardidae and Dudgeonei-dae. The position of Ratardidae is uncertain although some similarities were noted between that group and Metarbelinae (Holloway, 1986; Minet, 1991) and the family is provisionally included in Cossoidea in the present work. *Dudgeonea* (Dudgeoneidae) is also placed in Cossoidea (following Clench, 1957, 1959; Munroe, 1982; Common, 1990; Nielsen & Common, 1991), although the cossoid association of this genus has been questioned (Brock, 1971; Minet, 1982, 1983). Metarbelinae have been variously treated as a family, Metarbelidae, or as a subfamily of Cossidae.

Cossidae

The moths range from small to very large, and their wings often display a reticulate pattern. The ground colour of the wings is generally grey or brown, sometimes cream. The abdomen is long, and the bodies often contain large quantities of fat. The larvae are borers, predominantly in woody tissue - either in heartwood or bark - but sometimes they feed externally on roots and in stems of less woody plants (Stehr *in* Stehr, 1987). There are approximately 700 species (Schoorl, 1990).

Adult (Fig. 236)

External ocelli are present or absent, and chaetosemata are absent. The antennae may be simple, lamellate or prismatic, but are usually bipectinate. The proboscis is very short or absent, the maxillary palpi minute, and the labial palpi moderate or very short. On the prothorax, the patagia and parapatagia are protuberant, but sometimes secondarily reduced in height. The epiphysis is usually present. Tibial spurs are absent or have a formula of 0-2-4, or 0-2-2. The venation (e.g. Fig. 60) is almost complete. Vein M in the forewing is strong, and often forks within the cell. Frequently it also forks within the cell of the hindwing. Vein CuP is usually present in both wings (absent in Metarbelinae), although it is sometimes reduced, particularly in the hindwing. The wings are generally broader in females than in males. They are usually coupled by a frenulum and retinaculum, and by an overlap of lobes of the forewing or hindwing, but in Metarbelinae the frenulum and retinaculum are nearly always absent. The forewing-metathoracic locking device is present. Tympanal organs are absent. The ovipositor may be long and narrow or much shorter. (Munroe, 1982; Holloway, 1986.)

Immature stages

The egg (Common, 1990) varies in shape. It may be flat (Zeuzerinae and the Metarbelinae), or upright (Cossinae).

The larva (Stehr *in* Stehr, 1987; Holloway, 1986; Common, 1990) is cylindrical, and has a broad head with large mandibles. There is a distinctive prothoracic plate. A full complement of prolegs is present, each proleg being short or rudimentary. Crochets are arranged in an ellipse, a penellipse, or in transverse bands, and may be uni-, bi-, or triordinal (Stehr, 1987), or they may be arranged in a circle (Holloway, 1986).

The pupa (Mosher, 1916; Common, 1990) is long and approximately cylindrical, with two rows of spines on the abdomen and most abdominal segments movable. The pupa is protruded from the larval tunnel at eclosion.

Classification

The monophyly of Cossidae is based largely on wing shape and pattern, and also (Minet, 1986) on the presence of protuberant parapatagia on the prothorax. Schoorl (1990), who treated Metarbelidae as a separate family, based the monophyly of Cossidae on the relatively broad condition of the mesepimeron (when viewed dorsoventrally).

Chilecomadiinae (Schoorl, 1990) includes the genera *Chilecomadia* and *Rhizocossus*, which occur in Chile and Argentina. These genera were previously associated with *Pseudocossus* and *Dudgeonea* in Dudgeoneidae because all were thought to have tympanal or prototympanal organs. However, only *Dudgeonea* is actually so endowed (see Chapter 6). The two genera are grouped together on the basis of four rather inconspicuous specializations. The subfamily is difficult to characterize, but may be distinguished from Cossinae by the straight arrangement of veins R_3 and R_4 in the forewing of Chilecomadiinae compared with the sinuous course run by veins R_4 and R_5 in Cossinae. Unlike the situation in Cossidae, Chilecomadiinae do not have the tibia and first tarsomere of the hindleg swollen.

Cossinae are defined by four specialized characters (Schoorl, 1990). Segment 3 of the labial palpus is conical rather than ovate; veins R_3 and R_4 are sinuate; veins R_4 and R_5 are stalked; and the anal plate is only moderately long at the base of the hindwing. The antennae vary considerably in being filiform, unipectinate or uniserrate, bipectinate proximally or for their entire length (Holloway, 1986). Tibial spurs are present, and the hindtibia and first tarsomere are dilated. The male genitalia have a basal spine on the valva and, usually, a scobinate bulla uniting each half of the gnathos (Holloway, 1986). The frenulum is weaker in Cossinae than in Zeuzerinae. The subfamily occurs in Europe, Asia extending eastwards to Japan, New Guinea, Africa, and North and South America, with most species present in Asia (Schoorl, 1990).

A well known member of the subfamily is *Cossus cossus*, the goat moth (see Carter, 1984 for a summary of its biology), which is widely distributed in Europe. The common name is derived from the strong, goat-like odour of the larva caused by phenols the source of which is located in the mandibular gland (summary in Rothschild, 1985). The eggs of this species are laid in short rows or small clusters in crevices in the bark of trees. The larvae, which take two to four years to develop, bore between the bark and the wood of the host tree before tunnelling into the heartwood. They feed on a variety of trees, many of which bear commercial fruit. Often trees are killed by the larval activity.

Pseudocossinae (Schoorl, 1990), originally with the genera *Pseudocossus*, *Chilecomadia*, and *Rhizocossus* (Heppner, 1984), now include only the Madagascaran genus

Pseudocossus. There are five species, three of which are undescribed. The moths are medium-sized Cossidae with chestnut-brown wings marked with darker areas, and speckled with white. External ocelli are distinct, the antennal flagellomeres are prismatic, and in most species veins CuP and 1A + 2A converge and may meet distally.

Zeuzerinae include the Leopard moths with their typically spotted forewings. The subfamily is defined by several specialized characters (Schoorl, 1990). Among these are: the absence of a proximal pair of tibial spurs on the hindleg; a distinct anterio-dorsal process on the mesothoracic preepisternum; the extreme ventral position of the upper section of the mesothoracic parepisternal sulcus; and, at least in the male, an antenna that is bipectinate proximally with prismatic flagellomeres distally.

A general description is given by Holloway (1986). A large areole occurs in the forewing; in Cossinae this is much smaller. The anterior margin of the hindwing is developed into a convex lobe, and is presumably used in wing coupling in addition to the frenulum and retinaculum. In the male genitalia, the aedeagus is cleft at the apex and the scobinate bulla of the gnathos is absent. In the female, the ovipositor is long and slender as in typical members of Cossinae. In the larva, the prothoracic shield is particularly well sclerotized, and the prothorax is narrow and wedge-like (see Carter (1984) for illustration of *Zeuzera pyrina*, and Carter & Deeming (1980) for illustration of *Azygophleps albovittata*). The subfamily occurs in all areas, but most genera are found in Africa and the tropical part of Asia (Schoorl, 1990).

Several members of Zeuzerinae are pests. *Z. coffeae* damages coffee trees besides trees from many different families. The Palaearctic species *Z. pyrina* bores in the branches and sometimes trunks of many fruit trees and ornamentals, a habit that may kill the tree. *Azygophleps* (*sensu lato*) bores in stems of grasses, reeds, and rushes, as well as in leguminous plants. The group includes *Xyleutes* (see Common, 1990; Nielsen & Common, 1991), a widespread genus from both the Old and the New World, that may not be monophyletic. Some of the Australian species are very large. Enormous numbers of eggs are laid (more than 18 000 in *X. durvillei*). The larvae of this species form tunnels in the roots of *Acacia*. The witcheti grub (the larva of the Australian species *X. leucomochla*) feeds externally on the roots of *Acacia ligulata*.

Hypoptinae have been redefined by Schoorl (1990) to include those Cossidae where the third and fourth axillary sclerites of the hindwing are fused only anteriorly, and in which the fourth axillary sclerite lies, for most of its length, in the same dorsal plane as the third. As in Zeuzerinae, the upper section of the parepisternal sulcus is situated in a very ventral position. External ocelli are absent. Vein M lies close to CuA in both the forewing and the hindwing of most species, and CuP and 1A + 2A are linked by a cross vein in the forewing or virtually coalesce. Most species occur in South America, but the group is also represented in North America.

Metarbelinae (Janse, 1925; Holloway, 1986) are a predominantly Afrotropical group, but some species are found in the Oriental and Neotropical regions. They are often included in Cossoidea as a separate family (e.g., Common, 1990; Holloway, 1986). Minet (1986) included Metarbelinae in Cossidae, suggesting that the large parapatagium is a shared specialization. This character has not been established with certainty (Schoorl, 1990), but Metarbelinae are here included tentatively within Cossidae. However, Metarbelinae differ from other Cossidae in larval feeding habits and in the structure of the male and female genitalia.

The moths resemble other Cossidae in general pattern and colour. But Metarbelinae differ in lacking a frenulum (in all but a few species), lacking vein CuP in both the forewing and the hindwing, lacking vein M in the forewing, and differing in the structure of the ovipositor and the terminal abdominal segments. The ovipositor lobes bear dorsal, rounded expansions and a reduced ductus bursae and corpus bursae. The membrane between the seventh abdominal tergum and the genitalia is expanded. The antennae of both sexes are usually bipectinate throughout their length. A proboscis is absent. The legs bear at least the terminal pair of spurs.

Few details are known about the immature stages, but those available suggest that the larvae feed on the bark of trees rather than on heartwood. Janse (1925) found larvae of the Afrotropical species *Salagena tessellata* under the bark of pear trees causing considerable damage. Larvae of *Squamura maculata* (from Java, Sumatra, and Borneo) occur on many leguminous trees. The larva feeds at night on the cambium of the bark around the entrance of a tunnel in which it shelters during the day (Roepke, 1957).

Dudgeoneidae

The family includes *Dudgeonea*, occurring in Africa, Madagascar, India, S.E. Asia, and Australia and, possibly (Schoorl, 1990), *Acritocera*. Although currently assigned to a separate family and tentatively placed in Cossoidea, the affinities of *Dudgeonea* are not resolved (e.g., Schoorl, 1990).

The moths (e.g., Fig. 237) are medium-sized, with reddish brown forewings bearing silvery spots. External ocelli and chaetosemata are absent. The proboscis is absent, the maxillary palpi minute, and the labial palpi well developed and upturned. The wing venation is almost complete. A pair of tympanal organs, not dissimilar in structure to those of Pyralidae (Minet, 1983), occurs at the base of the abdomen. The ovipositor is fairly narrow.

The immature stages are unknown, but the larva of *Dudgeonea actinias* from Australia probably bores into *Canthium* (Rubiaceae) (summary in Common, 1990). (Munroe, 1982; Holloway, 1986; Common, 1990.)

Ratardidae

Ratardidae include about 10 species from the Oriental region, in particular from the Himalayan region of northern India, Taiwan to Borneo, Sumatra, and Java. The moths (e.g., Fig. 238) are medium to medium large in size, fairly drab, and with rounded wings. Wing colour is generally brown with white patches or white with brown markings. Ratardidae are known definitely only from females, and there is no information on juvenile stages. One species is diurnal (Heppner & Wang, 1987), but no comparative information is available for other species.

The phylogenetic position of the family is uncertain. Hering (1925) suggested that ratardids were related to Pterothysanidae (Calliduloidea, Chapter 12), while Brock (1971) and Holloway (1986) favoured placing them in Cossoidea. Bourgogne (1951) assigned them to Bombycoidea, while Heppner (1984) and Heppner & Wang (1987) followed Hering and associated them with Calliduliformes (= Calliduloidea). Minet (1987a) specifically considered them as representing the sister group of Callidulidae (including, in his classification, *Griveaudia* + Pterothysaninae (= Pterothysanidae)).

Characters shared by Ratardidae and Cossidae (Holloway, 1986) include the deep, narrow ovipositor lobes with dorsal expansions, a reduced ductus bursae and corpus bursae, and an expanded intersegmental membrane between tergum A7 and the female genitalia. A character shared by Ratardidae and Pterothysanidae (Calliduloidea) (Minet, 1987a) is the presence of long slender scales on the dorsal surface of the wings.

The larva is unknown.

TORTRICOIDEA

Tortricidae

Only one family, Tortricidae, is included. Tortricidae occur worldwide and are most diverse in temperate and tropical regions. There are over 5000 species (Horak, 1984), many of economic importance. Tortricidae are usually distinguishable from other moths of similar general size and appearance by the porrect labial palpi and the unscaled proboscis. Where labial palpi are upcurved, the terminal segment is never long and thin (Horak & Brown, 1991).

The moths are small to medium-sized. Most are fairly drab, and cryptic at rest; others exhibit spectacular colours. The shape of the forewings in many species is such that when they are folded, the outline of the moths, particularly females, resembles a bell. In some species, the wings are approximately rectangular and in others they are rounded apically. Many species have tufts of scales on the forewings and the thorax. The larvae are borers, leaf-tiers or rollers; rarely, they live and feed amongst leaf litter on the ground.

Adult (Fig. 239)

The head is generally rough-scaled. External ocelli are sometimes reduced, rarely vestigial, and occasionally (some Schoenotenini) absent. Chaetosemata are present. The antennae are generally filiform, but occasionally pectinate. There is no pecten on the scape. The base of the antennae may be notched or expanded and flattened in some males. The proboscis is unscaled. The maxillary palpi are primitively 4-segmented and very short; in most species they are not visible *in situ*.

The epiphysis is present, and the tibial spur formula is 0-2-4. Hairpencils or brushes of scales are present on the hindtibiae of males of several groups. The hindwings are about the same width as the forewings. The shape of the forewing varies considerably within the family and even between sexes within single species. When at rest, females of some Archipini, Epitymbiini, Sparganothini, and Atteriini, resemble, in outline, a bell. Both forewings and hindwings of males may have folds with associated scent scales. A costal fold at the base of the forewing of the male is found in many species and protects various scent glands and their distributing hair-pencils. In some species costal folds are vestigial or absent, a condition interpreted as a secondary reduction. The hindwing usually bears a cubital pecten of hairs in Olethreutinae and in Sparganothini. The groundplan wing venation is almost complete. In the forewing, the chorda (the base of R_{4+5}) and the stem of vein M are present in the discal cell of the most primitive species. Other members of the family, particularly Chlidanotinae and Tortricinae, lack these veins or have them in a vestigial condition. Primitively, all veins run separately from beyond the discal cell; specialization is by way of partial or sometimes entire fusion of certain veins. Vein CuP is usually present although often weak, but is absent from Cochylini (a tribe of Tortricinae). In the hindwing, the stem of vein M is never developed in the discal cell. Three branches of M are separated primitively, but varying degrees of fusion occur. Wing coupling is effected by a frenulum-retinacular system, and the forewing-metathoracic locking-device is present.

In the pregenital abdomen, the anterior corners of sternum A2 are elongated into a pair of anteriolateral processes (Kyrki, 1983a) (see Chapter 4). Some Tortricinae have a pair of dorsal pits or rarely a 'single pit' (i.e., fused pair) medially, on the tergum of A2, often that of A3 and, rarely, on subsequent terga as well. A pair of pits occurs on

tergum A2 of some Olethreutinae. The function of these structures is unknown. In the females of several groups, densely packed scales (the corethrogyne) are found on one or more of the terminal abdominal segments. In the female genitalia the anal papillae usually take the form of characteristic, well-developed pads. In some groups the ovipositor is enlarged, concave and bears modified, nail-like setae (floricomous condition). Rarely, the ovipositor is telescoped. In certain species a band-like sclerotization (cestum) runs along the wall of the ductus bursae.
(Horak, 1984; Horak, 1991.)

Immature stages

Despite considerable variation in shape, tortricid eggs (Powell & Common, 1985) are basically flat with the long axis horizontal and the micropyle at one end. The degree of flattening is often so great that the egg becomes scale-like. In *Scolioplecta comptana* (Phricanthini) the eggs are globose, shaped like a lemon.

The larva (Brown *in* Stehr, 1987) has six stemmata on each side of the head. On T1, the L-group is trisetose. Setae L_1 and L_2 are close together on A1-8. A full complement of prolegs, with crochets, is present. The crochets of the ventral prolegs are uniordinal, biordinal, or triordinal, and are arranged in a circle. On the anal prolegs the crochets are arranged in a continuous row. An anal comb or fork is usually present above the anus, but this structure is usually absent in borers. In Tortricidae the prongs of the anal combs are straight, whereas in Gelechiidae they are curved.

The pupa (Mosher, 1916) is adecticous, obtect, and has several movable abdominal segments. Dorsal spines are generally present on most abdominal segments, but in the Ceracini, the pupae of which are not extruded at adult emergence, spines are absent. Transverse grooves, or pits may occur near the anterior margin of abdominal segments A2 and A3. Their function is unknown. Pupation mostly takes place within the larval shelter.

Biology

Egg-laying habits (Powell & Common, 1985). Eggs may be laid singly or in batches. The flattened, circular eggs of Schoenotenini are laid singly or in rows of two or three, in Epitymbiini the flat eggs are weakly overlapped and laid in batches of 5-15. In most Archipini eggs are laid in masses formed of imbricated rows. In some Archipini and some Sparganothini, for example, they may be covered with an opaque or transparent colleterial fluid. In other instances, eggs are covered by corethrogyne scales from the abdomen. Scales may be affixed to the surface of an egg, so that it resembles a pincushion, and overlaid with a thatch of further scales (e.g., members of the *Euphona* group). Other Tortricidae arrange the scales as tufts that fringe or partly obscure the eggs. Scales from enlarged fringes on the costal margin of the hindwings are placed in clumps around the egg masses laid by females of *Cryptoptila*. Some species cover their eggs with debris using their floricomous ovipositors (e.g., as in Fig. 100).

Larval habits. The larvae are typically concealed feeders, often living within plant tissue (e.g., Roelofs & Brown, 1982; Brown *in* Stehr, 1987). Members of most of the tribes associated in the unnatural subfamily Tortricinae are foliage feeders, although the Epitymbiini feed on leaf-litter (Horak & Common, 1985), and the Cochylini in roots and seeds. Many Tortricinae are polyphagous. Most Olethreutinae are borers or gall-formers but there are many exceptions, such as Olethreutini, which feed mainly on foliage. Few Olethreutinae are polyphagous. Where known, larvae of Chlidanotinae are concealed feeders (e.g., in the cambium of *Pinus*, Heppner, 1982d).

The larva of *Semutophila saccharopa* (Olethreutinae) discharges from its anus a solution of sugar, amino acids and protein when stimulated by ants (Maschwitz *et al.*, 1986).

Pheromone studies. Sex attractants have been analysed for many Tortricidae (Roelofs & Brown, 1982) although, given the great number of species in the family, the information is necessarily fragmentary. Pheromone evolution in Lepidoptera may occur through modifying enzymes giving rise to variations of three main kinds: desaturation, chain-shortening and elongation, and acyl reduction (Roelofs & Brown, 1982). One generalization emerging from these studies is that while pheromones of Tortricinae are usually 14-carbon compounds, those of Olethreutinae are mostly 12-carbon compounds. The steps by which pheromones with similar components are synthesized may differ considerably, so caution is needed when using pheromones to indicate phylogenetic relationships.

Classification and diversity

The most recent classification (Horak & Brown, 1991) suggests that of the three subfamilies into which the Tortricidae are divided, two (Chlidanotinae and Olethreutinae) are almost certainly monophyletic, while Tortricinae are an unnatural group.

  Tortricinae (Horak & Brown, 1991) include groups with the least specialized male genitalia. Although the subfamily is unnatural, its members can be recognised relatively easily. The antennae nearly always bear two rows of scales on each flagellar segment, and the juxta and aedeagus are usually articulated rather than fused. Of the 11 constituent tribes, most are monophyletic, but the relationships of one tribe to another, and of Tortricinae to Chlidanotinae and Olethreutinae are unresolved. Phricanthini include 23 species found primarily in Australia and S.E. Asia. The hostplants, where known, all belong to the Dilleniaceae, and the larvae feed on foliage within a shelter, or on living bark. Most of the 300 or so Tortricini are leaf-rollers, but rarely they bore into plants. *Russograptis* is carnivorous and feeds on coccids. Schoenotenini, with approximately 200 species, are well developed in the Papuan region. Their larvae tie and roll leaves, bore into a variety of plant tissues, or sometimes mine in the first instar. The Cochylini (often classified as Phaloniidae), with about 600 species, include mostly oligophagous larvae usually feeding in flower-heads, seed-capsules, stalks and roots. They are rarely leaf-rollers or tiers. Mainly holarctic, the Cnephasiini include around 300 species. The larvae sometimes mine leaves in the first instar. Archipini, with well over 500 species, include several unrelated groups, although the core of the group is probably natural. *Archips* includes about 100 species and occurs through the Holarctic and Oriental regions as far east as Timor (Tuck, 1990). Many species are polyphagous. All larvae of Epitymbiini feed on leaf-litter as far as is known, nearly always that of *Eucalyptus*. The tribe is primarily Australian, and its core is monophyletic. The chemical structure of their pheromone components is unique. There are well over 100 species. Nearly all the 130 or so species of Sparganothini are found in the New World. The larvae are mostly leaf-rollers; those of *Synnoma lynosyrana* live in small groups in silken tubes. In the Atteriini the modified scales at the end of the abdomen of the female are used to form a fence around the eggs. The tribe, with 40 species, is Neotropical. The Euliini, with nearly all of the 240 species also from the Neotropics, may be unnatural. The Ceracini are almost certainly monophyletic. The tribe includes about 30 species of large and colourful Tortricidae centred in the Himalayan region. The larvae are leaf-rollers. Some are polyphagous, others feed on Pinaceae.

  Chlidanotinae (Tuck, 1981; Horak & Brown, 1991) include three tribes together constituting a monophyletic group based primarily on the unique possession of a deep, dorso-longitudinal invagination of each of the valvae. Into each invagination may fit a pencil of hairs the base of which is found on segment A8; but the hairpencils are often absent. The Polyorthini, with around 100 species, is primarily Neotropical and

Oriental/Australian. The pantropical Hilarographini include about 80 species. The glyphipterigid-like pattern on the wings led them to be misassociated, and only recently were they recognized as Tortricidae. The few known larvae are borers. Of the 37 species of Chlidanotini, one is found in Africa, two in New Caledonia, one in Norfolk Island and the rest in the Indo-Australian area. The apical area of the forewing is bent upwards at rest. Nothing is known of the immature stages.

Olethreutinae include six tribes (Horak & Brown, 1991), the most generalized of which include larvae that are leaf-rollers, seed-feeders, and stem-borers. With well over 3000 species, Olethreutinae are the largest of the three tortricid subfamilies. Characteristic of their members is the fusion of the aedeagus with the anellus and the juxta, and the presence of an 'excavation' at the base of the valva. The tribe Gatesclarkeanini includes about 60 species occurring particularly in S.E. Asia and the South Pacific. In the male genitalia of members of this tribe the valvae are usually asymmetrical. The Bactrini include over 100 species. The larvae bore in the stems of Cyperaceae and Juncaceae. Microcorsini are a small tribe occurring in South America, Africa and Madagascar, the Oriental and Australasian regions, and the eastern Palaearctic. About 1400 species are placed in the unnatural tribe Olethreutini. Olethreutini occur in all regions of the world, and include many serious pests. Eucosmini are also probably not a natural group. There are about 1000 species, mostly from the Holarctic. The polyphyletic Grapholitini, with about 600 species, have been associated on hindwing venation and reduction of structures in the male genitalia. This unnatural assemblage includes *Cydia pomonella*, the codling moth - a species that causes great damage to apple and stone fruits.

Phylogenetic relationships

The relationship of the Tortricidae to other superfamilies is unclear, but it has yet to be demonstrated that they should be united with Cossidae in a single superfamily (Forbes, 1923).

CASTNIOIDEA

Castniidae

The family includes around 150 species from the Neotropical, Oriental, and Australasian regions (Holloway *et al.*, 1987). Most occur in the Neotropics, an area with 30 genera (Miller, 1972, 1986). In Australia all Castniidae belong to *Synemon*, a genus with 34 species (Common, 1990). A small genus *Tascina* occurs in Borneo, Singapore, and the Philippines. Castniid moths are brightly coloured, medium-sized to large, diurnal Lepidoptera with clubbed antennae. Superficially they resemble certain butterflies (particularly Hesperiidae), and the larvae typically bore into stems or roots.

Adult (Fig. 240)

External ocelli are large and chaetosemata absent. The antennae are smooth and clubbed terminally, the club often extending into an apiculus or, sometimes, being flattened. Sensilla are present on the club, and sometimes flattened scales also occur (Miller, 1980). The proboscis may be well developed, slightly reduced or, rarely, strongly reduced, and the maxillary palpi are very small. The labial palpi are usually upturned. The epiphysis is present, and the tibial spur formula is 0-2-4. Sensory brushes (paronychia) occur on the pretarsus anterior to the claws. Both pairs of wings are broad. The wing venation is little reduced. The chorda is sometimes present, and

vein M occurs in the cell of the forewing and the hindwing and lies close to the cubital stem. Wing coupling is effected by a frenulum-retinaculum system, and the forewing-metathoracic locking-device is present. In both wings vein CuP may be present or absent. In the male genitalia, the aedeagus typically is curved, a characteristic feature of the family. In females, the ovipositor is usually elongated.
(Holloway, *et al.*, 1987; Miller, 1986; Common, 1990.)

Immature stages

The egg is spindle-shaped except for one species where it is said to be spherical. That of the Australian species *Synemon magnifica* bears numerous transverse ridges, and 4 longitudinal ridges (Common & Edwards, 1981), but the presence of 5 longitudinal ridges is typical of Castniidae.

The larva is prognathous, the head partly retracted into the prothorax, and the body cylindrical. On the thorax, the L group on T1 is trisetose. Patches of spinules in transverse patches occur dorsally on the abdomen of later instars. The prolegs are weakly developed. In early instars there are one or two uniserial bands of crochets, whereas in the mature larva crochets are either absent from the ventral prolegs or are multiserial. Anal prolegs of mature larvae are vestigial, and there are few, or no, crochets.

The pupa bears two rows of dorsal spines on A2-7 in males and A2-6 in females. The abdomen exhibits a high level of mobility.
(Miller, 1986; Miller *in* Stehr, 1987; Common & Edwards, 1981.)

Biology

The larvae, which are endophagous, bore into stems of their foodplants, or tunnel in the soil to feed on roots. *Synemon magnifica* from Australia lays its eggs near the base of the plant or even in the soil (Common & Edwards, 1981). First instar larvae of this species bore into the stem of the sedge *Lepidosperma viscidum* (Cyperaceae), but later instars form a silk-lined tunnel in the soil and feed on the rhizome before forming a system of tunnels in the rhizome. Pupation occurs in a vertical tunnel or 'chimney' to the soil surface. The pupa is protruded before adult emergence. Neotropical Castniidae bore into palms, bromeliads, sugercane, and bananas (all monocotyledons) (Miller, 1986; Miller *in* Stehr, 1987). Some species are of economic importance on sugar cane, bananas, and oil palms.

Classification

Two subfamilies are recognised (Miller, 1986) - Tascininae and Castniinae. Castniinae include two main clades, one represented by the Indo-Australian genus *Synemon*, the second by all other genera, which are confined to the Neotropics. The subfamily Tascininae occurs in the Oriental region.

Phylogenetic relationships

Castniidae may be most closely related to Sesioidea (Minet, 1986) (see below). With clubbed antennae and diurnal habits, the family is often said to be related to butterflies. However, these features are found widely within Lepidoptera, and there is no evidence of substance to suggest a close association. Many features of the larvae and pupae indicate a possible association with Cossidae (Miller, 1986).

SESIOIDEA

Families recently associated in this superfamily include Sesiidae, Brachodidae, Choreutidae, and Castniidae. However, the evidence for these inclusions is either provisional or unconvincing. The larvae of Sesiidae, Castniidae, and probably some Brachodidae are borers. Several characters may prove to be specializations of Sesiidae and Castniidae (Minet, 1986). In both families the adults are diurnal and have well developed external ocelli; the antenna is clubbed, and its apex bears a brush of long setae; the edges of the basisternum of the mesothorax are partially obscured by the edges of the parepisternum; the posterior tendons of the primary furcal arms overlap the laminae of the secondary furcal arms (Brock, 1971), and seta SD_2 is inserted near SD_1 and XD_2 on the prothorax of the larva. In tinthiine Sesiidae, the first two of these characters are absent, a condition presumed to represent a secondary reduction.

Brachodidae may be related to Sesiidae + Castniidae on the basis of two possible specializations (Minet, 1986): the anterior edge of the internal border of the ocular diaphragm is almost straight; and the posterior tendons of the primary arms of the metathoracic furca are attenuated, a situation rare in Apoditrysia. The second of these characters is more pronounced in Brachodidae than in Sesiidae and Castniidae. Although Choreutidae do not possess these characters in their groundplan, they do occur in the subfamily Millieriinae. However, their presence in this group is assumed to be homoplasious (Minet, 1986).

As these suggestions are provisional, Castniidae are placed in a separate superfamily (see above) following recent classifications (Brock, 1971; Miller *in* Stehr, 1987; Nielsen, 1989; Common, 1990). Sesiidae, Brachodidae, and Choreutidae are included here in Sesioidea, a placement following Heppner & Duckworth (1981) and Common (1990).

Sesiidae

The family comprises a cosmopolitan group of over 1000 species (Heppner & Duckworth, 1981). Most species occur in the tropics, but many are found in the Holarctic region. The moths are small to medium-sized diurnal Lepidoptera, typically mimicking certain Hymenoptera - sometimes to a striking degree. In most, the wings are extensively hyaline and the abdomen is banded.

Adult (Fig. 241)

Scales are appressed to the surface of the head. 'Chaetosemata' are present as a line of sensilla along the posterior margin of the head (Tinthiinae excepted), but they are absent from their usual position behind the ocelli. The external ocelli are large. The antennae are often swollen towards their apices before narrowing to form a setose tip, but sometimes they are filiform, pectinate or bipectinate. The proboscis is not scaled at its base, and the structure is occasionally reduced. Maxillary palpi are minute with 1 to 3 segments, and the labial palpi are upcurved.

The epiphysis is present, and the tibial spur formula is 0-2-4. Tufts of scales are sometimes present where the tibial spurs arise. The forewings are elongate and very narrow, particularly at the base. Although sometimes fully scaled, they are usually partially hyaline. Sometimes, scales are present at only the margin of the forewings. The forewing is elongated and narrow. In the forewing, veins R_4 and R_5 arise from a long stalk; CuP is absent, and $1A + 2A$ is short and lies in a longitudinal fold of the wing - sometimes with merely a vestige present at the base. The hindwing is broader than the forewing, and scaled or partly (sometimes considerably) hyaline. Veins $Sc + R_1$ and Rs run parallel and close together in a costal fold. Veins M_3 and CuA_1 are stalked. Of the anal veins, A_1 is usually indicated merely by a line of scales, A_2 is distinct and A_3 and A_4 are short but said to be distinct. In addition to the frenulum-retinacular wing coupling system, the anal area of the forewing is folded down to interlock with a corresponding costal fold of the hindwing, which is turned upwards

(Fig. 67). The forewing-metathoracic locking device is present. The abdomen tapers posteriorly and is sometimes narrowed basally. Frequently, it has bands of various colours. The ovipositor may be extensible or not extensible.
(Fibiger & Kristensen, 1974; Heppner & Duckworth, 1981.)

Immature stages

The larva is hypognathous to somewhat prognathous, and on T1 the L group is trisetose. Prolegs are reduced, although a full complement is present, and the crochets are uniordinal and arranged in transverse bands.

In the pupa a double row of dorsal spines is present on A2-7 in males and A2-6 in females. A single row occurs on A8-10. The pupa protrudes from the larval tunnel prior to adult emergence. There is no distinct cremaster. Most abdominal segments are mobile.
(Heppner & Duckworth, 1981.)

Biology

Eggs are laid singly in crevices of the foodplant, but females may drop eggs on the ground near the foodplant (Fibiger & Kristensen, 1974). Sesiid larvae chiefly bore in trunks, branches, or roots of trees, in woody shrubs and vines, occasionally in roots and stems of herbaceous plants and, rarely, in galls. Two species from Venezuela eat scale insects. Sesiidae may be of considerable economic importance, particularly where they feed on fruit bushes (e.g., raspberry and currant). A cocoon may or may not be formed. Pupation occurs within the tunnel constructed in the foodplant or in a silk-lined tunnel in the soil (Fibiger & Kristensen, 1974).

Wing transparency seems to have arisen independently on several occasions within the family for it is disparate in its distribution in Sesiinae, and achieved by different means (Kristensen, 1974). Transparency may be achieved by loss of deciduous scales shortly after emergence, by the presence of scales that are themselves transparent, or by the absence of scales from the outset. Unlike the usual condition in Heteroneura, the upper and lower lamellae of transparent scales make close contact leaving no internal lumen, and the upper lamella is not perforated.

Classification

Sesiidae are divided into three subfamilies, Tinthiinae, Paranthreninae, and Sesiinae (Heppner & Duckworth, 1981).

Tinthiinae include the most primitive members of the family with antennae that are not swollen distally and which may be filiform, pectinate, or bipectinate. The absence of a gnathos in the male genitalia may be a specialized character of that subfamily (Naumann, 1971).

The monophyly of the **Paranthreninae** has not been satisfactorily supported, and the assemblage may represent an unnatural group of less specialized Sesiinae, which is by far the largest of the subfamilies.

Sesiinae include species with the most derived characters such as a fused uncus and tegumen in the male genitalia.

Brachodidae

Brachodidae (previously, and in a more restricted sense, called Atychiidae) are tentatively included in Sesioidea. Represented in all zoogeographical regions except for the Nearctic, the family (Heppner & Duckworth, 1981) include less than 100 described species of small to medium-sized moths. Most species occur in the tropics. The forewings are often marked with metallic spots or bands, and the hindwings occasionally bear hyaline areas or spots.

Adult (Fig. 242)

Scales are usually closely appressed to the frons, but loosely appressed to the vertex, and are rarely long and hair-like. External ocelli are large, or, rarely, absent, and chaetosemata are absent. The antenna may be filiform, pectinate or bipectinate. The haustellum is not scaled at its base, and the maxillary palpi are typically minute and 1-3 segmented but, rarely, enlarged. The labial palpi are upcurved.

The hindwings are somewhat shorter and broader than the forewings. There is no stalking in the branches of veins Rs in the forewing, and CuP is reduced but sometimes present near the margin. Veins 1A + 2A are fused, but forked basally, and the other anal veins are usually absent. In the hindwing, 1A + 2A are usually fused except at the base of the wing. Wing coupling is effected by a frenulum-retinacular system, and the forewing-metathoracic locking device is present.
(Heppner & Duckworth, 1981.)

Immature stages

The larva is hypognathous. Prolegs vary in length, and crochets are well developed and arranged in two transverse bands.

The pupa bears dorsal spines on the abdomen, and protrudes from the cocoon prior to adult emergence. Most of the abdominal segments are mobile.
(Heppner & Duckworth, 1981.)

Biology

Brachodidae are typically diurnal, but at least one species is attracted to light. Larvae of Brachodinae bore into roots and pupate in the tunnel so formed. Phycodinae appear to be predominantly leaf-feeders, and pupate within well-developed silk cocoons on the surface of leaves.
(Heppner & Duckworth, 1981.)

Classification

Two subfamilies were recognized by Heppner and Duckworth (1981).

Brachodinae, with 73 species, occur mostly in the Oriental and Australasian regions, but they are pantropical. They lack the specialised scales on the labial palpi and the thorax found in Phycodinae, and the larvae are borers with reduced prolegs.

Phycodinae, with 23 species, are pantropical with most species occurring in the Old World. The labial palpi and the thorax have large, plate-like scales, and the larvae are external feeders with well developed prolegs.

Choreutidae

Choreutidae are diurnal, as in Sesiidae, Castniidae, and Brachodidae, but there are few, if any, other features to suggest a close association. Choreutid larvae are nearly all external feeders, unlike those of Sesiidae, Castniidae, and some Brachodidae. The family, which includes 350 described species, is found in all zoogeographical areas, and is best developed in the Orient and Australasia (Heppner & Duckworth, 1981). The moths are small, and have generally been placed in Glyphipterigidae on the basis of their wing pattern, diurnal habits and venation.

Adult (Fig. 243)

The scales on the frons are appressed to the head, those of the vertex loosely so. External ocelli are present and chaetosemata absent. Long setae are sometimes present ventrally on the filiform antennae of males. The maxillary palpi are minute, and the labial palpi upcurved. Scales are present at the base of the proboscis. The epiphysis is present on the foreleg, the tibial spur formula is 0-2-4, and scale tufts are present near the tibial spurs. The forewing often bears metallic spots. In the forewing venation, the radials are not stalked, and the chorda is nearly always present. The distal section of vein CuP is present. In the hindwing, M_3 is often stalked with CuA_1, and CuP is represented towards the margin. Wing coupling is effected by a frenulum-retinacular system, and the forewing-metathoracic locking device is present. The uncus is rarely present in the male genitalia and, in the female genitalia, although the ovipositor is generally not modified, it is sometimes floricomous. The ductus bursae is sometimes spiral.
(Heppner & Duckworth, 1981.)

Immature stages

Eggs are of the upright variety.
    The larva is hypognathous, rarely more or less prognathous. On T1, the L-group is trisetose. Prolegs are long and slender in most species but short in Millieriinae; they bear uniordinal, or sometimes biordinal, crochets usually arranged in a circle, but sometimes as a lateral penellipse.
    The pupa has dorsal spines arranged in a single row on segments A2-7 of most Choreutinae, but in Millieriinae the row is double, and there is considerable mobility in the abdominal segments. No distinct cremaster is present.
(Heppner & Duckworth, 1981.)

Biology

The eggs are usually deposited singly. Choreutid larvae do not bore; instead most are leaf-tiers, forming a loose tunnel. Those of a North American species (*Tebenna carduiella*) appear to feed on the inner surface of the hollow stems of the thistle *Carduus* (Asteraceae). A species from New Zealand (*Asterivora tillyardi*) also lives within stems. However, in neither case are the larvae modified for boring. Larvae of the monotypic European genus *Millieria* mine the leaves of *Aristolochia* (Aristolochiaceae), several larvae occurring in a single mine. Modifications to the larva are typical of leaf miners; prolegs are reduced and the head is predominantly prognathous (Heppner, 1982e). Other Millieriinae feed in monocotyledons, plants which are typically selected by Choreutidae in general.
    Pupation occurs in a silken cocoon composed of three layers, and attached to the surface of a leaf or leaf axil. The pupa is protruded from the cocoon before the adult emerges.

Classification

Three subfamilies are recognized (Heppner & Duckworth, 1981; Heppner, 1982e).

Most of the 66 species of **Brenthiinae** are tropical and occur mainly in the Nearctic, the Oriental, and the Australasian regions. The termen is rounded in the forewing and the chorda is absent.

Choreutinae, with 290 species, are present both in the temperate regions of the Holarctic region and pantropically. The number of species is high in the Indo-Australian region. The apex of the forewing tends to be pointed and the termen approximately quadrate. The chorda is only rarely vestigial.

Millieriinae, with 3 species, are found in Europe, Florida, and Chile. The proboscis is scaled basally, and the forewings are elongated with a rather pointed apex. The chorda is present or absent.

ZYGAENOIDEA

The composition of this superfamily has been disputed in recent years, although a consensus seems to be emerging that the view of Börner (1939), with some additions, should be accepted. Börner included Limacodidae, Zygaenidae, Megalopygidae, Epipyropidae, and Dalceridae, to which Common (1970) added Heterogynidae, Chrysopolomidae, and Cyclotornidae. Anomoeotidae and Himanopteridae are regarded as subfamilies of Zygaenidae. The monophyly of the superfamily as constituted above was challenged by Brock (1971) who restricted Zygaenoidea to Zygaenidae alone, placed Megalopygidae, Dalceridae, Limacodidae, and Chrysopolomidae in the Cossoidea, and Heterogynidae, Somabrachyidae, Himanopteridae, and Epipyropidae in other lower ditrysian superfamilies.

Common (1975) questioned the validity of including Limacodidae and Megalopygidae, which have exophagous larvae, in the Cossoidea, a superfamily with endophagous larvae. Minet (1986) considered that the monophyly of the various zygaenoid families was suggested by the markedly retractile larval head (at least in later instars), and by the fact that the abdominal spiracles on segment 2A of the pupa are usually covered by the wing cases when these are appressed to the abdomen. In Zygaenoidea, larval crochets are arranged in a mesoseries (but they are coronate in Epipyropidae), a feature seen extensively within macrolepidopterans but not among microlepidopteran families.

At present, the composition of Zygaenoidea is uncertain, and here I follow the classification adopted by Miller (*in press*) who included the following families: Zygaenidae, Dalceridae, Epipyropidae, Limacodidae, Megalopygidae (including *Somabrachys*), Heterogynidae, Chrysopolomidae, and Cyclotornidae.

Typically, the moths have a frenulum-retinacular system of wing coupling, although this is sometimes reduced, and the forewing-metathoracic locking device is present.

Zygaenidae

Zygaenidae (burnet and forester moths) include brightly coloured, mainly diurnal moths with about 800 species (Bourgogne, 1951). They occur in all zoogeographical regions, but are absent from New Zealand (Alberti, 1954). Although most species occur in the tropics, the family is well represented in the Palaearctic. The group has formed the subject of many biological studies (bibliography in Tremewan, 1988). Although Zygaenidae are placed among microlepidopterans and share numerous features with them, the presence, typically, of a free living larva, and the fact that the pupae sometimes escape from the cocoon prior to emergence resembles the conditions seen in macrolepidopterans.

Adult (Fig. 244)

External ocelli are usually present, and chaetosemata are usually present and very well developed. The antennae are filiform, clavate, weakly dentate, or bipectinate. The proboscis is typically well developed but sometimes reduced, the maxillary palpi are very small and 1- to 2-segmented, and the labial palpi are short and ascending. The epiphysis is present or absent, and the tibial spur formula is generally 0-2-2 or 0-2-4, although the medial pair of spurs on the hindleg may be vestigial or absent. The wings are narrow to broad, and the hindwing may be extended into a long tail. The wing venation is little reduced. In the forewing the veins are separate although veins R_3 and R_4 are sometimes stalked. Vein CuP is present in both forewing and hindwing. In the hindwing, $Sc + R_1$ is connected to Rs briefly beyond the middle of the cell, or it lies close to that vein. The ovipositor is fairly short. Sometimes (Zygaeninae) an additional pair of accessory glands (Petersen's glands) is present (Naumann, 1988). (Bourgogne, 1951; Alberti, 1954; Munroe, 1982; Tremewan, 1985.)

Immature stages

The eggs, which are ovoid and flattened with a micropyle at one end of the horizontal axis, are deposited in rows in a single layer or in several layers (Tremewan, 1985).

The larva (Stehr in Stehr, 1987) is brightly coloured - often yellow and black, short and broad, or sometimes slug-like. The head is retractile. The thorax and abdomen bear groups of secondary setae arranged in verrucae. Each thoracic leg bears one or two spatulate setae near the claw, and there is a conspicuous structure (possibly a gland) at the base of the prothoracic leg in many species. Prolegs are present on abdominal segments A3-6 and A10, and the crochets are arranged in a uniordinal mesoseries. In some genera, an anal fork is present. In Zygaeninae and Procridinae there is a gland adjacent to the posterior edge of the spiracle on A2 and A7.

The pupa (Tremewan, 1985) is obtect and adecticous. Spines are present on the abdominal terga.

Classification

The family was divided into three subfamilies by Bourgogne (1951) and into seven subfamilies by Alberti (1954). Four main groupings may exist: 1) Procridinae (widespread), 2) Chalcosiinae (Indo-Malaysia to E. Asia, rarely in the west Palaearctic) + Anomoeotinae (India) + Himantopterinae (Afrotropical and Indian), 3) Phaudinae (E. Asian and Indo-Malayan) + Charideinae (Afrotropical), and 4) Zygaeninae (western Palaearctic and east Asian, with some representatives in Australia).

Biology

Many aspects of the biology of Zygaenidae have been investigated (bibliography by Tremewan, 1988), probably because of the diurnal habits of these moths and their chemical defences.

The adults are mainly diurnal and feed at flowers. Flight, at least in the British species of *Zygaena*, depends on temperature and, to a lesser extent, sunshine

(Tremewan, 1985). The tissues of all stages in the life cycle of Zygaenidae release the highly toxic compound hydrogen cyanide, which is released after the breakdown of cyanoglucosides (summary in Rothschild, 1985). Cyanoglucosides appear not to be obtained from plants but are synthesised by the insect, a phenomenon that is possibly a zygaenid specialization. Zygaenidae feeding on cyanogenic plants are able to do so presumably because of their relative insensitivity to these normally toxic compounds. Larvae and adults are often aposematically coloured. Their tendency to live in colonial aggregations enhances their aposematic qualities.

The pupa either escapes from the cocoon prior to adult emergence, or protrudes from it. The cocoon is strong, elongate, and parchment-like.

Phylogenetic relationships

Zygaenidae may be most closely related to Heterogynidae. In both families the integument of the antennal scape is strongly pigmented.

Heterogynidae

This family includes only a few species in two genera. *Heterogynis* occurs in parts of Europe and North Africa, and *Janseola* in South Africa.

Adult (Fig. 245)

The male resembles some Zygaenidae in general appearance. External ocelli are absent. Chaetosemata are present, and particularly large in *Janseola*. The proboscis is absent in *Heterogynis*, and present but much reduced in *Janseola*; the maxillary and labial palpi are vestigial. The antennae are bipectinate. The frenulum is well developed. Females (unknown in *Janseola*) are apterous with their legs much reduced and mouthparts atrophied. The epiphysis is present in *Heterogynis* and absent from *Janseola*.
(Seitz, 1912.)

Immature stages

The larva (not known for *Janseola*) is similar to that of *Somabrachys*, but the additional pair of prolegs occurring in that genus is absent from *Heterogynis*. The head is retractile and the body short.
(Seitz, 1912.)

Biology

Females leave the cocoon to mate but they remain attached to it and lay their eggs therein. The larvae consume the female, which dies after oviposition, before becoming phytophagous.
(Bourgogne, 1951; Daniel & Dierl, 1966.)

Phylogenetic relationships

The family may be most closely related to Zygaenidae for reasons given under that family, above.

Megalopygidae

Moths of this family range in size from smallish to large. In general appearance, they are rather similar to Limacodidae. The abdomen is densely 'hairy'. Of the 230 species, placed in 23 genera, most are Neotropical while 11 are Nearctic (Miller, *in press*). The genus *Somabrachys* occurs in North Africa and related genera occur in South Africa (H. Geertsema, pers. comm.). The larvae are often extremely hairy. *Psycharium* is confined to the Western Cape of South Africa.

Adult (Fig. 246)

External ocelli are absent and chaetosemata are present. The palpi are very small or absent, and the proboscis rudimentary or absent. The antennae of the male are bipectinate, bipectinate only basally, or flattened and not pectinate. In the female they are weakly bipectinate or without pectinations. The tibial spurs are variously reduced or absent. The retinaculum of the male is subdorsal (at least in *Lagoa crispata*) and represented only by dense scaling (Braun, 1924). The wing venation is almost complete, and the stem of vein M is present in both forewing and hindwing. In the forewing of most species there is a terminal bifurcation composed of R_4 and R_5 rather than R_3 and R_4 (an arrangement similar to that in Limacodidae). The female of *Somabrachys* is apterous. In the male genitalia the gnathos is absent, and the valva is divided completely. In the female, there is a mass of hair scales situated on the expanded membrane between A7 and A8, which is deposited as a felt-like covering over the eggs.
(Hopp, 1934-35; Holloway *et al.*, 1987; Epstein, 1988.)

Immature stages

The eggs are round and laid in paired rows.
    The larvae are often extremely 'hairy' with a dense covering of long secondary setae, but some lack 'hair'. Three pairs of spined scoli are present on each side. Prolegs occur on abdominal segments A2 to A7, and on A10, a situation almost unique within the Lepidoptera. Crochets are present on A3-6 and A10, but absent from A2 and A7. On the ventral prolegs they are divided, or nearly divided, into two groups; on the anal prolegs they form a mesoseries.
    The pupa bears a flange or eye-plate (Mosher, 1916).
(Hopp, 1934-35; Holloway *et al.*, 1987; Stehr *in* Stehr, 1987.)

Biology

Eggs are covered by hair scales from the female. Many families of plants act as larval foodplants although the extent of the polyphagy of individual species is unclear, and the number of observations on the group is not extensive. The larvae have urticating setae, which may cause outbreaks of dermatitis, beneath the long hair-tufts.
    Pupation occurs in a leathery cocoon with a surrounding film of silk. Sometimes, an additional web covering several cocoons is found.
(Stehr *in* Stehr, 1987.)

Classification

    Three subfamilies: Trosiinae, Megalopyginae, and Aidinae were recognized by Hopp (1934-35) on the basis of the relationship of vein Sc to the stem of $R_1 + Rs$ in the forewing. The placement of the neotropical group Aidinae in Megalopygidae is questionable (Holloway, 1986) (see below).
    The old world genera *Somabrachys* and *Psycharium* are sometimes separated from Megalopygidae as Somabrachyidae. *Somabrachys* was included in Megalopygidae by Minet (1986) who noted that the larva bears ventral prolegs on segments A2-7, that the mesoseries of crochets on the prolegs are divided into two groups, and that the mouthparts of the adult are reduced. However, the first two of these characters may be

shared not only with Megalopygidae but with the megalopygid complex in general (Minet, 1986).

Phylogenetic relationships

Megalopygidae appear closely related to Limacodidae on the basis of several shared characters, in particular the stinging spines of the larva, the pupal eye-plate, and the general similarity of the shape of the wings (Holloway *et al.*, 1987). However, the family may not be monophyletic since the neotropical Aidinae possibly belong to Limacodidae on the basis of forewing venation and male and female genitalia (Holloway, 1986).

Limacodidae

Limacodidae are a family of predominantly tropical moths, but the group is represented in all the zoogeographical regions. There are about 1000 species (Epstein, 1988). The moths are small to medium-sized with broad rounded wings, 'hairy' legs and a 'hairy' body. The larvae are either covered with stinging spines (nettle caterpillars) and hairs or are smooth (gelatine caterpillars) (Cock *et al.*, 1987). Limacodidae are often referred to as 'slugs' because their larvae progress by means of peristaltic waves that pass along the sole-like ventral surface.

Adult (Fig. 247)

External ocelli and chaetosemata are absent. The antennae are usually bipectinate in males, at least for the basal half to two-thirds of their length, whereas in females they are simple and filiform. The proboscis is reduced or absent, the maxillary palpi are minute with 1-3 segments or absent, and the labial palpi are 2- or 3-segmented and, while usually short, may be large and bear hair-tufts.

The sclerotization of the mesothoracic epimeron is reduced (Epstein, 1988). The epiphysis is absent, and the tibial spur formula is 0-2-2 or 0-2-4. Wing coupling is effected by a frenulum-retinacular system. Venation is little reduced; vein M divides the cell in both wings, and CuP is present. In the forewing, veins R_3, R_4, and R_5 are stalked. In the hindwing $Sc + R_1$ fuses with Rs proximally.

The abdominal terga bear numerous detachable spines. An ampullate process projects externally from each side of segment A8 in the female (Holloway, 1986; Epstein, 1988). In the female, the ovipositor lobes are flattened and disc-shaped, and the ductus bursae is often spiralled with a sclerotized band running through its length. (Viette, 1980; Holloway, 1986; Cock *et al.*, 1987; Epstein, 1988; Munroe, 1982.)

Immature stages

The eggs are typically flat, scale-like structures with a thin shell.

The larvae are distinctive, characterized by a reduction of the thoracic legs, the absence of prolegs, and a sole-like ventrum. These modifications are associated with a slug-like mode of progression. Larvae may bear stinging spines and hairs, or they may be smooth. The head is concealed by prothoracic folds. The body armature varies: the groundplan arrangement may include a lateral row of spined scoli or tubercles just above the spiracles on each side, and a dorsolateral row on each side above the lateral scoli (Holloway, 1986). The lateral scolus on A1 is vestigial, and no scoli occur on the head. The reduction or partial or complete absence of scoli, or their enlargement, are interpreted as a modification of the groundplan arrangement. Suckers are often present.

The pupal appendages are free, and there is a sculptured plate or flange close to the eye (Mosher, 1916; Cock *et al*, 1987).

(Stehr *in* Stehr, 1987; Cock *et al.*, 1987; Epstein, 1988.)

Biology

The eggs may be laid singly or in shingle-like arrangements on leaves. Limacodid larvae are primarily plant feeders with first instars eating the epidermis, and subsequent stages the whole leaf. A wide variety of foodplants are accepted by the family, and most species appear to be polyphagous. Limacodidae are economically important defoliators of palms. Larvae hatched from eggs laid in batches tend to be gregarious, at least in the first instar. Pupation occurs in a hard, oval cocoon with a circular lid cut by the larva. The cocoon itself is constructed from silk hardened by saliva (Ishii *et al.*, 1984).

Adults are often observed in a distinctive resting posture with the body held at an angle to the substrate and the wings hanging straight down.

(Cock *et al.*, 1987; Epstein, 1988.)

Phylogenetic relationships

Limacodidae probably belong to a grouping also consisting of Megalopygidae, Dalceridae, and Epipyropidae (Miller, *in press*; Epstein, 1988). Possible specialized characters shared by these families include the absence of chaetosemata and externally visible ocelli; the reduction or absence of the maxillary palpi and the proboscis; bipectinate antennae; a slug-like motion in the larvae; and the general shape and appearance of the pupa and adult (Epstein, 1988).

Dalceridae

The family (Miller, *in press*) includes 84 species in 11 genera. The moths are medium-sized to small usually with a white, yellow or orange ground colour. The body is stout and hairy and the wings broad and rounded. Dalceridae are all Neotropical, although one species occurs up to Arizona. The larvae are slug-shaped and characterized by the gelatinous conical tubercles that cover the dorsum.

Adult (Fig. 248)

External ocelli and chaetosemata are absent. The proboscis and maxillary palpi are vestigial, and the labial palpi 2-segmented and upturned. The antennae are short, broadly bipectinate in males and narrowly bipectinate in females.

The epiphysis is absent as are tibial spurs. The wing venation is almost complete. In the forewing most of the stem of vein M is present in the discal cell. Stalking of the radial veins varies in extent within the family, but R_2 and R_3 are always stalked or fused as are R_4 and R_5 (a combination diagnostic within Zygaenoidea). In the hindwing also, most of the stem of vein M is present within the discal cell. In males of some species the frenulum and retinaculum are absent and the humeral angle of the hindwing somewhat expanded. The female frenulum is composed of a clump of long narrow scales.

Components of the male genitalia are much reduced or fused. The valvae are particularly reduced, typically as small, immobile processes fused to various sclerotizations of the diaphragma. In the female, the anal papillae are large and covered with setae and the anterior apophyses are absent. On each side of the anal papillae, and just anterior to them, occurs an 'accessory gland'. The two glands are not connected to each other, but each opens independently to the outside.
(Miller, *in press*.)

Immature stages

The eggs are small, oblong and smooth.
The larvae are characteristically slug-shaped, and covered with translucent, jelly-like tubercles dorsally. The head is concealed, and thoracic legs reduced in size. Prolegs are reduced to small lobes, but crochets are present on segments A2-A7 in the last two instars. The motion of the larvae is typically slug-like although the reduced prolegs appear to be functional to some extent at least in one species. The ventral surface is in the form of a mucous covered sole. Final instar larvae spin white or yellow cocoons, usually among clusters of leaves. The cocoon is typically composed of a loosely woven outer component and a tightly woven, fusiform inner layer.
The pupal exuviae are extruded at adult emergence.
(Stehr and McFarland *in* Stehr, 1987; Miller, *in press*.)

Biology

The eggs are laid singly. Adult dalcerids have a weak undulating flight, and they do not feed. At rest the wings are held loosely roof-like over the abdomen, the forelegs and midlegs are stretched out, and the antennae are held upright. Over 20 families of larval foodplants have been recorded. Dalceridae appear to be generalist herbivores with a preference for smooth-leaved trees and shrubs. As generalists they are sometimes pests on various crops (e.g., oil palms, cocoa and coffee). Pupation occurs in a hard, oval cocoon with a circular lid cut by the larva.
The lateral abdominal glands of the female (presumably homologous with those of Limacodidae) secrete a fluid that completely covers the eggs and dries rapidly to form a smooth layer. The ducts of the glands appear to empty into a cleft between the anal papillae.
(Miller, *in press*.)

Classification

A division into two subfamilies is supported (Miller, *in press*). In Acraginae an accessory cell is present in the forewing whereas in Dalcerinae it is absent. This serves to weakly distinguish the two taxa.

Phylogenetic relationships

The monophyly of Dalceridae is well supported on the basis of the stalking of vein R_2 with R_3 and of R_4 with R_5, the reduction of various components of the male genitalia, the gelatinous tubercles present on the larvae, and the presence of crochets on segments A2 and A7 as well as on A3-6 (a character that may be shared with Megalopygidae) (Miller, *in press*). However, relationships between Dalceridae and other zygaenoid (or possibly cossoid families) remain uncertain. Nevertheless, the several similarities shared by Dalceridae, Limacodidae, Megalopygidae, and Epipyropidae suggest that the four families form a monophyletic group (Miller, *in press*; Epstein, 1988).

Epipyropidae

The family includes about 40 species, some of which are undescribed, (Davis *in* Stehr, 1987). Epipyropidae are widespread, but occur mostly in the Old World tropics. The moths are small and dark with subtriangular forewings and shorter apically rounded wings, and the larvae are ectoparasitic on Homoptera.

Adult (Fig. 249)

External ocelli and chaetosemata are absent. Mouthparts are reduced, only minute labial palpi being visible. The antennae are short and bipectinate in males and females. Epiphyses and all tibial spurs are absent, and pretarsi are reduced. The forewing is approximately triangular and the hindwing rounded and much shorter than the forewing. Wing venation is little reduced. In the forewing, the chorda is often present, the base of M is present or vestigial within the cell, and vein CuP is visible near the wing margin. Typically, R_1 and the branches of Rs, M, and CuA arise directly from the cell. In the hindwing, CuP is present. Although a short frenulum occurs on the hindwing, the forewing lacks a retinaculum. The ovipositor is short. External glands of A8 are lacking.
(Krampl & Dlabola, 1983; Munroe, 1982.)

Immature stages

The eggs are smooth, and half as wide as long.
    Larval development is hypermetamorphic. The head of the larva is retracted into T1 and bears stylet-like mandibles. In later instars the thoracic legs are much reduced, but the tarsal claws are well developed. The abdominal prolegs are reduced, but bear uniordinal crochets arranged in a uniserial circle on A3-6, and a similar arrangement occurs on A10 except that the circle is broken posteriorly. All setae are reduced, and there are numerous secondary setae scattered over the body. First instar larvae are active and campodeiform, whereas later instars become progressively more stout and sessile. Secretion of a white waxy substance, which finally covers the larva is commenced by the third instar.
    The pupal appendages are weakly glued to the body, and the exuviae are partially extruded from the cocoon at adult emergence. The cocoon is oval or rosette-shaped.
(Davis *in* Stehr, 1987.)

Biology

All known larvae are parasitic on Homoptera, and the eggs are usually laid in batches on the foodplant of the host. The larvae are hypermetamorphic with first instars that seek the host, while later instars are sessile and restricted to the host. When on the body of the host, epipyropid larvae seem usually to feed on haemolymph released by the stylet-like mandibles. Final instar larvae of *Epimesophantia* damage not only superficial structures of the host, including the wax coating, but sometimes even penetrate the body where they presumably consume parts of internal organs (Krampl & Dlabola, 1983). In such instances epipyropid larvae kill their host; in situations where they feed only superficially they may not.
    Well developed tarsal claws are used for attachment to the host and the use of silk to provides a grasp for the crochets (Davis *in* Stehr, 1987).
    Pupation occurs in a silken cocoon attached to a plant (Davis *in* Stehr, 1987).

Phylogenetic relationships

Epipyropidae may be most closely related to Cyclotornidae on account of the similar ectoparasitic habits of the latter in the first instar (Davis *in* Stehr, 1987). Beyond this, the family appears to be related to Megalopygidae, Limacodidae, and Dalceridae.

Cyclotornidae

The family (Common, 1990) includes about a dozen species of fairly small, drab moths assigned to the genus *Cyclotorna* and restricted to Australia. They are mostly greyish, sometimes with dark spots on the forewing, but occasionally yellowish white or yellow and black. The body is relatively stout, and the wings are rounded but much narrower than in most other Zygaenoidea. The early instar larvae are external parasites of Homoptera and the later instars feed on ant brood.

Adult (Fig. 250)

External ocelli and chaetosemata are absent. The antennae are filiform and the scape is thickened and bears a pecten. The proboscis and maxillary palpi are absent, and the labial palpi are very short and drooping. The epiphysis is absent, and the tibial spur formula 0-2-4. The wing venation is little reduced. In the forewing, the chorda is present and vein M is forked in the cell. Vein CuP is present in both wings. The ovipositor is short.
(Common 1990.)

Immature stages

The eggs are oval and flattened.
   The first instar larva is flattened with a large head and long mandibles. Hypermetamorphosis occurs within the larval stage so that in later instars the head is relatively small and the body segments are extended laterally into a series of flanges. The thoracic legs have adhesive pads terminally and the prolegs bear crochets arranged in a uniordinal mesoseries. Fine secondary setae are also present.
   The pupa is flattened dorsoventrally.
(Common, 1990.)

Biology

Early larvae of *Cyclotorna monoentra* are external parasites of cicadellid leafhoppers. The leafhoppers appear to be located by the first instar larvae using trails made by attendant meat ants. After spinning a flat shelter beneath bark, cyclotornid larvae moult into the second phase. These later larvae are carried by the ants to their nests where they prey upon the ant brood. The larvae supply their ant hosts with an anal secretion. Final instars leave the nests of the ants, and pupal cocoons are spun in crevices of trees. The later instars may be brightly coloured. Similar larval habits seem to occur in other species.
(Dodd, 1912; Common, 1990.)

Chrysopolomidae

This family includes about 25 species of medium-sized Afrotropical moths. The adults are drab, and sometimes mostly whitish. Several species bear a small, but prominent, white spot on the forewing. They are rather limacodid-like with rounded wings and hairy bodies, and also resemble, superficially, certain Bombycoidea.

Adult (Fig. 251)

External ocelli and chaetosemata are absent. The proboscis is reduced. The antennae are bipectinate in the male, but filiform in the female. The epiphysis is absent, and the tibial spur formula 0-2-4. The wings are rounded, and the venation almost complete. In both forewing and hindwing the base of the medial vein is present in the cell. The frenulum is absent and the humeral area of the hindwing is slightly expanded. (Hering, 1937.)

Immature stages

The larvae have been likened to those of Limacodidae, but they lack spines (Pinhey, 1975). No other information is available.

Classification

The family was divided into two subfamilies (Ectropinae and Chrysopolominae) by Hering (1937) based on wing venation and the structure of the male genitalia.

Phylogenetic relationships

The limited comparisons that have been made between Chrysopolomidae and other Zygaenoidea suggest that the family is closely related to the Limacodidae.

IMMOIDEA

Immidae

The family Immidae was established by Heppner (1977) for the pantropical genus *Imma*, placed previously in Glyphipterigidae or Sesioidea. Eventually, Immidae were assigned to a separate superfamily because their origin and affinities were obscure (Common, 1979). Heppner included 238 described species in Immidae and elevated nine genera from synonymy with *Imma* so that there are now 10 genera in the family.

The wingspan ranges from about 16-40 mm in these smallish to medium-sized moths, which have often been confused with Noctuidae. The Australian species *Imma acosoma* rests with the wings folded and resembles certain Tortricidae (Common, 1979). The folded condition is presumably the arrangement typical in Immidae.

Adult (Fig. 252)

The front of the head is smooth-scaled, but scales are only loosely appressed to the vertex. External ocelli are rarely present and chaetosemata rarely absent. The antennae are usually filiform. The maxillary palpi are minute, and the proboscis is developed but not scaled at the base. The labial palpi are upcurved; in males, segment 2 on each palpus may bear a spined mesal projection at the apex.

The epiphysis is present, and the tibial spur formula is 0-2-4. Rarely, hyaline spots are present on both pairs of wings. The wing venation is little reduced. Vein CuP is present distally in both pairs of wings. In the hindwing, Rs is free throughout its length, although it may be stalked with M_1. The wings are coupled by a frenulum and retinaculum, and the forewing-metathoracic locking device is present.

The abdomen is relatively stout, and the sternal articulation with the thorax is 'tortricoid'. In the female, the ovipositor may be floricomous. Tympanal organs are absent.

(Heppner, 1982c; Common, 1990.)

Immature stages and biology

Information on the immature stages is restricted to the Australian species *Imma acosma*, which feeds on *Hymenanthera dentata* (Violaceae) and *I. vaticina*, which feeds on *Fenzlia obtusa* (Myrtaceae).

In the larva, the head is hypognathous. On the prothorax there are three prespiracular (L) setae. The prolegs are long and slender with the anal prolegs projecting posteriorly beyond the segment. Crochets are uniordinal and arranged in a mesoseries.

Pupation occurs in a cocoon spun by the larva across the depression of a leaf, so forming a cell for the pupa. Pupal appendages are firmly fused to each other and to the body. There are no dorsal spines, and the cocoon is not protruded prior to, or during, adult emergence. A cremaster is present. Abdominal segments A3 to A5 are movable. (Common, 1979.)

Phylogenetic relationships

Common (1979) noted that in some respects immids are typically microlepidopteran, for example in general larval form, the presence of three prespiracular setae on the prothorax and, in the adult, the absence of tympanal organs and the retention of vein CuP. However, he also pointed out that the group is advanced within the microlepidopterans, for example, crochets are arranged in a mesoseries, the larva is an external feeder, and the pupa is retained within the cocoon at eclosion.

COPROMORPHOIDEA

Copromorphidae and Carposinidae are placed provisionally in the same superfamily. Two characters possibly representing shared specializations have been suggested (Minet, 1986): the presence of raised scales on the upper surface of the forewing; and the fringing, by piliform scales, of the base of vein Cu on the hindwing. Both families have larvae that bore.

Copromorphidae

Copromorphidae are mostly smallish, relatively broad-winged, and generally brownish or green moths found mainly in Indo-Australia but also in the new World. There are about 60 species (Holloway *et al.*, 1987). In those few species for which information is available, the larvae are borers.

Adult (Fig. 253)

The head is rough-scaled. External ocelli are present or absent, and chaetosemata are absent. The antennae are simple, unipectinate or, rarely, bipectinate. The base of the proboscis is unscaled, the maxillary palpi 1- to 4-segmented and folded over the base of the proboscis, and the labial palpi are recurved or porrect. The epiphysis is present, and the tibial spur formula is 0-2-4. Raised scale tufts or ridges are usually present on the forewing. The wing venation is little reduced, but in the cell the chorda and the base of vein M are absent or vestigial, and vein CuP is reduced to a vestige. In the hindwing, a pecten usually occurs at the base of vein CuA, and vein CuP is present.

The wings are coupled by a frenulum-retinacular system, and the forewing-metathoracic locking device is present. At the base of the abdomen, sternum A2 has long, curved anteriolateral processes (Kyrki, 1983a). The ovipositor is not elongated. (Munroe, 1982.)

Immature stages

The larva is hypognathous and bears only primary setae. Group L on segment T1 is generally bisetose. On A8, the spiracle is sometimes borne on a stalk. The prolegs, present on A3-6 and A10, are slender, and the crochets are arranged in a uniordinal, mesal penellipse, or a circle. The pupa lacks tergal spines and the appendages are fused. (Heppner *in* Stehr, 1987.)

Biology

The larvae tunnel in bark, flowers, fruit, or galls. At least one species is a leaf-tier. Pupation occurs in the larval gallery or a cocoon in the soil or in a crevice. (Heppner *in* Stehr, 1987; Common, 1990.)

Carposinidae

The family, which is widely distributed, includes about 200 described species (Munroe, 1982) of small to small-medium moths. The wings are fairly broad compared with other microlepidopterans. Although most species occur in the Indo-Australian area (Diakonoff, 1989), the family is represented widely. The larvae bore in various plant parts; some are leaf miners in their early stages; many are fruit-feeders.

Adult (Fig. 254)

The frons is smooth-scaled and the vertex rough-scaled. External ocelli and chaetosemata are absent. The antennae are densely supplied with long cilia in males, but the cilia are short in females. The proboscis is short and not scaled at its base, and the maxillary palpi are usually 1-segmented and minute but sometimes an apical bud-like segment also occurs. The labial palpi are 3-segmented and recurved in males, but the apical segment may be longer and porrect in females. The epiphysis is present, and the tibial spur formula is 0-2-4. The upper surface of the forewing bears raised scale-tufts or ridges. Forewing venation is little reduced, but in the hindwing reduction is significant. In the forewing vein R_5 extends to the termen, and CuP is absent, and in the hindwing, 2 branches, or sometimes 1 branch, of vein M are absent, and a pecten occurs often at the base of CuA. Vein CuP is present. Wing coupling is effected by a frenulum-retinacular system, and the forewing-metathoracic locking device is present. The ovipositor is moderately long and telescoped. (Munroe, 1982; Davis, 1969; Diakonoff, 1989.)

Immature stages

In *Carposina niponensis*, the eggs are yellowish and spherical (summary in Davis, 1969). The larva (Heppner *in* Stehr, 1987) lacks secondary setae. The L group is bisetose on T1. Abdominal prolegs, present on A3-6 and A10, are slender, and the crochets are arranged in uniordinal circles. The pupa lacks abdominal spines, and has the appendages fused.

Biology

Eggs are deposited singly. Larvae bore in fruits, bark, twigs, stems, and galls, and some species are leaf miners. *Carposina niponensis* is a serious pest of apple and peach in Japan and the USA. Pupation occurs in the larval gallery or in a silken cocoon in the soil or litter.
(Heppner *in* Stehr, 1987; Diakonoff, 1989.)

SCHRECKENSTEINIOIDEA

The relationships of *Schreckensteinia* are uncertain (Minet, 1983; Kyrki, 1984). The genus appears not to belong to Yponomeutoidea, where it has often been placed, because the thoraco-abdominal articulation is of the tortricoid variety, and there are no pleural lobes on A8 of the male (Kyrki, 1984). The presence of 3 prespiracular setae on T1 of the larva and of dorsal spines on the abdomen of the pupa suggests that neither does it belong in Copromorphoidea (Kyrki, 1984). Since relationships of the genus were not established, it was assigned (Minet, 1983; 1986) to a separate superfamily - Schreckensteinioidea.

The moths are small to small-medium. Although there are few species, the genus is distributed widely. However, it is absent from Australia.

Adult (Fig. 255)

The head is smooth-scaled. External ocelli and chaetosemata are absent. There is no pecten on the antennal scape. The base of the proboscis lacks scales, the maxillary palpi are minute, and the labial palpi slender and upturned. The hindtibia bears stiff dorsal spines and an apical whorl of stiff spines. The wings are narrow and lanceolate. Vein R_5 of the forewing reaches the termen. Vein CuP is absent from both the forewing and the hindwing. The wings are coupled by a frenulum and a retinaculum. The ovipositor is short.
(Zimmerman, 1978; Minet, 1983.)

Immature stages

Eggs are of the upright variety.

The prespiracular group is trisetose on T1 of the larva. Prolegs are slender and bear few crochets.

On the pupa, the spiracles are projecting. Strong pointed teeth occur on certain of the abdominal segments. Tergal spines are present on the abdomen, the appendages are relatively free, and many segments of the abdomen are mobile. The pupa protrudes from the cocoon prior to adult emergence.
(Forbes, 1923; Minet, 1983.)

Biology

Larval foodplant records include the fruit racemes of Sumac (Anacardiaceae) for *Schreckensteinia erythriella*, and leaves of *Rubus* (Rosaceae) which are skeletonized by *S. festaliella* (Emmet, 1988).

Pupation occurs in a loosely woven network, not a dense cocoon. When the adult moth is at rest, the hindlegs are elevated.

URODOIDEA

Urodidae

The family Urodidae was erected by Kyrki (1988) for a group of 62 described species most of which had previously been assigned to Yponomeutidae. Most species occur in South and Central America. One species occurs in North America, one is found in Europe, and some others (undescribed) are present in the Oriental region. The wingspan ranges from 11-37 mm, and the moths are usually dark grey or brown often with blue or violet reflections.

Adult (Fig. 256)

Scales are more or less appressed to the head. External ocelli and chaetosemata are absent. The antennae are lamellate, particularly in males. The proboscis is not scaled at the base, and the maxillary palpi are reduced. The epiphysis is present, and the tibial spur formula is 0-2-4. The shape of the forewing varies from narrow to broadly oval. The venation of the forewing includes a strong chorda; vein R_5 runs to the costa or the apex; and CuP is reduced. A costal hairpencil is usually present in males. The anterior corners of abdominal sternum A2 are elongated and sclerotized, and sternal rods are well-developed, reduced, or absent. There are no metathoracic or abdominal tympanal organs. The apophyses of the female genitalia are elongated so forming a long ovipositor.
(Kyrki, 1988.)

Immature stages and biology

Larvae of only two species have been examined (Kyrki, 1988). Based on these, the family may be diagnosed by four main larval characters. The prolegs are long, narrow, and constricted medially, and bear a mesoseries of uniordinal and uniserial crochets. On segment A8, seta SV_1 lies almost at the same level as L_3. Microseta MXD_1 is absent from the prothorax. On the metathorax there is only a single MD seta. The larvae are free-living on leaves of broad-leaved trees.

The pupa (information derived from a single species, Kyrki, 1988) has abdominal segments A3-7 movable in males and A3-6 movable in females. Tergal spines are present. Pupation occurs in a meshed cocoon, from which the pupa is protruded prior to adult emergence.

Phylogenetic relationships

Urodus and its allies were excluded from Yponomeutoidea by Kyrki (1984) because of their tortricoid thoraco-abdominal articulation, the presence of three (rather than two) frenular bristles in the female, the arrangement of crochets on the larval prolegs in a mesoseries, the fact that abdominal setae L_1 and L_2 are found on a common pinaculum on the larva, and the spined condition of the abdominal terga of the pupa. The absence of the pleural lobes on A8 of the male is a further significant reason for the exclusion of the group. The actual relationships remain unresolved, although the grouping belongs within the primitive assemblage of Ditrysia with a pupa protruded from the cocoon prior to adult emergence (Kyrki, 1988). Despite having certain features (e.g., a free-living larva, long ventral prolegs, and a large meshed cocoon) in common with *Schreckensteinia* (Schreckensteinioidea), Kyrki was reluctant to unite the two taxa, preferring instead to leave Urodidae as a family of uncertain affinities within the Ditrysia. The family was listed as a separate superfamily, Urodoidea, by Nielsen (1989).

EPERMENIOIDEA

Epermeniidae

The family is widespread, and includes about 70 species (Holloway *et al.*, 1987) of very small, narrow-winged, rather drab moths.

Epermeniidae share no obvious specialized characters with any other group, and they have been placed in a separate superfamily (Minet, 1983). They are unlikely to belong to Yponomeutoidea because they bear 2 rather than 3 prespiracular setae on the prothorax of the larva, because the pupa is not protruded at adult emergence, and because pleural lobes are absent from segment A8 of adult males.

Adult (Fig. 257)

The head is smooth-scaled. External ocelli and chaetosemata are absent. The antennal scape bears an awning of scales but no pecten (bristles) (Dugdale, 1988). The base of the proboscis is not scaled, the maxillary palpi are 3-segmented and folded over the base of the proboscis, and the labial palpi are ascending. The epiphysis is present, and the tibial spur formula is 0-2-4. Stiff bristles extend along the length of the hindtibia. The wings are lanceolate. The forewing bears tufts of raised scales on its posterior margin. The wing venation is usually little reduced. However, in the forewing although the chorda may be present, the stem of vein M is absent, and vein R_4 or R_5 may be lost. Vein R_5 extends to the termen. In the hindwing the stem of vein M is also absent. Wing coupling is effected by a frenulum and retinaculum, and the forewing-metathoracic locking-device is present. Tergal spines on A2-7 are present or absent. The pleural area of segment A2 of the male sometimes bears projecting or eversible processes with long cilia. In the female, the anterior apophyses are forked at the base. (Gaedike, 1978; Minet, 1983; Dugdale, 1988; Common, 1990.)

Immature stages

Eggs are oval and of the flat variety.

The larva is hypognathous and cylindrical. Two L-group setae occur on T1, and seta SD_1 is almost directly above segment A1-8, a characteristic feature (Heppner *in* Stehr, 1987). A full complement of prolegs is present with uniordinal crochets arranged in a complete circle. The pupa lacks dorsal spines; abdominal segments A4-7 are movable in males and segments A3-6 movable in females.

The pupa is not protruded from the cocoon at adult emergence. (Minet, 1983; Heppner *in* Stehr, 1987.)

Biology

Eggs are laid singly. Larvae mainly bore in buds, fruit, and seeds, although some species are leaf miners in their early stages before becoming external leaf feeders (Heppner *in* Stehr, 1987). The foodplants often belong to Umbelliferae (Munroe, 1982). However, in Australia larvae of *Gnathifera eurybias* feed exposed on the leaves of *Exocarpos* (Santalaceae) (Common, 1990).

ALUCITOIDEA

Alucitoidea include Alucitidae and Tineodidae (Oxychirotidae) (Minet, 1986; Nielsen, 1989). The families are associated by the presence of V-shaped venulae on sternum A2, and by the transverse bands of spines situated on the adult abdominal terga.

Alucitidae

Alucitidae, known as many- or multi-plumed moths, are easily recognizable by the deep division of the forewing into 6 plumes and the division of the hindwings into 6 or 7 plumes. The family is small, with around 130 species (Heppner *in* Stehr, 1987) with most representatives occurring in temperate or warm regions. The moths are generally small to medium in size, although some larger species occur. The larvae bore in various plant substrates.

Adult (Fig. 258)

External ocelli are present or absent. The base of the proboscis is unscaled, the maxillary palpi are 3- to 5-segmented or absent, and the labial palpi are upturned or porrect. The forewing is divided into 6 and the hindwing into 6 or 7 feathery lobes. These divisions occur between veins R and M, between the branches of M, between M and CuA, between the branches of CuA, and between CuA and CuP. Wing coupling is effected by a frenulum-retinacular system. While in most species the aculei on the underside of the forewing base and the metascutum are absent, vestiges sometimes remain. The abdominal terga bear transverse patches of short spines. The ovipositor is not long.
(Munroe, 1982.)

Immature stages

The larva is semiprognathous. The L-group on T1 is bisetose. Prolegs are short, and the relatively few crochets are arranged in a uniordinal circle.

The pupa lacks abdominal spines and is not protruded prior to adult emergence. Pupation occurs in a cocoon on the ground or in galls.
(Munroe, 1982; Heppner *in* Stehr, 1987.)

Biology

Larvae bore into flowers, buds, seeds, leaves, shoots, and stalks. Pupation occurs in a cocoon on the ground or in galls.
(Munroe, 1982.)

Phylogenetic relationships

Uncertainty over relationships has led to certain authors placing Alucitidae in a separate superfamily, Alucitoidea (Leraut, 1980; Minet, 1983; Nielsen, 1989). They have been assigned to Pyraloidea (e.g., Bourgogne, 1951) and Copromorphoidea (Meyrick, 1928; Common, 1970), but are more likely to represent the sister group of Tineodidae (Minet, 1983; Nielsen, 1989).

Tineodidae

The family is small and occurs in Asia, New Guinea and, mainly, Australia. The moths are small to medium-sized. Usually, the wings are entire, but occasionally they may be deeply cleft and divided into 2 plumes. A few species superficially resemble agdistine Pterophoridae. The composition of the family is uncertain.

Adult (Fig. 259)

The head is smooth-scaled. External ocelli are sometimes present and chaetosemata are absent. The proboscis is well developed, the maxillary palpi small and 4-segmented, and the labial palpi moderately long and porrect. The epiphysis is present and the tibial spur formula is 0-2-4 or, rarely, 0-0-2. The wings are typically narrow and entire, or they are deeply cleft. Vein CuP is absent from both fore- and hindwing. Wing coupling is effected by a frenulum-retinacular system. The abdomen is long and slender, and a narrow band of spines occurs anteriorly on terga A3-6 or A2-7. (Munroe, 1982.)

Immature stages

The larva is semiprognathous, and lacks secondary setae. The crochets are uniordinal and arranged in a mesal penellipse on the ventral prolegs.
   The pupa lacks dorsal spines, bears hooked setae terminally, but a cremaster is not developed.
(Common, 1990.)

Biology

The larvae live between joined leaves or tunnel in fruits and seeds.
(Common, 1990.)

PTEROPHOROIDEA

Pterophoridae

A single family, Pterophoridae, is included. The moths, which are frequently distinctive with well developed clefts usually present in both the forewings and the hindwings, are called plume moths. The family is represented in all the main zoogeographical regions. There are about 500 species (Munroe, 1982).

Adult (Fig. 260)

The head is smooth-scaled. External ocelli and chaetosemata are absent. The proboscis is present and its base is not scaled, the maxillary palpi are vestigial, and the labial palpi slender or stout. The forewing is cleft, forming 2, or sometimes 3, usually deep divisions, but sometimes entire. The hindwing is cleft into 3 divisions, but is sometimes entire. Veins M_1 and M_2 are frequently reduced, and M_3 is frequently stalked with CuA. CuP may be present or reduced. In the hindwing, the venation is significantly reduced with M_1 and M_2 absent or vestigial, one branch of CuA often absent, and CuP absent. The wings are coupled by a frenulum-retinacular system, but the wing-thorax locking aculei are absent. In species with undivided wings the venation is little reduced. In species with cleft wings the venation is modified. In the forewing, the branches of vein Rs range from all present to mostly absent. The epiphysis is present and the tibial spur formula 0-2-4. The legs are typically long and slender. The abdomen is usually long and slender. In the male, the aedeagus is usually strongly curved. In the female, the ovipositor is fairly narrow but not attenuated, and the anterior apophyses frequently are absent. Tympanal organs are absent.
(Yano, 1963.)

Immature stages

Eggs are elongate-oval (Neunzig in Stehr, 1987).
   The larva (Yano, 1963) is cylindrical or slightly flattened. Only primary setae occur on the head, but primary and secondary setae are present on the body. The secondary

setae are variously modified. The prolegs are long and slender with uniordinal crochets arranged in a mesoseries. The L-group on the prothorax is trisetose.

The pupa (Yano, 1963) is slender and bears hooked setae ventrally on the caudalmost segments. Conspicuous spines often occur on the dorsal surface, and setae are widely distributed.

Biology

The larvae may feed on leaves, but often they bore into plant tissue, such as stems, flowers and fruits. The early stages of some species are leaf-miners, and others live in folded leaves.

At rest, the adult moths hold their forewings outstretched from the body with the lobes of the hindwings brought together and held under the forewing. (Yano, 1963.)

Classification

Three subfamilies are generally recognized (see Yano, 1963).

In **Agdistinae** the wings are entire, not cleft. Veins M_1 and M_2 are well developed in both forewing and hindwing.

In **Platyptiliinae** the forewing is bifid, rarely trifid, and the hindwing is trifid. Veins M_1 and M_2 are short and weak in both forewing and hindwing. The second lobe of the hindwing has 2 veins and the third lobe 2 veins.

In **Pterophorinae** the forewing is bifid and the hindwing trifid. Veins M_1 and M_2 are very short and weak in both forewing and hindwing. The second lobe of the hindwing has 2 veins and the third lobe 1 vein.

Phylogenetic relationships

The family has been placed often in Pyraloidea or Alucitoidea. However, there is no obvious reason for associating Pterophoridae with either of these superfamilies.

HYBLAEOIDEA

Hyblaeidae

The superfamily (Munroe, 1982) includes a single family, Hyblaeidae, with about 20 species. One species (in the genus *Torne*) occurs in the Neotropics, the others (in the genus *Hyblaea*) are found in tropical and subtropical areas of the Old World. The moths are of medium size, broad winged and stout bodied. Commonly the forewings are drab and the hindwings black with striking yellow or orange patches.

Adult (Fig. 261)

External ocelli are present, and chaetosemata absent. The proboscis is well developed and not scaled at its base; the maxillary palpi are short and 3- or 4-segmented, and the labial palpi are porrect and beak-like. The epiphysis is present, and the tibial spur formula is 0-2-0 in males and 0-2-4 in females. The hindtibia of the male bears a hairpencil, which fits into a projection from the hindcoxa. The wing venation is little reduced. In the forewing the branches of vein Rs arise independently from the cell, vein M_2 is closer to M_3 than to M_1, and vein CuP is absent. In the hindwing vein M_3 arises close to CuA_1, and CuP is weak basally but developed distally. Wing coupling is effected by a frenulum and a retinaculum, and the forewing-metathoracic locking

device is present. There are neither thoracic nor abdominal hearing organs. The ovipositor is fairly short.
(Munroe, 1982; Common, 1990.)

Immature stages

The egg is of the flat variety and is disc-like (Minet, 1986).

In the mature larva of *Hyblaea puera* (Singh, 1955), the head is semiprognathous. Secondary setae are absent. The L group on T1 is bisetose, and on the abdomen setae L_1 and L_2 are close together. On the ventral prolegs, the crochets are arranged in a triordinal, uniserial circle, and on the anal prolegs they are also triordinal but arranged in a mesal penellipse.

The pupa of *H. puera* (Singh, 1955) has the appendages fused to the body, and a cremaster is present. The pupa is not protruded at adult emergence.

Biology

Hyblaeid larvae live in silken galleries formed between tied leaves. *Hyblaea puera* defoliates teak (*Tectora grandis*: Verbenaceae) (Singh, 1955). The adults hold their antennae backwards over the wing when at rest, a feature also seen in some Tortricidae, particularly Olethreutinae (Minet, 1986), and Pyralidae.

Phylogenetic relationships

The absence of metathoracic tympanal organs, the form of the palpi, and various larval features, such as feeding habits and, perhaps, the arrangement of the crochets, demonstrate that Hyblaeidae do not belong in Noctuoidea where they have been, at times, misplaced. They share a number of similarities with Pyraloidea, a superfamily to which they are often assigned, but lack the abdominal tympanal organs of that family. Similarities between Hyblaeidae and Pyralidae mentioned by Singh (1955) are the bisetose L-group of T1 (but this character is found elsewhere, including in Noctuidae); the arrangement of crochets in a complete circle (the circle is not always complete in some Pyralidae); the proximity of setae L_1 and L_2 on the abdominal segments; and the absence of secondary setae (a feature also found in most Noctuidae besides many other groups). Given the uncertainties, and following Minet (1983; 1986), the family is placed in a separate superfamily. However, certain larval characters are similar in the Hyblaeidae and Pyralidae.

THYRIDOIDEA

Thyrididae

A single family, Thyrididae, is included in Thyridoidea with about 600 species of predominantly tropical, small-medium to medium sized moths (Munroe, 1982). The wings are broad and typically reticulated with a similar pattern on both forewing and hindwing. In some species the wings bear conspicuous hyaline patches. The larvae are often leaf rollers or stem borers.

Frequently associated with Pyraloidea, Thyatiridae lack abdominal tympanal organs and have a proboscis without basal scaling. Their affinities with Pyraloidea are therefore not established and they are therefore placed in a superfamily separate from Pyraloidea (Minet, 1986).

Adult (Fig. 262)

External ocelli are rarely present, but when occurring are very small. Chaetosemata are absent. The frons is occasionally extended into a prominent process. The compound eye sometimes bears interfacetal hairs. Typically minutely ciliate, the antenna may also be ciliate, dentate, or pectinate. The maxillary palpi are minute, the proboscis is often reduced or even vestigial, but when present lacks scales, and the labial palpi are usually 3-segmented and occasionally 2-segmented.

The epiphysis is rarely absent and the tibial spur formula is 0-2-2, 0-2-3 or 0-2-4. A hairpencil sometimes occurs on the hindtibia of the male. Wing pattern is usually reticulate, and the wings often bear hyaline spots or patches. The colour of the wings is often brown to red-brown. The wing venation is little reduced. In the forewing R_2-R_5 usually arise independently from the cell, but sometimes certain of these veins are united basally. In the hindwing, $Sc + R_1$ usually approaches Rs partway along the cell, but occasionally they fuse for a short distance. The metathoracic-forewing locking device is generally absent, but apparently (Kuijten, 1974) sometimes present. The wings are coupled by a frenulum-retinacular system; very rarely the frenulum is absent.

Tympanal organs occur at the bases of the wings in some Siculodinae (Minet, 1983), but they are absent from both the thorax and the abdomen. The ovipositor is short. (Whalley, 1971.)

Immature stages

Eggs are upright and ribbed (Common, 1990).

The larva is semiprognathous to hypognathous. Only two setae occur on each side of the thoracic segments T2 and T3 compared with most Pyralidae where there are three. Secondary setae are absent. Prolegs occur on A3-6 and on A10. Crochets on the prolegs are uni- to biordinal; on the ventral prolegs they are arranged in circles or broad ellipses.

The pupa lacks dorsal spines (Common, 1990). (Neunzig *in* Stehr, 1987.)

Biology

The larvae are concealed feeders living within stems or in rolled or tied leaves. Some induce the formation of galls. Adults rest in a distinctive position with the body steeply raised and the wings held outstretched. Many mimic dead leaves. Thyrididae seem mainly to be forest-dwellers. (Whalley, 1971.)

Classification

Four subfamilies were recognised by Whalley (1971). **Argyrotypinae** occur in Africa and Madagascar; **Pachythrinae** are pantropical; **Striglinae** (Whalley, 1976) are also pantropical; and **Siculodinae**, which are also widely distributed, include genera which may make the subfamily non-monophyletic.

Phylogenetic relationships

The pyralid-like general appearance of the larva seems to have been a major reason for the frequent association of Thyrididae with Pyralidae. However, no specialized characters have been found to support the inclusion of the two families within the same superfamily. Indeed, even the monophyly of the family Thyrididae has been questioned (Whalley, 1976).

PYRALOIDEA

The superfamily is taken to include the single family Pyralidae. Several other families have often been placed in Pyraloidea (particularly Hyblaeidae, Thyrididae, Pterophoridae, Oxychirotidae, and Tineodidae, but since they lack tympanal organs and appear to resemble Pyralidae only on primitive or convergent characters, they are excluded (Minet, 1982, 1983, 1986). Reduced tympanal organs have now been discovered in *Lathroteles*, and Lathrotelidae synonymized with Nymphulinae (see below) (Minet, 1991).

Pyraloidea are characterized by a pair of tympanal organs, the morphology of which is unique, situated ventrally on the first abdominal segment (Chapter 6).

Pyralidae

The family is probably the most speciose in the Lepidoptera with approximately 25,000 named species and possibly four times as many yet to be described (M. Shaffer, *pers. comm.*). Typically, Pyralidae are delicate moths, ranging in size from small to large. The larvae are generally concealed feeders (webbers, borers, miners). Although most species are phytophagous, considerable diversity is encountered in larval food substrate - including such peculiarities as coprophagy on the droppings of sloths and the consumption of wax of the combs of social Hymenoptera. Several species have aquatic larvae, and many are of economic importance.

Adult (Figs 263-267)

External ocelli and chaetosemata are present or absent. Generally large, the compound eyes are strongly reduced in certain diurnal species. The antennae are occasionally laminate, or uni- or bipectinate, but usually filiform. The proboscis is well developed and scaled at its base. The maxillary palpi are generally short and 4-segmented, but the number of segments may be reduced to 2 or 3. Associated with the maxillary palpi are various scales, which may form large flattened tufts, sprays, or pencils. The labial palpi are 3-segmented, vary in length and may be porrect or ascending.

The epiphysis is present on the foreleg, and the tibial spur formula is 0-2-4 or, rarely, 0-2-2. The wings are generally fairly broad, although they may be narrow in some species, and the shape is variable within the family. The wing venation is typically little reduced. In the forewing the chorda and the stem of vein M are absent. Vein R_3 is usually stalked with R_4 or occasionally even fused. Vein CuP is usually absent. In the hindwing, vein $Sc + R_1$ is fused with vein Rs to beyond the cell, or the two veins run very close together. Distally they 'diverge'. This character is diagnostic of the family. Vein CuP is usually present, and vein 3A is also found in association with the generally large anal area. Wing coupling is typically effected by a frenulum-retinacular system, but sometimes this is reduced. The forewing-metathoracic locking-device is present.

Tympanal organs, anterioventrally directed, are present at the base of the abdomen (see Chapter 6). The male genitalia are generally simple, but may bear complex ornamentations. In the female, the ovipositor generally comprises a pair of setose lobes.
(Munroe, 1972-[4]; Minet, 1982, 1985a.)

Immature stages

Eggs (Neunzig *in* Stehr, 1987) are scale-like, ovoid, or cylindrical, and of the flat variety.

The larva (Munroe, 1982; Neunzig *in* Stehr, 1987) is usually approximately cylindrical. Typically semiprognathous, the mouthparts are occasionally prognathous, particularly in leaf mining species. In the final instar the crochets of the ventral prolegs are usually biordinal or triordinal, but sometimes uniordinal. They are arranged in a circle, penellipse, or (Chrysauginae) in transverse bands. Filamentous gills are sometimes present.

The integument of the pupa (Mosher, 1916) is usually smooth or punctate, the appendages are fused, and the abdomen is not spined.

Biology

When scale-like, the eggs (Neunzig *in* Stehr, 1987) are laid in shingled masses, when ovoid or cylindrical they are deposited singly or in small groups. Sometimes they are glued together to form 'egg-sticks'. Eggs are covered by a transparent adhesive coating secreted as they are laid. In Schoenobiinae eggs are covered by a layer of wax in the form of numerous 'hairs' after they have been laid. The larvae are mainly phytophagous, and consume live plants, or dried plant material (e.g., seeds, hay, tobacco), but coprophagy and carnivory also occurs and honey, the wax of honey comb and chocolate are also eaten by some. Pyralid larvae are generally terrestrial, but a significant number are aquatic. Pupation typically takes place in silken shelters at the feeding site or in litter or soil on the ground.

Classification

Pyralidae are divided into two main lineages that have variously been termed Pyralinae and Crambinae, Pyraliformes and Crambiformes, and Pyralidae and Crambidae. This basic division was noted by Börner (1925) on the basis of the shape of the tympanum and the presence or absence of the ventral extension of the membrane (the praecinctorium) between the thorax and the abdomen. The differences were discussed in more detail by Kennel & Eggers (1933), and again by Minet (1982, 1985a), and derive further support from a study of larvae by Hasenfuss (1960). Here, following Munroe (1972-[4]), the names Pyraliformes and Crambiformes are used to reflect this division. By adapting this system of ranking, the family Pyralidae covers all those Lepidoptera with a characteristic tympanal organ. Although the monophyly of the two principal groups of Pyralidae is well supported, the various subfamilies of each group may not be monophyletic (Minet, 1985a). Descriptions of differences in the tympanal organs of the subfamilies were provided by Kennel & Eggers (1933). Further information was added by Minet (1980, 1982, 1985a) and Maes (1985). Minet (1982) recognized 25 subfamilies of Pyralidae, but this is a provisional arrangement.

Pyraliform classification and diversity

This group is clearly separable from the Crambiformes by the structure of the tympanal organs (e.g., Kennel & Eggers, 1933; Minet, 1982) and by certain characteristics of the larvae (Hasenfuss, 1960). In Pyraliformes the thoraco-abdominal intersegmental membrane is not extended into a praecinctorium, although medio-ventrally it is frequently covered with elongated scales. Furthermore, the tympanum and conjunctivum lie approximately in the same plane, and the scolopidia are inserted on the roof of the bulla tympani. The bulla tympani is said to be 'closed' in Pyraliformes since it surrounds the tracheal membrane. In the larva, seta SD_1 of A8 is surrounded by a ring of pigment. The Crambiformes have a well developed praecinctorium (sometimes secondarily reduced); the tympanum and the conjunctivum lie at an angle to each other, and the scolopidia are inserted onto a lateral projection (processus spiniforme) of the tympanic bulla rather than on its roof. The bulla tympani does not surround the tracheal membrane, and is said to be 'open'. Segment A8 of the larva is without a ring of pigment surrounding seta SD_1.

No specialized character convincingly supports the monophyly of the **Pyralinae** (Fig. 263), and the assemblage currently referred to the subfamily is probably not monophyletic (Minet, 1985a). Features by which the grouping may generally be recognized include a subcostal retinaculum and an extended posterior part of the tympanal frame rather than one that is distinctly trapezoidal or triangular (Minet, 1982, 1985a). In the hindwing, veins $Sc + R_1$ and Rs are often approximated rather than fused. Pyraline larvae are frequently associated with decaying plant material.

Endotrichinae are probably not closely related to typical Pyralinae, but show greater affinity to Chrysauginae; for example, in the strongly reduced maxillary palpi (Minet, 1985a).

Chrysauginae were considered to be monophyletic by Minet (1985a). The maxillary palpi are vestigial. In the males of some species, the frenulum and retinaculum are modified and function as a stridulatory organ (see Chapter 6). The adults of *Cryptoses* and *Bradypolicola* live in the fur of sloths, and the larvae are coprophagous on the droppings of these mammals.

Galleriinae are characterized by the reduction or loss of a number of structures (Whalley, 1964; Minet, 1985a), for instance external ocelli, chaetosemata, paraspinae (narrow, sharply pointed sclerotizations of the posterior part of the ligna tympani), proboscis, and maxillary palpi. Minet (1985a) considered that the monophyly of the group is established in particular by the loss of external ocelli, the modified labial palpi of the males, and the curved, rather than straight, ligna tympani. The larvae of many species feed on dried plant material, others are pests of palm trees and sugarcane (Holloway *et al.*, 1987), and some live on wax and debris in the combs of social Hymenoptera. The larvae of *Tirathaba parasitica* may feed on certain wood-boring larvae (Common, 1990). *Eldana saccharina*, an African sugarcane borer, lives primarily on sedges. Its invasion of sugarcane is associated with increased use of nitrogen and potassium fertilizers, and with monoculture (Atkinson, 1980).

Epipaschiinae occur mainly in the tropics or subtropics, and include moths that superficially resemble Noctuidae (Holloway *et al.*, 1987). Scale tufts are usually present on the upper surface of the forewing, and a fovea occurs in males. The proboscis is well developed. In some species the labial palpi are prominent and curved over the head. Tufts of erect scales occur on the thoraco-abdominal intersegmental membrane of most species (Minet, 1985a).

Phycitinae (Roesler, 1973; Neunzig, 1986) (Fig. 264) is a large subfamily of worldwide distribution, with most species occurring in the tropics. The forewings are generally mostly brown or grey, but a few are brightly coloured. The proboscis is usually well developed, but sometimes reduced, and the maxillary palpi of males may be developed and prominently scaled. The antennae are simple in females and sometimes pectinate in males. In the forewing, veins R_3 and R_4 are usually fused. Vein R_5 is absent. A cubital pecten is present in the hindwing. The prominent annular ring around the point of insertion of the scoloparium on the tympanum may be a derived character of Phycitinae (Minet, 1985a). The eggs are generally ovoid and flattened, and the chorion is often reticulate. Typically, eggs are laid singly or in small groups, but in some species that feed on cactus they are glued together and form a rod or 'egg-stick'. The larvae include many that feed on dried produce as varied as seeds, dried fruit, tobacco, and even chocolate. Well known is *Plodia interpunctella* (the Indian Meal Moth) and *Ephestia kuehniella* (the Mediterranean Flour Moth). However, most are tiers, webbers, and borers, with some (e.g., *Laetila coccidivora*) feeding on scale insects, or on larvae of Lymantriidae and Notodontidae (e.g., *Metoecis*). Also included in this subfamily is *Cactoblastis cactorum* used effectively in the control of *Opuntia*. Phycitinae generally pupate in silk lined chambers in the feeding site, or in litter or the soil.

Peoriinae (Shaffer, 1968) are distinguished from Phycitinae by the male and female genitalia. In particular, in females the ovipositor is strongly compressed, while in males the uncus bears prominent spicate processes, at least in American species. The moths resemble Phycitinae.

(Munroe, 1972-[4]; Minet, 1982; 1985a.)

Crambiform classification and diversity

Crambiformes are distinguished from Pyraliformes above.

In **Crambinae** (Bleszinsky, 1965) (Fig. 265), the wings are typically folded tightly around the body at rest and they are often longitudinally striped. These insects are frequently disturbed from grasslands, and are well known from lawns, but they are also associated with other monocotyledonous plants and may be pests of rice, maize, and sugarcane. The larvae are internal feeders of Gramineae. Several are pests of pastures. Scales found apically on the maxillary palpi making the palpi appear dilated; the labial palpi are prominent. A pecten of scales is present on the hindwing. Crambinae are not demonstrably monophyletic, although sizable subdivisions are probably natural (Minet, 1985a). The subfamily is cosmopolitan.

Midilinae (Munroe, 1970) are confined to the Neotropics. They are characterized by the small size of the tympanic bullae, which are weak and widely separated. The praecinctorium is present, but short and often hidden by scales (Minet, 1985a). The group is probably monophyletic.

Many **Schoenobiinae** are whitish or yellowish moths. Frequently, they rest among grasses, rushes, and sedges. The proboscis is vestigial. Vein CuP is present near the margin of the forewing. An anal tuft of hair-like scales on the female is used to cover the egg masses. The praecinctorium may be normal or exceptionally large and bilobed (Minet, 1985a). The presence of a pair of small 'tuberculate plates' on sternum A2 may be independently developed, and represent a specialized character of the subfamily (Minet, 1985a). The larvae tunnel in grasses, rushes, and sedges, and often respire underwater. *Scirpophaga* is a major pest of rice.

Cybalomiinae are characterized by a lateral division of sternum A2 in the adult, although the feature is not unique to the group (Minet, 1985a).

Linostinae (Munroe, 1959) include a single genus, *Linosta*, with three species from the Neotropics.

Nymphulinae (= Acentropinae) (Munroe, 1972-[4]; Speidel, 1984) (Fig. 266) include some of the most colourful Pyralidae. A number of species bear a row of metallic spots at the margin of the hindwing. The moths are often pale, and bear a complex pattern of transverse bands on the wings. The tympanic bullae are considerably reduced, and the praecinctorium is typically short, although it may also be very strongly reduced or lengthened. The scolopidia are attached either directly to the back of the bulla tympani or to dorsal rather than lateral projections. This character may be a specialization of the subfamily (Minet, 1985a). *Lathroteles*, which has been assigned to the Lathrotelidae, is now included in Nymphulinae. Like other Nymphulinae the tympanal organs are reduced (Minet, 1991). The larvae are predominantly aquatic, and many have gills. With some exceptions, larvae living in standing or slow moving water construct cases, eat leaves, and may or may not have gills. By contrast, larvae living on rocks in fast flowing water are web-spinners, feed on algae, and always bear gills (Lange, 1956). Gills are absent from *Nymphula*. During winter, the case, constructed from pieces of aquatic plants, is filled with water and respiration is cutaneous. In spring, the case is filled with air and respiration is spiracular. Larvae of Argyractini live under silken webs spun under stones in rapidly flowing streams, and bear tufted gills along the sides of the abdomen. In *Acentria ephemerella*, from the western Palaearctic and the Nearctic, a brachypterous morph of the *adult* female swims underwater aided by fringes of long hairs on the mid- and hindlegs (Berg, 1941). Respiration appears to be cutaneous.

Evergestinae are predominantly Holarctic, but a few genera are found in tropical and subtropical areas. A cubital pecten is present on the hindwing of all but a few species, and the praecinctorium is usually weakly bilobed. The larvae, where known, feed within webs on plants of Cruciferae or Capparidaceae (Munroe, 1972-[4]). North American species of *Orenaia* are arctic or alpine, and active by day. In common with many other diurnal moths that live in similar areas, they are small, dark, hairy, and have small eyes.

Musotiminae have greatly enlarged bullae tympani (Minet, 1985a). The larvae feed on ferns. The subfamily was included within Nymphulinae by Maes (1987).

Scopariinae adults often bear a tuft of scales on the discal zone of the forewing, but there is no cubital pecten on the hindwing of any but a few species. The subfamily has a worldwide distribution, and is based on the large, cosmopolitan genus *Scoparia*. Scopariine larvae typically feed on mosses, but some European species live on the roots of flowering plants, a Hawaiian species on the clubmoss *Lycopodium*, and an Australian species bores into tree-ferns (Munroe, 1972-[4]).

Pyraustinae (Munroe, 1976) include numerous species, most occurring in the tropics. This subfamily includes many species that probably belong in Odontiinae or Spilomelinae (Maes, 1987). The proboscis is usually present and the maxillary palpi are generally well developed. The praecinctorium is bilobed. Pyraustine eggs are often scale-like, and the larvae are leaf rollers and tiers, or borers. Many are pests, such as *Ostrina* a genus that feeds on a wide variety of crops such as corn, hemp, tomatoes, and fruit trees (Mutuura & Munroe, 1970). Pupation occurs in a flimsy cocoon or in the tunnel formed by the larva.

Odontiinae include very small to medium sized Pyralidae resembling Pyraustinae. However, the praecinctorium is not bilobed. Chaetosemata are absent and the compound eyes are often reduced. The larvae include leaf-webbers, miners, and seed-feeders. Although almost worldwide in their distribution, they are particularly well represented in the drier parts of the Holarctic.

Cathariinae include the single genus *Catharia* placed in a separate subfamily by Minet (1982). The species are found high in the Pyrenees Mountains, and the moths are mainly black. The structure of the tympanal organs is sexually dimorphic.

Dichogaminae include two genera, *Dichogama* and *Alatancusia* (Minet, 1982; Maes, 1987). The moths are relatively large Pyralidae. Although the group is represented in the Nearctic, it is almost exclusively Neotropical. The larvae feed on *Capparis* (Capparaceae) and live between spun leaves.

Spilomelinae (Fig. 267) (previously placed in Pyraustinae) were treated as a subfamily by Minet (1982) and Maes (1987) on the grounds that by so doing the Pyraustinae will become a more natural grouping. The subfamily is cosmopolitan, with most species occurring in the tropics, but heterogeneous. Spilomelinae resemble Pyraustinae, but the praecinctorium is only weakly bilobed in the former.

Noordinae (Minet, 1980) are a small subfamily with strongly modified tympanal organs. These structures are partially retracted into the thorax so that the tympana face backwards rather than in the usual anterio-ventral direction. The praecinctorium is reduced, and the vinculum of the male genitalia bears a pair of prominent, semimembranous structures anteriorly.

Glaphyriinae, a relatively small subfamily, is confined to the New World except for the genus *Hellula*, which also occurs elsewhere. External ocelli are generally present, and although there are no true chaetosemata a pair of prominent scale-tufts is found in their usual position (Munroe, 1972-[4]). The proboscis may be well developed or, less frequently, reduced, and the labial palpi are fairly short. The varied modes of life of the larvae were noted by Munroe (1972-[4]) and include predatory as well as phytopha-gous species. *Dicymdomia julianalis* eats larvae of Psychidae, and species of *Chalcoela* live in the nests of vespid wasps.

(Munroe, 1972-[4]; Minet 1981, 1985a; Maes, 1987.)

12

HIGHER DITRYSIA

The higher Ditrysia are also known as Macrolepidoptera, although the composition of this group is somewhat arbitrary and its monophyly doubtful. In the present work, ten superfamilies are included. The families Geometridae (Geometroidea) and Noctuidae (Noctuoidea) account jointly for about one quarter of all named Lepidoptera. The higher Ditrysia include the larger moths and the butterflies. Their larvae are typically external feeders on green plants with prolegs adapted for grasping aided by crochets arranged in a mesoseries; however, many have secondarily become borers, and carnivory and fungivory also occur. The pupae are obtect, with appendages fused to the pupal case, and only rarely do they project from the cocoon before adult eclosion. Phylogenetic relationships between the superfamilies are largely uncertain.

GEOMETROIDEA

Geometridae

The superfamily Geometroidea is taken to include a single family, Geometridae (Minet, 1983) (but see under *Phylogenetic relationships*, below, and Minet, 1991). Although Geometridae and several other families have tympanal organs at the base of the abdomen, the structure of these organs differs (Chapter 6).

The Geometridae are one of the largest families of moths, with an estimated figure of at least 20000 described species (Munroe, 1982). Their distribution is cosmopolitan. The moths range from small to large, with most being of moderate size. Generally, Geometridae are slender-bodied, broad-winged, and rather delicate macrolepidopterans, but there are many robust-bodied representatives. While some species are brightly coloured, most are cryptically patterned, frequently with the wings bearing several transverse wavy bands. The larvae are external feeders, and significant defoliators. Typically 'loopers', their form of movement is related to reduction of the number of prolegs.

Adult (Figs 268-270)

External ocelli are generally absent, and chaetosemata are present but small. The antennae are simple, serrate or bipectinate. The head is thickly scaled. The proboscis is well developed, and not scaled at its base, and the maxillary palpi are minute, 1- or occasionally 2-segmented. The labial palpi are upturned, but not strongly ascending.

The epiphysis is present, and the tibial spur formula is generally 0-2-4. The hindtibia of the male is sometimes swollen and bears a scent-brush lying in a groove. The wings are broad, but in females of a few species they are vestigial. In the forewing one or two areoles are usually formed by veins R_1 and Rs. Vein R_5 is stalked with R_3 and R_4. Vein M_2 usually arises closer to M_1 than to M_3, but may arise midway between M_1 and M_3. Vein CuP is absent. In the hindwing, vein Sc is usually bent strongly at its base, runs close to Rs and may fuse with it for part of its length. Otherwise it may be connected via R_1. It diverges before the end of the cell. Vein M_2 arises midway between M_1 and M_3 or near M_1. It may be absent or very weak. There is no forewing-metathoracic locking device, and the wings are rarely folded over the abdomen when the moth is at rest. A frenulum and a retinaculum are present but often reduced.

Tympanal organs (see Chapter 6) are present at the base of the abdomen other than in brachypterous females where the structures are lost. Each opens anteriorly. A tympanic handle or ansa (*Bügel* of Kennel & Eggers, 1933), which curves over the tympanum and from which the scolopidium is suspended, is unique to the family. In the female genitalia, the ovipositor is generally blunt.
(McGuffin, 1987; Common, 1990.)

Immature stages

The eggs are typically of the flat variety with the longitudinal axis lying parallel or almost parallel to the substrate. However, in some species they are upright with the longitudinal axis perpendicular to the substrate and attached by their posterior poles. The two kinds may even occur within the same genus (e.g., *Sterrha*, *Biston*) (Hinton, 1981). In certain species the eggs lie at an angle to the substrate, and in others they are not attached at all. In lateral view the eggs are usually ovoid, elliptical or bluntly elliptical, while when viewed from the ends, they are circular or oval (Salkeld, 1983). Sculpturing may be present or absent.

The larva (McGuffin, 1987; Stehr *in* Stehr, 1987) is generally cylindrical and slender-bodied (length about 10 times the width) or, less frequently, stout. It is rarely flattened. The cuticle is smooth or granular, and longitudinal stripes often run the length of the body. Various humps, filaments, or other processes are sometimes present. A full complement of primary setae is present, and a fourth SV seta occurs on the abdominal segments. The larvae are typically loopers with prolegs only on abdominal segments A6 and A10. However, some species have a full complement of prolegs, although some are reduced in size, or less than a full complement but more than the two pairs typical of Geometridae. Prolegs occurring on segments other than on A6 and A10 may lack crochets.

In the pupa (McGuffin, 1987), the palpi are not exposed and the cremaster is well developed.

Biology

The eggs are laid singly or in batches wrapped around twigs. They may be glued to plant substrate or simply dropped. The larvae feed on a wide range of plants, although usually they consume the leaves of trees and shrubs. They are often camouflaged, and frequently assume an outstretched twig- or stick-like position on the plant while holding on to the substrate with their prolegs on A6 and A10. Pupation occurs typically within a flimsy cocoon in debris, but some pupae are merely girdled. In several species of 'winter moths' the wings are virtually lost in the female. The moths are mated almost immediately after emergence from the pupa and lay their eggs soon after.

Classification

The family is currently divided into six subfamilies. The monophyly of all of them is in some doubt, and at least one (Oenochrominae *sensu lato*) is polyphyletic. The subfamilies themselves have frequently been divided into what are often rather loosely associated generic groupings. In general, the subfamily classification of the Geometridae is poorly resolved.

Archiearinae (Fig. 268) include 12 species of medium-sized Geometridae. The species are temperate or montane dwellers. They have a disjunct distribution, occurring extensively in the northern Palaearctic, and also in the southern Andes, and the alpine and subalpine regions of Tasmania. The moths are colourful diurnal Geometridae, with the front of the head bearing numerous hair-like scales. The compound eyes are oval. The moths lack an accessory tympanum on the metapostnotum the fenestra media being extremely narrow (Minet, 1983), a situation contrasting with other Geometridae. Eggs of *Archiearis* are of the upright variety. The larvae (at least in *Archiearis*) progress by 'looping' (Carter & Hargreaves, 1986) although a full complement of prolegs is present. However, the anterior pairs of prolegs are considerably reduced compared with those on A6.

In the Palaearctic, the males fly mainly in sunshine early in spring. Females of *Archiearis* tend to remain near the tops of their foodplant trees, aspen or birch. In Tasmania, males of *Acalyphes* also fly mainly in sunshine around pencil pines (*Arthrotaxis cupressoides*: Pinaceae), although this tree has not been established as the foodplant.

Ennominae (Fig. 269) (see McGuffin, 1987) is by far the most speciose subfamily with, in its present form, about half the number of species in the Geometridae. Ennominae are generally slender-bodied, medium to large Geometridae, but a group of robust-bodied species occurs mainly in southern Australia (e.g., McQuillan, 1981, 1985; McFarland, 1988). The subfamily is divided into numerous tribes, but the higher-level classification requires much resolution.

In the male, the antenna is simple or bipectinate whereas in the female it is usually simple. Both sexes are generally fully winged, but females of some species are almost wingless. Frequently, the colour pattern of the forewing and hindwing is similar. In males of several species, and in females of some, a blister-like fovea occurs near the base of the forewing between veins Cu and $1A+2A$. In the forewing, vein R_1 commonly anastomoses with Sc. Vein M_2 is typically weak or absent in the hindwing. A hairpencil, usually concealed in a deep groove, is often present on the tibia of the midleg. In several genera, widely scattered through the subfamily, sternum A3 of the male bears a transverse band of posteriorly directed setae (Holloway, *et al.*, 1987).

In some species (particularly from the southern hemisphere) the number of larval prolegs is not reduced to the extent typical of Geometridae. For example prolegs are present on segments A4-6 and A10 in the Australian genus *Thalania*, although they are somewhat reduced in size on A4 and A5. A similar situation is seen in the African genus *Callioratis* (Duke & Duke, 1988).

Oenochrominae (Fig. 270), as generally constituted (e.g., Prout, 1910), are an unnatural group of Geometridae largely encompassing those genera that cannot be placed in any other subfamily. Oenochrominae *sensu stricto* include the robust-bodied and almost exclusively Australasian moths belonging to *Oenochroma* or related genera (Scoble & Edwards, 1990). The 'slender bodied' Oenochrominae are probably not related to Oenochrominae *sensu stricto*. Even the composition of the Oenochrominae *sensu stricto* is uncertain, but the genera included are similar in general appearance, in the structure of the male genitalia, and in wing venation. The chief foodplants of Oenochrominae *sensu stricto* appear to be members of the Proteaceae and the Myrtaceae (McFarland, 1979; 1988). Larvae of Oenochrominae *sensu stricto* often bear two pairs of ventral prolegs as well as the anal pair. The extra pair, which occurs on segment A5, is much reduced.

Slender-bodied 'Oenochrominae' are distributed through the Orient, Africa, and Australasia with most species occurring in southern Australia. Many moths of this grouping are cryptic and well camouflaged.

Various brightly coloured moths are placed in 'Oenochrominae'. In Africa, *Aletis* and *Cartaletis* probably mimic certain butterflies, including the African Monarch, *Danaus chrysippus*. The South East Asian genera *Eumelea* and *Celerena* include bright yellow species which, like some other 'Oenochrominae', fly in forest understory by day (Holloway *et al.*, 1987).

Larvae of the Palaearctic genus *Alsophila* are defoliators of forest and fruit trees. *A. aescularia* is a European pest, and *A. pometaria* sometimes damages trees and bushes in North America. Unlike the typically smooth, slightly elongated eggs of Oenochrominae *sensu stricto*, which are laid singly or in small groups (McFarland, 1973b; 1988), those of *A. pometaria*, although smooth, are laid in large masses on bark or encircling twigs, and they are positioned with the micropyle at the top of each egg (Salkeld, 1983).

Geometrinae (see Ferguson, 1985) include the Emerald moths with about 1400 named species worldwide. Most are green and slender-bodied, but some are stout-bodied with a greyish, lichen-like or mottled appearance. The identity of the green pigment is unknown. It is, however, not a bile pigment, and appears not to fall into any established class of insect pigment (M.A. Cook, unpublished). The pigment predominates in Geometrinae but occurs in other Lepidoptera as well (M.A. Cook, unpublished). In the forewing, vein R_1 is usually separate from R_2, but anastomoses with Sc for a short distance. The accessory cell is often absent. Vein M_1 usually arises near the base of the stalk. In the hindwing, vein M_2 arises nearer M_1 than M_3. In the abdomen, two patches of deciduous spines are found on sternum A3, and a pair of large socii typically occur on the male genitalia (Holloway *et al.*, 1987).

The eggs (Salkeld, 1983; Ferguson, 1985; McFarland, 1988) are typically oblong and flattened on both sides (lozenge-like). They are generally exposed, not inserted into crevices, and are probably always dropped freely. They are usually laid singly or in twos, but in some robust-bodied species they are deposited in heaps or chains.

In the larvae (Ferguson, 1985; McFarland, 1988) the head is often bifid. Seta SV_4 is absent on A1, but present on A3-5. Only two pairs of prolegs (i.e., the typical geometrid number) are present. The habit of attaching plant fragments to the body, although occurring in a limited number of species, is found throughout the Geometridae.

Pupation occurs in a flimsy cocoon, usually on the foodplant, but sometimes on the ground among loose debris (not in the soil) (Ferguson, 1985; McFarland, 1988).

Sterrhinae (see McGuffin, 1967; Covell, 1983) are often called waves because of the numerous wavy fasciae on the wings. The moths are mainly small Geometridae and usually inconspicuous at rest with their cryptically coloured wings outstretched. They are rarely green, and both forewings and hindwings are concolorous with bands continuing from forewing to hindwing. Typically, a dark discal spot is present on each wing. Most species are nocturnal, but *Cyllopoda* and its relatives, with striking black and yellow markings, are active by day (Covell, 1983). Although most species are tropical, Sterrhinae are well represented in temperate regions especially the *Scopula*-group (Holloway *et al.*, 1987). In the forewing, one or two accessory cells are usually present. Vein $Sc+R_1$ of the hindwing fuses with Rs near the wing base for only a short distance before it diverges sharply.

The eggs (Salkeld, 1983) are flattened, and deposited loosely. They are variously ribbed and sculptured. The larvae (McGuffin, 1967; McFarland, 1988) are usually cryptic. Some have a bilobed head and a long, slender body, and only two pairs of prolegs are present. The larvae of most species feed on low herbs, but some (e.g., *Anisodes* and its relatives) are arboreal. Pupation generally occurs in a flimsy cocoon, but in *Anisodes* the pupae are girdled (Common, 1986) as are those of the related genus *Cyclophora*. However, the girdle traverses abdominal segment A3, rather than A1 or the thorax as in most butterflies.

Adults of some South East Asian members of *Scopula* feed on blood, from mammalian wounds, on sweat, and on tears (Bänziger & Fletcher, 1985). While the larvae of most species of the *Anisodes* group are arboreal, those of *Scopula* and *Idaea* include many herbaceous feeders.

Larentiinae (see Holloway *et al.*, 1987) are worldwide in distribution but most diverse in temperate regions and at high altitudes in the tropics. They are the dominant geometrid subfamily at high altitudes (McQuillan, 1986). A discal spot is frequently present on both forewing and hindwing. The forewings usually bear numerous fine transverse fasciae, usually arranged in multiple groups. Wings are much reduced in the females of some species. One or two areoles are generally present. In the hindwing, vein $Sc + R_1$ usually fuses with Rs for at least half the length of the cell, and vein M_2 is fully developed. In the male, a pair of coremata often extend from between A8 and the genitalia, and on A8 a pair of processes called octavals occur widely within the subfamily.

The eggs (Salkeld, 1983) are smooth, weakly or heavily sculptured, and usually oblong and variously flattened. The larvae (McFarland, 1988) are generally short and relatively stout. Some construct nests of leaves and silk, some resemble twigs or stems, and others hide in crevices. Certain species are concealed in ground litter by day and ascend to feed on the foliage of trees by night. Night-feeding is frequent in the group. In all but one of the Hawaiian species of *Eupithecia*, the larva is carnivorous and ambushes small insects and spiders (Montgomery, 1982). Prey is seized by the thoracic legs after a strike lasting about 1/12 second. Larvae of other *Eupithecia* feed primarily on flowers.

Phylogenetic relationships

The Geometridae may be most closely related to Uraniidae and Sematuridae (see below). Indeed, these families were reunited in Geometroidea by Minet (1991) based on two shared specializations. These are first that the larval spinneret is shorter than the prementum (the latter being measured along its midline), and second that the ventral arm of the tegula is fairly pointed or elongated.

URANIOIDEA

Uraniidae were removed from Geometroidea by Minet (1983) on the grounds that they shared no known defining characters with that group (but see under Geometroidea, under *'Phylogenetic relationships'*). Previously, Uraniidae were associated with Geometridae partly on the basis of the presence of abdominal tympanal organs. However, as shown by Kennel & Eggers (1933) the structure of these organs differ. Within Uraniidae occur some of the most spectacular of moths, iridescent and tailed although, with the inclusion of the Epipleminae, most members of the family are relatively drab. There are approximately 700 species, mostly Epipleminae, of this predominantly tropical group.

Epipleminae are included in Uraniidae due to the similarity of the tympanal organs, which exhibit only superficial differences from those of the latter. The moths range from small-medium to large.

Although they lack tympanal organs, Sematuridae and Epicopeiidae are included, tentatively, in Uranioidea.

Uraniidae

Adult (Figs 271, 272)

External ocelli are absent, and chaetosemata are large. In males, the antennae are thickened, and usually at least partly dentate. The proboscis is developed, maxillary palpi are 1-segmented, and the labial palpi are normally developed and ascending, or weakly ascending and short.

The epiphysis is present, and the tibial spur formula is usually 0-2-4 but sometimes 0-2-3. In the forewing, R_5 is stalked with M_1, or these veins arise from approximately the same point. Vein M_2 does not usually arise nearer M_3 than M_1. In the hindwing, $Sc + R_1$ diverges from Rs at the base. The forewing-metathoracic wing locking device is absent. A frenulum is present or absent; when absent the humeral angle of the wing is expanded.

The tympanal organs, which occur at the base of the abdomen are sexually dimorphic (see Chapter 6).
(Munroe, 1982; Common, 1990.)

Classification

Uraniinae (Fig. 271) (see Munroe, 1982; Common, 1990), which are often treated as a family, include about 100 species (Holloway *et al.*, 1987). The group is divided into Uraniini and Microniini, the former including the spectacular, iridescent, typically diurnal moths with prominently tailed hindwings. At times, they have been confused with papilionid butterflies. Well known is *Urania* from South America, *Chrysiridia* from Madagascar, and *Alcides* from Papua. *Lyssa*, from the Indo-Australian tropics, is nocturnal and not iridescent. Microniini are not iridescent, usually bear very short, pointed tails on the hindwing, are crepuscular or nocturnal, and are much more delicate. They are medium-sized moths with whitish wings bearing darker fasciae. Eyespots are usually present particularly at the base of the hindwing tails. The group is confined to the tropics or subtropics of the Old World. The frenulum is absent or vestigial.

Little is known about the larvae of most species. Prolegs are probably normally developed throughout the group. In the Indo-Australian tropics, uraniines may generally feed on members of Euphorbiaceae (Holloway *et al.*, 1987), and the same may be true of Neotropical species. In Africa, *Chrysiridia aroesus* feeds on mango (Anacardiaceae) and Indian Almond (Combretaceae) (Pinhey, 1975).

Adults are often migratory. In *Urania leilus*, both males and females migrate, and flights may be unidirectional dispersions in response to diminished food or other resources in the usual feeding areas (N. Smith, pers. comm.).

Epipleminae (Fig. 272) are widespread in the tropics, but poorly represented in temperate areas. There are about 600 species of these small-medium sized and generally darkish moths (Munroe, 1982). At rest, the hindwings are folded along the abdomen, but the forewings are extended horizontally and are usually rolled or curled (Pl. 1:8). A frenulum is present. The group has not been extensively studied, although Janse (1932) treated the species from South Africa and provided illustrations of the male genitalia.

The larva (Munroe, 1982; Stehr *in* Stehr, 1987), bears a few secondary setae in addition to the primaries. The head is hypognathous. Setae L_1 and L_2 are borne on the same pinaculum on A1-3, a distinctive feature of the group. Prolegs are normally developed, and crochets are arranged in a semicircular mesoseries in which, at least sometimes, they are uniordinal in the middle but biordinal at both ends. First instar larvae of some species inhabit a common web. Epiplemine larvae are more catholic in their choice of foodplants than are those of Uraniinae but, unlike the situation in Uraniinae, they have not been recorded from Euphorbiaceae (Holloway *et al*, 1987).

Sematuridae

The family includes a predominantly Neotropical group of species with about 35 species (Holloway *et al.*, 1987), most of which have tailed hindwings and resemble papilionid butterflies. The Afrotropical genus *Apoprogones*, which occurs in South Africa, is also included so the group has a disjunct distribution. Sematuridae lack tympanal organs, but have usually been associated with Uraniidae. Minet (1991) transferred them to a newly constituted superfamily Geometroidea (Geometridae, Uraniidae, and Sematuridae). As this suggestion requires further scrutiny before establishing the new findings into the classification, Sematuridae are retained in Uranioidea for the present.

Adult (Fig. 273)

External ocelli are absent, and chaetosemata are prominent. The proboscis and labial palpi are well developed, the latter being weakly upcurved with a porrect, often long, apical segment. The antennae, which are usually dilated towards the apex, narrow again before the tip, which is itself often recurved. The epiphysis is present, and the tibial spur formula is 0-2-4. A frenulum and retinaculum are present, although the frenulum is rather weak. The forewing-metathoracic locking-device is absent. In the forewing M_2 and M_3 arise close together. The first abdominal segment bears very large latero-tergal apodemes (Minet, 1991). Tympanal organs are absent.

Phylogenetic relationships

As in Geometridae and Uraniidae, the spinneret of the larva is short, and the ventral arm of the tegula is pointed or elongated (Minet, 1991). The absence of tympanal hearing organs means that the association of Sematuridae with Geometroidea, Uranioidea, or Drepanoidea is not straightforward. Their current inclusion in Uranioidea is, therefore, by no means established.

Epicopeiidae

The position of this group of superficially papilionoid-like moths from the Oriental region is not established, and it is here included in Uraniidae only tentatively. Since Epicopeiidae have no tympanal organs there is no comparison with those uniquely formed structures in Uraniidae. However, the venation is not dissimilar and the absence of tympanal organs may represent a secondary loss. The Epicopeiidae may be more appropriately placed in Drepanoidea (Minet, 1991), see under that superfamily, below). There are about 10 species. Certain species mimic troidine Papilionidae with the hindwings 'tailed'.

Structure and biology

In the adult (e.g., Fig. 274), external ocelli are absent and chaetosemata are present. The proboscis is developed, the maxillary palpi minute, and the labial palpi small and porrect. The antennae are bipectinate in both males and females. The epiphysis is present, and the tibial spur formula is 0-2-4. In the forewing, R_5 may be stalked with M_1, and M_2 and M_3 are well separated. Sometimes the base of vein M is weakly represented in the cell. The frenulum is weak. Tympanal organs are absent.

The larvae secrete wax, formed as granules or filaments, a habit possibly specialized and thus demonstrating the monophyly of a group of superficially disparate species.

Foodplants have been recorded among the Lauraceae and Ulmaceae. Pupation occurs in a weak cocoon on the ground.
(Munroe, 1982; Minet, 1983, 1986.)

DREPANOIDEA

Drepanidae

The superfamily Drepanoidea is composed of a single family, Drepanidae, of medium-sized moths including Drepaninae, Thyatirinae (= Cymatophorinae), and Cyclidiinae (Minet, 1983). These subfamilies are often treated as families. The form of the tympanal organs is characteristic (Kennel & Eggers, 1933; Gohrbandt, 1937; and see Chapter 6), strongly supporting the view that the group is monophyletic. The Australian genus *Hypsidia* represents an additional subgroup of Drepanidae (Scoble & Edwards, 1988) not yet assigned to any of the existing subfamilies.

Epicopeiidae have recently been included in Drepanoidea by Minet (1991) on the basis of the presence of a complete prespiracular sclerite connecting the anterior corner of the first abdominal sternum with the lateral bar of the first abdominal tergum in the adult; the concealed or slightly exposed foreleg femur in the pupa; the presence of setae on a flat and laterally delimited area of the larval mandible; and the occurrence, in the larva, of at least one secondary seta associated with seta L_3 on abdominal segments 1-8.

Adult (Figs 275-277)

External ocelli are usually absent and chaetosemata are reduced or absent. The antennae may be pectinate, lamellate, or filiform. The proboscis is present, vestigial, or absent, maxillary palpi are minute, and the labial palpi are short or long, and porrect or upcurved.

The epiphysis is rarely absent, and the tibial spur formula varies - 0-2-4, 0-2-2, 0-0-4, or 0-0-0. In the forewing, vein M_1 is separate from R_{4+5} or sometimes stalked with it. Vein M_2 often arises near, or at the same point as, M_3, but it may be approximately equidistant from M_1 and M_3. Vein CuP is absent. In the hindwing $Sc+R_1$ runs close and parallel to Rs to beyond the end of the cell, or may anastomose with it. Rarely, it diverges before the end of the cell. Vein CuP is absent. The forewing-metathorax locking device is present or absent. A frenulum-retinaculum system is present or absent. When present, the male frenulum is often clubbed.

Distinctive tympanal organs are present at the base of the first visible abdominal sternum (see Chapter 6). Both pleural and sternal components are present. The sternal components take the form of a pair of small oval structures, each composed of a small and a large chamber. The tympanum is situated *within* each sternal component between the two chambers. The pleural part is formed from a prominent fold supported by a large, 3-armed sclerite. The ovipositor is not elongated.

Immature stages are considered under the relevant subfamilies.
(Munroe, 1982; Common, 1990.)

Classification

Drepaninae with about 800 species (Holloway *et al.*, 1987), are known as hook-tips because of their typically, but not universally, falcate forewings. They occur chiefly in the tropics and subtropics of the Old World.

In the adult (Fig. 275) (Inoue, 1962; Watson, 1965a; Minet, 1985b), the proboscis is reduced or absent in Oretini (Oretinae of Watson), vestigial in *Nidara* (Nidarinae of Watson), but better developed, though sometimes reduced, in Drepanini. The antennae are frequently bipectinate. Unlike the situation in Thyatirinae, in Drepaninae vein M_2 arises very close to M_3 in the forewing. The forewing-metathorax locking device is absent, and the wings are held horizontal to the body when the moth is at rest. The epiphysis is typically present, but absent from *Nidara* where mid- and hindleg spurs are also lost. The male frenulum is usually present, rarely clubbed or, in Oretini, absent.

The eggs are flattened and pill-shaped (Sugi *et al.*, 1987). In the larva (Nakajima, 1970; Stehr *in* Stehr, 1987; Minet, 1985b), numerous secondary setae occur on Oretini, but few occur in Drepanini. A single verruca, or paired verrucae, may be present dorsally on the thorax. A supracoxal vesicle (probably a glandular structure) occurs on each side of the prothorax in Drepanini, but not in Oretini. Thoracic legs are developed. Prolegs are present on A3-6, but vestigial on A10 where they bear setae but no crochets. Crochets on the ventral prolegs are arranged as a uni- or occasionally biordinal mesoseries with a short uniordinal lateroseries. The anal prolegs are reduced and bear no crochets. Short to long processes extend from the suranal plate so that the larva appears tailed. Secondary setae, resembling primaries, are sometimes present.

The larvae are typically arboreal feeders. Pupation occurs in the ground, and the pupa is often covered with a bluish wax. Larvae of *Epicampoptera*, which are pests of coffee in Africa, have a strongly swollen thorax. At rest, drepanine larvae sometimes lie with the anterior part of the body curved around to run alongside the posterior part. In others, both the hind section and the fore section of the larva may be raised (Sugi, 1987).

Thyatirinae (= Cymatophorinae) are more robust than Drepaninae and appear more noctuid-like. They are generally more drab than Drepaninae, more often spotted, and the forewings are not falcate or only weakly falcate. The subfamily is small with little over 200 species (Munroe, 1982) and, although it is widespread, most species occur in the Palaearctic and Oriental regions. Thyatirinae are not represented in Australia and only four named species occur in Africa (Watson, 1965b; Lane, 1973).

In the adult (Fig. 276) (Forbes, 1936; Werny, 1966), the antennae are broad and prismatic or ciliate rather than pectinate. The male frenulum is typically clubbed. The forewing-metathoracic locking device is present and the wings are held roof-like over the abdomen at rest. Vein M_2 of the forewing does not usually arise so close to M_3 as it does in Drepaninae. In the male genitalia, the uncus is nearly always composed of three processes; one medial, and two (socii of Forbes, 1936; subunci of Werny, 1966) lateral.

The larva (Stehr *in* Stehr, 1987) is hypognathous. Prolegs are normal and bear crochets in a biordinal mesoseries, but anal prolegs are weak. Seta L_3 is situated well anterior to seta SV_4 on typical abdominal segments, and there are two additional setae - one dorsoposterior and the other ventroposterior to the spiracle.

The larvae are defoliators of trees and bushes, and either live among loosely rolled leaves (Forbes, 1923), or feed in more exposed situations. They bear tubercles and forked protuberances. In some species the end of the abdomen is raised at rest, as in certain Drepaninae.

Cyclidiinae (Inoue, 1962; Minet, 1985) include two genera, *Cyclidia* and *Mimozethes*, from the Orient and Japan, which were earlier assigned to either Drepaninae or Thyatirinae. *Cyclidia* includes large and usually white Drepanidae with darker markings, and *Mimozethes* includes medium-small, brownish species. In the forewing, although an anal fork is present, veins 1A and 2A continue independently beyond the fork so that they seem to form a cross. This character may unite these otherwise different looking genera. In both wings, vein M_2 arises from, or near to, the middle of the discal cell. A pair of long hair-tufts occur at the base of the abdomen. A pair of lateral arms arise from the base of the uncus, reminiscent of those in Thyatirinae. The larva bears numerous secondary setae, and ventral and anal prolegs are normally developed. (Adult e.g., Fig. 277.)

Genus *Hypsidia* includes six Australian species which remain unplaced in any of the drepanid subfamilies (Scoble & Edwards, 1988). Its members were previously assigned to either Pyralidae or Noctuidae, but the form of the tympanal organs is typically drepanid. In two species (both brightly coloured) external ocelli are present, an occurrence apparently unique within Drepanidae. In the others, more drab in appearance, the male frenulum is clubbed - a situation occurring in Thyatirinae, but rarely in Drepaninae. However, unlike Thyatirinae no locking mechanism exists between the forewing and the metathorax. A pair of lateral arms, reminiscent of those of Thyatirinae, arise from near the base of the uncus. However, in *Hypsidia* they are spatulate and bear large setae arranged in a comb, and they are not fused with the uncus. The life history is unknown.

AXIOIDEA

Axiidae

This small Mediterranean family includes fewer than 10 species. The moths are of medium size with extensive yellow-gold areas and silver markings. Their taxonomic affinities are obscure.

Adult (Fig. 278)

Chaetosemata are absent. The antennae are bipectinate in the male and serrate or bipectinate in the female. The proboscis is present and not scaled at its base, the maxillary palpi are vestigial, and the labial palpi short and porrect.

The epiphysis is absent, and the tibial spur formula is 0-2-4. The forewing is triangular, and veins R_2 to R_5 arise from an areole. Wing coupling is effected by a frenulum-retinacular system, and the forewing-metathoracic locking device is present. In the hindwing, $Sc+R_1$ approaches Rs.

The abdomen is fairly stout. At its base is a pair of squat apodemes; tergo-sternal sclerites are absent, and there are no venulae. There are no tympanal organs at the base of the abdomen. On A7 occurs a pair of specialized organs associated with the spiracle. Each of these structures, which is formed essentially from a sclerotized internal ring with an associated air sac, is covered by a prominent, protective operculum, which opens posteriorly. No associated sensory nerves or scoloparia have been discovered.
(Minet, 1983.)

Immature stages

The egg is of the flat variety with longitudinal ribs.

The larvae are without well developed hairs; the abdomen bears 4 pairs of ventral prolegs and a pair of enlarged anal prolegs.

The pupa is fairly squat, lacks spines on the dorsum, and does not project from the cocoon before emergence. Pupation takes place in the cocoon.
(Millière, 1862; [1864]; Minet, 1983.)

Phylogenetic relationships

The affinities of Axiidae have always been uncertain (e.g., Millière, 1862; [1864]), and they remain unresolved (Minet, 1983). The family has most frequently been placed in the Geometroidea (e.g., Bourgogne, 1951) because of the presence of organs thought to be tympanal structures on the abdomen. However, whereas in Geometridae the structures are found at the base of the abdomen, in Axiidae they occur on each side of A7. Furthermore, it is by no means established that the organs on A7 are tympanal because no associated sensory structures are apparent (Minet, 1983). For the same reasons, a possible relationship between Axiidae and Thyatirinae (Drepanidae) (e.g., Bourgogne, 1951) is doubtful.

CALLIDULOIDEA

The superfamily includes Callidulidae and Pterothysanidae (Munroe, 1982), groups that have been assigned to the same family by Minet (1986; 1990a). Here, the two taxa are retained as separate families. Minet (1987a) suggested that the form of the ovipositor in Callidulidae, Pterothysanidae, and Ratardidae (Chapter 11) may represent a specialized character indicative of the monophyly of the three groups. However, on balance, Ratardidae is placed in Cossoidea in this work.

Adult (Figs 279-280)

External ocelli are present (Callidulidae) or absent (Pterothysanidae). Chaetosemata are well developed. The antenna is not pectinate and its apex tends to be curved back dorsally. The proboscis is not scaled at its base, the maxillary palpi are reduced but always present, and the labial palpi are well developed and upturned.

The epiphysis is present and the tibial spur formula is 0-2-4 or 0-2-2. In the forewing the base of M_2 is very close to that of M_3. In the hindwing Sc+R closely approaches Rs and M_2 and M_3 arise together or with their bases approximated. The male frenulum is very short (occasionally divided in Pterothysanidae), but the retinaculum is absent. The forewing-metathoracic locking device is absent. Tympanal organs are absent.

In the female genitalia the anal papillae bear a pair of lateral lobes, sometimes not well developed.
(Munroe, 1982; Minet, 1990a.)

Callidulidae

The wings of these small to medium-sized tropical moths are usually dark brown with an orange band or patch. Callidulinae occur mainly in the Indo-Australian region, but fail to reach the Australian mainland, while Griveaudiinae occur in Madagascar (Seitz, 1924; Strand, 1911; Minet, 1990a). Most species of Callidulidae are diurnal, although some species are attracted to light, while Griveaudiinae are nocturnal and sexually dimorphic (Minet, 1990a). There are about 100 species (Munroe, 1982).

Adult (Fig. 279)

External ocelli are reduced. The antennae are filiform and slightly swollen distally. The proboscis is well developed. The epiphysis is present and the tibial spur formula is 0-2-4. In the forewing, R_2 to R_4 are stalked, and vein M_2 arises closer to M_3 than to M_1. In the hindwing, M_2 arises nearer M_1 than M_3. Vein CuP is absent from both wings. There are no narrow, erect scales arising from the surface of the wing, a character observed in Pterothysanidae (Minet, 1987a).
(Pagenstecher, 1902; Janse, 1932; Munroe, 1982; Minet, 1990a.)

Immature stages

(Based on a description of *Pterodecta felderi* by Tschistjakov and Belyaev (1987), from material collected in the Khasian region, Primary Territory (Minet, 1990a).)
    The egg is smooth, oval, and strongly flattened.
    The larval head is hypognathous. The prothoracic shield is wide. Secondary setae are absent. On each side of A1 is found a gland below seta L_2. Prolegs are present on A3-6 and A10. Crochets are arranged in a biordinal circle.
    The pupa is obtect with segments A8-10 fused.

Biology

Callidulidae larvae feed on ferns perhaps exclusively so (J.D. Weintraub, pers. comm.). Eggs are usually laid singly at the edge of leaflets. In their early stages, the larvae of *Cleosiris lycaenoides* live in rolled up leaflets, and later, the edges of leaflets are drawn together with silk (Barlow, 1982). Pupation occurs between the leaflets.
    The moths fly by day in the shade of understory vegetation in rainforest (Barlow, 1982; Holloway, et al., 1987). A few are attracted to light at dusk (Holloway et al., 1987). At rest, Callidulinae hold their wings vertically over the body similar to the position typical of butterflies. Griveaudiinae possibly hold their wings in a roof-like position (Minet, 1990a).

Classification

Two subfamilies occur (Minet, 1990a) - **Callidulinae** and **Griveaudiinae**. Griveaudiinae include the genus *Griveaudia* with three species.

Pterothysanidae

In general appearance, Pterothysanidae are geometrid-like with broad wings. They are fairly large moths, and there are about 12 species (Munroe, 1982). Most are oriental, but a single species is found in Madagascar. The genus *Hibrildes* from Africa has been transferred from this family to Bombycoidea by Minet (1986; 1990a).

Adult (Fig. 280)

External ocelli are absent; the antennae are filiform or pectinate; and the proboscis is short. In the forewing, R_3 and R_4 are stalked. The wings are often covered with long, erect, narrow scales. In *Pterothysanus*, from the Himalayas, the anal edge of the hindwing bears a prominent fringe of long hairs.
(Hering, 1926b; Munroe, 1982; Minet, 1990a.)

Immature stages

Nothing is known about the juvenile stages with the exception of a male pupa, described by Minet (1987b), of what is probably *Caloschemia monilifera*. The pupa is obtect, without dorsal spines, and with (at most) two mobile segments.

Biology

Pterothysanus is said to be diurnal (e.g., Seitz, 1915), while members of the family from Madagascar appear to be nocturnal (Minet, 1990a).

RHOPALOCERA: HEDYLOIDEA, HESPERIOIDEA, AND PAPILIONOIDEA

Butterflies are collectively known as the Rhopalocera, a term proposed by Duméril (1823), to refer to their clubbed antennae. Generally included in the group are skippers - Hesperioidea - and true butterflies - Papilionoidea. A third superfamily, Hedyloidea, without clubbed antennae, was added by Scoble (1986).

HEDYLOIDEA

Hedylidae

The group was considered by Guenée (1857) to be one of the 26 component families of the Phalénités, the equivalent of what, with some modification, are now called Geometridae. Prout (1910) regarded hedylids as a tribe ('Hedylicae') of Oenochrominae (Geometridae). The family is now considered to be one of the three superfamilies comprising the Rhopalocera or butterflies, despite the moth-like appearance of the species (Scoble, 1986; Scoble & Aiello, 1990).

There are 35 species (Scoble, 1990) in a single genus, *Macrosoma*. The family is restricted to tropical America including Cuba and Trinidad. The moths are medium-sized macrolepidopterans usually with greyish brown or brownish grey wings, but with a few species predominantly or entirely white or off-white. They are often marked with a white triangle on the costa of the forewing and exhibit a dark or chestnut brown apical area. Semi-translucent areas are frequently present medially on the wings. The egg is upright and ribbed, and resembles that of Pieridae. The larva is slender and bears horns and caudal processes, and the pupa is exposed and girdled.

Adult (Fig. 281)

The compound eyes are large, external ocelli are absent, and chaetosemata are present. Usually filiform, the antennae are bipectinate in a few species.

The epiphysis is present and the tibial spur formula is 0-2-2 or, rarely, 0-2-4. In males the foretarsus is 2-segmented, but not shortened, and the pretarsus is strongly reduced. Occasionally a scent brush is present. In the female the tarsus is 5-segmented and the pretarsus is not reduced. The forewing is fairly narrow with the apex usually weakly emarginated, and the hindwing is more rounded. Vein Sc is extended into a prominent expansion at the base of the forewing and forms a distinct fold. Very small tympanal organs are present at the base of the forewing formed within the swollen bases of veins Sc and Cu (Scoble, 1986; Minet, 1988a). The wings are typically coupled by means of a frenulum and a retinaculum: the frenulum in males takes the form of a strong spine. In a few species the frenulum and retinaculum are reduced or strongly reduced. The female bears a few weak costal bristles on the hindwing, and the retinaculum is absent. In the forewing, only veins R_4 and R_5 are stalked. Veins R_2 and R_3 run in a distinct, sinuous manner. The anepisternum of the mesothorax is not reduced, that is, there is no pronounced dorsal movement of the anapleural cleft. The aorta is expanded into a dorsal, horizontal chamber in the mesothorax, and the ascending and descending parts of the vessel run alongside each other (Minet, 1988b; Scoble & Aiello, 1990). The furcal apophyses of the metathorax are weakly sagittate in dorsal view. Metathoracic tympanal organs are absent.

Abdominal tympanal organs are absent. Tergum A1 is expanded into a prominent, membranous 'pouch', and the abdomen, particularly in males, tends to be down-

curved. A pre- and postspiracular bar are present at the base of the abdomen. The ovipositor is short.
(Scoble, 1986, 1990; Scoble & Aiello, 1990.)

Immature stages

The egg of *M. semiermis* (the only species for which such information is available) is upright and similar to eggs of several Pieridae, being slender and spindle-shaped with seven longitudinal ribs and about 30 crossribs (Aiello *in press*; Scoble & Aiello, 1990).
    The larva (Kendall, 1976; Scoble, 1986; Aiello, *in press*; Scoble & Aiello, 1990) of three species is known. The head is extended into a pair of long, anteriorly directed horns, and a pair of short caudal processes occur on segment A10. An anal comb is found terminally. A full complement of prolegs is present. Crochets on the ventral prolegs of the final instar are arranged in a penellipse or as transverse bands. The first instar lacks the cephalic horns of subsequent stages, and in second instars the horns are very small (Aiello *in press*). Various secondary setae are present on the body.
    The pupa (Kendall, 1976; Scoble, 1986; Aiello, *in press*; Scoble & Aiello, 1990) is girdled around segment A1 and there is no cocoon. A cremaster is present.

Biology

Although most adults have been collected at light, demonstrating nocturnal activity, at least some also fly by day. Adults of both sexes walk and rest with their forelegs folded under the thorax. At rest, the wings are outspread and slightly raised and, when the insect perches on a vertical support, the thorax is tilted backwards to such an extent that the posterior edges of the hindwings almost touch the substrate.
    The egg (known for *Macrosoma semiermis*) is glued to the edge of a leaf of *Byttneria aculiata* (Sterculiaceae) from which it projects (Aiello, *in press*; Scoble & Aiello, 1990). Larval behaviour of this species is described by Aiello (*in press*). First instars lie in a groove alongside a secondary vein. Much or most of the leaf is consumed by larger larvae, which lie alongside the midrib of leaves of the foodplant, and leave large holes in the leaf when they feed. When disturbed all instars snap back alongside or onto the midrib or secondary vein. Other hostplants recorded for Hedylidae (summary in Scoble, 1990) include *Croton* (Euphorbiaceae) (A.R. Pittaway, pers. comm.), *Byrsonima* (Malpighiaceae) (D.H. Janzen, pers. comm.), and *Hibiscus* (Malvaceae) (J. Mallet, pers. comm.).

Phylogenetic relationships

Several characters suggest an affinity with other butterflies, but many of these features occur only in certain groups. The egg strongly resembles that of Pieridae, and less strongly that of Nymphalidae. The larvae are like certain Apaturinae (Nymphalidae) in having horns and caudal processes, although this feature may be convergent as they do not apparently occur in the nymphalid groundplan. An anal comb occurs in Hedylidae, most Hesperiidae (excluding Megathyminae), in one species of Papilionidae, and in several Pieridae (Toliver *in* Stehr, 1987). Such a comb is also found in many microlepidopterans, and in some other macrolepidopterans. The location of the girdle over abdominal segment A1 is typical of its position in Pieridae. An exposed pupa is usual in Papilionoidea, although not in Hesperioidea. In general shape the hedylid pupa is not dissimilar from those of many Papilionidae (see illustrations of Igarashi, 1984).
    In the adult, the sagittate shape of the metathoracic furcal apophysis occurs in Hedyloidea, Hesperioidea, and Papilionoidea, but is weak in Hedyloidea. The configuration of the aorta in the thorax resembles that of Hesperioidea and Papilionoidea (except Papilionidae) (Minet, 1988b; Scoble & Aiello, 1990). The prominent dorsal

pouch on A1 is shared by Hedyloidea and Papilionoidea. A fusion of the tarsomeres in the foreleg of the male occurs in Hedyloidea and, usually, in Lycaenidae. Although the forelegs of Nymphalidae are greatly reduced, the tarsomeres are not fused. In male and female Hedylidae and in male Nymphalidae, the forelegs are folded under the thorax when perching and walking. Reduction of the pretarsus occurs in male Hedylidae and, to varying extents in Papilionidae and Lycaenidae.

Despite the many moth-like (primitive) characters displayed, the evidence suggests that either Hedylidae represent the sister group of Hesperiidae and Papilionidae, or, possibly, the sister group of Papilionoidea.

HESPERIOIDEA

Hesperiidae

A single family, with an estimated number of 3050 species, is included in Hesperioidea (Munroe, 1982). Their common name of skippers comes from their darting, irregular flight. They are generally included in the butterflies as a separate superfamily, but are occasionally treated as a family of Papilionoidea (e.g., Brock, 1971). Six subfamilies are accepted by Ackery (1984), although one of them, Megathyminae, is sometimes treated as a family. Hesperiidae are widespread, although they are absent from New Zealand. They are medium-sized, stout-bodied butterflies; although Megathyminae are slightly larger than other skippers. The monophyly of the group is supported by the presence of a distinct rim-like peripheral region of the compound eye (e.g., Minet, 1991).

Adult (Fig. 282)

The head is nearly always wider than the thorax, but narrower in Megathyminae. The periphery of the compound eye is demarcated from the central area (see, for example, Minet, 1991). External ocelli are absent, and chaetosemata are present. The antennal bases are widely separated, a feature related to the width of the head. Apically, each antenna is generally expanded into a club, which itself terminates in a short hook called an apiculus. Some skippers have one pair of chaetosemata on the top of the vertex, whilst in others there is an additional pair - one behind each antenna (Jordan, 1923b). The proboscis is well developed and not scaled at its base, maxillary palpi are absent, and labial palpi ascending.

The epiphysis is present. Tibial spurs are often absent from the midleg, but one or two pairs are found on the hindleg and the tibial spur formula is variable (0-2-4, 0-0-4, or 0-0-2). In the forewing, vein R_1 and all branches of Rs typically arise directly from the cell; there is no stalking. Vein CuP is absent from both wings. The frenulum and retinaculum are absent other than in males of the Australian skipper Euschemon where they are well developed. Although the anepisternum is reduced in the metathorax, it remains relatively prominent. The metafurcal apophysis is sagittate.

Abdominal tergum A1 of the abdomen is at most only weakly pouched in skippers. (Ehrlich, 1958b, 1960; Forbes, 1960; Scott, 1985.)

Immature stages

The eggs, which are laid singly, are upright. In many species they are somewhat wider than high; they may be smooth or ribbed.

In the larva (Toliver in Stehr, 1987), numerous secondary setae cover the head and body. In non-Megathyminae the head is larger than the prothorax giving rise to a prominent 'neck'. In Megathyminae the head is smaller than the prothorax. The prolegs of non-megathymines bear biordinal or triordinal crochets arranged in a circle.

In Megathyminae the crochets are biordinal and the circle that they form may be broken either mesally, laterally, or both. An anal comb is typically present in all skippers except Megathyminae.

The pupa may bear a horn on the head, and the proboscis is often long, even extending beyond the cremaster.

(Toliver *in* Stehr, 1987.)

Biology

The larvae are leaf-folders or tiers, or live in tunnels. A considerable range of angiosperm foodplants, both monocotyledonous and dicotyledonous, is accepted. Megathyminae larvae bore in the leaves, stems, and roots of Agavaceae. Larval shelters may be dusted with a waxy secretion from the abdominal glands. The pupa is concealed, or largely concealed, within the larval shelter. It is attached by a silk girdle within the weak cocoon.

Adult flight is rapid and often appears erratic. At rest, the wings may be held vertically above the body, sometimes with the head depressed, or horizontal and extended.

Classification

There are six subfamilies (summary and references in Ackery, 1984). The system dividing Hesperiidae into merely Megathyminae and 'Hesperiinae' (i.e., all non-megathymine Hesperiidae) may result in 'Hesperiinae' being non-monophyletic. **Pyrrhopyginae** are restricted to the tropics of the New World; **Pyrginae** occur in all the main zoogeographical regions; **Trapezitinae** are found in the New Guinea region and Australia; **Hesperiinae** (*sensu stricto*) are widely distributed; **Megathyminae** are restricted to Central America; and **Coeliadinae** occur in the Afrotropical, Oriental and Australasian regions.

Phylogeny

Hesperiidae are generally regarded as the sister group of Papilionoidea although other possibilities exist (see Kristensen, 1976; Scoble, 1986). Kristensen (1976) listed four possible shared specializations between Hesperioidea and Papilionoidea and noted two others worthy of consideration. 1) The mesal fusion of the dorsal lamellae of the secondary metathoracic furcal arms. This arrangement, noted as a prominent papilionoid (including Hesperiidae) feature by Brock (1971), should not be regarded as a unique specialization (J. Brock, pers. comm.). 2) The brain of butterflies is said to be characterized by the presence of very large optic lobes and a small deuterocerebrum (Ehnbom, 1948, see Kristensen, 1976). 3) The mesothoracic aorta has a horizontal chamber with the descending part of the aorta anterior to the chamber (Hessel, 1969). This feature also occurs in Hedyloidea and Papilionoidea, but it is secondarily modified in Papilionidae. 4) Ehrlich & Ehrlich (1963) found that the oblique lateral dorsal muscle in the mesothorax is twisted in Hesperioidea and, with the exception of Lycaenidae where the twist is considered to be secondarily absent, in Papilionoidea. The presence of clubbed antennae is often suggested as a shared specialization of hesperioids and papilionoids. However, mere swelling is an unconvincing character, at least taken in isolation, both because it occurs often in moths (e.g., many Zygaenidae, Castniidae, and many Sphingidae), and because the form of the apical club is not the same in Hesperioidea and Papilionoidea. The loss of a frenulum and a retinaculum in all but one hesperioid and in all papilionoids is another possible specialization, which would necessarily have to be interpreted as independently evolved because the loss is by no means unique within macrolepidopterans.

The sister group relationship of Hesperioidea and Papilionoidea is supported by Minet (1991) because A10 of the pupa bears a pair of anterior or anteriolateral tubercles; the antennal scape is inflated; the female lacks frenular bristles; and the adult tibiae bear numerous spines (at least in the perceived groundplans).

PAPILIONOIDEA

The Papilionoidea or true butterflies include over 14500 species (Munroe, 1982), many of which are very colourful. In size, they range from small to very large macrolepidopterans.

Four families were recognized by Ackery (1984), who reviewed the literature on the higher classification of butterflies: Papilionidae (swallowtails), Pieridae (whites or sulphurs), Lycaenidae (blues, hairstreaks, and coppers) (including Riodininae, the metalmarks), and Nymphalidae (including Libytheinae). Riodininae and Libytheinae are often treated as families.

Adult

The compound eyes are usually not hairy, but interfacetal hairs occur in some Lycaenidae and Nymphalidae (Yagi & Koyama, 1963). External ocelli are absent, and chaetosemata are present. The antennae are clubbed, but the clubs are not extended into an apiculus. Dorsal tentorial arms (Chapter 2) are sometimes present. The proboscis is usually well developed. Although maxillary palpi are well developed, sometimes even 2-segmented, they are generally minute and either 1-segmented or entirely absent. The labial palpi are 3-segmented, upturned or porrect, and sometimes considerably elongated (some Libytheinae), or reduced.

In the prothorax, the patagia are unsclerotized in Lycaenidae, most Pieridae, and many Papilionidae. Parapatagia are usually membranous. The anepisternal cleft differentiates the anepisternum from the pre-episternum in several butterflies, but is displaced dorsally. In most Nymphalidae and Lycaenidae the anepisternum virtually disappears (see also Brock, 1971). In these butterflies the parepisternum extends almost to the notum (Brock, 1971). A secondary sternopleural sulcus is found in Nymphalidae and Lycaenidae. It extends from the basisternal-parepisternal sulcus to a spur of the paracoxal suture. In all Papilionoidea the precoxal (paracoxal of Scott, 1985) sulcus fuses with the 'marginopleural' sulcus, unlike the condition in most Lepidoptera where the precoxal sulcus extends to the parepisternum. The apophyses of the metathoracic furca are sagittate, a condition that may be a specialization of Rhopalocera as a whole. In the mesothorax the aorta is expanded into a dorsally situated horizontal chamber, and the ascending and descending parts of the vessel run alongside each other (except in Papilionidae) (Hessel, 1966, 1969). The epiphysis is absent except in Papilionidae. The foreleg may be reduced, sometimes extremely so. Tarsal claws of the midlegs and hindlegs are bifid in Pieridae and some other butterflies. Except for a few female Papilionidae (Forbes, 1960), at least some stalking of the radial veins of the forewing occurs. Vein CuP is usually absent from both wings but is represented basally in Papilionidae. There is no frenular-retinacular system in true butterflies, wing-coupling is said to be effected by an amplexiform system made possible by the expanded humeral area. The forewing-metathoracic locking-device is absent. Minute tympanal organs are found in some Nymphalidae at the bases of certain wing veins (Chapter 6), but there are no metathoracic tympanal organs.

With few exceptions, the abdominal tergum A1 is strongly pouched. A prespiracular bar is present on A1 of all true butterflies except for Pieridae. A postspiracular bar is present or absent. There are no abdominal tympanal organs. In the female genitalia, the anterior apophyses are frequently reduced.

Immature stages

The egg is of the upright variety, and the shape and ribbing vary considerably between families.

In the larva, crochets are arranged in a circle in the first instar, but generally in a mesoseries in subsequent stages. A ventral neck gland is present.

The pupa is exposed, and is attached posteriorly to its substratum by a silken pad and, in many groups, a thoracic or abdominal silk girdle.

Papilionidae

The family includes 573 species (Collins & Morris, 1985) of predominantly tropical butterflies. They are generally large, sometimes (e.g., *Ornithoptera*) very large, and among the most spectacular of Lepidoptera. In many, the hindwings are 'tailed'.

The monophyly of the family is supported by four characters (Kristensen, 1976). A protrusible, forked osmeterium, occurs on the larval prothorax (Pl. 2:10). In the adult, the arolium and pulvilli of the pretarsus are reduced. In the forewing, vein 2A extends free to the margin whereas in other Papilionoidea, and most other Lepidoptera, it fuses with 1A after a short distance, and fails to reach the wing margin. The mesothoracic aorta lacks a horizontal chamber (unlike the condition found in Hesperioidea, Hedyloidea, and other Papilionoidea), but has ostiae (Hessel, 1966). The laterocervicalia are joined ventromedially, a feature absent from other butterflies (Ehrlich, 1958b; Miller, 1987c).

Adult (Fig. 283)

The maxillary palpi are minute and the labial palpi are porrect or upturned. The epiphysis is present and the foreleg is not reduced apart from the arolium and pulvilli. The tibial spur formula is 0-2-2 or 0-0-2. In the forewing R_4 and R_5 are typically stalked, CuP is usually present near the wing base, and there are two anal veins that are united at the base of the wing but then diverge. A humeral vein is present on the hindwing, and there is one anal vein. Sometimes scent brushes occur in an anal fold of the male hindwing. A sphragis, secreted by the male during mating, is found at the end of the abdomen of mated female Parnassiinae (see *Biology*).
(Munroe, 1982.)

Immature stages

The eggs are unsculptured and almost spherical.

In the larva, the presence of an osmeterium on the prothorax is unique to Papilionidae. The body is sometimes smooth, but numerous secondary setae typically occur on the body. The crochets are triordinal and, on the ventral prolegs, arranged in a mesoseries. A weak, biordinal lateroseries may also occur on the ventral prolegs. In some species various protuberances, filaments, or other adornments are found on the body.

The pupa is typically exposed and attached to its substrate by a silken girdle traversing the thorax, and by a cremaster. However, in Parnassiinae, pupation occurs in a loose cocoon on the ground.
(Munroe, 1982; Toliver *in* Stehr, 1987.)

Biology

The eggs are usually laid singly on the foodplant, but occasionally in small clusters. Sometimes they are deposited near the foodplant rather than upon it.

The osmeteria, which occur on the prothorax of the larva (Pl. 2:10), emit a pungent odour and are defensive. Chemical studies have shown that isobutyric acid and 2-methyl butyric acid, or a mixture of two sesquiterpenes, is released (Eisner *et al.*, 1971). The substances may deter predators (e.g., ants) and parasitoids (see Chapter 5). Many papilionid larvae are distasteful and aposematically patterned.

The adults are diurnal and often feed at flowers while on the wing. In many species the adults are sexually dimorphic. Usually the female is mimetic, and may be polymorphic with each morph mimicking a different model. The sphragis of Parnassiinae is better developed than in those other butterflies where it occurs. The structure is formed at the end of the abdomen of the female from a white substance, secreted by the male, which hardens during mating and prevents the female from being mated again.

Classification

The phylogeny of Papilionidae has been discussed by Ehrlich (1958b), Munroe (1961), and Hancock (1983). Miller (1987c) provided a detailed, cladistic study of Papilioninae, the subfamily with by far the most species. Three extant subfamilies are accepted by all four authors, to which Durden & Rose (1978) added the fossil Praepapilioninae. The foundation of papilionid classification was laid by authors particularly late last century and in the early part of the present century (Munroe, 1961; Miller, 1987c).

Baroniinae are represented by only one species (*Baronia brevicornis*), which occurs in the mountains of southwest Mexico. The antennae are short and unscaled, the legs are scaled, and the hindwing is not tailed. The mature larva is green, and bears a yellow stripe dorsally, and white transverse bands on each segment. Small black or yellow cephalic tubercles are present. The foodplant is *Acacia cymbispina* (Mimosaceae) (Vázquez & Pérez, 1961). The pupa is squat and ovoid, girdled, and pupates in an earthen cell in the ground. Two specialized characters of Baroniinae were noted by Miller (1987c): a lateral lobe of the male valva, and the absence of vein R_4 in the forewing.

The **Praepapilioninae** (see Durden & Rose, 1978) include a single extinct genus *Praepapilio* with two species from middle Eocene deposits of North America in Colorado. Although weak, vein CuP is present in both wings. Papilionid characters include (among others) the presence of a free vein 2A terminating on the posterior margin of the forewing. Very short hindwing 'tails' are present suggesting that these structures may represent the groundplan condition of the Papilionidae. In common with Papilioninae, *Praepapilio* has a basal spur (cubito-vannal vein) in the forewing, and, as in *Baronia* (Baroniinae) and *Parnassius* (Parnassiinae), lacks vein R_4 in the forewing (Miller, 1987c).

Parnassiinae, with about 50 species, occur mostly in the Holarctic region. The markings of the adults generally take the form of bands or rows of spots, which are often black. The hindwings are usually rounded, but sometimes 'tailed'. The antennae are of moderate length and may be scaled or unscaled. Hindwing tails are present or absent. The larvae are of the type termed red-tuberculate (Munroe, 1961), although in more specialized species red spots rather than tubercles are present. Primitively, members of the subfamily feed on Aristolochiaceae. The pupa may be attached by a girdle, or pupation may occur in a cocoon on the ground. The presence of a cocoon is probably specialized within this subfamily, and is assumed to be a character independently derived in Hesperioidea. Specialized characters of Parnassiinae (Miller, 1987c) include a sphragis (see above); a thin and heavily sclerotized aedeagus; a heavily sclerotized ostial region in the female genitalia; the elongate condition of the labial palpus; and an incurved middle discocellular vein (mdc, joining the bases of M_1 and M_2) in the forewing.

Most Papilionidae belong to **Papilioninae**, a subfamily with 509 species (Miller, 1987c), a widespread group with the greatest diversity in the tropics of the Old World. The monophyly of Papilioninae is based on five characters (Miller, 1987c) (two of which are homoplasious): a pseuduncus is present on tergum A8 of males; male-specific patches of scales are found along vein 2A on the underside of the hindwing ('anal brush' of various authors); the forewing has a basal spur (cubital-vannal vein); the metathorax bears a distinct meral suture with an internal lamella; and the larvae display a white saddle on the abdominal segments (a feature secondarily lost in many groups).

Pieridae

Pieridae include the 'white' and 'sulphur' butterflies, their predominant colours being white, yellow, and orange. There are approximately 1200 species (R.I. Vane-Wright pers. comm.). The family is widely distributed. The most easily visible of the six possibly specialized characters given by Kristensen (1976) (based mainly on Ehrlich, 1958b), is the presence of forked pretarsal claws on all the legs. Other possible specializations include the loss of a prespiracular bar from the first abdominal segment, and the presence of pterines in the wing-scales.

Adult (Fig. 284)

Maxillary palpi are absent or 1-segmented, and the labial palpi are porrect or ascending. The epiphysis is absent, and the tibial spur formula is 0-2-2. The pretarsal claws are bifid. In the forewing, one or more branches of Rs may be absent. In the hindwing, the humeral vein may be present or absent, and vein $Sc + R_1$ diverges from Rs at the base.
(Ehrlich, 1958b; Munroe, 1982; Ackery, 1984.)

Immature stages

The eggs (Döring, 1955) are upright and fusiform with vertical ridges and horizontal cross-ribs. They may be laid singly or in groups.
    The larva is cylindrical and covered with numerous short and fine secondary setae. There is no osmeterium. Each abdominal segment is divided, typically into six annulets. The crochets are arranged in a biordinal or triordinal mesoseries. An anal comb is present, but does not occur in all species.
    The pupa is attached to its substratum by a cremaster and a silken girdle typically around the first abdominal segment (Aiello *in press*) but, in *Pseudopontia*, the third.
(Toliver *in* Stehr, 1987.)

Biology

The eggs are usually laid singly on the foodplant, rarely in small batches. Larvae feed particularly on Brassicaceae and Fabaceae, but also on Capparidaceae and Loranthaceae and, indeed, other groups. The larvae of some species are cryptically coloured, others appear to be aposematic. Some adult Pieridae are migratory (e.g., see Chapter 3), sometimes moving considerable distances. In several species, the wings reflect light in the ultraviolet providing a means of inter- and intrasexual communication invisible to vertebrate predators (Chapter 3).

Classification

Four subfamilies were recognized by Ehrlich (1958b), one more than given by Klots (1933).

Pseudopontiinae are represented by a single white species, *Pseudopontia paradox*, mainly from West Equatorial Africa. In the hindwing, vein $Sc+R_1$ fuses secondarily with Rs before the middle of the wing, and M_2 is stalked with M_1. The valvae of the male genitalia are partly fused with each other along the ventral and lower distal margins. This feature also occurs in Dismorphiinae (but not in Pierinae or Coliadinae). The pupal girdle traverses A3 rather than A1.

Dismorphiinae, with about 100 species (Ackery, 1984), are mainly neotropical, but there is one small palaearctic genus. Vein $Sc+R_1$ is not secondarily fused with Rs, and M_2 arises from the cell. The larvae feed primarily on Fabaceae.

The **Pierinae** ('whites'), a subfamily with about 700 species, include most species of Pieridae. The group is cosmopolitan with its greatest diversity in the tropics, a situation also pertaining to Coliadinae ('sulphurs'). The humeral vein of the hindwing is well developed, $Sc+R_1$ is not secondarily fused with Rs, and M_2 arises from the cell. Pierinae larvae feed to a great extent on Brassicaceae, Capparidaceae and Loranthaceae.

In **Coliadinae**, the humeral vein of the hindwing is greatly reduced or absent, $Sc+R_1$ is not secondarily fused with Rs, and M_2 arises from the cell. The foodplants belong mainly to Fabaceae, Caesalpinaceae, and Mimosaceae.

Phylogenetic relationships

It has been suggested that Pieridae represent either the sister group of Papilionidae (Ehrlich, 1958b; Scott, 1985), or of Lycaenidae + Nymphalidae (Kristensen, 1976). This problem remains unresolved.

Nymphalidae

Nymphalidae, with an estimated 6000 species (Ackery, 1984), include about one third of all butterflies. In size, they range from small to large and exhibit considerable variation in colour and pattern.

Adult (Fig. 285-290)

The chaetosemata are elongated and run parallel to the eye margin. The maxillary palpi are 1-segmented and the labial palpi ascending. Two ventral grooves occur on most antennal segments. The epiphysis is absent. The forelegs are reduced and only rarely (female Libytheinae) used for walking. In males the pretarsus is lacking, and the number of tarsomeres are reduced. The tibial spur formula is 0-2-2 or 0-0-0. A humeral vein is usually present in the hindwing. In some Nymphalidae small tympanal organs occur in the swollen bases of certain wing veins (Chapter 6).
(Ehrlich, 1958a,b; Ackery, 1984.)

Immature stages

The eggs may be spherical, barrel-shaped, or conical.

The larvae are diverse. Frequently they bear humps, protuberances, scoli, or fleshy filaments. Secondary setae are abundant. Crochets usually form a triordinal mesoseries, but may be uni- or biordinal. In Libytheinae the crochets form a pseudocircle composed of a biordinal mesoseries with a few uniordinal crochets arranged in a lateroseries.

The pupa is usually suspended by the cremaster alone, but occasionally a weak cocoon is formed.
(Toliver *in* Stehr, 1987.)

Biology

The biology of Nymphalidae is extremely diverse. Typically, the eggs are laid in clusters. Early instar larvae are often gregarious and may construct shelters by means of silk and the leaves of their foodplant. A great diversity of plant families are utilized as foodplants (Ackery, 1988).
(Toliver *in* Stehr, 1987.)

Classification

The subfamily classification of the Nymphalidae was discussed by Ehrlich (1958b) and updated by Ackery (1984). Foodplant range is recorded by Ackery (1988). The taxonomic status of certain subfamilies is questionable (DeVries *et al.*, 1985). A revised classification of the family has been presented recently by Harvey *in* Nijhout (1991).

Brassolinae, with about 80 crepuscular, forest species, are Neotropical. Some are large, for example *Caligo*. The larvae are smooth, typically with a bifid 'tail', and feed exclusively on monocotyledons.

Amathusiinae (included in Morphinae as Amathusiini by Harvey *in* Nijhout, 1991), also with about 80 species, are strongly built butterflies from the Indo-Australian tropics. The larvae are covered with fine hairs, bear an anal fork and, usually, have cephalic processes. Amathusiinae and Brassolinae typically fly at dusk or in deep forest shade. Many are attracted to rotting bananas. The larvae feed on monocotyledons.

Satyrinae (Fig. 285), with around 1500 species (see Ackery, 1984) are polyphyletic (DeVries *et al.*, 1985). Satyrinae are often brown butterflies with eyespots frequently arranged in a row particularly near the margin of the hindwings, and with swollen bases to the wing veins. Typically, the larvae bear a pair of cephalic horns and a bifid 'tail'. Monocotyledons are the predominant foodplants. The group may be monophyletic if *Antirrhea* and *Caerois* are transferred to Morphinae (see below), and if Charaxinae are included (DeVries *et al.*, 1985).

Charaxinae, a group with an estimated number of species of between 300 and 400, are found throughout the tropics but rarely extend into more temperate regions (Ackery, 1984). The subfamily has 162 species in the Afrotropical region (Henning, 1988). They are typically forest dwellers. The butterflies are robust, fast flying, and colourful. They often feed on carrion, dung, and rotting fruit. The larvae feed on a variety of plants, but most charaxine tribes tend to be restricted to feeding on particular plant families or groups of families. The larval head is distinctive, usually bearing 1 or 2 pairs of horns.

Morphinae, usually regarded as containing only the genus *Morpho*, are monophyletic only with the inclusion of *Antirrhea* and *Caerois* - genera normally assigned to Satyrinae (DeVries *et al.*, 1985). The group is neotropical. *Morpho* butterflies are often brilliantly coloured, and feed on rotting, fleshy fruits. Most *Morpho* larvae feed on dicotyledonous plants, but some have monocotyledonous foodplants. *Antirrhea* and *Caerois* have been recorded feeding only on monocotyledons, mainly members of the Arecaceae.

Calinaginae are represented by a single genus, with uncertain affinities, from the Sino-Himalayan region. The larvae feed on members of the Moraceae.

Nymphalinae (Fig. 286) are almost certainly not monophyletic (Ackery, 1984; DeVries *et al.*, 1985), although there are undoubtedly many natural groups within the assemblage. There are about 3000 species (see Ackery, 1984), and the group is distributed throughout the world. Frequently, the forelegs of both sexes are reduced, and the larvae often bear spines. The biology of the group is diverse. Many species are migratory, for instance *Vanessa cardui*, and some are mimetic. In true Nymphalinae one series feeds mainly on Acanthaceae while the other is associated with Urticaceae.

Heliconiinae (Fig. 287) (included in Nymphalinae by Ehrlich, 1958b) are listed as a separate subfamily by Ackery (1984; 1988), who suggested, nevertheless, that they are probably no more than specialized members of the Argynnini. There are about 70 species of these New World butterflies, and most are Neotropical. The larvae feed on members of the Passifloraceae. Most adults feed on nectar, but some eat pollen (see Chapter 2). The larvae have 6 rows of spines arranged longitudinally; the head bears a pair of spines.

Acraeinae (Fig. 288) (see Ackery, 1984), with about 250 species, are a predominantly tropical subfamily with most species occurring in Africa. They are included, as tribe Acraeini, in Heliconiinae by Harvey *in* Nijhout (1991). The pretarsal claws are usually asymmetrical. The larvae bear 6 longitudinal rows of spines, with the spines often branched. The larval head is without processes, but the head of the pupa is forked. Within Acraeinae one group is typically associated with Passifloraceae and the other with Urticaceae, although further families are also utilized.

Danainae (milkweed butterflies) include about 150 species. Most are tropical and the greatest diversity occurs in the Oriental region. The monophyly of Danainae is supported (Ackery & Vane-Wright, 1984) by the presence of paired, sheathed, and eversible hairpencils on the abdomen of the male, and the strongly clubbed, spinose, 4-segmented pretarsus in the female. The hairpencils are everted by increased pressure of haemolymph and are withdrawn by retractor muscles. The tarsus of the female foreleg is 4-segmented and strongly clubbed. Larval instars beyond the first stage have between one and nine pairs of long fleshy filaments or tubercles dorsally. The larvae are usually aposematic with complete or incomplete stripes or hoops. The foodplants belong mostly to Asclepiadaceae, Apocyanaceae, and Moraceae.

In **Ithomiinae** (an adult of which is illustrated in Fig. 289) the larvae feed on Solanaceae and occasionally Apocynaceae. There are around 300 species (see Ackery, 1984), which are restricted to the Neotropics. Ackery & Vane-Wright (1984) pointed out that Ithomiinae are often associated with Danainae, but accepted three separate but related subfamilies (Danainae, Ithomiinae, and Tellervinae) given the unstable state of nymphalid classification. The antennae are scaled. The tarsus of the female foreleg is 4- or 5-segmented but not strongly clubbed.

Tellervinae includes the single genus *Tellervo* occurring from Maluku to the Solomon Islands. Males lack the secondary sexual characters so pronounced in Danainae and Ithomiinae. The tarsus of the female foreleg is 5-segmented and not strongly clubbed. The larvae, which feed on Apocynaceae, bear a single pair of fleshy filaments on the metathorax. The head is unornamented and a bifid 'tail' is lacking.

Libytheinae (Fig. 290) were included in Nymphalidae by Kristensten (1976), since their members share the specialized features of that family. The labial palpi are typically porrect and extended, a character that has led to libytheines being called 'snout' butterflies. The group is widespread, although there are only about 10 species. The larvae feed on members of the Ulmaceae with some records on Sapindaceae, Lauraceae, and Caprifoliaceae. They are smooth, lack processes on the head, and the 'tail' is not bifid.

Phylogenetic relationships

According to Kristensen (1976), Nymphalidae represent the sister group of Lycaenidae. One of the characters on which this suggestion is based is the reduction of the forelegs, at least in males. An alternative view is that they form part of a group also including Riodininae (as Riodinidae) and Libytheinae (as Libytheidae) (Robbins, 1988a).

Lycaenidae

Lycaenidae, with an estimated 6000 species (see Ackery, 1984), include the blues, hairstreaks, coppers, and metalmarks. Most species are tropical. The family was considered to be monophyletic by Kristensen (1976). Riodinidae (metalmarks) were included, as subfamily Riodininae, in Lycaenidae by Ehrlich (1958b) and Kristensen (1976), and are so treated here. But often they are given family status.

Adult (Figs 291, 292)

The eyes are hairy or not hairy, and they may be emarginated. The maxillary palpi are absent, and the labial palpi are usually well developed and ascending but occasionally reduced. The epiphysis is absent and the foreleg in males usually reduced, with the tarsal segments typically fused and the pretarsus reduced but in females normally developed. However, the legs are used for walking in most species, but not in Riodininae (Robbins, 1988b). Tibial spurs may be entirely absent, 0-0-2 or 0-1-1. The humeral vein is present or absent in the hindwing.
(Eliot, 1973; Munroe, 1982.)

Immature stages

The diameter of the egg usually exceeds that of its height, or the egg may be almost spherical. Often, lycaenid eggs are pitted or bear projections. In those of most species, parts of the highly porous chorion is perforated by numerous pores enabling it to act as a plastron when eggs are submerged in rain water (Downey & Allyn, 1981).

The larva (Downey *in* Stehr, 1987) is usually onisciform, resembling the shape of a woodlouse. Crochets are usually uniserial and tri- to multiordinal, and arranged in an interrupted mesoseries. The abdomen often bears a dorsal gland on segment A7, a pair of eversible organs on segment A8, and minute, round 'pore cupolas' associated particularly with the dorsal gland and the eversible organs.

The pupa is often girdled. Where not girdled, it may be positioned at an angle to the substrate or be suspended with the head downwards.

Classification and biology

Lipteninae are Afrotropical. The colour and pattern vary considerably and the butterflies may mimic those of other families. The larvae are broad and bear tufts of hair dorsally and laterally. They feed on lichen and microscopic fungi.

Poritiinae are Oriental. The egg is hexagonal and the larva long and thin with bundles of hair dorsally and hairs also on the side. The pupa lacks a girdle.

Liphyrinae are mainly African with a few species in the Oriental and Australasian regions. The proboscis is reduced. The larvae are aphytophagous (Cottrell, 1984), living on certain Homoptera and probably ant brood. The pupa lacks a girdle.

Miletinae occur mostly in Africa or the Orient, but some are present in the Holarctic region. The larvae are onisciform and probably always aphytophagous feeding on Homoptera, their secretions, and ant regurgitations. The pupa usually lacks a girdle.

Curetinae are Oriental with a few Palaearctic representatives. The larvae are long with the terminal tubercles permanently protruded.

Theclinae (Fig. 291) are represented in all the main zoogeographical areas. The hindwing usually bears one or more tails. The larvae are usually approximately onisciform. The pupae typically are girdled with their last segment usually dilated.

Lycaeninae occur mainly in the Holarctic region. The hindwing may or may not be 'tailed'. The larva is onisciform and the pupa girdled.

Polyommatinae are represented in all major biogeographical areas. The hindwing may be 'tailless' or may bear a filamentous 'tail'. The larva is onisciform and phytophagous or, at least in some instars, aphytophagous feeding on Homoptera, ant brood, or the regurgitations of ants. Saprophagy and cannibalism may also occur. The pupa may or may not be girdled.

Riodininae (Harvey, 1987), including *Styx*, have a worldwide distribution, but occur mainly in the Neotropics. Many species are brightly coloured and wing-shape is diverse within the group. The male forelegs are reduced with the tarsomeres fused and the pretarsi rarely bearing claws. They are not used for walking (Robbins, 1988b). The coxa extends as a spine-like structure to below the articulation point of the trochanter. The eggs are partially flattened to turban and dome-shaped (Downey & Allyn, 1980) and the larvae are often onisciform and frequently hairy. The larvae of some species are associated with ants (Cottrell, 1984; DeVries, 1988). The question of whether Riodininae should be treated as family Riodinidae is unresolved. (Adult e.g., Fig. 292.) (Eliot, 1973; Ackery, 1984; Downey *in* Stehr, 1987; Harvey *in* Stehr, 1987.)

Phylogenetic relationships

The Lycaenidae are thought to share a common ancestor with Nymphalidae by Ehrlich (1958b) and Kristensen (1976).

BOMBYCOIDEA

The superfamily is distinguished mainly by the absence or reduction of several adult structures. It includes most macrolepidopterans, other than butterflies, that lack metathoracic or abdominal tympanal organs. In Bombycoidea, the mouthparts, frenulum, and retinaculum are usually reduced or absent. External ocelli and chaetosemata are always absent. The presence of one or a pair of protuberances on segment A8 of the larva of many Bombycoidea, and the frequent occurrence of a stalked R_2 and R_3 in those groups with a full complement of radial veins lends some very provisional support for the monophyly of the superfamily, or at least for a major section of it (Holloway, 1987a). Most bombycoid moths are stout-bodied, densely scaled, and have broad wings.

The composition of the group accepted here follows Franclemont (1973) except that Sphingidae are included (e.g., Brock, 1971; Minet, 1986) as are Mimallonidae and also some smaller families found outside North America. The order in which the families are presented follows W. Nässig (pers. comm.).

Bombycoidea fall into two groups - those with densely hairy larvae, and those with larvae that are naked or spiny, but not densely hairy. Families with hairy larvae include Lasiocampidae, Anthelidae, Lemoniidae, Eupterotidae, and Apatelodidae. Families without densely hairy larvae include Carthaeidae, Cercophanidae, Oxytenidae, Saturniidae, Brahmaeidae, Endromidae, Bombycidae, Sphingidae and Mimallonidae. Members of the second group bear a dorsal protuberance on A8 of the larva at least in early instars, and often paired horns or filaments on some or all thoracic segments. The protuberance takes different forms (e.g., a horn or a scolus) and may not be homologous (Nässig, 1988).

Morphological details are not presented under the superfamily Bombycoidea because the monophyly of the grouping remains in question.

Lasiocampidae

The family includes the eggar and lappet moths with about 2200 described species (Holloway *et al.*, 1987). The group is most diverse in the Old World tropics, absent from New Zealand, but otherwise of worldwide occurrence. The moths are medium to large, and of a robust and hairy appearance. They are generally cryptically coloured and patterned. Both forewings and hindwings are usually broad, however there are some Old World genera in which the forewings are relatively narrow, the hindwings relatively small, and with the abdomen extending well behind the wings when at rest. Sometimes the female is brachypterous. Sexual dimorphism is typical.

Adult (Fig. 293)

The antennae are bipectinate to the apex in both sexes. The proboscis is vestigial or absent, and the maxillary palpi absent. The labial palpi are small to large, and often porrect or upturned.

The epiphysis is present in males but reduced in or absent from females. In the forewing, vein R_5 is usually stalked with M_1. Vein M_2 is stalked with M_3 or at least arises near its base. In the hindwing, Sc and Rs are separate at the base of the wing, but they meet again to form a humeral cell. Vein M_2 usually arises from the base of M_3. Vein CuP is absent from both the fore- and the hindwing. The tibial spur formula is 0-2-2, and the spurs are very short. The frenulum is absent, and wing-coupling is assisted by an expanded humeral area in the hindwing, which is supplied with one or more humeral veins. The forewing-metathoracic locking device is present.

In the male genitalia, the uncus and gnathos are frequently reduced. Sometimes the valva is reduced (Kuznetzov & Stekolnikov, 1985). A terminal tuft is frequently present on the abdomen of females.
(Franclemont, 1973; Munroe, 1982; Common, 1990.)

Immature stages

Eggs are of the flat variety with the micropyle situated on the side.

The larva is hairy, often densely so, and may be cylindrical or somewhat flattened. The flattened variety bears lateral tufts (lappets) on each segment, which gives it a characteristic appearance. Crochets are arranged in biordinal mesoseries.

The pupa is stout, often hairy, and bears a few anal hooks.
(Franclemont, 1973; Munroe, 1982; Stehr *in* Stehr, 1987.)

Biology

The eggs are laid singly, in small clusters, or in compact masses. They may be covered by scales from the terminal tuft of the female or, in some groups, by spumaline - a frothy substance released from the accessory glands of the female. Lasiocampidae are often sexually dimorphic. The female is generally sluggish, and in some species may be apterous. The resting posture of many lappet moths is characteristic, with the hindwings extended beyond the front margin of the forewing (e.g., *Gastropacha quercifolia*). The larvae may be solitary or colonial. The best known colonial species are the Tent caterpillars of the temperate genus *Malacosoma* in which larvae spin large webs or tents in which they feed gregariously (Stehr & Cook, 1968). Tent caterpillars may cause serious damage to forest trees. In a number of species, the secondary hairs on the larvae are urticating.

Pupation occurs within a dense, silken cocoon.

Classification

The family was divided into seven subfamilies by Aurivillius (1927) based on details of wing venation. This classification was questioned by Franclemont (1973) who suggested that the North American species, at least, are more appropriately divided into two groups. In one group, the male genitalia appear generalized and the humeral cell of the hindwing is small or very small. In the other, the male genitalia are strongly modified, with the reduction or loss of some components and a hook-like aedeagus, and a small to very large humeral cell in the hindwing.

Anthelidae

The family, which includes about 100 species (Holloway *et al.*, 1987) of medium-sized to large moths, occurs in Australia with a few species in New Guinea.

Adult (Fig. 294)

The antennae are bipectinate to the apex in males; in females they are usually bipectinate, but may be pectinate or dentate. The proboscis is rarely developed, maxillary palpi are vestigial, and the labial palpi porrect.

The epiphysis is present in the male, and usually reduced or absent in females. The tibial spur formula is either 0-2-4 or 0-2-2, with the spurs short. The wings are generally broad and rounded, but in a number of species they are narrower. Occasionally, they are vestigial in females. In the forewing venation one or two areoles may be formed by R_1 and R_s. Vein M_1 may arise at the same point as R_5 or be stalked with it. Vein M_2 arises nearer the base of M_3 than M_1. Vein CuP is absent. In the hindwing $Sc + R_1$ is either free from Rs, or R_1 may connect Rs with Sc. Vein M_2 arises nearer the base of M_3 than M_1, and CuP is absent. A frenulum-retinacular system is present in males (sometimes reduced), but is reduced or absent in females. The forewing-metathoracic locking-device is absent.

There are no thoracic or abdominal tympanal organs.
(Common, 1990.)

Immature stages

Anthelinae larvae, even first instars, bear long, often branched, secondary setae arising from verrucae. However, in Munychryiinae the setae are short, unbranched and not arranged on verrucae (Common & McFarland 1970). A full complement of prolegs is present, and crochets are arranged in a uniordinal, biordinal, and multiordinal mesoseries. In the first instar of *Munychryia* the anal prolegs are greatly enlarged.

The pupa is stout or very stout, bears a few hooked anal setae (Anthelinae) or lacks setae.

Biology

Several Anthelinae feed on *Eucalyptus*, *Acacia*, and *Pinus*. Grasses are also recorded as foodplants (Common, 1990). The larva of *Munychryia* (Munychryiinae) feed on *Casuarina* (Casuarinaceae) (Common & McFarland, 1970). The cocoon, which is double-walled, bears protective irritant spines incorporated by the larva during construction (Common, 1990).

Classification

The family is divided into two subfamilies (Common & McFarland, 1970), Anthelinae and Munychryiinae. A proboscis occurs in Munychryiinae. It is well developed and coiled in *Munychryia* but short in *Gephyroneura*. The differences in larval characteristics between the two subfamilies were noted above.

Lemoniidae

About 20 species are currently placed in this heterogeneous family (Holloway *et al.*, 1987). Two genera, probably unrelated, are included - *Lemonia* and *Sabalia*. *Spiramiopsis*, sometimes placed within this family, is provisionally assigned to Brahmaeidae.

Lemonia (Fig. 295) is a Palaearctic genus with about 12 species of yellow-brown, medium-sized moths found in temperate Asia and Europe. They resemble Lasiocampidae but may belong to Eupterotidae (W. Nässig, *pers. comm.*). The larva is hairy, resembling that of Lasiocampidae, and feeds on Compositae. However, it lacks silk glands, so no cocoon is spun.

Sabalia includes about six species, all from tropical Africa. It has been placed in various bombycoid families and even treated as a separate family. (Munroe, 1982.)

Eupterotidae

The family includes over 300 species (Holloway *et al.*, 1987) of medium to large moths from the Old World tropics of Asia, Australasia, and Africa. In general appearance the moths are broad-winged and drab, and the larvae bear long secondary setae.

Adult (Fig. 296)

The antennae are bipectinate to the apex in both sexes. The proboscis is weakly developed or vestigial, maxillary palpi absent, and labial palpi short. The epiphysis is sometimes present, and the tibial spur formula is 0-2-2. Typically, the wings bear many narrow, crenulate fasciae. These are particularly conspicuous distal of the generally strong postmedial, straight fascia. In the forewing, veins R_2 and R_3 may be fused. The branches of Rs are variously arranged and vein M_1 may arise from the same point as Rs or close to it. In the hindwing, Rs and M_1 are stalked and M_3 is associated with CuA_1. In both wings the cell is small. Vein CuP is absent from both wings. The frenulum and retinaculum are often present in males, although weak; in the female the frenulum is usually absent. The forewing-metathoracic locking-device is absent. (Forbes, 1955; Holloway, 1987a; Munroe, 1982; Common, 1990.)

Immature stages

The egg (see Holloway, 1987a), which is of the upright variety, is a smooth, dome-shaped structure.

The larva (Holloway, 1987a; Common, 1990) bears very long secondary setae arranged in bands around its cylindrical body. The crochets are biordinal and arranged in a mesoseries.

The pupa is ovoid and bears a cremaster of varied form (Holloway, 1987a).

Biology

Eggs are generally laid in clusters on the lower surfaces of leaves and bark.

Both larvae and adults are active at night. In Asia, Eupterotidae are typically forest dwelling and fly close to the forest floor. The larvae of some species live communally in webs, particularly in early instars.

Pupation takes place in the litter or soil surface in a weak or dense cocoon in which secondary setae from the larvae are often incorporated.

Classification

Apatelodidae (see below) were included as a subfamily, Apatelodinae, of Eupterotidae by Forbes (1955). He suggested that there appeared to be no great difference between the adults and those of Eupterotinae, and stated that the larvae of the two groupings resembled each other.

Apatelodidae

The family is small and confined to the New World (Franclemont, 1973), but it is uncertain as to whether it forms a natural group (W. Nässig, *pers. comm.*). The moths are moderately small to moderately large, and there are around 250 species (Holloway *et al.*, 1987). The larvae are hairy (Franclemont, 1973).

Adult (Fig. 297): structure and classification

The antennae are pectinate in both sexes. The proboscis is much reduced or absent, and the frenulum well developed, particularly in males. Vein M_2 is closer to M_1 than to M_2 in both forewing and hindwing. Segment A8 may be modified. (Franclemont, 1973.)

Immature stages

The larvae of some species are very hairy, others less hairy. A pencil of dark setae projects from T2 and T3 in *Apatelodes*. Middorsal tufts of setae arise posteriorly on segments A1-8. Crochets are arranged in a biordinal mesoseries. (Franclemont, 1973; Stehr *in* Stehr, 1987.)

Biology

The larvae are external feeders on various shrubs and trees (Stehr *in* Stehr, 1987). Pupation occurs in the ground in North American species, but many neotropical species spin cocoons like those of Lasiocampidae and Bombycidae (Franclemont, 1973).

Classification

In Apatelodinae the antennae are long and the abdomen unmodified, whereas in Epiinae the antennae are short and abdominal segment A8 is modified. In some Epiinae this abdominal modification is such that segment A8 takes over the function of the valvae and uncus. (Franclemont, 1973.)

Carthaeidae

The family includes a single species - *Carthaea saturnioides* - from western Australia, which has been assigned to various families during its taxonomic history. Its bombycoid affinities were recognized by Common (1966). *C. saturnioides* is a large, broad-winged moth.

Adult (Fig. 298)

The antennae are bipectinate in males and dentate in females. The proboscis is well developed and coiled, the maxillary palpi are short and 3-segmented, and the labial palpi are prominent and 3-segmented.

The epiphysis is present, and the tibial spur formula is 0-2-4. The wings are broad, and the hindwing hairy. A large eyespot is present on each wing. In the forewing, all branches of veins Rs are present. Vein M_1 is not stalked with Rs. In the hindwing, Sc approaches Rs before one half the length of the cell, from which point R_1 joins Sc and Rs. Vein CuP is absent from both wings. A retinaculum is present in males but absent from females. The frenulum is well developed in males and reduced to a number of presumably functionless bristles in females, although its base is present. The forewing-metathoracic locking-device is absent.

The abdomen is stout and hairy. Tympanal organs are absent from both the metathorax and the abdomen.
(Common, 1966.)

Immature stages

The eggs (Common, 1990) are of the flat variety, broadly oval in shape, and pale orange yellow.

The penultimate- and final-stage larval instars (Common, 1966) bear a full complement of prolegs, and the crochets of the ventral prolegs are arranged in a biordinal mesoseries. The anal prolegs are very large. Abdominal segments A1-8 each have a prominent lateral eyespot associated with the spiracle. In the final instar these are particularly well defined. Secondary setae are short, and not numerous. The primary setae of the penultimate instar are clubbed, and the scoli of seta D_1 of segment A8 fuse to form a prominent dorsal horn. While numerous scoli are present on the penultimate instar, they are absent from the final instar, and so, therefore, is the dorsal horn of A8. However, A8 is humped dorsally.

The pupa (Common, 1966) is well sclerotized, and tapers posteriorly into a cremaster with hooked setae.

Biology

When disturbed, adults have been seen to protract their forewings thus displaying the eyespots, and move the hindwings rhythmically from side to side. The larvae feed on Proteaceae, and are aposematically coloured black or brown, and orange. Pupation probably occurs in debris on the ground in a flimsy cocoon of silk and sand.
(Common, 1966.)

Phylogenetic relationships

The affinities of *Carthaea* are uncertain, the family exhibiting several primitive characters but no character by which it may be associated with another bombycoid family. *Carthaea* may represent a primitive group of those Bombycoidea with naked larvae.

Cercophanidae

This family (Jordan, 1924), includes 10 described species of medium-sized bombycoid moths (e.g., as in Fig. 299) confined to the coastal areas of Chile with one genus from the Andes. Cercophanidae probably represent an unnatural grouping distinguished from Saturniidae by their possession of bipectinate, not quadripectinate, antennae, and by the presence of a crossvein from the discal cell to vein Sc of the hindwing.

There are two subfamilies (Jordan, 1924). In Cercophaninae the proboscis is vestigial while in Janiodinae it is well developed, particularly in the male. Larvae of Cercophaninae are known from the final instar of two species. There are no prominent tubercles, and A10 bears a short, backwardly directed peak rather than a horn. A variety of dicotyledonous families serve as foodplants. Pupation occurs in a hard cocoon, which may be open at one or both ends. The juvenile stages of Janiodinae are unknown.

The presence of the crossvein in the hindwing is probably a primitive feature, so the monophyly of the family is in doubt. It was considered to be most closely related to Saturniidae (Jordan, 1924; Michener, 1952).

Oxytenidae

The family (Jordan, 1924) includes 34 described species of medium-sized bombycoid moths (e.g., Fig. 300) occurring in tropical South America (Munroe, 1982). *Asthenidia* superficially resembles microniine Uraniidae in being pale with prominent dark fasciae aligned diagonally across the wings, and having hindwings with short tails. The other two genera (*Oxytenis* and *Homopteryx*) include much darker moths. The proboscis is well developed and bears large carinate papillae, and the antennae in both males and females are bipectinate with the branches arising ventrally. Some of the branches of vein Rs are absent from the forewing, a feature often encountered in Saturniidae.

At rest, the anterior part of the body of the larva (Jordan, 1924) curves around so that it lies close to the abdomen. This posture, together with the 'oily' surface texture, makes the larva resemble a bird dropping. Six rows of small tubercles are present in the first instar, but there are fewer in the last stage. The mesothorax is enlarged laterally, and bears an eyespot dorsolaterally on each side. A caudal horn, bifurcate at its apex, is present on the abdomen. All known foodplants belong to Rubiaceae. The pupa is concealed within a leaf the margins of which are held together by silk. There is no cocoon (Jordan, 1924).

Oxytenidae are most closely related to Saturniidae (Michener, 1952); there is no evidence that they represent the sister group of Cercophanidae.

Saturniidae

The family includes some of the largest moths. There are around 1300 described species (Holloway *et al.*, 1987), and the group is of worldwide distribution with most species occurring in the tropics of the New World (Lemaire, 1978; 1980; 1988).

Adult (Figs 301, 302)

Primitively, the eyes are large, but they are reduced in various members of the family. The antennae are longer than the thorax in primitive species but have become short to very short in some groups. Generally, the antennae of males are quadripectinate (doubly bipectinate), although not always to the apex; exceptionally they are bipectinate. In females, antennae range from simple to quadripectinate, but when pectinate the rami are always shorter than in males. The proboscis is usually absent or rudimentary, but rarely it is larger and probably functional. The maxillary palpi are vestigial. The labial palpi are often reduced in both size and number of segments, and sometimes the reduction is extreme.

The epiphysis may be present or absent. Tibial spurs may be absent, or have the formula 0-2-2, or 0-2-4. The wings are broad with considerable variation in detail of shape. They may be rounded or falcate, and typically, both wings bear an eyespot or a window. In some species the hindwing is extended into a tail, which is sometimes very long. The discal cell may be open or closed. In the forewing, one vein of Rs may be lost, and vein M may arise from about midway between M_1 and M_3 or its base may be closer to M_1. In the hindwing, the bases of M_1 and M_2 arise close together. Wing coupling is generally amplexiform with the humeral angle expanded in the hindwing. The frenulum and retinaculum are lacking. The forewing-metathoracic locking-device is absent.

The male abdomen is relatively small when compared with the size of the wings. (Michener, 1952; Ferguson, 1971-2; Holloway, 1987a.)

Immature stages

Egg are ovoid, usually smooth, and slightly flattened. They may be transparent or opaque whitish.

The larva may be setose, granulose, or smooth. Even in first instars, numerous secondary setae may occur. Scoli are prominent on the first instar. In mature larvae these are generally reduced to warts or short tubercles, but the presence of scoli in some species is probably universal. The dorsal pair of scoli of the meso- and metathorax are often more prominent than those on the rest of the body. A full complement of prolegs is present. The crochets are arranged in a biordinal mesoseries. In many species there is a caudal projection on abdominal segment A8, but in some a pair of scoli, which may be partially fused, occur.

The pupa rarely bears setae, and a cremaster may be present or absent. (Godfrey *et al.*, in Stehr, 1987.)

Biology

The eggs may be laid singly, in small groups, or in masses around twigs. They are generally stuck to the substrate with an adhesive substance.

Pupation generally occurs in a strong silken cocoon attached to the foodplant or suspended from it. Some Saturniidae lack a cocoon and pupate under the surface of the ground.

Adult habits were characterized by Janzen (1984). In contrast with Sphingidae, Saturniidae are non-feeding, short-lived moths with lightweight fast-flying males adapted almost exclusively to finding heavy-bodied females. Adaptations in females are strongly related to ensuring that the eggs are laid on the correct foodplant. The fastest and longest flying species have narrow, sphingid-shaped wings. Females tend to deposit eggs, often in large batches, on scented plants. These characteristics reduce the time taken to oviposit, and reduce the chance of predation. In *Hylesia lineata* the eggs are laid all at once, in a ball. Some species lay unfertilized, but viable, eggs.

Adults, particularly of diurnal species, frequently display the eyespots on the wings as a device to startle predators. Larvae tend to be defended by their urticating hairs, and by spines that may deliver toxic substances.

Classification

The Saturniidae are divided into seven subfamilies (Michener, 1952). Six of these subfamilies are probably monophyletic. The seventh, Saturniinae, represent an unnatural grouping.

Arsenurinae (= Rhescyntinae) (Michener, 1952; Lemaire, 1980), with about 60 species, are exclusively Neotropical and include species with many primitive saturniid characters such as quadripectinate antennae, galeae that are not strongly reduced, and vein M_2 arising midway between M_1 and M_3. The ground colour of the moths is mostly grey or brown. In most species, the hindwings are extended into prominent tails. First instar larvae bear a pair of horn-like scoli subdorsally on the metathorax, and a single dorsal scolus on segment A8. In last instars, the scoli are lost and usually replaced by protuberances. There is no cocoon, pupation taking place in the soil.

Ceratocampinae (= Citheroniinae) (Michener, 1952; Ferguson, 1971-2; Lemaire, 1980), with 170 species, are exclusively American, and mostly Neotropical. The bodies of these moths are large relative to the size of the wings, and the insects are the most sphingiform of all Saturniidae. The proboscis varies from being minute to relatively long. The male antenna is quadripectinate for part of its length then simple to the apex. In the female, the antenna is like that of the male, or entirely simple. The discal wing 'ocelli' are small in most species. Vein M_1 is stalked with the radials in the forewing, and vein M_2 arises from near the middle of the discal cell. On the larvae, there are usually two pairs of prominent scoli on the meso- and metathorax; a single scolus (the caudal horn) occurs mid-dorsally on A8; and a shorter mid-dorsal scolus occurs on A9. Occasionally a pair of short, unfused scoli, are found on A8. In final instars thoracic and abdominal scoli may be present or absent; often they are reduced. Pupation takes place in a cell in the ground and there is no cocoon. The pupa often bears many small spines, some of them strong, which assist it to move through the soil just before eclosion.

Agliinae (Michener, 1952) include two, or possibly three (W. Nässig, pers. comm.) species. Two are from S.E. China and the third from Japan. The larva bears large thoracic horns in early instars, but these structures are absent from the last stage. Pupation occurs in a cocoon.

Hemileucinae (Fig. 301) (Michener, 1952) are confined to the New World, and are best developed in the Neotropics. There are approximately 600 species. Brachypterous females are found in a few species. The eggs are somewhat elongated, and in *Coloradia* and *Hemileuca* often laid in bands encircling twigs of the foodplants. The larvae bear large tubercles, often with sharp spines, with a defensive function. The larvae of *Lonomia* deliver a powerful anticoagulant by way of the spines. The pupa is enclosed in a cocoon, the strength of which varies from being formed from a few strands to a relatively well constructed structure. The cocoon is situated in debris on the ground rather than in the ground. In *Hylesia*, the female moth bears urticating hairs. Although the subfamily is associated with a wide variety of foodplants, individual species are generally fairly host-specific - a characteristic atypical of most saturniid groups.

Ludiinae (Michener, 1952), a small Ethiopian subfamily, include fairly small Saturniidae often with irregular hyaline patches on the wings instead of eyespots. The tubercles of the larvae bear sharp spines and long hairs. Pupation occurs in a cocoon.

Salassinae (Michener, 1952) include a single genus from southern Asia with characters more primitive than those exhibited by Saturniinae.

Saturniinae (Fig. 302) (Attacinae) (Michener, 1952; Ferguson, 1971-2; Lemaire, 1978) are probably an unnatural grouping. They are worldwide in distribution. The moths are usually large. The larvae are generally stout and sluggish. Scoli may be elongated or much reduced and bear spines or hairs, which deliver toxic and urticating substances. With few exceptions, pupation occurs in a dense silken cocoon attached to the foodplant. Some species (e.g., *Samia cynthia*) have been used for commercial silk production. The subfamily is best developed in the tropics of the Old World, but it occurs throughout the world.

Brahmaeidae

The family includes around 20 species (Nässig, 1980; Holloway *et al.*, 1987) in four genera (Holloway, 1987a) of medium- to large-sized moths found in Africa, the tropics of the Orient and the Far East, and in southern Europe.

Adults and larvae are distinctive. The wings are broad, and bear complex markings of numerous wavy fasciae and a row of submarginal 'ocelli', which are sometimes incompletely formed.

Adult (Fig. 303)

The antennae are bipectinate in both sexes, but in some species a cone-like projection is present ventrally on each flagellomere suggesting a reduced tripectinate condition (Sauter, 1986). The proboscis is absent or well developed and functional, and the labial palpi are prominent, upturned structures. The epiphysis is present. All the radial veins are generally present in the forewing. The stem of vein M is present in the cell of the hindwing, and may also be found in that of the forewing. Vein R_1 is connected, in the hindwing, to Sc by a crossvein. The frenulum is absent, as is the forewing-metathoracic locking-device.
(Mell, 1928; Sauter, 1986; Munroe, 1982.)

Immature stages and biology

The larvae are sometimes brightly coloured and, in their early instars, bear long, paired, fleshy filaments usually on the dorsum of T2, T3, A8 and A9/10. In the final instar these structures are absent, but a pair of eyespots occur in larger species, which are used, together with a threat posture, for defence. A mediodorsal caudal horn is present on A8. The crochets are arranged in a mesoseries. The larvae feed on Oleaceae and Asclepiadaceae. There is no cocoon, and pupation takes place on the ground in crevices (such as under stones), but not deep in the soil.
(Mell, 1928; Sauter, 1986; Munroe, 1982.)

Classification

The monotypic, South African genus *Spiramiopsis* is included in Brahmaeidae following Jordan (1923c) although its affinities are uncertain. The larva is not hairy, but long paired dorsal processes occur in the meso- and metathorax, and a caudal horn is found on abdominal segment A8. The genus should certainly be excluded from Lemoniidae (W. Nässig, *pers. comm.*), where it has been placed.

Endromidae

The family includes a single species - *Endromis versicolora*, known in Britain as the Kentish Glory. It occurs in the *Betula* forests of Europe, its range extending through Asia to southern Siberia and the Far East.

Adult structure, immature stages and biology

The single species (Fig. 304) (Seitz, 1911) is a medium-sized bombycoid, sexually dimorphic, and brown with white and dark brown fasciae and other marks. The proboscis is not developed, the maxillary palpi absent, and the labial palpi short. In the forewing, vein M_2 arises closer to M_3 than to M_1. In the hindwing, vein M_2 arises from near the middle of the cell or closer to M_3 than to M_1. The frenulum and retinaculum are absent. The forewing-metathoracic locking-device is absent.

The larva (Seitz, 1911; Carter & Hargreaves, 1986) has a broad body but narrows strongly towards the head. It rests characteristically, with the anterior part of the body raised. The head and body are green marked with yellowish white bands and black spots. The larva is not hairy, nor does it bear protuberances other than a short horn-like hump on abdominal segment A8. In the early stages it is gregarious. Pupation takes place in a dark brown cocoon in the litter or just under the surface of the soil. The foodplant is usually *Betula* but the larva may feed on other deciduous trees.

Classification

The genus *Mirina* from temperate Asia was included in Endromidae on the basis of superficial similarity and in the absence of anywhere else to place it (Seitz, 1911). The larva bears long scoli on the dorsum of the meso- and metathorax, a situation similar to the condition seen in Brahmaeidae, a number of Saturniidae, and in *Ceratomia* (Sphingidae) (Minet, 1986). Based on the findings of a study of the musculature of the male genitalia by Kutznetsov & Stekolnikov (1985), *Mirina* was placed in a monotypic family, Mirinidae by Minet (1986).

Bombycidae

The family (*sensu* Holloway, 1987a) includes around 60 species of relatively small, Old World bombycoids. Most species occur in the tropics of the orient and southeastern Asia. There are two groupings in the Oriental Bombycidae (designated as Bombycinae and Oberthueriinae by Kutznetsov & Stekolnikov (1985), or the *Bombyx* group or the *Mustilia* group of Holloway, 1987a).

Adult (Fig. 305)

In males, the antennae are bipectinate throughout their length; in females they may be either bipectinate, weakly bipectinate, or filiform. The mouthparts are virtually absent, but minute labial palpi occur in *Bombyx mori*. The epiphysis is present or absent, and the tibial spurs are absent or have the formula 0-2-2. The frenulum is usually absent but occasionally present and short. The forewing-metathoracic locking-device is absent. The forewing is often falcate. The branches of Rs are variably arranged, and sometimes one branch is absent. Vein M_1 is stalked with Rs. Vein CuP is sometimes present in the fore- and hindwings. The dorsum of the hindwing is pleated and often excavated.
(Franclemont, 1973; Holloway, 1987a.)

Immature stages

Eggs are of the flat variety with the micropyle on the side of each egg. Typically they look like flattened spheres.
   Superficially, the larvae appear smooth, but they are usually covered with short secondary setae. A full complement of prolegs is present. A caudal horn or hump occurs on abdominal segment A8 in most species.
(Holloway, 1987a.)

Biology

As in certain other bombycoids, the adults of some Bombycidae rest in a characteristic fashion with the forewing held at right angles to the body and the pleats of the dorsum of the hindwing folded over the posterior edge of the forewing. Within Bombycidae this habit may be characteristic of the *Bombyx* group (Bombycinae) (Holloway, 1987a).

Pupation occurs in a dense cocoon, and cocoons of *Bombyx mori* form the basis of the silk industry (see Chapter 7). *Bombyx mori* was also used in the pioneering experiments on pheromones (see Chapter 6).

Classification

In the *Bombyx*-group, vein CuP is present in both fore- and hindwings. In the forewing vein R_s is 4-branched, and a vestige of M often occurs in the cell. The tergum and sternum of segment A8 of the male abdomen are strongly modified. In the *Mustilia*-group neither CuP nor the base of M is present. In the forewing, there are only 3 distinct Rs branches. The tergum and sternum of segment A8 of the abdomen of the male is not strongly modified.

The monotypic genus *Dalailama* from Tibet is provisionally included in Bombycidae (Grünberg, 1911), but its true affinities are uncertain, and it has not been taken into account in the comments on structure (above). The antenna is pectinate to beyond its middle but serrated towards its apex, and the colour pattern is distinct and unlike that typical of Bombycidae.

Sphingidae

Hawkmoths are often included in a superfamily of their own, but here they are placed in Bombycoidea. The group includes about 1050 species (D'Abrera, 1986), and although best developed in the tropics it is represented worldwide. The most conspicuous aspect of sphingid biology is the capacity of the moths for fast, long-distance, and often migratory flight. The moths are appropriately shaped with narrow wings and a fusiform abdomen. Many species have evolved the ability to hover in front of flowers in a manner resembling that of hummingbirds.

Adult (Fig. 306)

The antennae are usually filiform but occasionally bipectinate in the male. Sometimes the flagellum is swollen distally and hooked at the apex. The proboscis is usually well developed, and sometimes extremely long, but it may be reduced and non-functional. The maxillary palpi are very small and 1-segmented, and the labial palpi are prominent, ascending, and appressed to the head. On segment 1 of the labial palpus, the inner surface may bear fine, dense setae the presence of which is of taxonomic importance. Segment 2 is sometimes swollen and acts, together with the pilifer, as a non-tympanate hearing organ.

The epiphysis is present, and the tibial spur formula is usually 0-2-4 or occasionally 0-2-2. The midleg often bears a row of slender setae, known as the midtarsal comb, on the first tarsal segment. The forewings, typically, are considerably longer than broad, and the hindwings only about twice as long as broad. In the forewing, veins R_2 and R_3 are stalked or fused. Vein M_1 arises from the same point as Rs or is stalked with it. In the hindwing R_1 diverges from a point about halfway along the cell and meets Sc. CuP is absent from both fore- and hindwings. Wing coupling is nearly always effected by means of a frenulum and retinaculum, but the system is occasionally reduced. The forewing-metathoracic locking-device is absent.

The abdomen is robust, but narrows posteriorly. There are no metathoracic or abdominal tympanal organs.

(Rothschild & Jordan, 1903; Hodges, 1971; Derzhavets, 1984; Holloway, 1987a.)

Immature stages

The eggs are of the flat variety, translucent greenish, and rounded or slightly flattened. The surface is not strongly sculptured.

The larvae are medium-sized to large. Secondary setae may be present, but they are not conspicuous. A prominent dorsal horn usually occurs on abdominal segment A8, although the structure is sometimes rudimentary. Rarely, a complete set of scoli or spines are present on the body. A full complement of prolegs is present, and the crochets are arranged in a biordinal mesoseries. Prominent eyespots may occur on the thorax.

The pupa is fusiform, and the cremaster prominent. To accommodate the long adult tongue, the pupal proboscis is often looped or coiled away from the body and the adult tongue turned back on itself to terminate in the usual position between the legs. Two types of extended pupal tongues occur (I.J. Kitching, *pers. comm.*). In one, the structure loops down from the front of the head. In the other, it extends directly anterior from the head before turning posteriorly.

Biology

Adults are generally nocturnal or crepuscular, and there are few truly diurnal species. Both males and females are generally long lived and strong fliers. Sphingidae, in contrast with Saturniidae, feed over a long period, possess appropriate flight mechanisms, and have females that lay eggs singly, not in large batches (Janzen, 1984). Since their flight often involves finding and revisiting flowers, Sphingidae also require a suitably developed neural capacity (Janzen, 1984). Many members of the group suck nectar from flowers while at rest, but others - notably *Macroglossum*, the Hummingbird Hawkmoths - feed while hovering in front of flowers. In Smerinthini, the proboscis is reduced and non-functional, and the mode of life described above does not occur.

Aposematic markings are well known in the larvae. For example, the Elephant Hawkmoths (e.g., *Deilephila elpenor*) bear three pairs of eyespots on the anterior section of the abdomen. These markings are displayed in a threatening way when the head and thorax are drawn in and the abdomen bulges. The central American species *Eumorpha labruscae* is said to mimic a snake, the eyespots on the inflated abdomen representing the eyes of the reptile, and a line that ends with the ecdysial cleavage line of the head its forked tongue (Rothschild, 1985). *Hemeroplanes* (see Pl. 2:9) is also a snake-mimic. Larvae of most members of the family are however cryptically coloured.

The adult of *Acherontia atropos* (the Death's Head Hawkmoth) makes a high-pitched squeaking sound (see Chapter 6), which is said to mimic the piping of a queen bee and to give the moth protection as it robs hives of honey. An ability to hear has been demonstrated in Choerocampini (Macroglossinae) (Roeder, 1972) (see Chapter 6), members of which have non-tympanate, cephalic hearing organs.

Classification

The family is divided into two subfamilies, Sphinginae and Macroglossinae (e.g., Carcasson, 1976; Hodges, 1971; Derzhavets, 1984).

Sphinginae (Fig. 306) are characterized by the absence of the patch of short, sensory setae on the inner surface of the first segment of the labial palpus, the presence of symmetrical male genitalia, and female genitalia with the genital plate composed of both lamella antevaginalis and lamella postvaginalis. Although diverse, the larvae all have paired lateral, oblique lines that run from the anterio-ventral margin to the posterio-dorsal margin of a segment. The caudal horn is generally well developed. Sphinginae larvae are usually more granulose or minutely spinose than are those of Macroglossinae. In the pupa, the proboscis is often partly looped away from the body. There are two tribes.

Sphingini include over 150 species of moderate to very large Sphingidae. Most species occur in the New World, but the tribe is also found in the Old World.

Smerinthini, with over 200 species, occur mainly in the Ethiopian and Oriental regions, but some are Holarctic. The hindwing often bears an ocellate spot. The proboscis is typically reduced. In the larva, besides oblique segmental stripes, longitudinal stripes sometimes run along the body. The proboscis of the pupa is not looped away from the body, but is entirely fused with it.

In **Macroglossinae**, there is a patch of short sensory hairs, on an otherwise naked patch, on the inner surface of segment 1 of the labial palpus; the male genitalia may be symmetrical or asymmetrical; and often the genital plate is reduced and situated posterior to the ostium. The caudal horn of the larva is often reduced to a short squat structure, and it is within Macroglossinae that eyespots on the larvae tend to be encountered on the abdomen. There are four tribes.

The Dilophonotini include over 130 species. Most occur in the Neotropics, but the group is also represented in the Holarctic. The genitalia of both the male and the female are often asymmetrical. The larva is often brightly coloured and bears black, white, and yellow bands. In the Holarctic genus *Hemaris*, areas of the wings lose their scales shortly after the moths make their maiden flights.

The Philampelini, with 20 species, are predominantly Neotropical. The very slender caudal horn occurring in the first instar larva is replaced with a button-like structure in the last instar.

The Choerocampini, of worldwide distribution, include about 160 species. They are characterized by the modification of the labial palpi and pilifers into hearing organs (Chapter 6). The larvae often bear a row of ocellate spots dorsolaterally.

The Macroglossini are an unnatural grouping representing those Macroglossinae that do not fit into the other tribes. There are about 225 species. The larvae often bear ocellate spots on the thorax.

(Rothschild & Jordan, 1903; Hodges, 1971; Derzhavets, 1984; Holloway, 1987a.)

Mimallonidae (= Lacosomidae)

There are about 200 species (Munroe, 1982) in the family. Mimallonidae are mainly confined to the Neotropics, although four species are found in North America. The family was given superfamily rank by Franclemont (1973), but a more inclusive definition of Bombycoidea is accepted in the present work. Since Lasiocampidae and Sphingidae are placed in Bombycoidea it seems reasonable to include Mimallonidae. Holloway (1987a) pointed out that in both Mimallonidae and other Bombycoidea veins R_2 and R_3 are stalked and arise independently from the cell in the forewing. Scott (1986) suggested that mimallonids are pyraloids on the basis of certain larval similarities. However, Mimallonidae lack the tympanal organs so characteristic of Pyraloidea, and several other characters establish Mimallonidae within the bombycoid complex (Minet, 1991).

Adult (Fig. 307)

External ocelli and chaetosemata are absent. The antennae are pectinate. Both maxillary and labial palpi are vestigial, and the proboscis is almost entirely absent. The body bears a deep vestiture and the wing scaling is dense. The forewings are commonly falcate. Veins R_2 and R_3 are stalked, a character seen in many bombycoids (Holloway, 1987a). The frenulum is present but reduced, or absent. The forewing-metathoracic locking-device is absent. The tibial spur formula is 0-2-2 or 0-2-4. Tympanal organs are absent.

(Franclemont, 1973; Munroe, 1982.)

Immature stages and biology

In the larva (Stehr *in* Stehr, 1987), the L-group is bisetose on T1, and on the abdomen L1 and L2 lie close together. On A3-6, the prolegs are short, with crochets arranged into an unbroken transverse oval, a situation said to be unique within Lepidoptera. Prolegs on A10 are also short with crochets similarly arranged. Early instars live between two leaves in a web. In later instars the larvae construct portable cases composed of leaves and silk and, in some tropical species, frass. The cases are open at both ends (Franclemont, 1973).

The pupa is heavily sclerotized, obtect, and typically bombycoid (Franclemont, 1973).

NOCTUOIDEA

The Noctuoidea are by far the largest superfamily of Lepidoptera with the number of named species estimated at little under 42 000. The group may be recognized by the presence of metathoracic tympanal organs, the structure and position of which are unique. Noctuoidea are distributed globally. Although the moths range in size from small to very large, they are typically robust-bodied. Chaetosemata are always absent. Wing coupling is effected by a frenulum-retinacular system, and the forewing-metathoracic locking device is usually present. The eggs of Noctuoidea are of the upright variety. The larvae are cylindrical with prolegs bearing crochets arranged in a mesoseries. Although most species are external feeders on plants, many bore into plant tissue.

The way in which the Noctuoidea should be divided remains unresolved.

Oenosandridae

Oenosandra, an Australian genus with three species (Common, 1990), was placed in a separate family by Miller (1991) who was unable to assign it to any other family of Noctuidae. Previously it had generally been included in Thaumetopoeinae (Notodontidae). The adult has a nodular sclerite in the tympanal organ (unlike the situation in Doidae and Notodontidae), a posterior- rather than ventral-facing tympanum, and a prespiracular counter tympanal hood on the first abdominal segment. The genus *Discophlebia*, with five species, is closely related according to Common (1990).

Adult (Fig. 308) (based on *Oenosandra*)

The front of the head is protruded. The antennae are bipectinate in the male and filiform in the female. The proboscis is short and the labial palpi very short. The tibial spur formula is 0-2-4. The wings are fairly elongated, with the forewing venation trifid, vein M_2 not being approximated to M_3. Metathoracic tympanal organs are present, with a nodular sclerite and a posteriorly facing tympanum. (Miller, 1991.)

Immature stages and biology

The eggs of *Oenosandra boisduvali*, which are laid in groups, are approximately cylindrical, bear weak longitudinal grooves, have rounded ends, and are partially covered with scales from the female moth (McFarland, 1973b). Those of *Discophlebia catocalina* are laid in rows and small groups. They lack any scale covering and are prominently ribbed (McFarland, 1973b).

The larva (Miller, 1991) lacks the cervical gland. There is no covering of secondary setae, and dorsal tubercles are absent. The prolegs bear biserial crochets arranged in two rows of equal length, and there are three SV setae, or fewer than three, on the lateral surfaces of the prolegs. The larvae feed on *Eucalyptus* leaves at night and are often found in groups under loose bark by day (Common, 1990). *Discophlebia catocalina* also feeds on *Eucalyptus*, and young larvae, at least, are also rather gregarious (Common, 1990).

Doidae

The family was established (Donahue & Brown *in* Stehr, 1987) for a small group of Neotropical species belonging to two or three genera. The group is mainly Mexican, with two species occurring in the southwestern U.S.A, but was excluded from Notodontidae because the male genitalia appear to lack socii, and the larva has only a single SV seta, rather than two setae, on A1 (Miller, 1991).

Adult (Fig. 309) (based on *Doa* alone)

The antenna in both sexes is bipectinate. The proboscis is of moderate length and the labial palpus fairly long. The tibial spur formula is 0-2-4 and the tarsal claws are simple. A metathoracic tympanal organ, similar in structure to that of Notodontidae is present. In the male genitalia, socii are probably absent and the valva is simple. (Miller, 1991.)

Immature stages and biology

Eggs of *Doa ampla* (Donahue & Brown *in* Stehr, 1987) are yellow, oblong, covered with fine punctuations, and usually laid in a cluster of parallel rows.

The larva (Donahue & Brown *in* Stehr, 1987; Miller, 1991) lacks secondary setae on the head, and the mandibles are serrated. A cervical gland is absent. Long spicules cover the integument, but secondary setae are confined to the region of L_3 on all abdominal segments. The thoracic segments are somewhat humped dorsally. Prolegs occur on segments A3-6 and A10, and the crochets are biordinal.

All recorded foodplants belong to the Euphorbiaceae.

Phylogenetic relationships

Although Doidae have been assigned to various families, evidence strongly suggests that they are closely related to Notodontidae (Miller, 1991).

Notodontidae

Notodontidae (as defined here) include around 2500-3000 species (Holloway *et al.*, 1987) of medium-sized to large moths. The family is found worldwide, and has often been treated as superfamily Notodontoidea. Thaumetopoeidae, Dioptidae, and Thyretidae have either been included within Notodontidae, or treated as families within Notodontoidea (e.g., Kiriakoff, 1963). In the present work, Thaumetopoeidae and Dioptidae are treated as subfamilies of Notodontidae, while Thyretidae are placed in Arctiidae as Thyretinae (Holloway, 1983).

The monophyly of the Notodontidae is well supported (Miller, 1991). In the adult, the tips of the tibial spurs have serrated edges; and there is a teardrop-shaped, swollen area (bulla) of the metascutal region above the tympanum. In the larva, the cutting edge of the mandible is primitively smooth; there are two MD setae on segment A1; and a seta or a verruca occurs on the anteriolateral corner of the anal shield.

Adult (Figs 110-112)

External ocelli are usually reduced or absent, and chaetosemata are absent. The male antennae are typically bipectinate, often to the apex, while those of the female are weakly bipectinate or simple. The proboscis is well developed, reduced, or absent, the maxillary palpi are very small or absent, and the labial palpi porrect, upturned, or vestigial.

The epiphysis is present in males, but occasionally absent from the female. The tips of the tibial spurs are typically sclerotized with their margins serrated, and the tibial spur formula is 0-2-2 or 0-2-4. In the forewing, vein R_2 is often stalked with R_{3-5}, and vein M_2 does not arise nearer M_3 than to M_1. There is no accessory cell. The metathoracic tympanal organ is characterized by a ventrally directed tympanum. The nodular sclerite is absent. A teardrop-shaped bulla occurs on the metascutal area above the tympanum.

A counter tympanal cavity with a postspiracular hood-like projection occurs at the base of the abdomen. Sometimes a cteniophore, a membranous pleural outpushing, covers a pleural cavity on both sides of abdominal segment A4 in males (Jordan, 1923a). Sternum A8 of the male is frequently modified for grasping the female. A membranous, pleated sacculus, with the pleats enclosing long androconial scales, occurs widely in the genitalia of males.
(Holloway, 1983; Common, 1990; Miller, 1991.)

Immature stages

The egg is of the upright variety, hemispherical or almost spherical in shape, and not ribbed (Döring, 1955; Common, 1990).

The larva (Godfrey & Appleby *in* Stehr, 1987; Common, 1990; Miller, 1991) is extremely variable, but often bears spines, tubercles, and dorsal humps. Secondary setae are present or absent from the head. The cutting edge of the mandible is primitively smooth, but is serrated in some groups. The body may be smooth, or may bear dense secondary setae. Unlike the condition in other Noctuoidea, the MD group is bisetose (rather than unisetose) on A1. A midventral cervical ('neck') gland occurs in many species. When extruded, the gland is dorso-ventrally compressed and strongly bifid unlike the condition in other Noctuoidea (Miller, 1991). Ventral prolegs are present, usually fully developed, and bear homoideous, uniordinal crochets arranged in a mesoseries. The anal prolegs are diverse, ranging from vestigial (a condition frequently encountered) to situations in which they are extended horizontally and where they may be elongated and bear eversible glands distally - the stemapodiform condition. Dorsally, segments A1-8 are unmodified or may bear tubercles, humps, or both on one or more segments.

There is no certain way to distinguish the pupa of Notodontidae from that of Noctuidae and some Bombycoidea. However, in Notodontidae visible labial palpi are absent or minute, and the maxillae are very short and fail to extend to the distal wing pad margin (Godfrey & Appleby *in* Stehr, 1987).

Biology

The larvae (Godfrey & Appleby, *in* Stehr, 1987; Miller, 1991) frequently feed on the leaves of trees or bushes rather than herbs, first instars tending to be skeletonizers later instars consuming most of the leaf. However, Dioptinae are secondarily herb feeders. They may be solitary or may aggregate depending on species or instar. Although some Notodontidae are cryptically coloured, several are able to spray formic acid and ketones from the cervical gland at predators and parasitoids. Others rear and lash their stenopodiform anal prolegs, adopting various postures thought to be threatening (e.g., as in Pl. 2:8).

Classification

Eight subfamilies were recognised by Miller (1991) representing major monophyletic groups within the family. The subfamilial placement of some other genera is uncertain.

Thaumetopoeinae (Fig. 310) include about 100 species occurring in Africa, the Mediterranean and southern Europe to northern India and Australasia (Holloway *et al.*, 1987). The group represents the sister group of all other Notodontidae (Miller, 1991). The proboscis is reduced or absent, and the labial palpi are small and often strongly reduced. The antennae in both sexes are pectinate to the apex. The epiphysis is sometimes absent from females. The frenulum in the female consists of more than 20 bristles. The larva bears secondary setae on the head, including the mandibles, and body. The prothoracic gland is absent. Anal prolegs are not reduced. The larvae are processionary, colonial, and live in silken nests; at night they move in procession to feed.

Pygaerinae include, with certainty, only the genus *Clostera* (Miller, 1991), a more comprehensive group having yet to be established. *Clostera* contains 43 species and is represented in all the main zoogeographical regions. The proboscis is short and the labial palpi long. The frenulum of the female consists of fewer than 15 bristles. The tarsal claws are bifid and the hindtibia bears 2 pairs of spurs. The male genitalia have a pleated, saccular scent organ. The larva bears secondary setae on the head and the body, and the mandibular cutting edge is smooth.

Notodontinae (Fig. 311) include at least 200 species, but the composition of the subfamily remains uncertain and the number of species is likely to be much greater (Miller, 1991). Included in Miller's concept of the subfamily are the Cerurinae of previous authors. The male antennae are pectinate to the apex. The proboscis is short and broad or absent, and the labial palpi are often small. The frenulum of the female is composed of more than 20 bristles. The tarsal claws are not bifid, and the epiphysis is usually elongate and flattened. The sacculus is usually not pleated. The larva lacks secondary cephalic setae, but secondary setae are either scattered on the body or absent; the body is never hairy. The anal prolegs are smaller than the ventral prolegs and sometimes stemapodiform.

Phalerinae are of uncertain composition (Miller, 1991) and thus of uncertain distribution. In the broad sense they are represented in Africa, Indo-Australia, the Palaearctic and the Nearctic. The male antenna is pectinate or ciliate. The proboscis is of moderate length as are the labial palpi. The frenulum consists of fewer than 10 bristles in the female. The tarsal claws are bifid, and the hindtibiae bear 2 pairs of spurs. The sacculus is very weakly pleated or not pleated. The larva lacks secondary cephalic setae, but the body may or may not bear them. The anal prolegs are smaller than the ventral prolegs.

Dudusinae have been broadened considerably in concept (Miller, 1991) and include representatives worldwide. The male antenna is pectinate to the apex or to below the apex and the female antenna is pectinate or ciliate. The proboscis is longer than the thorax, and the labial palpi are of moderate length. The tarsal claws are bifid, rarely simple, and the tibial spur formula is 0-2-4 or 0-2-2. The frenulum consists of fewer than 10 bristles in the female. The sacculus is faintly pleated or lacks pleats. The larva typically lacks secondary setae on the head, and the cutting edge of the mandible is either smooth or serrate. The anal prolegs are small and frequently stemapodiform.

Heterocampinae occur mainly in the New World but are also found in the Old World. However, the composition of the subfamily remains uncertain (Miller, 1991). The male antenna is pectinate, usually to below its tip whereas the female antenna is simple or pectinate. The proboscis is long or short and the labial palpi are of moderate length. The tarsal claws are usually bifid but sometimes simple or absent. The frenulum in the female consists of from 2 to 10 bristles. The sacculus may be pleated or not pleated. The larva usually lacks secondary setae on the head and the mandible has a smooth cutting edge. The anal prolegs are reduced and sometimes stemapodiform.

Nystaleinae include about 300 species (S.J. Weller *in* Miller, 1991) and occur in the Neotropics with three genera extending into North America. The male antenna is pectinate to below the tip whereas that of the female is ciliate. The proboscis is usually long but sometimes reduced. The tarsal claws are bifid and the tibial spur formula 0-2-4. The frenulum in the female consists of 2 or 3 bristles. The sacculus is usually pleated and often very large. The larval head lacks secondary setae and the mandibular cutting surface is smooth. The anal prolegs are small but not stemapodiform.

Dioptinae (Fig. 312), with 400 species, occur in the Neotropics with one species in the west coast of the U.S.A. (Miller, 1987b, 1991). The moths are delicate, brightly coloured and diurnal. The male antenna may be pectinate to the apex or may lack pectinations, and the female antenna is either pectinate or ciliate. The proboscis is long and the labial palpi of moderate length. The tarsal claws are bifid and the tibial spur formula 0-2-4. The frenulum in the female consists of fewer than 10 bristles. The sacculus is either pleated or not pleated. The larval head lacks secondary setae and the mandibular cutting edge is either serrated or smooth. The prolegs are small and tubular or stemapodiform.

Lymantriidae

The family includes about 2700 described species of medium-sized macrolepidopterans (Ferguson, 1978) called Tussock moths. Most occur in the Old World tropics. The larvae are hairy with tufts and hairpencils characteristically arranged on the body. Frequently polyphagous and arboreal, they sometimes defoliate large areas of forest.

Adult (Fig. 313)

External ocelli are absent. The antennae are usually bipectinate in both sexes, but in the male the pectinations are longer and terminate in 1 to 3 long spinules. The proboscis is strongly reduced or absent, the maxillary palpi are 1-segmented or absent, and the labial palpi are moderately developed.

The epiphysis is absent and the tibial spur formula is 0-2-2 or 0-2-4. The wings are usually broad, but strongly reduced in the female of some species. In the forewing, an areole is present or absent. Occasionally, one branch of Rs may be absent. In the hindwing, vein Sc diverges from Rs basally but approximates to it (or connects with it), via R_1, towards the middle of the cell. Vein M_2 arises much nearer the base of M_3 than M_1. Metathoracic tympanal organs are present.

A prespiracular counter-tympanal hood is present at the base of the abdomen. Frequently, what is probably a tymbal organ occurs on segment A3 of males (Zerny & Beier, 1936; Dall'Asta, 1988). The male genitalia are fairly simple. In the female genitalia the ovipositor lobes are usually large, fleshy, and setose and occasionally elongated. Terminally, large tufts of hair-like scales occur in the female. (Ferguson, 1978.)

Immature stages

The egg is of the upright variety, spherical, hemispherical, or subcylindrical.

The larva sometimes bears brightly coloured markings and dense tufts of secondary setae (sometimes urticating) and spines arising from verrucae. The cervical gland is lacking. Usually, characteristic hairpencils extend especially from T1, A8 and A9. Prolegs occur on A3-6 and A10 with the crochets arranged as a homoideous, uniordinal mesoseries. A single middorsal gland, usually yellow or red, is situated on each of segments A6 and A7.

The pupa is stout and hairy with setae mainly arising from the verrucal scars of the larva. A cremaster is present.

(Ferguson, 1978; Godfrey *in* Stehr, 1987.)

Biology

The eggs are often laid in large masses and covered by, or mixed with, terminal hair of the female. A frothy secretion from the female, which hardens when exposed to air, coats the eggs in many species. Rarely, as in *Dasychira*, the eggs are deposited singly or in small groups. The larvae feed on the leaves mainly of trees and shrubs, but in some cases herbaceous plants. They tend to be polyphagous (e.g., see Holloway *et al.*, 1987), and many are serious pests (e.g., members of the genera *Lymantria* and *Euproctis*). The secondary setae of *Euproctis chrysorrhoea* are urticating and can lead to dermatitis in man (see Chapter 7).

The resting position of the adult is characteristic, with males often assuming a roughly triangular posture with the wings appressed to the substratum and the 'hairy' forelegs extended forwards (Ferguson, 1978). Sometimes the wings are held in a roof-like position. Females often tend towards flightlessness.

Classification

Studies of the family have been regional, and an understanding of the classification of the family requires analytical examination (Ferguson, 1978). In North America, at least, Lymantriidae fall into two groupings one including *Orgyia* and its relatives and the other *Lymantria* and its relatives.

Phylogenetic relationships

The metathoracic tympanal organs show that Lymantriidae are noctuoids, but their closest relatives within that superfamily are uncertain.

Arctiidae

This large family includes about 11 000 species (Watson & Goodger, 1986) of medium-sized to fairly large Noctuoidea. Many are brightly coloured, aposematic or mimetic species. About 6000 species occur in the Neotropics (Watson & Goodger, 1986) but the family is represented in all the main zoogeographical regions. The composition of the family is controversial and here a fairly inclusive definition is adopted (see *Classification*, below). The larvae are typically densely setose with setae arising from verrucae ringing each segment of the body.

Arctiidae are best defined by the presence of a tymbal organ (see Chapter 6) on the metepisternum of the adult (sometimes lost), and the presence of two subventral (SV) setae on the meso- and metathorax of the larva (Kitching, 1984). Two other possible specializations include the swollen condition of the base of vein Sc in the hindwing, and the presence of a pair of glands situated anterio-dorsally between the ovipositor lobes (Holloway, 1988).

Adult (Figs 314-317)

External ocelli are present or absent. The antennae are bipectinate, ciliated, or simple. The proboscis is usually reduced, but sometimes well-developed, the maxillary palpi are minute, 1-segmented structures, and the labial palpi are typically short, but occasionally long.

The epiphysis is present, and the tibial spur formula is 0-2-4 or 0-2-2. In the forewing, vein Sc is usually separate. Veins R_{2-5} are usually stalked, and sometimes form an areole. Vein M_2 arises much nearer M_3 than M_1, or occasionally it is absent. In the hindwing, $Sc+R_1$ is often swollen basally and usually fuses with Rs to near the middle of the cell. A tymbal organ, sometimes absent or reduced, is present on each metepisternum. Tympanal organs occur in the metathorax, and the tympanic membranes are directed backwards. The forewing-metathorax locking device is sometimes lacking.

The abdomen bears a prespiracular tympanal hood. A pair of glands, usually dichotomously branched, arises dorsally and anteriorly from between the ovipositor lobes.
(Munroe, 1982; Holloway, 1988; Common, 1990.)

Immature stages

The egg is of the upright variety and usually hemispherical. The surface is usually reticulated.

The larva typically bears from 3-5 conspicuous verrucae above the coxa on the meso- and metathorax. The abdomen bears numerous secondary setae arising from verrucae on the body and also from the prolegs. Prolegs occur on A3-6 and A10 and usually bear heteroideous crochets arranged in a mesoseries. Typically, the planta are considerably extended.

The pupa is stout, sometimes brightly coloured, and the cremaster is weak or absent.
(Habeck *in* Stehr, 1987.)

Biology

Eggs are usually laid in clusters, but sometimes singly. The larvae are typically exposed feeders, sometimes living gregariously in webs. A wide variety of foodplants is accepted ranging from epiphytic algae, or the algal component of lichens, to grasses and deciduous trees and shrubs. The secondary setae of the larvae are often barbed. Secondary plant substances, mainly pyrrolizidine alkaloids, but also cardenolides, are frequently sequestered by larvae and stored by pupae and adults (Rothschild, 1985). Many species are aposematic.

The tymbal organs on the metepisternum of the adults are sound producing organs involved in defence against bats (see Chapter 6).

Classification

Great uncertainty exists over the subfamily classification of the Arctiidae. Here, five subfamilies are tentatively accepted, but changes are expected (I.J. Kitching, pers. comm.).

Arctiinae (Fig. 314) (tiger moths) include robust, mainly large Arctiidae with broad forewings and relatively broad hindwings. In the hindwing $Sc+R_1$ is swollen at its base (Nielsen & Common, 1991). The subfamily may not be monophyletic (Holloway,

1988). The species are frequently colourful with red, black and yellow predominating on the wings. Frequently, the abdomen is banded. Some genera (e.g., *Rhodogastria*, *Utetheisa*, and *Spilosoma*) are very widely distributed. A detailed study of the genitalia of Palaearctic and Oriental Arctiinae was presented by Kôda (1987). There are around 2000 species. The larvae (Munroe, 1982; Holloway *et al.*, 1987) feed on a wide variety of plants, particularly herbs, and are often polyphagous. They are densely setose, with setae arranged in tufts from tubercles, and often referred to as woolly bears.

In **Lithosiinae** (Fig. 315) (Munroe, 1982; Holloway *et al.*, 1987), the forewing is narrow but the hindwing is broader and more rounded. The group is widely distributed but mainly tropical. There is an estimated number of over 2000 species. In the forewing, vein R_1 often merges with Sc for most of its length and frequently one of the veins arising from the cell is lost. The larva has a less dense covering of secondary setae than in other Arctiidae. The crochets are arranged as a uniordinal mesoseries. Sometimes the prolegs have extended lobes laterally. Lithosiinae larvae feed on epiphytic algae or on the algal component of lichens (Habeck *in* Stehr, 1987).

Ctenuchinae (*sensu lato*) (Figs 316, 317) are often treated as family Ctenuchidae (= Euchromiidae, Amatidae) and are here taken to include also Syntominae and Euchromiinae (e.g., Watson *et al.*, 1980). They share the specializations of Arctiidae, although in many the metepisternal tymbal organs are absent. It has been suggested (Minet, 1986) that the tymbal organs were never gained in these species and that consequently Ctenuchinae are an unnatural grouping made natural only by removing, as Syntomidae, the species without tymbal organs from Ctenuchinae (and, indeed, from Arctiidae as a whole). However, since the species lacking tymbal organs have typical arctiid dorsal glands it is perhaps more likely that the tymbal organs have been secondarily lost (Holloway, 1988). Moreover, veins Sc and Rs are fused in the hindwing in Ctenuchinae (*sensu lato*), (that is, including Ctenuchinae, Syntominae, and *Euchromia* - the Ctenuchidae of Watson *et al.*, 1980), which is a possible specialization for this assemblage. Ctenuchinae (*sensu lato*) (Munroe, 1982; Holloway *et al.*, 1987) are widely distributed, but best developed in the Neotropics. New World species are narrow-winged and often metallic. In the Old World, the hindwings are usually reduced, the abdomen banded, and the wings often bear white or yellowish semitransparent patches against a black background. The antennae are often white-tipped. Many species are diurnal and thought to mimic wasps. The proboscis is well-developed. There are about 2000 species. Foodplants include grasses and members of Apocynaceae (Habeck *in* Stehr, 1987) mosses and lichens (Bell *in* Holloway, 1988). The larvae (Habeck *in* Stehr, 1987) bear dense brushes of secondary setae mainly on verrucae, with the setae not as long as in typical Arctiinae. The crochets are uniordinal and heteroideous in derived species, but homoideous in primitive members of the group. It is not possible to distinguish with certainty larvae of Ctenuchinae from those of other Arctiidae. Pupation occurs in a cocoon of silk and larval setae. Ctenuchinae (*sensu lato*) may not be monophyletic, and changes in its composition are likely (I.J. Kitching, pers. comm.).

Thyretinae have been placed in Notodontidae on account of their ventrally directed tympanum (Kiriakoff, 1960), but they are now assigned to Arctiidae because they have a tymbal organ (A. Watson *in* Holloway *et al.*, 1987) and typically arctiid larvae (J.E. Rawlins *in* Holloway, 1988). The subfamily (Kiriakoff, 1960; Munroe, 1982) includes about 200 species of medium-sized macrolepidopterans from mainland Africa. The forewings are narrow and the hindwings reduced. The moths are often brightly coloured, mainly red and orange, and some resemble wasps. The larvae (Pinhey, 1975; Munroe, 1982) bear numerous secondary setae, at least sometimes arranged in tufts.

Pericopinae are confined to the New World and occur mainly in the tropics. There are over 300 species (Munroe, 1982), most of which are fairly delicate, but large, Arctiidae. Typically, they are brightly coloured and often sexually dimorphic. Some are apparently aposematic, others mimic various other Lepidoptera. In the hindwing, the subcostal and radial veins are fused for only a short distance. The larvae (Habeck *in* Stehr, 1987) are usually brightly coloured and have only 3 verrucae above the coxa on the mesothorax (as in Ctenuchinae, but not as in Arctiidae). The setae are barbed and, except for those on the prolegs, arise from verrucae. The crochets are arranged in a heteroideous mesoseries. Foodplant families include Fabaceae, Asclepiadaceae, Apocyanaceae, Boraginaceae, and Compositae. The pupa is often brightly coloured and develops in a flimsy cocoon.

Phylogenetic relationships

The composition of the Arctiidae is now fairly well established with the identification of specialized characters (above). The classification within the family, however, is uncertain, as is the relationship of the family to other Noctuoidea.

Noctuidae

The Noctuidae, with an estimated 21000 described, valid species (Holloway *et al.*, 1987), are worldwide in distribution, and particularly well developed in the tropics. The adults are usually robust-bodied, small to very large macrolepidopterans. Most are relatively drab, grey-brown moths, but many are brightly coloured. The larvae of certain species are serious pests, sometimes causing extensive loss of crops as defoliators, borers into plant tissue, or cutworms.

The concept of the family is uncertain. It appears likely that the Nolinae complex and the Pantheinae will be excluded from the family in the future (I.J. Kitching pers. comm.). The subfamily classification is also unresolved, and a new synthesis is in preparation (I.K. Kitching & J.E. Rawlins).

Adult (Figs 318-321)

External ocelli are usually present and chaetosemata are absent. With few exceptions, the proboscis is well developed. The antennae are typically filiform or serrated, but sometimes they are pectinate. The maxillary palpi are minute and 1-segmented, and the labial palpi are porrect or ascending.

The epiphysis is present, and the tibial spur formula is 0-2-4. In the forewing, an areole is usually present. Vein M_1 arises either from the point at which R_{4+5} arises, or from close to the upper corner of the cell. In the hindwing, $Sc+R_1$ is fused for a short length with Rs before it diverges. Vein M_2 may arise about halfway between M_1 and M_3 causing the cubital system to appear 3-branched (the trifine condition), or the base of vein M_2 approximates that of M_3, so that the cubital system appears 4-branched (quadrifine condition). The wings are coupled by a frenulum and a retinaculum. The metathoracic-forewing locking device is present, and the wings are folded over the body when the moth is at rest.

At the base of the abdomen there is a swelling termed a counter-tympanal hood, typically postspiracular, sometimes prespiracular, and sometimes double. A secondary 'tympanal hood', situated anterio-lateral to the counter-tympanal hood, is actually a flattened pouch termed a pleural pouch (Kitching, 1987). The ovipositor is usually composed of short, fleshy lobes, but is sometimes wedge-shaped for laying eggs into crevices.

(Munroe, 1982; Kitching, 1984; Common, 1990.)

Immature stages

The egg (e.g., Döring, 1955; Salkeld, 1984) is of the upright variety, usually domed with conspicuous vertical ribs, and sometimes with weaker horizontal ridges.

The larva (Godfrey *in* Stehr, 1987) is cylindrical, typically lacks secondary setae, and, usually, has a full complement of well-developed prolegs with crochets arranged in a homoideous mesoseries. However, there are many exceptions. For example, the prolegs may be reduced in number or size. The crochets are usually uniordinal, less frequently biordinal. The body is variously marked, often with longitudinal stripes.

The pupa (Mosher, 1916) bears hooked setae on the cremaster (unlike the condition in Arctiidae). Larval setae are not incorporated into the cocoon.

Biology

The eggs may be laid singly or in clusters. The larvae are typically phytophagous, sometimes carnivorous, and, rarely, coprophagous. Plant feeding habits vary considerably within the family. Some species are external feeders on leaves, others bore into fruits, stems, and roots. Leaf-mining is rare. Various species (cutworms) cut leaves and stems. Those species with reduced prolegs are called semi-loopers, a term reflecting their mode of locomotion. A few species construct nests from folded leaves of their foodplant.

Classification

The history of noctuid classification is involved and contentious (Kitching, 1984). The family has long been divided into trifine Noctuidae and quadrifine Noctuidae on the basis of the appearance of the cubital vein system (see above), but although the trifines are very likely to be monophyletic, the quadrifines are not (Kitching, 1984).

Both Herminiinae and Aganainae are here placed within Noctuidae although they are frequently treated as having family rank. Herminiinae were excluded from Noctuidae because of their prespiracular (rather than postspiracular) counter tympanal hood. Aganainae were frequently excluded because the prespiracular counter tympanal hood was said to be double. However, both these statements are dubious (I.J. Kitching pers. comm.). Moreover, both groups share with Noctuidae two characters that may be specialized (Minet, 1986). First, the anterio-lateral extensions of abdominal sternum A2 are each united with a corresponding latero-tergal sclerite by a dorsally enlarged, postspiracular bar. The ventral part of this connection is occasionally lost secondarily. Second, the counter tympanal cavities are so close to one another that they appear to be largely contiguous.

Pantheinae (Fig. 318) are a poorly defined and probably unnatural group of quadrifines with eyes bearing long interfacetal hairs (Holloway, 1985a). They also have a reduced tympanal hood. Secondary setae are present on the larval head and (as in many Acronictinae) on the body. They are likely to be removed from Noctuidae into a separate family in the future (I.J. Kitching, pers. comm.).

The Nolinae complex includes Nolinae (Fig. 319), Chloephorinae, and Sarrothripinae. All have a boat-shaped cocoon and a bar-shaped retinaculum. It is by no means certain that the three subfamilies are monophyletic, nor even that the complex is correctly placed in Noctuidae. Nolinae are cosmopolitan and include small to very small Noctuidae in which external ocelli are not present, and the larvae bear tufted setae. The moths are fairly drab, and have been placed in various non-noctuid families, most frequently Arctiidae, or assigned to a family of their own. Nolinae and Sarrothripinae also have tufts of raised scales on the cell of the forewing, although this feature is found outside these groups in other Noctuidae (Kitching, 1984). The first pair of ventral prolegs is absent from the larvae. Chloephorinae include small Noctuidae with most species occurring in the Old World tropics. They are arguably subordinate to Sarrothripinae (Kitching, 1984). Chloephorinae differ from Sarrothripinae in lacking scale-tufts on the cell of the forewing. The abdominal sternum is modified into a pair of pouches (possibly sound-producing organs) in at least some species (Holloway, 1988).

Most Sarrothripinae occur in the Old World tropics. The moths are generally small to medium-sized and rather drab Noctuidae. They bear raised scales on the cell of the forewing as in Nolinae. Larvae of some species bore into pods of Fabaceae. Several South African Sarrothripinae have been reared from flower and seed-pod galls on *Acacia karoo* (Fabaceae) (C.H. Scholtz & M.J. Scoble, unpublished).

Aganainae (=Hypsinae) have been included either in Arctiidae, a family to which they bear a strong resemblance in colour pattern, or they have been treated as a separate family. The absence of tymbal organs from the metathorax indicates that aganaines are not arctiids. However, the presence of a bar-shaped retinaculum suggests a possible relationship. The labial palpi of Aganainae are long and upwardly pointed (e.g., Common, 1970). The group is primarily tropical. Eggs of Aganainae are laid in clusters and the larvae are sparsely covered with long hairs and are often brightly coloured. Pupation occurs in a weak cocoon.

Herminiinae include mainly relatively small, drab Noctuidae occurring in the New World tropics. Herminiinae have been variously assigned to Hypeninae, treated as an independent noctuid subfamily, or given full family status. Unlike other Noctuidae, they have a pre-spiracular tympanal hood. Herminiinae (see Owada, 1987) have large, and porrect or upturned labial palpi with the second segment being sickle-shaped or long and blade-like, and the third segment long and acute or blade-like. Secondary sexual organs in the male are diverse in form and location. Antennal modifications, confined to the male, include, in some species, a series of thickened flagellar segments. Since long scales may be present on the flagellomeres, the function of the thickenings may be olfactory. Alternatively, the modifications may be used in some way to hold the female during mating. Various tufts and brushes are found on the labial palpi; sometimes the tufts are exposed, sometimes they are concealed. Various specialized scales are contained within costal folds of the forewing. In the foreleg, complex modifications occur often so that scale tufts may be concealed. The larvae are characterized by the presence of five pairs of well developed prolegs, compared with related subfamilies where the first two pairs are nearly always reduced or lost (Owada, 1987). Typically, the larvae feed on dead leaves of dicotyledonous angiosperms or conifers, but occasionally they consume live plant tissue (e.g., see Owada, 1987).

Hypeninae (including Rivulinae) may not be monophyletic. They have long palpi and 'lashed' eyes (Forbes, 1954), but these features also occur outside the subfamily. The moths are small, brown or grey Noctuidae and *Hypena* and its relatives have long, porrect palpi, giving the moths their colloquial name of 'snouts'. The larva has setae borne on raised chalazae (Gardner, 1946). The larvae are green semiloopers, which generally feed high up on plants (Forbes, 1954). However, larvae of *Osericana gigantalis* from the Afrotropical region, feed on grass in termite nests (Pinhey, 1975). The subfamily is widely distributed. *Hypena* is represented worldwide except for South America.

Catocalinae (Fig. 320) are here taken to include Ophiderinae since the distinction between them cannot be upheld (e.g., Berio, 1959; Kitching, 1984). Catocalinae, as accepted here, include about 10 000 species, a figure approaching half that of the total number in Noctuidae. There exist several monophyletic groups within Catocalinae *sensu lato*, and *Calpe* and its relatives are likely to be moved to their own subfamily (I.J. Kitching, pers. comm.). Most Catocalinae (e.g., Holloway *et al.*, 1987; Common, 1990) have broad wings and upturned labial palpi. In many, the tibiae are spined, and a hairpencil occurs on the midtibia. The larvae are semiloopers, with one or more pairs of prolegs reduced. Often, the pupa has a whitish bloom. The subfamily includes several fruit-piercing moths and the skin-piercing and blood-sucking moths (see Chapter 2).

Euteliinae (Holloway, 1985a; Kitching, 1987; Holloway *et al.*, 1987) are mainly tropical, and tend to be associated with savanna and semi-arid areas. The moths are often colourful. Adult resting posture is characteristic: the wings are folded longitudinally, slanted backwards, and held well down against the substrate, while the abdomen is curved sharply upwards. A pair of characteristic lenticular flanges occur on the basal sternum (Holloway, 1985a). The larvae are smooth and bear four pairs of prolegs. Their foodplants frequently belong to Anacardiaceae (including mango, *Mangifera*: Anacardiaceae). Pupation occurs in a cocoon in the soil or in rotten wood.

Stictopterinae (Holloway, 1985a; Holloway *et al.*, 1987), a tropical subfamily of predominantly forest dwellers, are almost certainly monophyletic. The forewings of the adults tend to be reticulate and somewhat metallic, and the hindwings typically have a large translucent patch medially to basally. The frenulum of the female is generally reduced to a single spine. The ovipositor lobes of the female bear numerous setae, and the margins of the lobes appear 'frayed'. At rest the wings are folded at an acute angle or rolled around the body, an arrangement resembling a broken twig. The larva has a complete set of prolegs and pupates in a cocoon in the soil, and the cremaster of the pupa is Y-shaped. Two lineages were recognized by Holloway (1985a). In the *Stictoptera/Gyrtona* group, the valvae of the males are divided longitudinally into a pair of slender arms, and costal processes are generally absent. The limited records available suggest that this group may be associated with plants of the family Guttiferae. In the other group (*Lophoptera* and its relatives), sternum A8 of the female is more strongly developed than the corresponding tergum, and the apophyses are short and broad rather than long and slender. There is an association with Dipterocarpaceae. A well-corroborated sister group relationship exists between Stictopterinae and Euteliinae (Kitching, 1987).

Acronictinae have a trifine wing venation, but lack the characteristic hairpencils at the base of the male abdomen in trifine Noctuidae. Larvae of most species have secondary setae on the body. The moths are greyish, cryptically coloured Noctuidae. Although Acronictinae are cosmopolitan, most species occur in temperate regions.

Plusiinae (Eichlin & Cunningham, 1978; Holloway, 1985a; Kitching, 1987) are distributed worldwide. They are found in tropical, temperate and even arctic regions. Well represented in Africa and the eastern Palaearctic, they are apparently not rich in South America. Most species occur in tropical montane and temperate areas. They frequently bear metallic silver, and sometimes gold markings on the forewings. Many species fly during the day and night, and adults have often been seen visiting flowers. Plusiinae seem not to be strongly attracted to light. Most larvae are semiloopers with prolegs on segments A3 and A4 vestigial or missing. Generally, they are green with white stripes. Pupation occurs in a cocoon usually spun between leaves. The larvae of many species are pests (as defoliators) of crops, vegetables, greenhouse plants and ornamentals (Eichlin & Cunningham, 1978). A very great diversity of plants serve as plusiine hosts.

Traditionally, Plusiinae have been regarded as quadrifines with lashed eyes, but several other related groups have lashed eyes, and one plusiine (*Mouralia*) is trifine. However, four characters are both unique to the subfamily and do not revert to the primitive state in the more advanced members (Kitching, 1987). These characters include the presence of a convex occiput; few apical styloconic sensilla on the proboscis; a proboscis with semicircular strengthening bars on the most apical portion; and anal papillae with dorsoventral rows of setae in the female genitalia. Several other characters, more easily visible but prone to exceptions, also occur. An eversible dorsal pouch is found between the anal papillae and A8. The structure is lost from Abrostolini. Triangular pilifers, a long thin proboscis with smooth styloconic sensilla, and a long thoracic crest are features that although found throughout Plusiinae, occur in at least one other related group. Other characters present in the plusiine groundplan are both present in certain related groups and are also prone to transformation within the subfamily. Of these, vein M_2 of the hindwing is half as strong as M_3; flanges are present on the sacculus of the valva; and a clavus is borne on a peninsula on the sacculus.

Trifine Noctuidae have been traditionally associated by the so-called trifid venation, where vein M_2 of the hindwing is lost. However, there are exceptions where the quadrifid venation occurs. Stronger support for the monophyly of the trifines comes from the presence of hairpencils at the base of the abdomen arising on sclerotized levers and associated with scent glands (see Chapter 6) (Holloway, 1989; Lafontaine & Poole, *in press*). A further character uniting the trifines is the presence of a corona of stout setae at the apex of the male valva (Lafontaine & Poole, *in press*). Both these characters are often secondarily reduced or lost.

Acontiinae are small to very small, often robust Noctuidae with broad, rounded wings. The group is not monophyletic. However, within Acontiinae, tribe Acontiini is likely to be so on the basis of the enlargement and sclerotization of the alula and the consequent reduction or loss of the tympanal hood, and the presence of a pair of anal hair-masses (Kitching, 1984). *Eublemma* (Eublemmini) is a large genus of small moths found mainly in the tropics of the Old World. Larvae of a few species prey on scale insects and other Homoptera, for example the widespread species *E. scitula*, which occurs in Africa, and parts of Europe and Asia. Those of *Catoblemma* from Australia also feed on scale insects. Larvae of *Enispa*, also from Australia, live in the webs of spiders and feed on detritus and the remains of insects caught in spider webs (Common, 1990). The adults of some species of *Acontia* resemble bird droppings.

Agaristinae include about 300 species and are predominantly tropical. The subfamily is probably monophyletic, and has often been given family status (Kitching, 1984). Characters that appear to be specializations include a counter-tympanum several times the size of the tympanal membrane; a reduction or absence of the counter-tympanal hood; and adults that are often diurnal, usually brightly coloured, and with clubbed antennae. The larvae are also usually colourful. In general appearance, Agaristinae resemble Arctiidae (tiger moths) and are often called false-tigers.

Adults and larvae of many species are probably aposematic. Several species are mimetic.

Stiriinae are probably closely related to Heliothinae (Hardwick, 1970; Matthews, 1991). The evidence for this close relationship, according to Kitching (1984), is the occurrence of a clawed foretibia, and the choice of flowers and young fruits as a food source by the larvae. Stiriinae (*sensu* Matthews, 1991) occur in North America, Africa, and the Palaearctic. The monophyly of the group is supported by the strongly flattened, scale-like form of the larval spinneret (Matthews, 1991).

Heliothinae are best characterized by larval features (Matthews, 1991): the presence of spiny skin and the transverse, as opposed to vertical, alignment of setae L_1 and L_2 on the prothorax (i.e., L_1 and L_2 are approximately in a horizontal line). The larvae feed almost exclusively on flowers, fruits, and seeds. The group includes many very important pests, for example, those of the corn-earworm complex (*Helicoverpa*, see Hardwick, 1965). *H. armigera* feeds in pods and buds of beans and peas, bores in tobacco stems, ears of corn, in cotton, maize and numerous other plants - many of which are of commercial importance. The species is widespread. The North American heliothine genera were studied by Hardwick (1970) who emphasized the importance of the male genitalia in their diagnosis.

Amphipyrinae are probably not monophyletic, but consist of trifine Noctuidae that lack hairy or 'lashed' eyes, tibial spines, and spiny larvae (Kitching, 1984; Holloway, 1989). The moths vary considerably in colour and size. Larvae of the genus *Spodoptera* are serious pests on a large number of farm crops and garden plants.

Cuculliinae are probably polyphyletic, although some of the constituent tribes are likely to be monophyletic (Kitching, 1984). Cuculliinae are predominantly temperate, although the type genus (*Cucullia*) is mainly represented in hotter areas. The moths are often cryptically coloured. Apart from *Cucullia*, these insects tend to be spring or autumnal fliers (Holloway *et al.*, 1987).

Hadeninae are Noctuidae with hairy eyes and unspined tibiae, but there is no convincing specialized character demonstrating their monophyly. The subfamily is cosmopolitan. Glottulini are peculiar and include species with brightly coloured larvae that feed on bulbs of Amaryllidaceae (see Kitching, 1984). Larvae of *Diaphone eumelea* from South Africa feed on flower buds, fruits and ovaries of *Thuranthos macranthum* (Liliaceae), and the moths pollinate the night-opening flowers of this plant (Stirton, 1976).

Noctuinae (Fig. 321) are a grouping the monophyly of which has not been adequately demonstrated (Kitching, 1984). However, two features seem to be diagnostic of the group. One is that, at rest, the moths hold their wings horizontally over the body, each in the same plane, rather than roof-like. The other is the presence of rows of spines on the hindtibiae. The subfamily includes generally small to medium-sized Noctuidae. The forewings are typically narrow with the termen squared off. Significant pests include the cutworms (e.g., *Agrotis*), noctuine larvae that feed on stems near the ground consequently destroying all the aerial parts of the plant.

Figs 186–193. *186, MICROPTERIGIDAE:* Micropterix schaefferi; *187, AGATHIPHAGIDAE:* Agathiphaga vi-tiensis; *188, HETEROBATHMIIDAE:* Heterobathmia *spp; 189, ERIOCRANIIDAE:* Eriocrania sparmannella; *190, ACATHOPTEROCTETIDAE:* Acanthopteroctetes unifascia; *191, LOPHOCORONIDAE:* Lophocorona pediasia; *192, NEOPSEUSTIDAE:* Neopseustis meyricki; *193, MNESARCHAEIDAE:* Mnesarchaea acuta. *(Scales: 3 mm.)*

Figs 194–201. *194, HEPIALIDAE:* Sthenopis argenteomaculatus; *195, HELIOZELIDAE:* Antispila pfefferella; *196, ADELIDAE:* Nemophora barbarella; *197, CRINOPTERIGIDAE:* Crinopteryx barbatella; *198, INCURVAR-IIDAE:* Incurvaria praelatella; *199, CECIDOSIDAE:* Cecidoses eremita; *200, PRODOXIDAE:* Tegeticula yuccasella; *201, NEPTICULIDAE:* Stigmella aurella. (*Scales: 194, 10 mm; 195-200, 3 mm; 201, 1 mm.*)

Figs 202–209. *202, OPOSTEGIDAE:* Opostega auritella; *203, PALAEPHATIDAE:* Palaephatus fusciterminus; *204, TISCHERIIDAE:* Tischeria complanella; *205, TINEIDAE:* Trichophaga cuspidata; *206, ERIOCOTTIDAE:* Eriocottis fuscanella; *207, ACROLOPHIDAE:* Acrolophus doeri; *208, PSYCHIDAE:* Oiketicoides febretta; *209, GALACTICIDAE:* Homodaula anisocentra. (*Scales: 3 mm.*)

Figs 210–217. *210, GRACILLARIIDAE:* Acrocercops resplendens; *211, BUCCULATRICIDAE:* Bucculatrix ala-ternella; *212, ROESLERSTAMMIIDAE:* Roeslerstammia erxlebella; *213, DOUGLASIIDAE:* Tinagma dryadis; *214, YPONOMEUTIDAE:* Yponomeuta paurodes; *215, YPSOLOPHIDAE:* Ochsenheimeria bubalella; *216, HELIODINIDAE:* Heliodines nodosella; *217, LYONETIIDAE:* Lyonetia ledi. (*Scales: 210-212, 214-217, 3 mm; 213, 1 mm.*)

Figs 218–225. *218, GLYPHIPTERIGIDAE:* Glyphipteryx haworthana; *219-225, OECOPHORIDAE: 219, DEPRESSARIINAE:* Depressaria multifidae; *220, ETHMIINAE:* Ethmia nigroapicella; *221, PELEOPODINAE:* Peleopoda obiterella; *222, AUTOSTICHINAE:* Autosticha demotica; *223, XYLORYCTINAE:* Uzucha humeralis; *224, STENOMATINAE:* Stenoma lutifica; *225, OECOPHORINAE:* Tonica effractella. (*Scales: 3 mm.*)

Figs 226–233. *226, LECITHOCERIDAE:* Timyra crassella; *227, ELACHISTIDAE:* Elachista argentella; *228, PTEROLONCHIDAE:* Pterolonche pulverulenta; *229, COLEOPHORIDAE:* Coleophora trifolii; *230, AGONOX-ENIDAE:* Agonoxena argaula; *231, BLASTOBASIDAE:* Blastobasis decolorella; *232, MOMPHIDAE:* Mompha laspeyrella; *233, SCYTHRIDIDAE:* Scythris punctivittella. (*Scales: 3 mm.*)

Figs 234–241. *234, COSMOPTERIGIDAE:* Cosmopterix sapporensis; *235, GELECHIIDAE:* Teliopsis bagriotella; *236, COSSIDAE,* Xyleutes eucalypti; *237, DUDGEONEIDAE:* Dudgeonea actinias; *238, RATARDIDAE:* Ratarda furvivestita; *239, TORTRICIDAE:* Dynatocephala omophaea; *240, CASTNIIDAE:* Xanthocastnia evalthe; *241, SESIIDAE:* Aegeria tibiale. (*Scales: 234, 235, 239, 3 mm; 236, 240, 10 mm; 237, 238, 241, 5 mm.*)

Figs 242–249. *242, BRACHODIDAE:* Atychia appendiculata; *243, CHOREUTIDAE:* Choreutis punctosa; *244, ZYGAENIDAE:* Zygaena occitanica; *245, HETEROGYNIDAE:* Heterogynis *sp.;* *246, MEGALOPYGIDAE:* Podalia pedacia; *247, LIMACODIDAE:* Parasa consocia; *248, DALCERIDAE:* Acraga moorei; *249, EPIPYROPIDAE:* Heteropsyche *sp.* (*Scales: 3 mm.*)

Figs 250–257. *250, CYCLOTORNIDAE:* Cyclotorna monocentra; *251, CHRYSOPOLOMIDAE:* Chrysopoloma isabellina; *252, IMMIDAE:* Imma albifasciata; *253, COPROMORPHIDAE:* Copromorpha efflorescens; *254, CARPOSINIDAE:* Peragrarchis *sp.; 255, SCHRECKENSTEINIIDAE:* Schreckensteinia festaliella; *256, URODIDAE:* Urodus parvula; *257, EPERMENIIDAE:* Epermenia chaerophyllella. *(Scales: 3 mm.)*

Figs 258–265. *258, ALUCITIDAE:* Alucita hexadactyla; *259, TINEODIDAE:* Anomima phaeochroa; *260, PTEROPHORIDAE:* Stenoptilia pnemonanthes; *261, HYBLAEIDAE:* Hyblaea flavifasciata; *262, THYRIDIDAE:* Macrogonia igniaria; *263-265, PYRALIDAE: 263, PYRALINAE:* Pyralis manihotalis; *264, PHYCITINAE:* Phycita coronatella; *265 CRAMBINAE:* Crambus magnificus. (*Scales: 3 mm.*)

Figs 266–273. *266, 267, PYRALIDAE: 266, NYMPHULINAE:* Aulacodes heptopis; *267, SPILOMELINAE:* Spoladea recurvalis; *268–270, GEOMETRIDAE: 268, ARCHIEARINAE:* Archiearis parthenias; *269, ENNOMINAE:* Epimecis hortaria; *270, OENOCHROMINAE (s. str.):* Oenochroma vinaria; *271–273: URANIIDAE: 271, URANIINAE:* Chrysiridia ripheus; *272, EPIPLEMINAE:* Epiplema himala; *273, SEMATURIDAE:* Sematura empedocles. *(Scales: 266–270, 272, 3 mm; 271, 273, 10 mm.)*

Figs 274–281. *274, EPICOPEIIDAE:* Épicopeia *sp.; 275-277: DREPANIDAE: 275, DREPANINAE:* Drepana arcuata; *276, THYATIRINAE:* Thyatira batis; *277, CYCLIDIINAE:* Cyclidia fractifasciata; *278, AXIIDAE:* Axia margarita; *279, CALLIDULIDAE:* Callidula *sp.; 280, PTEROTHYSANIDAE:* Pterothysanus lanaris; *281, HEDYLIDAE:* Macrosoma conifera. (*Scales: 274, 277, 10 mm; 275, 276, 278-281; 3 mm.*)

Figs 282–289. *282, HESPERIIDAE:* Calpodes ethlias; *283, PAPILIONIDAE:* Iphiclides podalirius; *284, PIERI-DAE:* Colias eurythme; *285-289, NYMPHALIDAE: 285, SATYRINAE:* Ethope himachala; *286, NYMPHALINAE:* Vanessa atalanta; *287, HELICONIINAE:* Euides isabella; *288, ACRAEINAE:* Acraea zetes; *289, ITHOMIINAE:* Patricia dercyllidas. (*Scales: 5 mm.*)

Figs 290–297. *290, LIBYTHEINAE (NYMPHALIDAE):* Libythea carineata; *291-292, LYCAENIDAE: 291, THE-CLINAE:* Jalmenus evagoras; *292, RIODININAE:* Ancyluris aristodorus; *293, LASIOCAMPIDAE:* Entometa apicalis; *294, ANTHELIDAE:* Anthela ocellata; *295, LEMONIIDAE:* Lemonia dumi; *296, EUPTEROTIDAE:* Eupterote undans; *297, APATELODIDAE:* Apatelodes pandarioides. (*Scales: 5 mm.*)

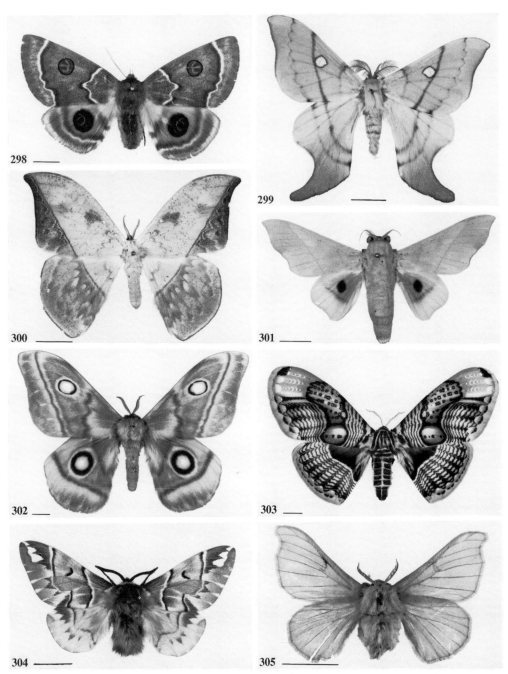

Figs 298–305. *298, CARTHAEIDAE:* Carthaea saturnioides; *299, CERCOPHANIDAE:* Cercophana venusta; *300, OXYTENIDAE:* Oxytenis modestia; *301, 302, SATURNIIDAE: 301, HEMILEUCINAE:* Syssphinx molina; *302, SATURNIINAE:* Imbrasia cytheraea; *303, BRAHMAEIDAE;* Brahmaea wallichii; *304, ENDROMIDAE:* Endromis versicolor; *305, BOMBYCIDAE:* Bombyx mori. (*Scales: 10 mm.*)

Figs 306–313. *306, SPHINGIDAE:* Manduca calapagensis; *307, MIMALLONIDAE:* Mimallo amilia; *308, OENOSANDRIDAE,* Oenosandra; *309, DOIDAE:* Doa raspa; *310-312, NOTODONTIDAE: 310, THAUMETO-POEINAE:* Anaphe; *311, NOTODONTINAE:* Pheosia rimosa; *312, DIOPTINAE:* Phanoptis cyanomelas; *313, LYMANTRIIDAE:* Aroa discalis. (*Scales: 306, 10 mm; 307-313, 5 mm.*)

Figs 314–321. 314-317, *ARCTIIDAE*: 314, *ARCTIINAE*, Aphyle affinis; 315, *LITHOSIINAE*, Eilema lurideola; 316, 317, *CTENUCHINAE*, (316) Syntomis marjana; (317) Ctenucha refulgens; 318-321, *NOCTUIDAE*: 318, *PANTHEINAE*: Panthea coenobita; 319, *NOLINAE*: Nola cucullatella; 320, *CATOCALINAE*: Othreis fullonia; 321, *NOCTUINAE*, Agrotis segetum. (*Scales: 5 mm.*)

REFERENCES

Ackery, P.R. 1984. Systematics and faunistic studies on butterflies. *In* Vane-Wright, R.I. and Ackery, P.R. (Eds). *The Biology of Butterflies.* pp. 9-21. Academic Press, London.

Ackery, P.R. 1988. Hostplants and classification: a review of nymphalid butterflies. *Biological Journal of the Linnean Society* 33: 95-203.

Ackery, P.R. & Vane-Wright, R.I. 1984. *Milkweed butterflies: their cladistics and biology.* ix + 425 pp. British Museum (Natural History), London.

Acocks, J.P.H. 1975. Veld types of South Africa. *Memoirs of the Botanical Survey of South Africa* 40: 128 pp. + 5 maps. 2nd edition.

Adamski, D. & Brown, R.L. 1989. Morphology and systematics of North American Blastobasidae [Lepidoptera: Gelechioidea]. *Mississippi Agricultural and Forestry Experiment Station Technical Bulletin* 165: iii + 70 pp.

Adler, P.H. 1982. Soil- and puddle-visiting habits of moths. *Journal of the Lepidopterists' Society* 36: 161-173.

Agassiz, D.J.L. 1985. Douglasiidae. *In* Heath J. & Emmet, A.M. (Eds). *The Moths and Butterflies of Great Britain and Ireland,* 2: 408-409. Harley Books, Colchester.

Agee, H.R. 1971. Ultrasound production by wings of adults of *Heliothis zea. Journal of Insect Physiology* 17: 1267-1273.

Aiello, A. *In press.* Nocturnal butterflies in Panama. Hedylidae (Lepidoptera: Rhopalocera). *In* Quintero, D. and Aiello, A. (Eds). *Insects of Panama and Mesoamerica: Selected studies.* Oxford University Press.

Alberti, B. 1954. Über die stammesgeschichtliche Gliederung der Zygaenidae nebst Revision einiger Gruppen (Insecta, Lepidoptera). *Mitteilungen aus dem Zoologischen Museum der Humboldt-Universität Berlin* 30: 115-480.

Allyn, A.C., Barbier, M., Bois-Choussy, M. & Rothschild, M. 1981. Have the bile pigments in the wing membranes of butterflies special function? *Antenna* 5: 29-31.

Allyn, A.C., Rothschild, M., & Smith, D.S. 1982. Microstructure of blue/green and yellow pigmented wing membranes in Lepidoptera with remarks concerning the function of pterobilins. *Bulletin of the Allyn Museum* 75: 20 pp.

Aplin, R.T. & Rothschild, M. 1972. Poisonous alkaloids in the body tissues of the Garden Tiger moth (*Arctia caja* L.) and the Cinnabar moth (*Tyria* (= *Callimorpha*) *jacobaeae* L.) (Lepidoptera). *In* Vries, A. de and Kochva, K. (Eds). *Toxins of animal and plant origin.* pp. 579-595. Gordon and Breach, London.

Arms, K., Feeny, P. & Lederhouse, R.C. 1974. Sodium: stimulus for puddling behaviour by Tiger Swallowtail butterflies, *Papilio glaucus. Science* 185: 372-374.

Arocha-Pinaǹgo, C.L. & Layrisse, M. 1969. Fibrinolysis produced by contact with a caterpillar. *Lancet* 7599: 810-812.

Atkinson, P.R. 1980. On the biology and natural host-plants of *Eldana saccharina* Walker (Lepidoptera: Pyralidae). *Journal of the Entomological Society of southern Africa* 43: 171-194.

Aurivillius, C. 1927. Family: Lasiocampidae. *In* Seitz, A.A. (Ed.). *The Macrolepidoptera of the World* 14: 205-281.

Bailey, W.J. 1978. Resonant wing systems in the Australian whistling moth *Mecatesia* (Agarisidae [*sic*], Lepidoptera). *Nature* 272: 444-446.

Baker, R.R. 1978. *The Evolutionary Ecology of Animal Migration.* xvii + 1012 pp. Hodder & Stoughton, London.

Baker, R.R. 1984. The Dilemma: When and how to go or stay. *In* Vane-Wright, R.I. and Ackery, P.R. (Eds). *The Biology of Butterflies.* pp. 279-296. Academic Press, London.

Bänziger, H. 1970. The piercing mechanism of the fruit-piercing moth *Calpe* [*Calyptra*] *thalictri* Bkh. (Noctuidae) with reference to the skin-piercing blood-sucking moth *C. eustrigata* Hmps. *Acta Tropica* 27: 54-88.

Bänziger, H. 1971. Extension and coiling of the lepidopterous proboscis - a new interpretation of the blood-pressure theory. *Mitteilungen der schweizerischen Entomologischen Gesellschaft* 43: 225-239.

Bänziger, H. 1973. Biologie der lacriphagen Lepidopteren in Thailand und Malaya. *Revue Suisse de Zoologie* 79: 1381-1469.

Bänziger, H. 1975. Skin-piercing blood-sucking moths I: ecological and ethological studies on *Calpe eustrigata* (Lepid., Noctuidae). *Acta Tropica* 32: 125-144.

Bänziger, H. 1979. Skin-piercing blood-sucking moths II: studies on a further 3 adult *Calyptra* [*Calpe*] sp. (Lepid., Noctuidae). *Acta Tropica* 36: 23-37.

Bänziger, H. 1980. Skin-piercing blood-sucking moths III: feeding act and piercing mechanism of *Calyptra eustrigata* (Hmps.) (Lep., Noctuidae). *Mitteilungen der schweizerischen entomologischen Gesellschaft* 53: 127-142.

Bänziger, H. 1982. Fruit-piercing moths (Lep., Noctuidae) in Thailand: a general survey and some new perspectives. *Mitteilungen der schweizerischen entomologischen Gesellschaft* 55: 213-240.

Bänziger, H. 1983. Lachryphagous Lepidoptera recorded for the first time in Laos and China. *Mitteilungen der schweizerischen entomologischen Gesellschaft* 56: 73-82.

Bänziger, H. 1988a. Lachryphagous Lepidoptera recorded for the first time in Indonesia (Sumatra) and Papua New Guinea. *Heterocera Sumatrana* 2: 133-144.

Bänziger, H. 1988b. The heaviest tear drinkers: ecology and systematics of new and unusual notodontid moths. *Natural History Bulletin of the Siam Society* 36: 17-53.

Bänziger, H. 1988c. Unsuspected tear drinking and anthropophily in thyatirid moths, with similar notes on sphingids. *Natural History Bulletin of the Siam Society* 36: 117-133.

Bänziger, H. 1989a. A persistent tear drinker: notodontid moth *Poncetia lacrimisaddicta* sp. n., with notes on its significance to conservation. *Natural History Bulletin of the Siam Society* 37: 31-46.

Bänziger, H. 1989b. Skin-piercing blood-sucking moths V: Attacks on man by 5 *Calyptra* spp. (Lepidoptera, Noctuidae) in S and SE Asia. *Mitteilungen der schweizerischen entomologischen Gesellschaft* 62: 215-233.

Bänziger, H. & Fletcher, D.S. 1985. Three new zoophilous moths of the genus *Scopula* (Lepidoptera: Geometridae) from South-east Asia. *Journal of Natural History* 19: 851-860.

Barlow, H.S. 1982. *An Introduction to the Moths of South East Asia*. ix+305 pp. + pls 1-50. The Malayan Nature Society, Kuala Lumpur, and E.W. Classey, Faringdon, U.K.

Barth, F.G. 1985. *Insects and Flowers: the biology of a partnership*. ix+297 pp. Allen & Unwin, London and Sydney.

Barton Browne, L., Soo Hoo, C.F., van Gerwen, A.C.M., & Sherwell, I.R. 1969. Mating flight behaviour in three species of *Oncopera* moths (Lepidoptera: Hepialidae). *Journal of the Australian Entomological Society* 8: 168-172.

Becker, V.O. 1977. The taxonomic position of the Cecidosidae (Brèthes) (Lepidoptera). *Polskie Pismo Entomologiczne* 47: 79-86.

Bengtsson, B.A. 1984. The Scythrididae (Lepidoptera) of Northern Europe. *Fauna Entomologica Scandinavica* 13: 1-137.

Bennett, A.L. 1989. On the mechanism of ultrasound production in *Galleria mellonella* (L.) (Lepidoptera: Pyralidae). *Journal of the entomological Society of southern Africa* 52: 317-323.

Berenbaum, M. 1983. Coumarins and caterpillars: a case for coevolution. *Evolution* 37: 163-179.

Berg, K. 1941. Contributions to the biology of the aquatic moth *Acentropus niveus* (Oliv.). *Videnskabelige Meddelelser fra Dansk naturhistorisk Forening* 105: 59-139 + pl. 2.

Berio, E. 1959. Studi sulla systematica delle cosiddette 'Catocalinae' e 'Othreinae' (Lepidoptera, Noctuidae). *Annali del Museo Civico di Storia Naturale di Genova* 71: 276-327.

Berry, R.J. 1990. Industrial melanism and peppered moths (*Biston betularia* (L.)). *Biological Journal of the Linnean Society* 39: 301-322.

Betts, C.R. & Wootton, R.J. 1988. Wing shape and flight behaviour in butterflies (Lepidoptera: Papilionoidea and Hesperioidea): a preliminary analysis. *Journal of Experimental Biology* 138: 271-288.

Birch, M.C. 1970a. Structure and function of the pheromone-producing brush-organs in males of *Phlogophora meticulosa* (L.) (Lepidoptera: Noctuidae). *Transactions of the Royal Entomological Society of London* 122: 277-292.

Birch, M.C. 1970b. Pre-courtship use of abdominal brushes by the nocturnal moth, *Phlogophora meticulosa* (L.) (Lepidoptera: Noctuidae). *Animal Behaviour* 18: 310-316.

Birch, M.C. 1970c. Persuasive scents in moth sex life. *Natural History* 79: 34-39, 72.

Birch, M.C. 1972. Male abdominal brush-organs in British noctuid moths and their value as a taxonomic character. *The Entomologist* 105: 185-205.

Birch, M.C. 1974. Aphrodisiac pheromones in insects. *In* Birch, M.C. (Ed.). *Pheromones. Frontiers of Biology* 32: 115-134. North Holland/Amsterdam.

Birch, M.C. & Haynes, K.F. 1982. *Insect Pheromones.* (*Studies in Biology* 147) 60 pp. Edward Arnold, London.

Birket-Smith, J. & Kristensen, N.P. 1974. The skeleto-muscular anatomy of the genital segments of male *Eriocrania* (Insecta. Lepidoptera). *Zeitschrift für Morphologie der Tiere* 77: 157-174.

Blest, A.D. 1957a. The function of eyespot patterns in the Lepidoptera. *Behaviour* 11: 209-255.

Blest, A.D. 1957b. The evolution of protective displays in the Saturnioidea and Sphingidae (Lepidoptera). *Behaviour* 11: 257-309.

Blest, A.D. 1963. Relations between moths and predators. *Nature* 197: 1046-1047.

Blest, A.D. 1964. Protective display and sound production in some New World arctiid and ctenuchid moths. *Zoologica New York* 49: 161-181.

Blest, A.D., Collett, T.S. and Pye, J.D. 1963. The generation of ultrasonic signals by a New World arctiid moth. *Proceedings of the Royal Society of London* (B) 158: 196-207.

Bleszinsky, S. 1965. Crambinae. *In* Amsel, H.G., Gregor, F., and Reisser, H. (Eds). *Microlepidoptera Palaearctica* 1: xlvii+553 pp. + 133 pls. Georg Fromme, Vienna.

Bocharova-Messner, O.M. & Aksyuk, T.S. 1981. Formation of a tunnel by the wings of butterflies (Lepidoptera, Rhopalocera) during their flight. *Doklady Akademii Nauk Soyuza Sovetskikh Sotsialisticheskikh Respublik* (Leningrad) 260: 1490-1493. [In Russian.]

Bodenheimer, F.S. 1951. *Insects as Human Food.* 352 pp. Junk, The Hague.

Bogner, F., Boppré, M., Ernst, K.-D., and Boeckh, J. 1986. CO_2 sensitive receptors on labial palps of *Rhodogastria* moths (Lepidoptera: Arctiidae): physiology, fine structure and central projection. *Journal of Comparative Physiology* A 158: 741-749.

Boppré, M. 1984a. Redefining 'pharmacophagy'. *Journal of Chemical Ecology* 10: 1151-1154.

Boppré, M. 1884b. Chemically mediated interactions between butterflies. *In* Vane-Wright, R.I. and Ackery, P.R. *The Biology of Butterflies.* pp. 259-275. Academic Press, London.

Boppré, M. & Schneider, D. 1989. The biology of *Creatonotos* (Lepidoptera: Arctiidae) with special reference to the androconial system. *Zoological Journal of the Linnean Society* 96: 339-356.

Boppré, M. & Vane-Wright, R.I. 1989. Androconial systems in Danainae (Lepidoptera): functional morphology of *Amauris, Danaus, Tirumala* and *Euploea. Zoological Journal of the Linnean Society* 97: 101-133.

Börner, C. 1925. Lepidoptera, Schmetterlinge. *In* Brohmer, P. (Ed.). *Fauna aus Deutschland.* pp. 358-387.

Börner, C. 1939. Die Grundlagen meines Lepidopterensystems. *Verhandlungen VII Internationalen Kongresses für Entomologie* 2: 1372-1424.

Bourgogne, J. 1946. Un type nouveau d'appareil génital femelle chez les Lépidoptères. *Annales de la Société Entomologique de France* 115: 69-80.

Bourgogne, J. 1949. L'appareil génital femelle de quelques Hepialidae (Lépidoptères). *Bulletin de la Société Zoologique de France* 74: 284-291.

Bourgogne, J. 1951. Ordre des Lépidoptères. *In* Grassé, P.-P. *Traite de Zoologie* 10: 174-448 + pls. 1-3.

Bradley, J.D. 1951. Notes on the family Arrhenophanidae (Lepidoptera: Heteroneura), with special reference to the morphology of the genitalia, and descriptions of one new genus and two new species. *The Entomologist* 84: 178-185.

Bradley, J.D. 1953. On *Harmaclona* Busck, 1914 (Lep., Arrhenophanidae), with descriptions of a new species, of a supplementary wing-coupling device present, and some new synonymy. *The Entomologist* 86: 61-66 + pl. 2.

Bradley, J.D. 1966. A comparative study of the coconut flat moth (*Agonoxena argaula* Meyr.) and its allies, including a new species (Lepidoptera, Agonoxenidae). *Bulletin of Entomological Research* 56: 453-472.

Braun, A.F. 1919. Wing structure of Lepidoptera and the phylogenetic and taxonomic value of certain persistent trichopterous characters. *Annals of the Entomological Society of America* 12: 349-366 + 1 pl.

Braun, A.F. 1924. The frenulum and its retinaculum in the Lepidoptera. *Annals of the Entomological Society of America* 17: 234-256.

Braun, A.F. 1948. Elachistidae of North America (Microlepidoptera). *Memoirs of the American Entomological Society* 13: 110 pp. + 26 pls.

Braun, A.F. 1963. The genus *Bucculatrix* in America north of Mexico (Microlepidoptera). *Memoirs of the American Entomological Society* 18: iii + 208 pp. + 45 pls.

Braun, A.F. 1972. Tischeriidae of America north of Mexico. *Memoirs of the American Entomological Society* 28: 148 pp.

Brock, J.P. 1971. A contribution towards an understanding of the morphology and phylogeny of the Ditrysian Lepidoptera. *Journal of Natural History* 5: 29-102.

Brock, J.P. 1990. Pupal protrusion in some bombycoid moths (Lepidoptera). *Entomologist's Gazette* 41: 91-97.

Brooks, D.R. 1979. Testing the context and extent of host-parasite coevolution. *Systematic Zoology* 28: 299-307.

Brower, L.P. 1977. Monarch migration. *Natural History* 86 (6): 40-53.

Brower, L.P. 1984. Chemical defence in butterflies. *In* Vane-Wright, R.I. and Ackery, P.R. (Eds). *The Biology of Butterflies*. pp. 109-134. Academic Press, London.

Brower, L.P. & Brower, J. van Z. 1964. Birds, butterflies, and plant poisons: a study in ecological chemistry. *Zoologica (New York)* 49: 137-159.

Brower, L.P., Brower, J.v.Z., & Cranston, F. 1965. Courtship behaviour of the Queen butterfly, *Danaus gilippus berenice* (Cramer). *Zoologica (New York)* 50: 1-40 + pls 1-7.

Brower, L.P & Glazier, S.C. 1975. Localization of heart poisons in the Monarch butterfly. *Science* 188: 19-25.

Brower, L.P., Walford, P., & Calvert, W.H. *In press*. Lipids in the annual life history cycle of the Monarch butterfly, *Danaus plexippus* L. *In* Malcolm, S.B. and Zaluchi, M. (Eds). *Biology and Conservation of the Monarch Butterfly*. Los Angeles County Museum, Los Angeles.

Brown, K.S. 1984. Adult-obtained pyrrolizidine alkaloids defend ithomiine butterflies against a spider predator. *Nature* 309: 707-709.

Brown, K.S. 1991. Conservation of neotropical environments. *In* Collins, N.M. and Thomas, J.A. (Eds). *The Conservation of Insects and their Habitats (15th Symposium of the Royal Entomological Society of London)*. pp. 349-404. Academic Press, London.

Brown, K.S. & Neto, J.V. 1976. Predation on aposematic ithomiine butterflies by Tanagers (*Pipraeidea melanonota*). *Biotropica* 8: 136-141.

Buckner, C.H. 1969. The common shrew (*Sorex araneus*) as a predator of the winter moth (*Operophtera brumata*) near Oxford, England. *Canadian Entomologist* 101: 370-375.

Busck, A. & Böving, A. 1914. On *Mnemonica auricyanea* Walsingham. *Proceedings of the Entomological Society of Washington* 16: 151-163 + pls 9-16.

Büttiker, W. & Whellan, J.A. 1966. Records of eye-frequenting moths from Rhodesia. *Rhodesia Agricultural Journal* 2306: 4 pp.

Callahan, P.S. 1975. Insect antennae with special reference to the mechanism of scent detection and the evolution of the sensilla. *International Journal of Insect Morphology and Embryology* 4: 381-430.

Calvert, W.H. & Brower, L.P. 1986. The location of Monarch butterfly (*Danaus plexippus* L.) overwintering colonies in Mexico in relation to topography and climate. *Journal of the Lepidopterists' Society* 40: 164-187.

Carcasson, 1976. *Revised Catalogue of the African Sphingidae (Lepidoptera) with descriptions of the East African species.* 2nd Edn. 148 pp. + 17 pls. E.W. Classey, Faringdon, England.

Carey, B., Cameron, P.S., Cerda, L., & Garda, R. 1978. Ciclo estacional de un minador subcortical de coigue (*Nothofagus dombeyi*). *Turrialba* 28: 151-153.

Carpenter, G.D.H. 1938. Audible emission of defensive froth by insects. *Proceedings of the Zoological Society* A 108: 243-252 + 2 pls.

Carter, D.J. 1984. Pest Lepidoptera of Europe with special reference to the British Isles. *Series Entomologica* 31: 431 pp. Junk; Dordrecht, Boston and Lancaster.

Carter, D.J. & Deeming. J.C. 1980. *Azygophleps alborittata* Bethune-Baker (Lepidoptera: Cossidae) attacking groundnuts in Northern Nigeria, with descriptions of the immature and imaginal stages. *Bulletin of Entomological Research* 70: 399-405.

Carter, D.J. & Dugdale, J.S. 1982. Notes on collecting and rearing *Micropterix* (Lepidoptera: Micropterigidae) larvae in England. *Entomologist's Gazette* 33: 43-47.

Carter, D.J. & Hargreaves, B. 1986. *A field guide to Caterpillars of Butterflies and Moths in Britain and Europe*. 296 pp. Collins.

Carter, D.J., Kitching, I.J. & Scoble M.J. 1988. The adaptable caterpillar. *The Entomologist* 107: 68-78.

Casey, T.M. 1981. Behavioral mechanisms of thermoregulation. *In* Heinrich, B. (Ed.). *Insect Thermoregulation*. pp. 79-114. John Wiley & Sons.

Chapman, R.F. 1982. *The Insects: structure and function*. 3rd Edn. xiv+919 pp. Hodder & Stoughton.

Chapman, T.A. 1896. On the phylogeny and evolution of the Lepidoptera from a pupal and oval standpoint. *Transactions of the Entomological Society of London* 1896: 567-587.

Chapman, T.A. 1917. *Micropteryx* entitled to ordinal rank; order Zeugloptera. *Transactions of the Entomological Society of London* 1917: 310-314 + pls 81-92.

Chauvin, J.T. & Chauvin, G. 1980. Formation des reliefs externes de l'oeuf de *Micropteryx* [sic] *calthella* L. (Lepidoptera: Micropterigidae). *Canadian Journal of Zoology* 58: 761-766.

Chauvin, G. & Faucheaux, M. 1981. Les pieces buccales et leurs recepteurs sensoriels chez l'imago de *Micropterix calthella* L. (Lepidoptera: Micropterigidae). *International Journal of Insect Morphology and Embryology* 10: 425-439.

Chew, F.S. & Robbins, R.K. 1984. Egg-laying in butterflies. *In* Vane-Wright, R.I. and Ackery, P.R. (Eds). *The Biology of Butterflies*. pp. 65-79. Academic Press, London.

Clarke, C.A., Mani, G.S., Wynne, G. 1985. Evolution in reverse: clean air and the peppered moth. *Biological Journal of the Linnean Society* 26: 189-199.

Clench, H.K. 1957. Cossidae from Chile (Lepidoptera). *Mitteilungen der Münchner entomologischen Gesellschaft* 47: 122-142 + pls 4-6.

Clench, H.K. 1959. On the unusual structure and affinities of the Madagascan genus *Pseudocossus*. *Revue Française d'Entomologie* 26: 44-50.

Cock, M.J.W., Godfray, H.C.J. & Holloway, J.D. 1987. *Slug and Nettle Caterpillars*. x+270 pp. + 36 pls. C.A.B. International, Wallingford.

Collins, C.T. & Watson, A. 1983. Field observations of bird predation on neotropical moths. *Biotropica* 15: 53-60.

Collins, N.M. & Morris, M.G. 1985. *Threatened Swallowtail Butterflies of the World*. The IUCN Red Data Book. vii+401 pp. + pls 1-8. IUCN; Gland and Cambridge.

Common, I.F.B. 1954. A study of the ecology of the adult Bogong moth, *Agrotis infusa* (Boisd.) (Lepidoptera: Noctuidae), with special reference to its behaviour during migration and aestivation. *Australian Journal of Zoology* 2: 223-263.

Common, I.F.B. 1966. A new family of Bombycoidea (Lepidoptera) based on *Carthaea saturnioides* Walker from western Australia. *Journal of the Entomological Society of Queensland* 5: 29-36.

Common, I.F.B. 1969a. A wing-locking or stridulatory device in Lepidoptera. *Journal of the Australian Entomological Society* 8: 121-125.

Common, I.F.B. 1969b. A new genus *Perthida* for the Western Australian Jarrah leaf miner *P. glyphopa* sp. n. and *Tinea phoenicopa* Meyrick (Lepidoptera: Incurvariidae). *Journal of the Australian Entomological Society* 8: 126-130.

Common, I.F.B. 1970. Lepidoptera. *In* Mackerras, I.M. (Ed.). *The Insects of Australia*. pp. 765-866. Melbourne University Press, Melbourne.

Common, I.F.B. 1973. A new family of Dacnonypha (Lepidoptera) based on three new species from southern Australia, with notes on the Agathiphagidae. *Journal of the Australian Entomological Society* 12: 11-23.

Common, I.F.B. 1975. Evolution and classification of the Lepidoptera. *Annual Review of Entomology* 20: 183-203.

Common, I.F.B. 1979. The larva and pupa of *Imma acosma* (Turner) and *I. vaticina* Meyrick (Lepidoptera: Immidae), and the taxonomic relationships of the family. *Journal of the Australian Entomological Society* 18: 33-38.

Common, I.F.B. 1980. The systematic position of *Hypertropha* (Lepidoptera) and related Australian genera. *Entomologica scandinavica* 11: 17-31.

Common, I.F.B. 1986. Exposed pupae in some Australian moths. *News Bulletin of the Entomological Society of Queensland* 13: 120-123.

Common, I.F.B. 1990. *Moths of Australia.* vi+535 pp. + 32 pls. E.J. Brill, and Melbourne University Press.

Common, I.F.B. & Bellas, T.E. 1977. Regurgitation of host-plant oil from a foregut diverticulum in the larvae of *Myrascia megalocentra* and *M. bracteatella* (Lepidoptera: Oecophoridae). *Journal of the Australian Entomological Society* 16: 141-147.

Common, I.F.B. & Edwards, E.D. 1981. The life history and early stages of *Synemon magnifica* Strand (Lepidoptera: Castniidae). *Journal of the Australian Entomological Society* 20: 295-302.

Common, I.F.B. & McFarland, N. 1970. A new subfamily for *Munychryia* Walker and *Gephyroneura* Turner (Lepidoptera: Anthelidae) and the description of a new species from Western Australia. *Australian Entomological Society* 9: 11-22.

Comstock, J.H. 1893. *A contribution to the classification of the Lepidoptera.* pp. 37-113 + 3 pls. *The Wilder Quarter-Century Book.* Ithaca, New York.

Comstock, J.H. 1918. *The Wings of Insects.* xvii+430 pp. + 10 pls. New York.

Conner, W.E., Eisner, T., Van der Meer, R.K., Guerrero, A., and Meinwald, J. 1981. Precopulatory sexual interaction in an arctiid moth (*Utetheisa ornatrix*): role of a pheromone derived from dietary alkaloids. *Behavioural Ecology and Sociobiology* 9: 227-235.

Cook, M.A. & Scoble, M.J. 1992. Tympanal organs of geometrid moths: a review of their morphology, function, and systematic importance. *Systematic Entomology* 17.

Cott, H.B. 1940. *Adaptive coloration in animals.* xxxii+508 pp. + 49 pls. Methuen, London.

Cottrell, C.B. 1984. Aphytophagy in butterflies: its relationship to myrmecophily. *Zoological Journal of the Linnean Society* 80: 1-57.

Covell, C.V. 1983. The state of our knowledge of the Neotropical Sterrhinae (Geometridae). *Second Symposium on Neotropical Lepidoptera Arequipa, Peru 1983.* Supplement 1: 17-23.

Cowie, R.J. & Hinsley, S.A. 1988. Feeding ecology of Great Tits (*Parus major*) and Blue Tits (*Parus caeruleus*), breeding in suburban gardens. *Journal of Animal Ecology* 57: 611-626.

Culvenor, C.C.J. & Edgar, J.A. 1972. Dihydropyrrolizine secretions associated with coremata of *Utetheisa* moths (family Arctiidae). *Experientia* 28: 627-628.

D'Abrera, B. 1986. *Sphingidae Mundi, Hawkmoths of the World,* 226 pp. E.W. Classey, Faringdon.

Dahm, K.H., Meyer, D., Finn, W.E., Reinhold, V., Röller, H. 1971. The olfactory & auditory mediated sex attraction in *Achroia grisella* (Fabr.). *Naturwissenschaften* 58: 265-266.

Dall'Asta, U. 1988. The tymbal organs of the Lymantriidae (Lepidoptera). *Nota Lepidotera* 11: 169-176.

Dalton, S. 1975. *Borne of the wind.* 160 pp. Chatto & Windus, London.

Damman, H. 1986. The osmaterial [*sic*] glands of the swallowtail butterfly *Eurytides marcellus* as a defence against natural enemies. *Ecological Entomology* 11: 261-265.

Daniel, F. & Dierl, W. 1966. Zur Biologie und Anatomie von *Heterogynis penella* (Hbn.) (Lep.). *Zoologischer Anzeiger* 176: 449-464.

Davis, D.R. 1964. Bagworm moths of the western hemisphere. *Bulletin of the United States National Museum* 244: 233 pp.

Davis, D.R. 1967 A revision of the moths of the subfamily Prodoxinae (Lepidoptera: Incurvariidae). *Bulletin of the United States National Museum* 255: 170 pp.

Davis, D.R. 1969. A revision of the American moths of the family Carposinidae (Lepidoptera: Carposinoidea). *Bulletin of the United States National Museum* 289: 1-105.

Davis, D.R. 1972. *Careospina quercivora*, a new genus and species of moth infesting live oaks in California. *Proceedings of the Entomological Society of Washington* 74: 121-128.

Davis, D.R. 1975a. A review of the West Indian moths of the family Psychidae with descriptions of new taxa and immature stages. *Smithsonian Contributions to Zoology* 166: 1-66.

Davis, D.R. 1975b. Review of Ochsenheimeriidae and the introduction of the Cereal Stem Moth *Ochsenheimeria vacculella* into the United States (Lepidoptera: Tineoidea). *Smithsonian Contributions to Zoology* 192: 20 pp.

Davis, D.R. 1975c. Systematics and zoogeography of the family Neopseustidae with the proposal of a new superfamily (Lepidoptera: Neopseustoidea). *Smithsonian Contributions to Zoology* 210: 45 pp.

Davis, D.R. 1978. A revision of the North American moths of the superfamily Eriocranioidea with the proposal of a new family, Acanthopteroctetidae (Lepidoptera). *Smithsonian Contributions to Zoology* 251: 1-131.

Davis, D.R. 1986. A new family of monotrysian moths from Austral South America (Lepidoptera: Palaephatidae), with a phylogenetic review of the Monotrysia. *Smithsonian Contributions to Zoology* 434: 202 pp.

Davis, D.R. 1989. Generic revision of the Opostegidae, with a synoptic catalog of the world's species (Lepidoptera: Nepticuloidea). *Smithsonian Contributions to Zoology* 478: 97 pp.

Davis, D.R. 1990. Neotropical Microlepidoptera XXIII. First report of the family Eriocottidae from the New World, with descriptions of new taxa. *Proceedings of the Entomological Society of Washington* 92: 1-35.

Davis, D.R., Clayton, D.H., Janzen, D.M., & Brooke, A.P. 1986. Neotropical Tineidae, II: biological notes and descriptions of two new moths phoretic on spiny pocket mice in Costa Rica (Lepidoptera: Tineoidea). *Proceedings of the Entomological Society of Washington* 88: 98-109.

Davis, D.R. & Faeth, S.H. 1986. A new oak-mining eriocraniid moth from southeastern United States (Lepidoptera: Eriocraniidae). *Proceedings of the Entomological Society of Washington* 88: 145-153.

Davis, D.R. & Heppner, J.B. 1987. New discoveries concerning *Ischnuridia* a remarkable genus of Indo-Australian Tineidae (Lepidoptera). *Tinea* 12, (supplement): 145-151.

Davis, D.R. & Milstrey, E.G. 1988. Description and biology of *Acrolophus pholeter* (Lepidoptera: Tineidae), a new moth commensal from Gopher tortoise burrows in Florida. *Proceedings of the Entomological Society of Washington* 90: 164-178.

Davis, D.R. & Nielsen, E.S. 1980. Description of a new genus and two new species of Neopseustidae from South America, with discussion of phylogeny and biological observations (Lepidoptera: Neopseustoidea). *Steenstrupia* 6: 253-289.

Davis, D.R. & Nielsen, E.S. 1985. The South American neopseustid genus *Apoplania* Davis: a new species, distribution records and notes on adult behaviour (Lepidoptera: Neopseustina). *Entomologica scandinavica* 15: 497-509.

DeFoliart, G.R. 1989. The human use of insects as food and as animal feed. *Bulletin of the Entomological Society of America.* 35: 22-35.

Dempster, J.P. 1983. The natural control of populations of butterflies and moths. *Biological Reviews of the Cambridge Philosophical Society* 58: 461-481.

Denis, J.-R. & Bitsch, J. 1973. Structure céphalique dans les ordres des insectes. *In* Grassé, P.-P. *Traité de Zoologie* 8 (Fasc. 1): 101-593. Paris.

Denis, M. & Schiffermüller, I. 1775. *Ankündung eines systematischen Werkes von den Schmetterlingen der Wienergegend* 324 pp. + 3 pls. Vienna.

Derzhavets, Y.A. 1984. Review of the classification of the Sphingidae (Lepidoptera) with a list of the species of the USSR. *Entomologicheskoye Obozreniye* 3: 604-620. [In Russian: English translation in Entomological Review 63: 6-28.]

Dethier, V.G. 1963. *The Physiology of Insect Senses.* ix+266 pp. Methuen.

DeVries, P.J. 1988. The larval ant-organs of *Thisbe irenea* (Lepidoptera: Riodinidae) and their effects upon attending ants. *Zoological Journal of the Linnean Society* 94: 379-393.

DeVries, P.J., Kitching, I.J. & Vane-Wright, R.I. 1985. The systematic position of *Antirrhea and Caerois*, with comments on the classification of the Nymphalidae (Lepidoptera). *Systematic Entomology* 10: 11-32.

Diakonoff, A. 1989. Revision of the Palaearctic Carposinidae with descriptions of a new genus and new species (Lepidoptera: Pyraloidea). *Zoologische Verhandelingen* 251: 1-155.

Dickens, J.C. & Eaton, J.L. 1973. External ocelli in Lepidoptera previously considered to be anocellate. *Nature* 242: 205-206.

Dierl, W. 1970. Compsoctenidae: ein neues Taxon von Familienstatus (Lepidoptera). *Veröffentlichungen der Zoologischen Staatssammlung Museum München* 14: 1-41 + 2 pls.

Dodd, F.P. 1912. Some remarkable ant-friend Lepidoptera of Queensland. With supplement by E. Meyrick. *Transactions of the Entomological Society of London* 1911: 577-590 + pl. 48.

Doesburg, P.H., van. 1966. Über valväre Stridulation bei Schwärmer (Lepidoptera Sphingidae). *Zoologische Mededeelingen* 41: 161-170.

Döring, E. 1955. *Zur Morphologie der Schmetterlingseier.* pp. 154 + pls 1-61. Akademie-Verlag, Berlin.

Douglas, M.M. 1986. *The Lives of Butterflies.* University of Michigan Press, Ann Arbor xv + 241 pp. + 16 pls.

Downes, J.A. 1968. A nepticulid moth feeding at the leaf-nectaries of Poplar. *Canadian Entomologist* 100: 1078-1079.

Downes, J.A. 1973. Lepidoptera feeding at puddle-margins, dung, and carrion. *Journal of the Lepidopterist's Society* 27: 89-99.

Downey, J.C. & Allyn, A.C. 1975. Wing-scale morphology and nomenclature. *Bulletin of the Allyn Museum* 31: 32 pp.

Downey, J.C. & Allyn, A.C. 1978. Sounds produced in pupae of Lycaenidae. *Bulletin of the Allyn Museum* 48: 14 pp.

Downey, J.C. & Allyn, A.C. 1980. Eggs of Riodinidae. *Journal of the Lepidopterists' Society* 34: 133-145.

Downey, J.C. & Allyn, A.C. 1981. Chorionic sculpturing in eggs of Lycaenidae. Part I. *Bulletin of the Allyn Museum* 61: 1-29.

Dugdale, J.S. 1974. Female genital configuration in the classification of Lepidoptera. *New Zealand Journal of Zoology* 1: 127-146.

Dugdale, J.S. 1975. The insects in relation to plants. *In* Kuschel, G. (Ed.). Biogeography and ecology in New Zealand. *Monographiae Biologicae* 27: 561-589.

Dugdale, J.S. 1988 (volume dated 1987). *Thambotricha vates* Meyrick, reassigned to Epermeniidae (Lepidoptera: Epermenioidea). *New Zealand Journal of Zoology* 14: 375-383.

Duke, A.J.H. & Duke, N.J.S. 1988. Life cycle of a little-known south African geometer moth, *Callioratis abraxas* Felder (Lepidoptera: Geometridae). *Journal of the Entomological Society of southern Africa* 51: 144-146.

Dumbleton, 1952. A new genus of seed-infesting micropterigid moths. *Pacific Science* 6: 17-29.

Duméril, A.M.C. 1823. *Considérations générales sur la classe des insects* xii + 272 + pls 1-60. Leorault, Paris.

Dumortier, B. 1963. Morphology of sound emission apparatus in Arthropoda. *In* Busnel, R.-G. (Ed.). *Acoustic behaviour of animals.* Elsevier, Amsterdam.

Dunning, D.C. 1968. Warning sounds of moths. *Zeitschrift für Tierpsychologie* 25: 129-138.

Dunning, D.C. & Roeder, K.D. 1965. Moth sounds and the insect-catching behavior of bats. *Science* 147: 173-174.

Durden, C.J. & Rose, H. 1978. Butterflies from the middle Eocene: the earliest occurrence of fossil Papilionoidea (Lepidoptera). *The Pearce-Sellards Series Texas Memorial Museum* 29: 25 pp.

Dyar, H.G. 1894. A classification of lepidopterous larvae. *Annals of the New York Academy of Science* 8: 194-232.

Eassa, Y.E.E. 1953. The development of imaginal buds in the head of *Pieris brassicae* Linn. (Lepidoptera). *Transactions of the Royal Entomological Society* 104: 39-50 + 1 pl.

Eastham, L.E.S. & Eassa, Y.E.E. 1955. The feeding mechanism of the butterfly *Pieris brassicae* L. *Philosophical Transactions of the Royal Society of London* (B) 239: 1-43.

Eggers, F. 1919. Das thoracale bitympanale Organ einer Gruppe der Lepidoptera Heterocera. *Zoologische Jahrbücher abteilung für Anatomie und Ontogenie der Tiere* 41: 273-376 + pls 20-24.

Ehnbom, K. 1948. Studies on the central and sympathetic nervous system and some sense organs in the head of neuropteroid insects. *Opuscula Entomologica* (Lund) Supplement 8: 162 pp.

Ehrlich, P.R. 1958a. The integumental anatomy of the Monarch butterfly *Danaus plexippus* L. (Lepidoptera: Danaidae). *The University of Kansas Science Bulletin* 38: 1315-1349.

Ehrlich, P.R. 1958b. The comparative morphology, phylogeny and higher classification of the butterflies (Lepidoptera: Papilionoidea). *The University of Kansas Science Bulletin* 39: 305-370.

Ehrlich, P.R. 1960. The integumental anatomy of the Silver-spotted skipper, *Epargyreus clarus* Cramer (Lepidoptera: Hesperiidae). *Microentomology* 24: 3-23.

Ehrlich, P.R. & Ehrlich, A.H. 1963. The thoracic and basal abdominal musculature of the butterflies (Lepidoptera: Papilionoidea). *Microentomology* 25: 91-126.

Ehrlich, P.R. & Raven, P.H. 1965. Butterflies and plants: a study in coevolution. *Evolution* 18: 586-608.

Eichlin, T.D. & Cunningham, H.B. 1978. The Plusiinae (Lepidoptera: Noctuidae) of America North of Mexico, emphasizing genitalic and larval morphology. *Technical Bulletin. United States Department of Agricultural Research Service* 1567: 222 pp.

Eisner, T. 1980. Chemistry, defence, and survival: case studies and selected topics. *In* Locke, M. and Smith, D.S. (Eds). *Insect Biology in the Future*. pp. 847-878. Academic Press.

Eisner, T., Kluge, A.F., Ikeda, M.I., Meinwald, Y.C. & Meinwald, J. 1971. Sesquiterpenes in the osmeterial secretion of a papilionid butterfly, *Battus polydamus. Journal of Insect Physiology* 17: 245-250.

Eisner, T. & Meinwald, Y.C. 1965. Defensive secretions of a caterpillar (*Papilio*). *Science* 150: 1733-1735.

Eliot, J.N. 1973. The higher classification of the Lycaenidae (Lepidoptera): a tentative arrangement. *Bulletin of the British Museum (Natural History)* Entomology 28: 373-505 + pls 1-6.

Ellington, C.P. 1984a. The aerodynamics of flapping animal flight. *American Zoologist* 24: 95-105.

Ellington, C.P. 1984b. The aerodynamics of hovering insect flight (I-VI). *Philosophical Transactions of the Royal Society of London* 305: 1-181 + 7 pls.

Eltringham, H. 1925. On a new organ in certain Lepidoptera. *Transactions of the Entomological Society of London* 1925: 7-9 + pl. 2.

Eltringham, H. 1938. Appendix. *In* Carpenter, G.D.H. Audible emission of defensive froth by insects. *Proceedings of the Zoological Society* A 108: 251-252 + pl. 2.

Emmet, A.M. 1976. Nepticulidae; Heliozelidae. *In* Heath, J. (Ed.). *The Moths and Butterflies of Great Britain and Ireland* 1: 171-267, 300-306 + pls 1-7, 9, 11, 12; and pls 8, 13. Blackwell Scientific Publications, & The Curwen Press.

Emmet, A.M. 1985. Lyonetiidae. *In* Heath, J. and Emmet, A.M. (Eds). *The Moths and Butterflies of Great Britain and Ireland* 2: 212-239 + pls 1, 2, 9, 10.

Emmet, A.M. (Ed.). 1988. *A Field Guide to the smaller British Lepidoptera*. 2nd edition. 288 pp. British Entomological and Natural History Society, London.

Emmet, A.M. 1991. *The scientific names of the British Lepidoptera - their history and meaning*. 288 pp. Harley Books, Colchester.

Epstein, M.E. 1988. An overview of Slug caterpillar moths (Lepidoptera: Limacodidae) with emphasis on genera in the New World *Parasa* group. PhD thesis. University of Minnesota.

Erhardt, A. & Thomas, J.A. 1991. Lepidoptera as indicators of change in the semi-natural grasslands of lowland and upland Europe. *In* Collins, N.M. and Thomas, J.A. (Eds). *The Conservation of Insects and their Habitats (15th Symposium of the Royal Entomological Society of London)*. pp. 213-236. Academic Press, London.

Exner, S. 1891. Die Physiologie der facettirten Augen von Kresben und Insecten. Deuticke, Leipzig and Wien.

Fabricius, I.C. 1775. *Systema entomologiae*. xxvii+832 pp. Leipzig.

Feinsinger, P. 1983. Coevolution and pollination. *In* Futuyma, D.J. & M. Slatkin (Eds). *Coevolution*. pp. 282-310. Sinauer Associates Inc., Massachusetts.

Feltwell, J. 1990. *The Story of Silk.* 233 pp. Alan Sutton, Stroud.

Fenton, M.B. & Fullard, J.H. 1979. The influence of moth hearing on bat echolocation strategies. *Journal of Comparative Physiology* A 132: 77-86.

Fenton, M.B. & Fullard, J.H. 1981. Moth hearing and the feeding strategies of bats. *American Scientist* 69: 266-275.

Ferguson, D.C. 1971-2. Bombycoidea (Saturniidae). *In* Dominick, R.B. *et al.* (Eds). *The Moths of America North of Mexico* 20 (2A & B): 275 pp. + pls 1-22. E.W. Classey and R.B.D. Publications, London.

Ferguson, D.C. 1978. Noctuoidea (Lymantriidae). *In* Dominick, R.B. *et al.* (Eds). *The Moths of America North of Mexico* 22 (2): x+110 pp. + pl. A + pls 1-8. E.W. Classey and The Wedge Entomological Research Foundation, London.

Ferguson, D.C. 1985. Geometroidea (Geometridae). *In* Dominick, R.B. *et al.* (Eds). *The Moths of America North of Mexico* 18 (1): xv+131 pp. + 4 pls. The Wedge Entomological Research Foundation, Washington.

Fibiger, M. & Kristensen, N.P. 1974. The Sesiidae (Lepidoptera) of Fennoscandia and Denmark. *Fauna Entomologica Scandinavica* 2: 1-91.

Fletcher, D.S. 1979. Geometroidea. *In* Nye, I.W.B. (Ed.). *The Generic Names of Moths of the World* 3: x+243 pp. + 2 pls.

Forbes, W.T.M. 1923. The Lepidoptera of New York and Neighboring States. *Memoirs of the Cornell University Agricultural Experiment Station* 68: 729 pp.

Forbes, W.T.M. 1936. The classification of the Thyatiridae (Lepidoptera). *Annals of the Entomological Society of America* 29: 779-800 + 2 pls.

Forbes, W.T.M. 1954. The Lepidoptera of New York and Neighboring States: III. *Memoirs of the Cornell University Agricultural Experiment Station* 329: 433 pp.

Forbes, W.T.M. 1955. The subdivisions of the Eupterotidae (Lepidoptera). *Tijdschrift voor Entomologie* 98: 85-132.

Forbes, W.T.M. 1960. Lepidoptera of New York and Neighboring States: IV. *Memoirs of the Cornell University Agricultural Experiment Station* 371: 188 pp.

Forbes, W.T.M. & Franclemont, J.G. 1957. The striated band (Lepidoptera, chiefly Arctiidae). *Lepidopterist's News* 11: 147-150.

Fracker, S.B. 1915. The classification of lepidopterous larvae. *Illinois Biological Monographs* 2: iv+169 pp.

Franclemont, J.G. 1973. Mimallonoidea; Bombycoidea (in part) in Dominick, R.B., *et al.* (Eds). *Moths of America North of Mexico* 20 (1): 86 pp. + 11 pls. E.W. Classey and R.B.D Publications, London.

Free, J.B. 1970. *Insect Pollination of Crops* xi+544 pp. Academic Press, London and New York.

Freeman, T.N. 1972. The coniferous feeding species of *Argyrestia* in Canada (Lepidoptera: Yponomeutidae). *Canadian Entomologist* 104: 687-697.

French, D.D. & Smith, V.R. 1983. A note on the feeding of *Pringleophaga marioni* Vietti [*sic*] larvae at Marion Island. *South African Journal of Antarctic Research* 13: 45-46.

Fullard, J.H. 1979. Behavioral analyses of auditory sensitivity in *Cycnia tenera* Hübner (Lepidoptera: Arctiidae). *Journal of Comparative Physiology* A 129: 85-95.

Fullard, J.H. 1982a. Echolocation assemblages and their effects on moth auditory systems. *Canadian Journal of Zoology* 60: 2572-2576.

Fullard, J.H. 1982b. Cephalic influences on a defensive behavior in the dogbane tiger moth, *Cycnia tenera. Physiological Entomology* 7: 157-162.

Fullard, J.H. 1984a. Listening for bats: pulse repetition rate as a cue for a defensive behavior in *Cycnia tenera* (Lepidoptera: Arctiidae). *Journal of Comparative Physiology* A 154: 249-252.

Fullard, J.H. 1984b. Acoustic relationships between tympanate moths and the Hawaiian hoary bat (*Lasiurus cinereus semotus*). *Journal of Comparative Physiology* A 155: 795-801.

Fullard, J.H. & Barclay, R.M.R. 1980. Audition in spring species of arctiid moths as a possible response to differential levels of insectivorous bat predation. *Canadian Journal of Zoology* 58: 1745-1750.

Fullard, J.H., Fenton, M.B. & Simmons, J.A. 1979. Jamming bat echolocation: the clicks of arctiid moths. *Canadian Journal of Zoology* 57: 647-649.

Gaedike, R. 1974. Revision der paläarktischen Douglasiidae (Lepidoptera). *Acta Faunistica Entomologica Musei Nationalis Pragae* 15: 79-102.

Gaedike, R. 1978. Versuch der phylogenetischen Gliederung der Epermeniidae der Welt (Lepidoptera). *Beiträge zur Entomologie* 28: 201-209.

Gardner, J.C.M. 1946. On larvae of the Noctuidae. *Transactions of the Royal Entomological Society of London* 96: 61-72.

Gaston, K.J. 1991. The magnitude of global insect species richness. *Conservation Biology* 5: 283-296.

Gatehouse, A.G. 1987. Migration: a behavioural process with ecological consequences. *Antenna* 11: 10-12.

Gauld, I.D. 1988. Evolutionary patterns of host utilization by ichneumonoid parasitoids (Hymenoptera: Ichneumonidae and Braconidae). *Biological Journal of the Linnean Society* 35: 351-377.

Gauld, I.D. & Bolton, B. (Eds) 1988. *The Hymenoptera.* xi+332 pp. + 10 col. pls. British Museum (Natural History) and Oxford University Press.

Gerasimov, A.M. 1952. Insects-Lepidoptera. Part 1. Caterpillars. *Zoological Institute of the Academy of Sciences USSR*, Fauna SSSR, New Series 56 (1): 338 pp. [In Russian.]

Ghiradella, H. 1984. Structure of iridescent lepidopteran scales: variations on several themes. *Annals of the Entomological Society of America* 77: 637-645.

Ghiradella, H. & Radigan, W. 1976. Development of butterfly scales. II. Struts, lattices and surface tension. *Journal of Morphology* 150: 279-297.

Gibbs, G.W. 1979. Some notes on the biology and status of the Mnesarchaeidae (Lepidoptera). *New Zealand Entomologist* 7: 2-9.

Gilbert, L.E. 1972. Pollen feeding and reproductive biology of *Heliconius* butterflies. *Proceedings of the National Academy of Sciences U.S.A.* 69: 1403-1407.

Gilbert, L.E. 1984. The biology of butterfly communities. *In* Vane-Wright, R.I. and Ackery, P.R. (Eds). *The Biology of Butterflies.* pp. 41-54. Academic Press, London.

Glendinning, J.I. *In press.* Crawling off the ground: a defense against mouse attack in overwintering Monarch butterflies. *In* Malcolm, S.B. and Zaluchi, M. (Eds). *Biology and Conservation of the Monarch Butterfly.* Los Angeles County Museum, Los Angeles.

Gohrbandt, I. 1937. Das Tympanalorgan der Drepaniden und der Cymatophoriden. *Zeitschrift für wissenschaftliche Zoologie* 149: 537-600.

Gozmany, L. 1978. Lecithoceridae. *In* Amsel, H.G., Gregor, F., and Reisser, H. (Eds). *Microlepidoptera Palaearctica* 5: xxviii+306 pp. + 93 pls. Frome & Co, Vienna.

Graham, M.W.R. de V. 1950. Postural habits and colour pattern evolution in Lepidoptera. *Transactions of the Society for British Entomology* 10: 217-232.

Grassé, P.-P. 1975. Les couleurs physiques ou structurales du tégument des insectes. *In* Grassé, P.-P. (Ed.). *Traité de Zoologie* 8: 185-198.

Greene, E. 1989. A diet-induced developmental polymorphism in a caterpillar. *Science* 243: 577-700.

Grehan, J.R. 1989. Larval feeding habits of the Hepialidae (Lepidoptera). *Journal of Natural History* 23: 803-824.

Grimes, L.R. & Neunzig, H.H. 1968a. Morphological survey of the maxillae in last stage larvae of the suborder Ditrysia (Lepidoptera): Palpi. *Annals of the Entomological Society of America* 79: 491-509.

Grimes, L.R. & Neunzig, H.H. 1986b. Morphological survey of the maxillae in last-stage larvae of the suborder Ditrysia (Lepidoptera): mesal lobes (laciniogaleae). *Annals of the Entomological Society of America* 79: 510-526.

Grodnitsky, D.L. & Kozlov, M.V. 1985. Functional morphology of the wing apparatus and flight peculiarities in primitive moths (Lepidoptera: Micropterigidae, Eriocraniidae). *Zoologichesky Zhurnal* 64: 1661-1671.

Grünberg, K. 1911. Family: Bombycidae. *In* Seitz, A.A. (Ed.). *The Macrolepidoptera of the World* 2: 189-192.

Guenée, A. 1857. Species général des Lépidoptères. *In* Boisduval, J.A. & Guenée, A. *Histoire Naturelle des Insectes* 10: 584 pp. Paris.

Gwynne, D.T. & Edwards, E.D. 1986. Ultrasound production by genital stridulation in *Syntonarcha iriastis* (Lepidoptera: Pyralidae): long-distance signalling by male moths? *Zoological Journal of the Linnean Society* 88: 363-376.

Haber, W.A. *In press*. Seasonal migration of Monarchs in Costa Rica, Central America. *In* Malcolm, S.B. and Zaluchi, M. (Eds). *Biology and Conservation of the Monarch Butterfly*. Los Angeles County Museum, Los Angeles.

Haber, W.A, & Frankie, G.W. 1989. A tropical hawkmoth community: Costa Rican dry forest Sphingidae. *Biotropica* 21: 155-172.

Hamilton, K.G.A. 1972. The insect wing. Part II. *Journal of the Kansas Entomological Society* 45: 54-58.

Hancock, D.L. 1983. Classification of the Papilionidae: a phylogenetic approach. *Smithersia* 2: 48 pp.

Handschin, E. 1946. Silk Moths. *Ciba Review* 53: 1902-1936.

Hannemann, H.J. 1956a. Die Kopfmuskulatur von *Micropteryx calthella* (L.) (Lep.). Morphologie und Funktion. *Zoologische Jahrbücher Abteilung für Anatomie und Ontogenie der Tiere* 75: 177-206.

Hannemann, H.J. 1956b. Über ptero-tarsale Stridulation und einige andere Arten der Lauterzeugung bei Lepidopteren. *Deutsche Entomologische Zeitschrift* (Neue Folge) 3: 14-27.

Harborne, J.B. 1982. *Introduction to ecological biochemistry*. 2nd edition. xi+278 pp. Academic Press.

Hardwick, D. 1965. The corn earworm complex. *Memoirs of the Entomological Society of Canada* 40: 247 pp.

Hardwick, D. 1970. A generic revision of the North American Heliothidinae (Lepidoptera: Noctuidae). *Memoirs of the Entomological Society of Canada* 73: 59 pp.

Hart-Dyke, Z. 1949. *So spins the silkworm*. xii+165 pp. Rockliff, London.

Harvey, D.J. 1987. *The higher classification of the Riodinidae (Lepidoptera)*. 216 pp. PhD thesis. University of Texas at Austin.

Hasbrouck, F.F. 1964. Moths of the family Acrolophidae in America North of Mexico (Microlepidoptera). *Proceedings of the United States National Museum* 114: 487-706.

Hasenfuss, I. 1960. *Die Larvalsystematik der Zünsler (Pyralidae)*. 263 pp. Akademie-Verlag, Berlin.

Hawkins, B.A & Lawton, J.H. 1987. Species richness for parasitoids of British phytophagous insects. *Nature* 326: 788-790.

Haynes, K.F. & Birch, M.C. 1986. Temporal reproductive isolation between two species of plume moths (Lepidoptera: Pterophoridae). *Annals of the Entomological Society of America* 79: 210-215.

Heath, J. 1962. The eggs of *Micropteryx* (Lep., Micropterygidae). *Entomologist's Monthly Magazine* 97: 179-180 + 1 pl.

Heath, J. 1976. Micropterigidae; Eriocraniidae. *In* Heath, J. (Ed.). *The Moths and Butterflies of Great Britain and Ireland*. 1: 151-155, 156-165 + pls 8, 9. Blackwell Scientific Publications, and The Curwen Press.

Heinrich, B. 1981. Ecological and evolutionary perspectives. *In* Heinrich, B. (Ed.). *Insect Thermoregulation* pp. 235-302. John Wiley & Sons.

Heinrich, C. 1916. On the taxonomic value of some larval characters in the Lepidoptera. *Proceedings of the Entomological Society of Washington* 18: 154-164.

Heinrich, C. 1918. On the Lepidoptera genus *Opostega* and its larval affinities. *Proceedings of the Entomological Society of Washington* 20: 27-38.

Hemming, F. 1937. *Hübner*. 1: xxxiv+605 pp. London.

Hennig, W. 1966. *Phylogenetic Systematics*. xiii+263 pp. Urbano, Chicago and London.

Hennig, W. 1981. *Insect Phylogeny*. xxii+514 pp. John Wiley & Sons.

Henning, S.F. 1988. *The Charaxinae Butterflies of Africa*. 457 pp. Aloe Books, Johannesburg.

Heppner, J.B. 1977. The status of the Glyphipterigidae and a reassessment of relationships in ypon. meutoid families and ditrysian superfamilies. *Journal of the Lepidopterists' Society* 31: 124-134.

Heppner, J.B. 1981. Revision of the new genus *Diploschiza* (Lepidoptera: Glyphipterigidae) for North America. *The Florida Entomologist* 64: 309-336.

Heppner, J.B. 1982a. Synopsis of the Glyphipterigidae (Lepidoptera: Copromorphoidea) of the World. *Proceedings of the Entomological Society of Washington* 84: 38-66.

Heppner, J.B. 1982b. A world catalogue of genera associated with the Glyphipterigidae auctorum (Lepidoptera). *Journal of the New York Entomological Society* 89: 220-294.

Heppner, J.B. 1982c. Review of the family Immidae, with a world checklist (Lepidoptera: Immoidea). *Entomography* 1: 257-279.

Heppner, J.B. 1982d. Synopsis of the Hilarographini (Lepidoptera: Tortricidae) of the World. *Proceedings of the Entomological Society of Washington* 84: 704-715.

Heppner, J.B. 1982e. Millieriinae, a new subfamily of Choreutidae, with new taxa from Chile and the United States (Lepidoptera: Sesioidea). *Smithsonian Contributions to Zoology* 370: 27 pp.

Heppner, J.B. 1984. Pseudocossinae: a new subfamily of Cossidae (Lepidoptera). *Entomological News* 95: 99-100.

Heppner, J.B. & Duckworth, W.D. 1981. Classification of the superfamily Sesioidea (Lepidoptera: Ditrysia). *Smithsonian Contributions to Zoology* 314: 144 pp.

Heppner, J.B. & Wang, H.Y. 1987. A rare moth, *Ratarda tertia* Strand (Lepidoptera: Ratardidae), from Palin, Taiwan. *Journal of the Taiwan Museum* 40: 91-94.

Hering, [E.] M. 1925. Die Familie der Ratardidae (Lep.). *Mitteilungen aus dem Zoologischen Museum in Berlin* 12: 151-158 + 1 pl.

Hering, [E.] M. 1926a. *Biologie der Schmetterlinge* vi+480 pp. 13 pls. Berlin.

Hering, [E.] M. 1926b. Family: Pterothysanidae. *In* Seitz, A. *The Macrolepidoptera of the World* 14: 123-125.

Hering, [E.] M. 1937. Revision der Chrysopolomidae (Lep.). *Annals of the Transvaal Museum* 17: 233-257 + 1 pl.

Hering, E.M. 1951. *Biology of the leaf miners.* iv+420 pp. Junk, The Hague.

Hering, E.M. 1958. Die Tegula der Lepidoptera, ihre Funktion und taxonomische Verwertbarkeit. *Proceedings of the tenth International Congress of Entomology* 2: 303-312.

Herrich-Schäffer, G.A.W. 1843-1856. *Systematische Bearbeitung der Schmetterlinge von Europa*, vols 1-6. Regensburg.

Hessel, J.H. 1966. A preliminary comparative anatomical study of the mesothoracic aorta of the Lepidoptera. *Annals of the Entomological Society of America* 59: 1217-1227.

Hessel, J.H. 1969. The comparative morphology of the dorsal vessel and accessory structures of the Lepidoptera and its phylogenetic implications. *Annals of the Entomological Society of America* 62: 353-370.

Heyneman, A.J. 1983. Optimal sugar concentrations of floral nectars - dependence on sugar intake efficiency and foraging costs. *Oecologia (Berlin)* 60: 198-213.

Hill, J.E. & Smith, J.D. 1984. *Bats. A natural history.* 243 pp. British Museum (Natural History), London.

Hinton, H.E. 1946a. On the homology and nomenclature of the setae of lepidopterous larvae, with some notes on the phylogeny of the Lepidoptera. *Transactions of the Royal Entomological Society of London* 97: 1-37.

Hinton, H.E. 1946b. A new classification of insect pupae. *Proceedings of the Zoological Society of London* 116: 282-382.

Hinton, H.E. 1947. The dorso cranial areas of caterpillars. *Annals and Magazine of Natural History* 14: 843-852.

Hinton, H.E. 1948. Sound production in lepidopterous pupae. *The Entomologist* 81: 254-269.

Hinton, H.E. 1955. Sound producing organs in Lepidoptera. *Proceedings of the Royal Entomological Society of London* (C) 20: 5-6.

Hinton, H.E. 1956. The larvae of the species of Tineidae of economic importance. *Bulletin of Entomological Research* 47: 251-346.

Hinton, H.E. 1958. The phylogeny of the panorpoid orders. *Annual Review of Entomology* 3: 181-206.

Hinton, H.E. 1981. *Biology of Insect Eggs* 1-3: xxiv+1125 pp. Pergamon Press.

Hodges, R.W. 1962. A revision of the Cosmopterigidae of America North of Mexico, with a definition of the Momphidae and Walshiidae (Lepidoptera: Gelechioidea). *Entomologica Americana* 42: 1-171.

Hodges, R.W. 1966. Review of the New World species of *Batrachedra* with descriptions of three new genera (Lepidoptera: Gelechioidea). *Transactions of the American Entomological Society* 92: 585-651 + pls 29-53.

Hodges, R.W. 1971. Sphingoidea, Hawkmoths. *In* Dominick, R.B. *et al.* (Eds). *The Moths of America North of Mexico* 21: xii+158 pp. + 14 pls. E.W. Classey and R.B.D. Publications, London.

Hodges, R.W. 1974. Gelechioidea: Oecophoridae (in part). *In* Dominick, R.B. *et al.* (Eds). *The Moths of America North of Mexico*: Oecophoridae (in part) 6(2): 142 pp. + 1 b & w and 7 colour pls. E.W. Classey and The Wedge Entomological Research Foundation, London.

Hodges, R.W. 1978. Gelechioidea (in part): Cosmopterigidae. *In* Dominick, R.B. *et al.* (Eds). *The Moths of America North of Mexico* 6 (1): 166 pp. + 6 colour pls. E.W. Classey and The Wedge Entomological Research Foundation.

Hodges, R.W. 1986. Gelechioidea: Gelechiidae (part), Dichomeridinae. *In* Dominick, R.B. *et al.* (Eds). *The Moths of America North of Mexico* 7(1): xiii+195 pp. The Wedge Entomological Research Foundation, Washington.

Hodkinson, I.D. & Casson, D. 1991. A lesser predilection for bugs: Hemiptera (Insecta) diversity in tropical rain forests. *Biological Journal of the Linnean Society* 43: 101-109.

Hoegh-Guldberg, O. 1972. Pupal sound production of some Lycaenidae. *Journal of Research on the Lepidoptera* 10: 127-147.

Holloway, J.D. 1977. The Lepidoptera of Norfolk Island. *Series Entomologica* 13: 291 pp. Junk, The Hague.

Holloway, J.D. 1980. Insect surveys - an approach to environmental monitoring. *Atti XII Congresso Nazionale Italiano di Entomologia Roma* 1: 239-261.

Holloway, J.D. 1983. The Moths of Borneo: Family Notodontidae. *Malayan Nature Journal* 37: 107 pp.+98 b & w figs and 9 col. pls.

Holloway, J.D. 1984. The larger moths of the Gunung Mulu National Park; a preliminary assessment of their distribution, ecology, and potential as environmental indicators. *The Sarawak Museum Journal* 30 (2): 149-190 + pl. xii.

Holloway, J.D. 1985a. The Moths of Borneo, 14. *Malayan Nature Journal* 38: 157-317 + 339 figs. and 8 col. pls.

Holloway, J.D. 1985b. Moths as indicator organisms for categorizing rain-forest and monitoring charges and regeneration processes. *In*: Chadwick, A.C. and Sutton, S.L. (Eds). *Tropical Rain-Forest: The Leeds Symposium* pp. 235-242. Special Publications of the Leeds Philosophical and Literary Society.

Holloway, J.D. 1986. The Moths of Borneo, 1. *Malayan Nature Journal* 40: 166 pp. + 9 pls.

Holloway, J.D. 1987a. *The Moths of Borneo* 3: 199 pp. + 9 b & w pls and 20 col. pls. Southdene, Kuala Lumpur, Malaysia.

Holloway, J.D. 1987b. Bracken-feeding Geometridae in the genus *Idiodes* Guenée, 1857, and allied taxa in the tribe Lithinini (Lepidoptera). *Tinea* 12 (Supplement): 242-248.

Holloway, J.D. 1988. *The Moths of Borneo* 6, 101 pp. + 168 figs, + 6 col. pls. The Malayan Nature Society and Southdene, Kuala Lumpur.

Holloway, J.D. 1989. *The Moths of Borneo* 12. *The Malayan Nature Journal* 42: 57-226 + 404 figs + 8 pls.

Holloway J.D., Bradley, J.D. & Carter, D.J. 1987. Lepidoptera. *In* Betts, C.R. (Ed.). *CIE Guides to Insects of Importance to Man* 1: 262 pp. CAB International, Wallingford.

Holloway, J.D. & Hebert, P.D.N. 1979. Ecological and taxonomic trends in macrolepidopteran host plant selection. *Biological Journal of the Linnean Society* 112: 229-251.

Hopp, W. 1934-5. Family: Megalopygidae. *In* Seitz, A. *The Macrolepidoptera of the World* 6: 1071-1101.

Horak, M. 1984. Assessment of taxonomically significant structures in Tortricinae (Lep., Tortricidae). *Mitteilungen der Schweizerischen Entomologischen Gesellschaft* 57: 3-64.

Horak, M. 1991. Morphology. *In* van der Geest, L.P.S. and Evenhuis, H.H. (Eds). *Tortricoid Pests, their Biology, Natural Enemies and Control.* pp. 1-22. Elsevier, Amsterdam.

Horak, M. & Brown, R.L. 1991. Taxonomy and phylogeny. *In* van der Geest, I.P.S. and Evenhuis, H.H. (Eds). *Tortricoid Pests, their Biology, Natural Enemies and Control*. pp. 23-48. Elsevier, Amsterdam.

Horak, M. & Common, I.F.B. 1985. A revision of the Australian genus *Epitymbia* Meyrick, with remarks on the Epitymbiini (Lepidoptera: Tortricidae). *Australian Journal of Zoology* 33: 577-622.

Humphries, C.J., Cox, J.M. & Nielsen, E.S. 1986. *Nothofagus* and its parasites: a cladistic approach to coevolution. *In* Stone, A.R. and Hawksworth, D.L. (Eds). *Coevolution and Systematics*. pp. 55-76. The Systematics Association and the Clarendon Press, Oxford.

Hunter, M.D. 1987. Sound production in larvae of *Diurnea fagella* (Lepidoptera: Oecophoridae). *Ecological Entomology* 12: 355-357.

Huxley, J. 1975. The basis of structural colour variation in two species of *Papilio*. *Journal of Entomology* A 50: 9-22.

Huxley, J. 1976. The coloration of *Papilio zalmoxis* and *P. antimachus* and the discovery of Tyndall blue in butterflies. *Proceedings of the Royal Society of London* B 193: 441-453.

Huxley J. & Barnard, P.C. 1988. Wing-scales of *Pseudoleptocerus chirindensis* Kimmins (Trichoptera: Leptoceridae). *Zoological Journal of the Linnean Society* 92: 285-312.

Igarashi, S. 1984. The classification of the Papilionidae mainly based on the morphology of their immature stages. *Tyô to Ga* 34: 41-96.

Inoue, H. 1962. Lepidoptera: Cyclidiidae [and] Drepanidae. *Insecta Japonica* 2: 1-54 + pls 1-3.

Ishii, S., Inokuchi, T., Kanazawa, J. & Tomizawa, C. 1984. Studies of the cocoon of the oriental moth, *Monema flavescens*, (Lepidoptera: Limacodidae) 3. Structure and composition of the cocoon in relation to hardness. *Japanese Journal of Applied Entomology and Zoology* 28: 269-273.

Issiki, S.T. & Stringer, H. 1932. On new oriental genera and species of the Hepialoidea (Lepidoptera Homoneura). Part I: Descriptions of new genera and species. Part II: Observations on the systematic position. *Stylops* 1: 71-72; 73-80.

Jäckh, E. 1955. Schutzvorrichtung zum Bau des Verpuppungskokons bei Arten der Gattung *Bucculatrix* Z. und *Lyonetia* Hb. (Lep., Lyonetiidae). *Zeitschrift der Wiener Entomologischen Gesellschaft* 40: 118-121 + pls 6-9.

James, D.G. 1984. Migration and clustering phenology of *Danaus plexippus* in Australia. *Journal of the Australian Entomological Society* 23: 199-204.

Jander, U. 1966. Untersuchungen zur Stammesgeschichte von Putzbewegungen von Tracheaten. *Zeitschrift für Tierpsychologie* 23: 799-844.

Janse, A.J.T. 1925. A revision of the South African Metarbelinae. *South African Journal of Natural History* 5: 61-100 + pls 4-8.

Janse, A.J.T. 1932. *The Moths of South Africa* 1: xi+376 pp. + pls 1-15. Pretoria.

Janse, A.J.T. 1942-8. *The Moths of South Africa* 4: xxv+185 pp. + pls 47-94. Pretoria.

Janse, A.J.T. 1945. On the history and life-history of *Leto venus* Stoll. *Journal of the Entomological Society of southern Africa* 8: 154-156.

Janzen, D.H. 1980. When is it coevolution? *Evolution* 34: 611-612.

Janzen, D.H. 1983. (Ed.). *Costa Rican Natural History*. xi+816 pp. University of Chicago Press, Chicago and London.

Janzen, D.H. 1984. Two ways to be a tropical big moth: Santa Rosa saturniids and sphingids. *Oxford Surveys in Evolutionary Biology* 1: 85-140.

Janzen, D.H. 1987. Insect diversity of a Costa Rican dry forest: why keep it, and how? *Biological Journal of the Linnean Society* 30: 343-356.

Jayewickreme, S.H. 1940. A comparative study of the larval morphology of leaf-mining Lepidoptera in Britain. *Transactions of the Royal Entomological Society of London* 90: 63-105.

Jobling, B. 1936. On the stridulation of the females of *Parnassius mnemosyne* L. *Proceedings of the Entomological Society of London* A 11: 66-68.

Johannsmeier, M.F. 1976. *A morphological study of the proboscis and observations on the feeding habits of fruit-piercing moths*. M.Sc. thesis, University of Pretoria. 95 pp. + 14 figs.

Johnson, C.G. 1969. *Migration and Dispersal of Insects in Flight*. xxii+763 pp. Methuen, London.

Jordan, K. 1898. Contributions to the morphology of Lepidoptera. *Novitates Zoologicae* 5: 374-415.

Jordan, K. 1923a. On the comb-bearing flap present on the fourth abdominal segment in the males of certain Notodontidae. *Novitates Zoologicae* 30: 153-154.

Jordan, K. 1923b. On a sensory organ found on the head of many Lepidoptera. *Novitates Zoologicae* 30: 155-158 + pl. 2.

Jordan, K. 1923c. A note on the families of moths in which R² (= vein 5) of the forewing arises near the centre or from above the centre of the cell. *Novitates Zoologicae* 30: 163-166.

Jordan, K. 1924. On the Saturnoidean families Oxytenidae and Cercophanidae. *Novitates Zoologicae* 31: 135-193 + pls vi-xxi.

Kammer, A.E. 1981. Physiological mechanisms of thermoregulation. *In* Heinrich, B. (Ed.). *Insect Thermoregulation.* pp. 115-158. John Wiley & Sons.

Kammer, A.E. & Bracchi, J. 1973. Role of the wings in the absorption of radiant energy by a butterfly. *Comparative Biochemical Physiology* A 45: 1057-1064.

Karsch, F., 1898. Giebt es ein System der recenten Lepidopteren auf phylogenetischer Basis? *Entomologische Nachrichten* 24: 296-303.

Karsholt, O. & Nielsen, E.S. 1984. A taxonomic review of the stem moths, *Ochsenheimeria* Hübner, of northern Europe (Lepidoptera: Ochsenheimeriidae). *Entomologica scandinavica* 15: 233-247.

Karsholt, O., Kristensen, N.P., Kaaber, S., Larsen, K., Nielsen, E.B., Palm, E., Schnack, K., Skou, P. & Skule, B. 1985. Catalogue of the Lepidoptera of Denmark. *Entomologiske Meddelelser* 52: 163 pp. [In English and Danish.]

Kayser, H. 1985. Pigments. *In* Kerkut, G.A. and Gilbert, L.I. *Comprehensive Insect Physiology, biochemistry and pharmacology* 10: 367-415. Pergamon Press.

Kellogg, V.A. 1894. The taxonomic value of the scales of the Lepidoptera. *The Kansas University Quarterly* 3: 45-89.

Kendall, R.O. 1976. Larval foodplants and life history notes for eight moths from Texas and Mexico. *Journal of the Lepidopterists' Society* 30: 264-271.

Kennel, J. & Eggers, F. 1933. Die abdominalen Tympanalorgane der Lepidopteren. *Zoologische Jahrbücher, Abteilung für Anatomie und Ontogenie der Tiere* 57: 1-104 + 6 pls.

Kettlewell, H.B.D. 1955. Recognition of appropriate backgrounds by the pale and black phases of Lepidoptera. *Nature* 175: 943-944.

Kettlewell, H.B.D. 1973. *The Evolution of Melanism* xxiv+423 pp. + 19 pls. Clarendon Press, Oxford.

Kingsolver, J.G. 1988. Thermoregulation, flight, and the evolution of wing pattern in pierid butterflies: the topography of adaptive landscapes. *American Zoologist* 28: 899-912.

Kingsolver J.G. & Daniel, T.L. 1979. On the mechanics and energetics of nectar feeding in butterflies. *Journal of Theoretical Biology* 76: 167-179.

Kiriakoff, S.G. 1960. Lepidoptera. Fam. Thyretidae. *In* Wytsman, P. (Ed.). *Genera Insectorum* 214E: 66 pp. + 2 pls. Brusselles.

Kiriakoff, S.G. 1963. The tympanic structures of the Lepidoptera and the taxonomy of the order. *Journal of the Lepidopterists' Society* 17: 1-6.

Kitching, I.J. 1984. An historical review of the higher classification of the Noctuidae (Lepidoptera). *Bulletin of the British Museum (Natural History) (Entomology)* 49: 153-234.

Kitching, I.J, 1987. Spectacles and silver Ys: a synthesis of the systematics, cladistics and biology of the Plusiinae (Lepidoptera: Noctuidae). *Bulletin of the British Museum (Natural History) (Entomology)* 54: 75-261.

Klots, A.B. 1933. A generic revision of the Pieridae (Lepidoptera). *Entomologica Americana* 12: 139-242 + pls 5-13.

Klots, A.B. 1970. Lepidoptera. *In* Tuxen, S.L. (Ed.). *Taxonomist's Glossary of genitalia in insects*, 2nd edn. pp. 115-130. Munksgaard, Copenhagen.

Kobayashi, Y. & Ando, H. 1981. The embryonic development of the primitive moth *Neomicropteryx nipponensis* Issiki (Lepidoptera, Micropterigidae): morphogenesis of the embryo by external observation. *Journal of Morphology* 169: 49-59.

Kobayashi, Y. & Gibbs, G.W. 1990. The formation of the germ rudiments and embryonic membranes in the mnesarchaeid moth, *Mnesarchaea fusilella* (Lepidoptera: Mnesarchaeidae). *Bulletin of the Sugadaira Montane Research Centre* 11: 107-109.

Kôda, N. 1987. A generic classification of the subfamily Arctiinae of the Palaearctic and Oriental regions based on the male and female genitalia (Lepidoptera, Arctiidae), Part I. *Tyô to Ga* 38: 153-237.

Kozhantschikov, I.V. 1956. Psychidae - *Fauna SSSR* 62: 1-516. [English translation 1969 by Israel Programme for Scientific Translations.]

Krampl, F. & Dlabola, J. 1983. A new genus and species of Epipyropid moth from Iran ectoparasitic on a new *Mesophantia* species, with a revision of the host genus (Lepidoptera, Epipyropidae; Homoptera, Flatidae). *Acta Entomologica Bohemoslovaca* 80: 451-472 + pls 1-2.

Kristensen, N.P. 1967. Erection of a new family in the lepidopterous suborder Dacnonypha. *Entomologiske Meddelelser* 35: 341-345.

Kristensen, N.P. 1968a. The skeletal anatomy of the heads of adult Mnesarchaeidae and Neopseustidae (Lep., Dacnonypha). *Entomologiske Meddelelser* 36: 137-151.

Kristensen, N.P. 1968b. The anatomy of the head and the alimentary canal of adult Eriocraniidae (Lep., Dacnonypha). *Entomologiske Meddelelser* 36: 239-315.

Kristensen, N.P. 1970. Morphological observations on the wing scales in some primitive Lepidoptera (Insecta). *Journal of Ultrastructure Research* 30: 402-410.

Kristensen, N.P. 1974. On the evolution of wing transparency in Sesiidae (Insecta, Lepidoptera). *Videnskabelige Meddelelser fra Dansk Naturhistorisk* 137: 125-134.

Kristensen, N.P. 1976. Remarks on the family-level phylogeny of butterflies (Insecta, Lepidoptera, Rhopalocera). *Zeitschrift für Zoologische Systematik und Evolutionsforschung* 14: 25-33.

Kristensen, 1978a. A new familia of Hepialoidea from South America, with remarks on the phylogeny of the suborder Exoporia (Lepidoptera). *Entomologica Germanica* 4: 272-294.

Kristensen, N.P. 1978b. Observations on *Anomoses hylecoetes* (Anomosetidae), with a key to the hepialoid families (Insecta: Lepidoptera). *Steenstrupia* 5: 1-19.

Kristensen, N.P. 1984a. The pregenital abdomen of the Zeugloptera (Lepidoptera). *Steenstrupia* 10: 113-136.

Kristensen, N.P. 1984b. Studies on the morphology and systematics of primitive Lepidoptera (Insecta). *Steenstrupia* 10: 141-191.

Kristensen, N.P. 1984c. The male genitalia of *Agathiphaga* (Lepidoptera, Agathiphagidae) and the lepidopteran ground plan. *Entomologica scandinavica* 15: 151-178.

Kristensen, N.P. 1984d. The larval head of *Agathiphaga* (Lepidoptera, Agathiphagidae) and the lepidopteran ground plan. *Systematic Entomology* 9: 63-81.

Kristensen, N.P. 1984e. Skeletomuscular anatomy of the male genitalia of *Epimartyria* (Lepidoptera: Micropterigidae). *Entomologica scandinavica* 15: 97-112.

Kristensen, N.P. 1986. Structure and phylogeny of the lowest Lepidoptera - Glossata: some recent advances. *Fifth European Congress in Lepidopterology Budapest* [Abstracts]: p. 27.

Kristensen, N.P. 1990a. The trunk integument of zeuglopteran larvae: one of the most aberrant arthropod cuticles known (Insecta: Lepidoptera). *Bulletin of the Sugadaira Montane Research Centre* 11: 101-102.

Kristensen, N.P. 1990b. Morphology and phylogeny of the lowest Lepidoptera-Glossata: recent progress and unforeseen problems. *Bulletin of the Sugadaira Montane Research Centre* 11: 105-106.

Kristensen, N.P. & Nielsen, E.S. 1979. A new subfamily of micropterigid moths from South America. A contribution to the morphology and phylogeny of the Micropterigidae, with a generic catalogue of the family (Lepidoptera: Zeugloptera). *Steenstrupia* 5: 69-147.

Kristensen, N.P. & Nielsen, E.S. 1980. The ventral diaphragm of primitive (non-ditrysian) Lepidoptera. A morphological and phylogenetic study. *Zeitschrift für zoologische Systematik und Evolutionsforschung* 18: 123-146.

Kristensen, N.P. & Nielsen, E.S. 1981a. Intrinsic proboscis musculature in non-ditrysian Lepidoptera-Glossata: Structure and phylogenetic significance. *Entomologica scandinavica* Supplement 15: 299-304.

Kristensen, N.P. & Nielsen, E.S. 1981b. Double-tube proboscis configuration in neopseustid moths (Lepidoptera: Neopseustidae). *International Journal of Insect Morphology and Embryology* 10: 483-486.

Kristensen, N.P. & Nielsen, E.S. 1982. South American micropterigid moths: two new genera of the *Sabatinca*-group (Lepidoptera: Micropterigidae). *Entomologica scandinavica* 13: 513-529.

Kristensen, N.P. & Nielsen, E.S. 1983. The *Heterobathmia* life history elucidated: Immature stages contradict assignment to suborder Zeugloptera (Insecta, Lepidoptera). *Zeitschrift für zoologische Systematik und Evolutionsforschung* 21: 101-124.

Kudrna, O. 1986. Aspects of the conservation of butterflies in Europe. *In* Kudrna, O. (Ed.). *Butterflies of Europe* 8: 323 pp. Aula-Verlag, Wiesbaden.

Kuijten, P.J. 1974. On the occurrence of a hitherto unknown wing-thorax coupling mechanism in Lepidoptera. *Netherlands Journal of Zoology* 24: 317-322.

Kuroko, H. 1964. Revisional studies on the family Lyonetiidae of Japan (Lepidoptera). *Esakia* 4: 1-61 + 17 pls.

Kuroko, H. 1990. Preliminary ecological notes on *Ogygioses caliginosa* Issiki & Stringer (Palaeosetidae). *Bulletin of the Sugadaira Montane Research Centre* 11: 103-104.

Kuznetzov, V.I. & Stekol'nikov, A.A. 1985. Comparative and functional morphology of the male genitalia of the Bombycoid moths (Lepidoptera, Papilionomorpha: Lasiocampoidea, Sphingoidea, Bombycoidea) and their systematic position. *Proceedings of the Zoological Institute USSR Academy of Sciences* 134: 3-48. [In Russian with an English summary.]

Kyrki, J. 1983a. Adult abdominal sternum II in ditrysian tineoid superfamilies - morphology and phylogenetic significance (Lepidoptera). *Annales Entomologici Fennici* 49: 89-94.

Kyrki, J. 1983b. *Roeslerstammia* Zeller assigned to Amphitheridae, with notes on the nomenclature and systematics of the family (Lepidoptera). *Entomologica scandinavica* 14: 321-329.

Kyrki, J. 1984. The Yponomeutoidea: a reassessment of the superfamily and its suprageneric groups (Lepidoptera). *Entomologica scandinavica* 15: 71-84.

Kyrki, J. 1988. The systematic position of *Wockia* Heinemann, 1870, and related genera (Lepidoptera: Ditrysia: Yponomeutidae auct.). *Nota lepidopterologica* 11: 45-69.

Kyrki, J. 1990. Tentative reclassification of holarctic Yponomeutoidea (Lepidoptera). *Nota lepidopterologica* 13: 23-42.

Kyrki, J. & Itämies, J. 1986. Immature stages and the systematic position of *Orthotelia sparganella* (Thunberg) (Lepidoptera: Yponomeutoidea). *Systematic Entomology* 11: 93-105.

Lafontaine, J.D. & Poole, R.W., in press. Noctuoidea: Noctuidae (part), Plusiinae. *In* Dominick, R.B., *et al.* (Eds). *The Moths of America North of Mexico*. The Wedge Entomological Research Foundation, Washington.

Laithwaite, E., Watson, A. & Whalley, P.E.S. 1975. *The Dictionary of Butterflies and Moths in colour.* xlvi + 296 pp. Michael Joseph, London.

Land, M.F. 1985. The eye: optics. *In* Kerkut, G.A. and Gilbert, L.I. (Eds). *Comprehensive insect physiology, biochemistry and pharmacology* 6: 225-275. Pergamon Press.

Landry, J.-F. 1991. Systematics of Nearctic Scythrididae (Lepidoptera: Gelechioidea): phylogeny and classification of supraspecific taxa, with a review of described species. *Memoirs of the Entomological Society of Canada* 160: 341 pp.

Lane, M.A. 1973. A new genus and species of Ethiopian Thyatiridae (Lepidoptera). *Journal of Natural History* 7: 267-272.

Lange, W.H. 1956. Aquatic Lepidoptera. *In* Usinger, R.L. (Ed.). *Aquatic Insects of California* pp. 271-288. University of California Press, Berkeley and Los Angeles.

Larsen, T.B. 1981. Butterflies as prey for *Orthetrum austenti* (Kirby) (Anisoptera: Libellulidae). *Notulae odonatologicae* 1: 130-133.

Latreille, P.A. 1796. *Précis des caractères génériques des insectes.* xiii + 201[7] pp. Bordeaux.

Lemaire, C. 1978. Les Attacidae Americains. Attacinae. 238 pp. + 49 pls. C. Lemaire, Neuilly-sur-Seine, France.

Lemaire, C. 1980. Les Attacidae Americains. Arsenurinae. 199 pp. + 75 pls. C. Lemaire, Neuilly-sur-Seine, France.

Lemaire, C. 1988. Les Saturniidae Americains. Ceratocampinae. 479 pp. + 64 pls. Museo Nacional de Costa Rica, San José.

Leraut, P. 1980. Liste systématique et synonymique des lépidoptères de France, Belgique et Corse. *Alexanor* (Supplement) 334 pp.

Link, E. 1909. Über die Stirnaugen der Neuroptera und Lepidopteren. *Zoologische Jahrbücher Abteilung für Anatomie und Ontogenie der Tiere* 27: 213-242 + pls 15-17.

Linnaeus, C. 1758. *Systema Naturae* (Edn. 10). 823 pp. Stockholm.

Lloyd, J.E. 1974. Genital stridulation in *Psilogramma menephron* (Sphingidae). *Journal of the Lepidopterists' Society* 28: 349-351.

Lorenz, R.E. 1961. Biologie und Morphologie von *Micropterix calthella* (L.). *Deutsche Entomologische Zeitschrift* (Neue Folge) 8: 1-23.

Luff, M.L. 1964. [Untitled.] *Proceedings of the Royal Entomological Society of London* C 29: 6.

Lyonet, P. 1832. Recherches sur l'anatomie et les métamorphoses de différentes espèces d'insectes. *Mémoires du Muséum d'Histoire Naturelle* 18-22. Paris 1829-1832.

McCabe, T.L. & Wagner, D.L. 1989. The biology of *Sthenopis aurafus* (Grote) (Lepidoptera: Hepialidae). *Journal of the New York Entomological Society* 97: 1-10.

McColl, H.P. 1969. *The sexual scent organs of male Lepidoptera.* MSc thesis. 214 pp. University College of Swansea, Wales.

McFarland, N. 1973a. Notes on describing, measuring, preserving and photographing the eggs of Lepidoptera. *Journal of Research on the Lepidoptera* 10: 203-214.

McFarland, N. 1973b. Egg photographs depicting 40 species of southern Australian moths. *Journal of Research on the Lepidoptera* 10: 215-247.

McFarland, N. 1979. Annotated list of larval foodplant records for 280 species of Australian moths. *Journal of the Lepidopterists' Society* 33 suppl: 72 pp.

McFarland, N. 1988. *Portraits of South Australian Geometrid Moths* iv + 400 pp. Allen Press, Kansas.

McGuffin, W.C. 1967. Guide to the Geometridae of Canada (Lepidoptera). 1. Subfamily Sterrhinae. *Memoirs of the Entomological Society of Canada* 50: 67 pp.

McGuffin, W.C. 1987. Guide to the Geometridae of Canada (Lepidoptera). *Memoirs of the Entomological Society of Canada* 138: 182 pp.

McIver, S.B. 1985. Mechanoreception. *In* Kerkut, G.A. and Gilbert, L.I. (Eds). *Comparative insect physiology, biochemistry and pharmacology* 6: 71-132.

McQuillan, P.B. 1981. A review of the Australian moth genus *Thalania* (Lepidoptera: Geometridae: Ennominae). *Transactions of the Royal Society of South Australia* 105: 1-23.

McQuillan, P.B. 1985. A review of the Australian autumn gum moth genus *Mnesampela* Guest (Lepidoptera: Geometridae, Ennominae). *Entomologica scandinavica* 16: 175-202.

McQuillan, P.B. 1986. Trans-Tasman relationships with the highland moth (Lepidoptera) fauna. *In* Barlow, B.A. (Ed.). *Flora and Fauna of Alpine Australasia* pp. 263-276.

Maes, K. 1985. A comparative study of the abdominal tympanal organs in Pyralidae (Lepidoptera). I. Description, terminology, preparation technique. *Nota lepidopterologica* 8: 341-350.

Maes, K.V.N. 1987. *Een vergelijkende, morfologisch-systematische studie van de Pyralidae (Lepidoptera) met bijzondere aandacht voor de Pyraustinae.* PhD thesis. Rijksuniversiteit, Gent.

Malcolm, S.B., Cockrell, B.J., & Brower, L.P. *In press.* Milkweed cardenolides as labels of the Monarch's spring remigration strategy. *In* Malcolm, S.B. and Zaluchi, M. (Eds). *Biology and Conservation of the Monarch Butterfly.* Los Angeles County Museum, Los Angeles.

Mallet, J. 1984. Sex roles in the ghost moth *Hepialus humuli* (L.) and a review of mating in the Hepialidae (Lepidoptera). *Zoological Journal of the Linnean Society* 79: 67-82.

Marti, O.G. & Rogers, C.E. 1988. Anatomy of the ventral eversible gland of Fall Armyworm, *Spodoptera frugiperda* (Lepidoptera: Noctuidae), larvae. *Annals of the Entomological Society of America* 81: 308-317.

Maschwitz, U., Dumpert, K. & Tuck, K.R. 1986. Ants feeding on anal exudate from tortricid larvae: a new type of trophobiosis. *Journal of Natural History* 20: 1041-1050.

Matsuda, R. 1965. Morphology and evolution of the insect head. *Memoirs of the American Entomological Institute* 4: 334 pp.

Matsuda, R. 1970. Morphology and evolution of the insect thorax. *Memoirs of the Entomological Society of Canada* 76: 431 pp.

Matsuda, R. 1976. *Morphology and evolution of the insect abdomen.* viii + 534 pp. Pergamon Press, Headington.

Matthews, M. 1987. The African species of *Heliocheilus* Grote (Lepidoptera: Noctuidae). *Systematic Entomology* 12: 459-473.

Matthews, M. 1991. Classification of the Heliothinae. *Bulletin of the Natural Resources Institute* 44: vi + 195 pp.

May, M.L. 1985. Thermoregulation. pp. 507-552. *In* Kerkut, G.A. and Gilbert, L.I. (Eds). *Comparative Insect Physiology, Biochemistry and Pharmacology* 4: 507-552. Pergamon Press.

Mazanec, Z. 1974. Influence of Jarrah leaf miner on the growth of Jarrah. *Australian Forestry* 37: 32-42.

Mazanec, Z. 1983. The immature stages and life history of the Jarrah leaf miner, *Perthida glyphopa* Common (Lepidoptera: Incurvariidae). *Journal of the Australian Entomological Society* 25: 149-159.

Mazokhin-Porshnyakov, G.A. & Kazyakina, V.I. 1984. Structure of the organs of vision in the larvae of *Lymantria dispar* L. (Lepidoptera, Lymantriidae). *Entomological Review* 63: 26-32. [English translation from *Entomologicheskoye Obrozreniye*.]

Mell, R. 1922. *Beiträge zur Fauna Sinica* 2. (Biologie und Systematik der südchinesischen Sphingiden). 331 pp. Friedländer, Berlin.

Mell, R. 1928. Family Brahmaeidae. *In* Seitz, A. (Ed.). *The Macrolepidoptera of the World* 10: 521-522.

Meyrick, E. 1895. *A handbook of British Lepidoptera*. vi + 843 pp. Macmillan, London and New York.

Meyrick, E. 1920. Descriptions of South African Micro-Lepidoptera. *Annals of the South African Museum* 17: 273-318.

Meyrick, E. 1928. *A revised Handbook of British Lepidoptera*. vi + 914 pp. Watkins and Doncaster, London.

Michener, C.D. 1952. The Saturniidae (Lepidoptera) of the western hemisphere. *Bulletin of the American Museum of Natural History* 98: 339-501.

Mikkola, K. 1974. Timing of swarming in *Hepiolus* [sic.] *humuli* L. (Lep., Hepiolidae [sic.]): a summary. *Entomologiske Meddelelser* 42: 76-77.

Mikkola,K. 1984. On the selective forces acting in the industrial melanism of *Biston* and *Oligia* moths (Lepidoptera: Geometridae and Noctuidae). *Biological Journal of the Linnean Society* 21: 409-421.

Miller, J.S. 1987a. Host-plant relationships in the Papilionidae (Lepidoptera): parallel cladogenesis or colonization? *Cladistics* 3: 105-120.

Miller, J.S. 1987b. A revision of the genus *Phryganidia* Packard with description of a new species (Lepidoptera: Dioptidae). *Proceedings of the Entomological Society of Washington* 82: 303-321.

Miller, J.S. 1987c. Phylogenetic studies in the Papilioninae (Lepidoptera: Papilionidae). *Bulletin of the American Museum of Natural History* 186: 365-512.

Miller, J.S. 1991. Cladistics and classification of the Notodontidae (Lepidoptera: Noctuoidea) based on larval and adult morphology. *Bulletin of the American Museum of Natural History* 204: 230 pp.

Miller, J.Y. 1972. Review of the central American *Castnia inca* complex (Castniidae). *Bulletin of the Allyn Museum* 6: 1-13.

Miller, J.Y. 1980. Studies in the Castniidae. III *Microcastnia. Bulletin of the Allyn Museum* 60: 15 pp.

Miller, J.Y. 1986. *The taxonomy, phylogeny, and zoogeography of the neotropical moth subfamily Castniinae (Lepidoptera: Castnioidea: Castniidae)*. xv + 571 pp. PhD thesis, University of Florida.

Miller, S.E. *In press*. Systematics of the Neotropical moth family Dalceridae (Lepidoptera). *Bulletin of the Museum of Comparative Zoology of Harvard*.

Millière, P. 1862. Iconographie et Description de Chenilles et Lépidoptères inédits. *Annales de la Société Linéenne de Lyon* (NS) 6: [385]-446 pls 1-6.

Millière, P. 186[4]. Iconographie et Description de Chenilles et Lépidoptères inédits. *Annales de la Société Linéenne de Lyon* (NS) 10: [217]-244 pls. 41-44.

Minet, J. 1980. Création d'une sous-famille particulière *Noorda* Walker, 1859, et définition d'un nouveau genre parmi les Odontiinae. *Bulletin de la Société Entomologique de France* 85: 79-87.

Minet, J. 1982 (volume dated 1981). Les Pyraloidea et leurs principales divisions systématique. *Bulletin de la Société Entomologique de France* 86: 262-280.

Minet, J. 1983. Étude morphologique et phylogénétique des organes tympaniques des Pyraloidea. 1. Généralités et homologies. (Lep. Glossata). *Annales Société Entomologique de France* 19: 175-207.

Minet, J. 1984. Contribution a l'analyse phylogénétique des Néolépidoptères (Lepidoptera, Glossata). *Nouvelle Revue d'Entomologie* 2: 139-149.

Minet, J. 1985a. Étude morphologique et phylogénétique des organes tympaniques des Pyraloidea. 2 - Pyralidae; Crambidae, première partie. (Lepidoptera Glossata.) *Annales Société Entomologique de France* 21: 69-86.

Minet, J. 1985b. Définition d'un nouveau genre au seine des Drepanoidea paléarctiques (Lep. Drepanoidea). *Entomologica Gallica* 1: 291-304.

Minet, J. 1986. Ébauche d'une classification modern de l'ordre des Lépidoptères. *Alexanor* 14: 291-313.

Minet, J. 1987a. Les Ratardidae proposés comme groupe-frère des Callidulidae (Lep. Ditrysia). *Bulletin de la Société entomologique de France* 92: 39-44.

Minet, J. 1987b. Description d'une chrysalide de Pterothysaninae (Lep. Callidulidae). *Nouvelle Revue d'Entomologie* 4: 312.

Minet, J. 1988a. Structure and evolution of tympanic organs in Lepidoptera. *Proceedings of the XVII International Congress of Entomology* p. 78.

Minet, J. 1988b. The major ditrysian lineages and their interrelationships. *Proceedings of the XVII International Congress of Entomology* p. 79.

Minet, J. 1990a (volume dated 1989). Nouvelles frontières, géographiques et taxonomiques, pour la famille des Callidulidae (Lepidoptera, Calliduloidea). *Nouvelles Revue d'Entomologie* 6: 351-368.

Minet, J. 1990b. Remaniement partiel de la classification des Gelechioidea, essentiellement en fonction de caractères pré-imaginaux. *Alexanor* 16: 239-255.

Minet, J. 1991. Tentative reconstruction of the ditrysian phylogeny (Lepidoptera: Glossata). *Entomologica scandinavica* 22: 69-95.

Minnich, D.E. 1925. The reactions of the larvae of *Vanessa antiopa* Linn. to sounds. *Journal of Experimental Zoology* 42: 443-469.

Minnich, D.E. 1936. The responses of caterpillars to sounds. *Journal of Experimental Zoology* 72: 439-453.

Mitter, C. & Brooks, D.R. 1983. Phylogenetic aspects of coevolution. In Futuyma, D.J. and M. Slatkin (Eds). *Coevolution*. pp. 65-98. Sinauer, Sunderland, Massachusetts.

Montgomery, S.L. 1982. Biogeography of the moth genus *Eupithecia* in Oceania and the evolution of ambush predation in Hawaiian caterpillars (Lepidoptera: Geometridae). *Entomologia Generalis* 8: 27-34.

Moriuti, S. 1977. Yponomeutidae s. lat. (Insecta: Lepidoptera). *Fauna Japonica*. 327 pp. + 95 pls. Keigaku Publishing Co., Tokyo.

Moriuti, S. 1978. Amphitheridae (Lepidoptera): four new species from Asia, *Telethera blepharacma* Meyrick new to Japan and Formosa and *Sphenograptis* Meyrick transferred to the family. *Bulletin of the University of Osaka Prefecture* B 30: 1-17.

Morris, R.B. 1975. Iridescence from diffraction structures in the wing scales of *Callophrys rubi*, the Green Hairstreak. *Journal of Entomology* A 49: 149-154.

Mosher, E. 1916. A classification of the Lepidoptera based on characters of the pupa. *Bulletin of the Illinois State Laboratory of Natural History* 12: 15-159 + pls 19-27.

Munroe, E. 1959. Revision of the genus *Linosta* Möschler (Lepidoptera: Pyralidae) with characterization of the subfamily Linostinae and a new subfamily. *Canadian Entomologist* 91: 485-488.

Munroe, E. 1961. The classification of the Papilionidae (Lepidoptera). *Canadian Entomologist* (Supplement) 17: 1-51.

Munroe, E. 1970. Revision of the subfamily Midilinae (Lepidoptera: Pyralidae). *Memoirs of the Entomological Society of Canada* 74: 1-94.

Munroe, E. 1972-[4]. Pyraloidea: Pyralidae (in part). *In* R.B. Dominick *et al.* (Eds). *The Moths of America north of Mexico.* 13 (1): xx+304 pp.+ 23 pls. E.J. Classey and R.B.D. Publications, London.

Munroe, E. 1976. Pyraloidea: Pyralidae (in part). *In* Dominick, R.B., *et al. The Moths of America North of Mexico.* 13 (2): xvii+78 pp. + 12 pls. E.W. Classey and The Wedge Entomological Research Foundation, London.

Munroe, E.G. 1982. Lepidoptera. *In* Parker, S.B. (Ed.). *Synopsis and Classification of Living Organisms* 2: 612-651. McGraw-Hill.

Mutuura, A. 1972. Morphology of the female terminalia in Lepidoptera, and its taxonomic significance. *Canadian Entomologist* 104: 1055-1071.

Mutuura, A. & Munroe, E. 1970. Taxonomy and distribution of the European Corn Borer and allied species: genus *Ostrinia* (Lepidoptera: Pyralidae). *Memoirs of the Entomological Society of Canada* 71: 112 pp.

Nachtigall, W. 1976. Wing movements and the generation of aerodynamic forces by some medium-sized insects. *In* Rainey, R.C. *Insect Flight. Symposia of the Royal Entomological Society of London* 7: 31-47. Blackwell Scientific Publications.

Nakajima, H. 1970. A contribution to the knowledge of the immature stages of Drepanidae occurring in Japan. (Lepidoptera). *Tinea* 8: 167-184 + pls 34-53.

Nässig, W. 1980. Ein Beitrag zur Kenntnis der Saturniidae und Brahmaeidae des Iran und der Türkei (Lepidoptera). Teil 1: Brahmaeidae. *Nachrichten entomologische Verhandlung Apollo* (Neue Folge) 1: 77-91.

Nässig, W.A. 1988. Wehrorgane und Wehrmechanismen bei Saturniidenraupen (Lepidoptera, Saturniidae). *Verhandlungen des Westdeutscher Entomologentag* pp. 253-264. [In German with English summary.]

Nässig, W.A. & Lüttgen, M. 1988. Notes on genital stridulation in male hawkmoths in South East Asia (Lep., Sphingidae). *Heterocera Sumatrana* 2: 75-77.

Naumann, C. 1971. Untersuchungen zur Systematik und Phylogenese der holarktischen Sesiiden (Insecten, Lepidoptera). *Bonner Zoologische Monographie* 1: 190 pp.

Naumann, C.M. 1988. The internal female genitalia of some Zygaenidae (Insecta, Lepidoptera): their morphology and remarks on their phylogenetic significance. *Systematic Entomology* 13: 85-99.

Neunzig, H.H. 1986. Pyraloidea: Pyralidae (in part). *In* Dominick, R.B., *et al.* (Eds). *The Moths of America North of Mexico* 15 (2): xii+113 pp. The Wedge Entomological Research Foundation, Washington.

Nichols, S.W., Tulloch, G.S., and Torre-Bueno, J.R. 1989. *The Torre-Bueno glossary of entomology,* revised edition of *A glossary of entomology* (1937). xvii+840 pp. The New York Entomological Society and the American Museum of Natural History, New York.

Nielsen, A. 1980. A comparative study of the genital segments and the genital chamber in female Trichoptera. *Biologiske Skrifter udgivet af det Kongelige Danske Videnskabernes Selskab* 23(1): 200 pp.

Nielsen, E.S. 1978. On the systematic position of the genus *Eriocottis* Zeller, 1847, with remarks on the phylogeny of primitive Tineoidea (Lepidoptera). *Entomologica scandinavica* 9: 279-296.

Nielsen, E.S. 1980. A cladistic analysis of the Holarctic genera of adelid moths (Lepidoptera: Incurvarioidea). *Entomologica scandinavica* 11: 161-178.

Nielsen, E.S. 1981. A taxonomic revision of the species of *Alloclemensia* n. gen. (Lepidoptera: Incurvariidae s.str.). *Entomologica scandinavica* 12: 271-294.

Nielsen, E.S. 1982a. Incurvariidae and Prodoxidae from the Himalayan area (Lepidoptera: Incurvarioidea). *Insecta Matsumurana* 26: 187-200.

Nielsen, E.S. 1982b. The maple leaf-cutter moth and its allies: a revision of *Paraclemensia* (Incurvariidae s. str.). *Systematic Entomology* 7: 217-238.

Nielsen, E.S. 1982c. Review of the higher classification of the Lepidoptera, with special reference to lower heteroneurans. *Tyô to Ga* 33: 98-101.

Nielsen, E.S. 1985a. Primitive (non-ditrysian) Lepidoptera of the Andes: diversity, distribution, biology and phylogenetic relationships. *Journal of Research on the Lepidoptera* (Supplement) 1: 1-16.

Nielsen, E.S. 1985b. A taxonomic review of the adelid genus *Nematopogon* Zeller (Lepidoptera: Incurvarioidea). *Entomologica scandinavica* (Supplement) 25: 66 pp.

Nielsen, E.S. 1985c. The monotrysian heteroneuran phylogeny puzzle: a possible solution. *Proceedings of the 3rd Congress of European Lepidopterology*: pp. 138-143.

Nielsen, E.S. 1987. The recently discovered primitive (non-ditrysian) family Palaephatidae (Lepidoptera) in Australia. *Invertebrate Taxonomy* 1: 201-229.

Nielsen, E.S. 1988. The peculiar Asian Ghost Moth genus *Bipectilus* Chu & Wang: taxonomy and systematic position (Lepidoptera: Hepialidae s. str.). *Systematic Entomology* 13: 171-195.

Nielsen, E.S. 1989. Phylogeny of major lepidopteran groups. *In* Ferholm, B., Bremer, K., and Jörnvall, H. (Eds). *The Hierarchy of Life*. pp. 281-294. Elsevier.

Nielsen, E.S. & Common, I.F.B. 1991. Lepidoptera (moths and butterflies) Chapter 41. *In* Naumann, I.D. (Ed.). *The Insects of Australia* (2nd edn) 2: 817-915. Melbourne University Press, Carlton, Victoria & University College of London Press, London.

Nielsen, E.S. & Davis, D.R. 1981. A revision of the Neotropical Incurvariidae s. str., with the description of two new genera and two new species (Lepidoptera: Incurvarioidea). *Steenstrupia* 7: 25-57.

Nielsen, E.S. & Davis, D.R. 1985. The first southern hemisphere prodoxid and the phylogeny of the Incurvarioidea (Lepidoptera). *Systematic Entomology* 10: 307-322.

Nielsen, E.S. & Kristensen, N.P. 1989. Primitive Ghost Moths. Morphology and taxonomy of the Australian genus *Fraus* Walker (Lepidoptera: Hepialidae *s. lat.*). *Monographs on Australian Lepidoptera* 1: xvii+286 pp. CSIRO, Australia.

Nielsen, E.S. & Scoble, M.J. 1986. *Afrotheora*, a new genus of primitive Hepialidae from Africa (Lepidoptera: Hepialoidea). *Entomologica scandinavica* 17: 29-54.

Nielsen, E.T. 1961. On the habits of the migratory butterfly *Ascia monuste* L. *Biologiske Meddelelser* 23: 1-81.

Nieukerken, E.J. 1986. Systematics and phylogeny of Holarctic genera of Nepticulidae (Lepidoptera, Heteroneura: Monotrysia). *Zoologische Verhandelingen*: 236: 93 pp.

Nieukerken, E.J. & Dop, H. 1987. Antennal sensory structures in Nepticulidae (Lepidoptera) and their phylogenetic implications. *Zeitschrift für Zoologische Systematik und Evolutionsforschung* 25: 104-126.

Nijhout, H.F. 1978. Wing pattern formation in Lepidoptera: a model. *Journal of Experimental Zoology* 206: 119-136.

Nijhout, H.F. 1981. The colour patterns of butterflies and moths. *Scientific American* 245 (5): 104-115.

Nijhout, H.F. 1985. The developmental physiology of colour patterns in Lepidoptera. *In* Berridge, M.J., Treherne, J.E., and Wigglesworth, V.B. (Eds). *Advances in Insect Physiology* 18: 181-247. Academic Press.

Nijhout, H.F. 1991. *The development and evolution of butterfly wing patterns.* xvi+297 pp., 8 pls. Smithsonian Institution Press, Washington and London.

Nilsson, D.-E. 1989. Optics and evolution of the compound eye. *In* Stavenga, D.G. and Hardie, R.C. (Eds). *Facets of Vision*. pp. 30-73. Springer-Verlag.

Nilsson, D.-E., Land, M.F., and Howard, J. 1984. Afocal apposition optics in butterfly eyes. *Nature* 312: 561-563.

Norris, M.J. 1934. Contributions towards the study of insect fertility III. Adult nutrition, fecundity, and longevity in the genus *Ephestia* (Lepidoptera, Phycitidae). *Proceedings of the Zoological Society of London* 1934: 333-360.

Norris, M. J. 1936. The feeding-habits of adult Lepidoptera Heteroneura. *Transactions of the Royal Entomological Society of London* 85: 61-90.

Ochsenheimer, F. 1807-35. *Die Schmetterlinge von Europe* 1-10. Leipzig.

O Dell, T.M., Shields, K.S., Mastro, V.C., & Kring, T.J. 1982. The epiphysis of the gypsy moth, *Lymantria dispar* (Lepidoptera: Lymantriidae): structure and function. *Canadian Entomologist* 114: 751-761.

Oiticica J. 1947. Sobre a genitália das fœmeas de Hepialidae. *Summa Brasiliensis Biologiae* 1: 384-428.

Oord, W. 1981. Baltvlucht van *Nemophora degeerella* (Linnaeus) (Lep., Incurvariidae). *Entomologische Berichten* 41: 80.

Oudemans, J.T. 1903. Étude sur la position de repos chez les Lepidoptera. *Verhandelingen der Koninklijke Akademie van Wetenschappen, Amsterdam* (2) 10: 90 pp. + 11 pls.

Owada, M. 1987. *A taxonomic study on the subfamily Herminiinae of Japan* (Lepidoptera, Noctuidae). 208 pp. National Science Museum, Tokyo.

Pagenstecher, A. 1902. Callidulidae. *In* Schulze, F.E. *Das Tierreich*. 24 pp. R. Friedländer und Sohn, Berlin.

Paterson, H.E.H. 1985. The recognition concept of species. *In* Vrba, E.S. (Ed.). *Species and Speciation. Transvaal Museum Monograph* 4: 21-29. Pretoria.

Pavan, M. 1981. Utilità delle formiche del gruppo *Formica rufa*. *Collana verde* 57: 99 pp.+5 maps. (2nd Edn.) Ministero dell'Agricoltura delle Foreste, Roma.

Petersen, G. 1978. Zur systematischen Stellung der Gattung *Crinopteryx* Peyerimhoff, 1871 (Lepidoptera: Incurvariidae). *Beiträge zur Entomologie* 28: 217-220.

Peterson, A. 1962. *Larvae of Insects. An Introduction to Nearctic species*. Part I. 4th Edn. 315 pp. Columbus, Ohio.

Philpott, A. 1924. The wing coupling apparatus in *Sabatinca* and other primitive genera in Lepidoptera. *Report of the Australian Association for the Advancement of Sciences* 16: 414.

Philpott, A. 1925. On the wing coupling apparatus of the Hepialidae. *Transactions of the Entomological Society of London* 1925: 331-340.

Pinhey, E. 1975. *Moths of Southern Africa* 273 pp. + 63 pls. Tafelberg, Cape Town.

Pivnick, K.A. & McNeil, J.N. 1985. Effects of nectar concentration on butterfly feeding: measured feeding rates for *Thymelicus lineola* (Lepidoptera: Hesperiidae) and a general feeding model for adult Lepidoptera. *Oecologia* 66: 226-237.

Pliske, T.E. 1975a. Attraction of Lepidoptera to plants containing pyrrolizine alkaloids. *Environmental Entomology* 4: 455-473.

Pliske, T.E. 1975b. Courtship behavior and use of chemical communication by males of certain species of ithomiine butterflies (Nymphalidae: Lepidoptera). *Annals of the Entomological Society of America* 68: 935-942.

Pliske, T. E. & Eisner, T. 1969. Sex pheromone of the Queen Butterfly: biology. *Science* 164: 1170-1172.

Pocock, R.I. 1903. Notes on the commensalism subsisting between a gregarious spider, *Stegodyphus* sp., and the moth *Batrachedra stegodyphobius*, Wlsm. *The Entomologist's Monthly Magazine* 39: 167-170.

Portier, P. 1949. La biologie des lépidoptères. *In* Lechevalier, P. (Ed.) *Encyclopédie Entomologique* A 23: 643 pp. Paris.

Powell, J.A. 1971. Biological studies on moths of the genus *Ethmia* in California (Gelechioidea). *Journal of the Lepidopterists' Society* 25 (Supplement 3): 67 pp.

Powell, J.A. 1973. A systematic monograph of New World ethmiid moths (Lepidoptera: Gelechioidea). *Smithsonian Contributions to Zoology* 120: 302 pp.

Powell, J.A. 1980. Evolution of larval food preferences in Microlepidoptera. *Annual Review of Entomology* 25: 133-159.

Powell, J.A. & Common, I.F.B. 1985. Oviposition patterns and egg characteristics of Australian tortricine moths (Lepidoptera: Tortricidae). *Australian Journal of Zoology* 33: 179-216.

Proctor, M. & Yeo, P. 1973. *The pollination of flowers*. 418 pp. + 56 pls. Collins, London.

Prout, L.B. 1910. Lepidoptera Heterocera. Fam. Geometridae. Subfam. Oenochrominae. *In* Wytsman, P. (Ed.). *Genera Insectorum* 104: 120 pp. + 2 pls.

Rankin, M.A., McAnelly, & M.L. Bodenhamer, J.E. 1986. The oogenesis-flight syndrome revisited. *In* Danthanarayana, W. (Ed.). *Insect flight: dispersal and migration*. pp. 27-48. Springer-Verlag, New York, Heidelberg, and Berlin.

Rath, O. vom. 1887. Über die Hautsinnesorgane der Insecten. *Zoologischer Anzeiger* 10: 627-631, 645-649.

Rawlins, J.E. 1984. Mycophagy in Lepidoptera. *In* Wheeler Q. and Blackwell, M. (Eds). *Fungus-insect relationships*. pp. 382-423. New York: Columbia University Press.

Richards, A.G. 1933. Comparative skeletal morphology of the noctuid tympanum. *Entomologica Americana* 13: 1-43 + pls 1-20.

Rindge, F.M. 1986. Generic descriptions of New World Lithinini (Lepidoptera, Geometridae). *American Museum Novitates* 2838: 1-68.

Robbins, R.K. 1980. The lycaenid 'false head' hypothesis: historical review and quantitative analysis. *Journal of the Lepidopterists' Society* 34: 194-208.

Robbins, R.K. 1988a. Comparative morphology of the butterfly foreleg coxa and its trochanter (Lepidoptera) and its systematic implications. *Proceedings of the Entomological Society of Washington* 90: 133-154.

Robbins, R.K. 1988b. Male foretarsal variation in Lycaenidae and Riodinidae, and the systematic placement of *Styx infernalis* (Lepidoptera). *Proceedings of the Entomological Society of Washington* 90: 356-368.

Robbins, R.K. 1989. Systematic implications of butterfly leg structures that clean the antennae. *Psyche* 96: 209-222.

Robinson, G.S. 1979. Clothes-moths of the *Tinea pellionella* complex: a revision of the world's species (Lepidoptera: Tineidae). *Bulletin of the British Museum (Natural History) (Entomology)* 38: 57-128.

Robinson, G.S. 1980. Cave-dwelling tineid moths: a taxonomic review of the world species (Lepidoptera: Tineidae). *Transactions of the British Cave Research Association* 7: 83-120.

Robinson, G.S. 1986. Fungus moths: a review of the Scardiinae (Lepidoptera: Tineidae). *Bulletin of the British Museum (Natural History) (Entomology)* 52: 37-181.

Robinson, G.S. 1988a. The systematic position of *Thermocrates epischista* Meyrick (Lepidoptera: Tineidae) and the biology of the Dryadaulinae. *Nota lepidopterologica* 11: 70-79.

Robinson, G.S. 1988b. Keratophagous moths in tropical forests - investigations using artificial birds' nests. *The Entomologist* 107: 34-45.

Robinson, G.S. 1988c. A phylogeny for the Tineoidea (Lepidoptera). *Entomologica scandinavica* 19: 117-129.

Robinson, G.S. & Nielsen, E.S. *In press*. Tineid genera of Australia. *Monographs on Australian Lepidoptera* 2.

Roeder, K.D. 1965. Moths and ultrasound. *Scientific American* 212 (4): 94-102.

Roeder, K.D. 1972. Acoustic and mechanical sensitivity of the distal lobe of the pilifer in choerocampine Hawkmoths. *Journal of Insect Physiology* 18: 1249-1264.

Roeder, K.D. & Treat, A.E. 1970. An acoustic sense in some hawkmoths (Choerocampinae). *Journal of Insect Physiology* 16: 1069-1086.

Roelofs, W.L. & Brown, R.L. 1982. Pheromones and evolutionary relationships of Tortricidae. *Annual Review of Ecology and Systematics* 13: 395-422.

Roepke, W. 1957. The cossids of the Malaya region (Lepidoptera: Heterocera). *Verhandelingen der Koninklijke Nederlandse Akademie van Wetenschappen* (Natuurkunde) 52: 1-60 + 4 pls.

Roesler, R.U. 1973. Phycitinae. *In* Amsel, H.G., Gregor, F., and Reisser, H. (Eds). *Microlepidoptera Palaearctica* 4: xvi + 752 pp; 137 pp. + 170 pls.

Rothschild, M. 1985. British aposematic Lepidoptera. *In* Heath, J. and Emmet, A.M. (Eds). *The Moths and Butterflies of Great Britain and Ireland* 2: 9-62.

Rothschild, M., Keutmann, H., Lane, N.J., Parsons, J., Prince, W. & Swales, S.S. 1979. A study on the mode of action and composition of a toxin from the female abdomen and eggs of *Arctia caja* (L.) (Lep. Arctiidae): an electrophysiological, ultrastructural and biochemical analysis. *Toxicon* 17: 285-306.

Rothschild, W. & Jordan, K. 1903. A revision of the lepidopterous family Sphingidae. *Novitates Zoologicae* (Supplement) 9: cxxxv + 972 pp. + 67 pls.

Salkeld, E.H. 1983. A catalogue of the eggs of some Canadian Geometridae (Lepidoptera), with comments. *Memoirs of the Entomological Society of Canada* 126: 271 pp.

Salkeld, E.H. 1984. A catalogue of the eggs of some Canadian Noctuidae (Lepidoptera). *Memoirs of the Entomological Society of Canada* 127: 167 pp.

Sanderford, M.V. & Conner, W.E. 1990. Courtship sounds of the Polka-Dot Wasp Moth *Syntomeida epilais*. *Naturwissenschaften* 77: 345-347.

Sattler, K. 1967. Ethmiidae. *In* Amsel, H.G., Gregor, H.G., and Reisser, H. (Eds). *Microlepidoptera Palaearctica* 2: 185 pp. + 106 pls. Georg Fromme, Vienna.

Sattler, K. 1991a. Der 'Achselkamm' der Lepidoptera und seine Funktion. *Deutsche Entomologische Zeitschrift* 38: 7-11.

Sattler, K. 1991b. A review of wing reduction in Lepidoptera. *Bulletin of the British Museum (Natural History)* (Entomology) 60: 243-288.

Sattler, K. & Tremewan, W.G. 1978. A supplementary catalogue of the family-group and genus-group names of the Coleophoridae (Lepidoptera). *Bulletin of the British Museum (Natural History)* (Entomology) 37: 73-96.

Sauter, W. 1986. Zur Morphologie von *Acanthobrahmaea europaea* (Hartig, 1963) und zur systematischen Gliederung der Brahmaeidae (Lepidoptera): Dactyloceratinae, subfam. n. *Nota lepidopterologica* 9: 262-271.

Schneider, D. 1984a. Insect Olfaction - our research endeavour. *In* Dawson, W.W. and Enoch, J.M. (Eds). *Foundations of Sensory Science* pp. 381-418.

Schneider, D. 1984b. Pheromone biology in the Lepidoptera: overview, some recent findings and some generalizations. *In* Bolis, L., Keynes, R.D., and Maddrell, S.H.P. (Eds). pp. 301-313. *Comparative Physiology of Sensory Systems*. Cambridge University Press.

Schneider, D., Boppré, M., Zweig, J., Hansen, K., & Diehl, E.W. 1982. Scent organ development in *Creatonotos* moths: regulation by pyrrolizidine alkaloids. *Science* 215: 1264-1265.

Schoorl, J.W. 1990. A phylogenetic study on Cossidae (Lepidoptera: Ditrysia) based on external adult morphology. *Zoologische Verhandelingen* 263: 1-138.

Schrank, P. 1802. *Fauna Boica* 2(2): 412 pp. Ingolstadt.

Schultz, H. 1914. Das Pronotum und die Patagia der Lepidopteren. *Deutsche Entomologische Zeitschrift* 1914: 17-42 + 11 pls.

Schulze, P. 1911. Die Nackengabel der Papilionidenraupen. *Zoologische Jahrbücher für Anatomie und Ontogenie der Tiere* 32: 181-244 + pls 12-14.

Schwanwitsch, B.N. 1924. On the groundplan of wing-pattern in nymphalids and certain other families of rhopalocerous Lepidoptera. *Proceedings of the Zoological Society of London* B, 34: 509-528.

Scoble, M.J. 1980. A new incurvariine leaf-miner from South Africa, with comments on structure, life-history, phylogeny, and the binominal system of nomenclature (Lepidoptera: Incurvariidae). *Journal of the Entomological Society of southern Africa* 43: 77-88.

Scoble, M.J. 1982. A pectinifer in the Nepticulidae (Lepidoptera) and its phylogenetic implications. *Annals of the Transvaal Museum* 33: 123-129.

Scoble, M.J. 1983. A revised cladistic classification of the Nepticulidae (Lepidoptera) with descriptions of new taxa mainly from South Africa. *Transvaal Museum Monograph* 2: 105 pp.

Scoble, M.J. 1986. The structure and affinities of the Hedyloidea: a new concept of the butterflies. *Bulletin of the British Museum (Natural History)* (Entomology) 53: 251-286.

Scoble, M.J. 1990. An identification guide to the Hedylidae (Lepidoptera: Hedyloidea). *Entomologica scandinavica* 21: 121-158.

Scoble, M.J. 1991. Classification of the Lepidoptera. *In* Emmet, A.M. and Heath, J. (Eds). *The Moths and Butterflies of Great Britain and Ireland* 7(2): 11-45. Harley Books, Colchester.

Scoble, M.J. & Aiello, A. 1990. Moth-like butterflies (Hedylidae: Lepidoptera): a summary, with comments on the egg. *Journal of Natural History* 24: 159-164.

Scoble, M.J. & Edwards, E.D. 1988. *Hypsidia* Rothschild: a review and a reassessment (Lepidoptera: Drepanoidea, Drepanidae). *Entomologica scandinavica* 18: 333-353.

Scoble, M.J. & Edwards, E.D. 1990. *Parepisparis* Bethune-Baker and the composition of the Oenochrominae (Lepidoptera: Geometridae). *Entomologica scandinavica* 20: 371-399.

Scoble, M.J. & Scholtz, C.H. 1984. A new, gall-feeding moth (Lyonetiidae: Bucculatricinae) from South Africa with comments on larval habits and phylogenetic relationships. *Systematic Entomology* 9: 83-94.

Scott, J.A. 1985. The phylogeny of butterflies (Papilionoidea and Hesperioidea). *Journal of Research on the Lepidoptera* 23: 241-281.

Scott, J.A. 1986. On the monophyly of the Macrolepidoptera, including a reassessment of their relationship to Cossoidea and Castnioidea, and a reassignment of Mimallonidae to Pyraloidea. *Journal of Research on the Lepidoptera* 25: 30-38.

Seitz, A. 1911. Family: Endromididae. *In* Seitz, A.A. (Ed.). *The Macrolepidoptera of the World* 2: 193-194.

Seitz, A. 1912. Family: Heterogynidae. *In* Seitz, A. (Ed.). *The Macrolepidoptera of the World* 2: 349-350.

Seitz, A. 1915. Genus *Pterothysanus* Wkr. *In* Seitz, A. (Ed.). *The Macrolepidoptera of the World* 10: 277.

Seitz, A. 1924. Family: Callidulidae. *In* Seitz, A. *The Macrolepidoptera of the World* 10: 491-496.

Sellier, R. 1973. Recherches en microscopie électronique par balayage, sur l'ultrastructure de l'appareil androconial alaire dans le genre *Argynnis* et dans les genres voisins (Lep. Rhopalocères Nymphalides). *Annales de la Société Entomologique de France* 9: 703-728.

Shaffer, J.C. 1968. A revision of the Peoriinae and Anerastiinae (auctorum) of America north of Mexico (Lepidoptera: Pyralidae). *Bulletin of the United States National Museum* 280: 124 pp.

Shannon, R.C. 1928. Zoophilous moths. *Science* 58: 461-462.

Sharplin, J. 1963a. A flexible cuticle in the wing basis of Lepidoptera. *Canadian Entomologist* 95: 96-100.

Sharplin, J. 1963b. Wing base structure in Lepidoptera I. Fore wing base. *Canadian Entomologist* 95: 1024-1050.

Sharplin, J. 1963c. Wing base structure in Lepidoptera II. Hind wing base. *Canadian Entomologist* 95: 1121-1145.

Sharplin, J. 1964. Wing folding in Lepidoptera. *Canadian Entomologist* 96: 148-149.

Shepard, H.H. 1930. The pleural and sternal sclerites of the lepidopterous thorax. *Annals of the Entomological Society of America* 23: 237-260.

Shreeve, T.G. 1990. The behaviour of butterflies. *In* Kudrna, O. *Butterflies of Europe* 2: 480-511. Aula-Verlag, Wiesbaden.

Sibatani, A., Ogata, M., Okada, Y., & Okagaki, H. 1954. Male genitalia of Lepidoptera: morphology and nomenclature: I, Divisions of the valvae in Rhopalocera, Phalaenidae (= Noctuidae) and Geometridae. *Annals of the Entomological Society of America* 47: 93-106.

Sick, H. 1937. Die Tympanalorgane der Uraniden und Epiplemiden. *Zoologische Jahrbücher. Abteilung für Anatomie und Ontogenie der Tiere* 63: 351-398.

Silberglied, R.E. 1977. Communication in the Lepidoptera. *In* Sebeok, T.A. (Ed.). *How Animals Communicate.* pp. 362-402. Indiana University Press, Bloomington and London.

Silberglied, R.E. 1979. Communication in the ultraviolet. *Annual Reviews of Ecology and Systematics* 10: 373-398.

Silberglied, R.E. 1984. Visual communication and sexual selection among butterflies. *In* Vane-Wright, R.I. and Ackery, P.R. (Eds). *The Biology of Butterflies*, pp. 207-223, Academic Press.

Silberglied R.E. & Taylor, O.R. 1978. Ultraviolet reflection and its behavioral role in the courtship of the Sulfur butterflies *Colias eurytheme* and *C. philodice* (Lepidoptera, Pieridae). *Behavioral Ecology and Sociobiology* 3: 203-243.

Silow, C.A. 1976. Edible and other insects of mid-western Zambia. Studies in Ethno-Entomology II. *Occasional Papers of the Institutionen för Allmän och Jämförande Etnografi vid Uppsala Universitet* 5: 223 pp.

Singh, B. 1955. Description and systematic position of larva and pupa of the Teak defoliator, *Hyblaea puera* Cramer (Insecta, Lepidoptera, Hyblaeidae). *Indian Forest Records* (Entomology) 9: 1-16.

Skalski, A.W. 1990. An annotated review of fossil records of lower Lepidoptera. *Bulletin of the Sugadaira Montane Research Centre* 11: 125-128.

Smithers, C.N. 1977. Seasonal distribution and breeding status of *Danaus plexippus* (L.) (Lepidoptera: Nymphalidae) in Australia. *Journal of the Australian Entomological Society* 16: 175-184.

Someren, V.G.L., van. 1922. Notes on certain colour patterns in Lycaenidae. *Journal of the East Africa and Uganda Natural History Society* 17: 18-21 + 1 pl.

Sonnenschein, M. & Häuser, C.L. 1990. Presence of only eupyrene spermatozoa in adult males of the genus *Micropterix* Hübner and its phylogenetic significance (Lepidoptera: Zeugloptera, Micropterigidae). *International Journal of Insect Morphology and Embryology* 19: 269-276.

Southwood, T.R.E. 1962. Migration of terrestrial arthropods in relation to habitat. *Biological Reviews* 37: 171-214.

Southwood, T.R.E. 1973. The insect/plant relationship - an evolutionary perspective. *Symposia of the Royal Entomological Society of London* 6: 3-30.

Spangler, H.G. 1985. Sound production and communication by the Greater Wax Moth (Lepidoptera: Pyralidae). *Annals of the Entomological Society of America* 78: 54-61.

Spangler, H.G. 1988. Moth hearing, defense, and communication. *Annual Review of Entomology* 33: 59-81.

Spangler, H.G., Greenfield, M.D. & Takessian, A. 1984. Ultrasonic mate calling in the lesser wax moth. *Physiological Entomology* 9: 87-95.

Speidel, W. 1977. Ein Versuch zur Unterteilung der Lepidopteren in Unterordnungen. *Atalanta* 8: 119-121.

Speidel, W. 1984. Revision der Acentropinae des palaearktischen Faunengebietes (Lepidoptera, Crambidae). *Neue Entomologische Nachrichten* 12: 157 pp.

Stehr, F.W. (Ed.). 1987. *Immature Insects*. ix+754 pp., Kendall/Hunt, Dubuque, Iowa.

Stehr, F.W. & Cook, E.F. 1968. A revision of the genus *Malacosoma* Hübner in North America (Lepidoptera: Lasiocampidae): Systematics, biology, immatures and parasites. *Bulletin of the United States National Museum* 276: 321 pp.

Stekol'nikov, A.A. 1967. Functional morphology of the copulatory apparatus in the primitive Lepidoptera and general evolutionary trends in the genitalia of the Lepidoptera. *Entomological Review* 46: 400-409. [English translation of *Entomologicheskoye Obozreniye* 46: 670-689.]

Stekol'nikov, A.A. & Kuznetsov, V.I., 1986. Origin and general trends of the evolution of the skeleton and musculature of the male genitalia in Lepidoptera. *Entomological Review* 65: 130-145. [English translation of *Entomologicheskoye Obozreniye* 65: 59-73.]

Steward, R.C. 1985. Evolution of resting behaviour in polymorphic industrial melanic moth species. *Biological Journal of the Linnean Society* 24: 285-293.

Stirton, C.H. 1976. *Thuranthos*: notes on generic status, morphology, phenology and pollination biology. *Bothalia* 12: 161-165.

Stork, N.E. 1988. Insect diversity: facts, fiction and speculation. *Biological Journal of the Linnean Society* 35: 321-337.

Strand, E. 1911. Family: Callidulidae. *In* Seitz, A. *The Macrolepidoptera of the World* 2: 207-208.

Strong, D.R., Lawton, J.H. & Southwood, T.R.E. 1984. *Insects on plants: Community patterns and mechanisms*. vi+313 pp. Blackwell Scientific Publications, Oxford.

Sugi, S. (Ed.). 1987. *Larvae of Larger Moths in Japan*. 453 pp. Kodansha.

Surlykke, A. & Gogala, M. 1986. Stridulation and hearing in the noctuid moth *Thecophora fovea* (Tr.). *Journal of Comparative Physiology* A 159: 267-273.

Swihart, S.L. 1967. Hearing in butterflies (Nymphalidae: *Heliconius*, *Ageronia*). *Journal of Insect Physiology* 13: 469-476.

Tautz, J. & Markl, H. 1978. Caterpillars detect flying wasps by hairs sensitive to airborne vibration. *Behavioural Ecology and Sociobiology* 4: 101-110.

Thien, L.B., Bernhardt, P., Gibbs, G.W., Pellmyr, O., Bergström, G., Groth, I., & McPherson, G. 1985. The pollination of *Zygogynum* (Winteraceae) by a moth, *Sabatinca* (Micropterigidae): An ancient association? *Science* 227: 540-543.

Thorpe, W.H. 1928. Note on *Hyponomeuta cognatella* Hübn. feeding on the honey-dew of *Aphis rumicis* Linn. *Entomologist's Monthly Magazine* 64: 46.

Tillyard, R.J. 1918. The panorpoid complex. Part i - the wing-coupling apparatus, with special reference to the Lepidoptera. *Proceedings of the Linnean Society of New South Wales* 43: 285-319 + 2 pls.

Tillyard, R.J. 1923. On the mouth-parts of the Micropterygoidea (Order Lepidoptera). *Transactions of the Entomological Society of London* 1923: 181-206.

Tinbergen, N. 1951. *The Study of Instinct*. 228 pp. Oxford University Press, Oxford.

Traugott-Olsen, E. & Nielsen, E.S. 1977. The Elachistidae (Lepidoptera) of Fennoscandia and Denmark. *Fauna Entomologica Scandinavica* 6: 299 pp.

Tremewan, W.G. 1985. Zygaenidae. *In* Heath, J. and Emmet, A.M. *The Moths and Butterflies of Great Britain and Ireland* 2: 74-123 + pls 4-6.

Tremewan, W.G. 1988. *A Bibliography of the Zygaeninae* (*Lepidoptera: Zygaenidae*). 188 pp. Harley Books, Colchester.

Tschistjakov, Y.A. & Belyaev, E.A. 1987. The immature stages of *Pterodecta felderi* (Bremer) and systematic position of the family Callidulidae (Lepidoptera). *Tinea* (Supplement) 12: 285-289.

Tuck, K.R. 1981. A new genus of Chlidanotini (Lepidoptera, Tortricidae) from New Caledonia, with a key to genera and check-list of species. *Systematic Entomology* 6: 337-346.

Tuck, K.R. 1990. A taxonomic revision of the Malaysian and Indonesian species of *Archips* Hübner (Lepidoptera: Tortricidae). *Entomologica scandinavica* 21: 179-196.

Turner, J.R.G. 1977. Butterfly mimicry: the genetical evolution of an adaptation. *In*: Hecht, M.K., Steere, W.C., and Wallace, B. (Eds). *Evolutionary Biology* 10: 163-206. Plenum Publishing Cooperation.

Turner, J.R.G. 1984. Mimicry: The palatability spectrum and its consequences. *In* Vane-Wright R.I. and Ackery, P.R. (Eds). *The Biology of Butterflies.* pp. 141-161. Academic Press, London.

Tuskes, P.M. & Smith, N.J. 1984. The life history and behaviour of *Epimartyria pardella* (Micropterigidae). *Journal of the Lepidopterists' Society* 38: 40-46.

Tutt, J.W. 1904. *A Natural History of the British Lepidoptera.* 4: xvii+535 pp. London and Berlin.

Tweedie, M.W.F. 1988. Resting posture in the Lepidoptera. *British Journal of Natural History* 1: 1-8.

Tweedie, M.W.F. & Emmet, A.M. 1991. Resting posture in the Lepidoptera. *In* Emmet, A.M. and Heath, J. (Eds). *The Moths and Butterflies of Great Britain and Ireland* 7(2): 46-60.

Urquhart, F.A. 1960. *The Monarch Butterfly.* xv+361 pp. University of Toronto Press.

Urquhart, F.A. 1976. Found at last: the Monarch's winter home. *National Geographic* 150: 160-173.

Urquhart, F.A. 1987. *The Monarch Butterfly: International Traveler.* xxii+232 pp., 24 pls. Nelson-Hall, Chicago.

Vane-Wright, R.I. 1976. A unified classification of mimetic resemblances. *Biological Journal of the Linnean Society* 8: 25-56.

Vane-Wright, R.I. 1979. The coloration, identification and phylogeny of *Nessaea* butterflies (Lepidoptera: Nymphalidae). *Bulletin of the British Museum (Natural History)* (Entomology) 38: 27-56.

Vane-Wright, R.I. 1981. Only connect. *Biological Journal of the Linnean Society* 16: 33-40.

Varí, L. 1961. South African Lepidoptera. Volume 1, Lithocolletidae. *Transvaal Museum Memoir* 12: 238 pp. + 112 pls.

Vázquez, G.L. & Pérez, R.H. 1961. Observaciones sobre la biología de *Baronia brevicornis* Salv. (Lepidoptera: Papilionidae: Baroniinae). *Anales del Instituto de Biologia Universidad de Mexico* 32: 295-311.

Vielmetter, W. 1954. Die Temperaturregulation des Kaisermantels in der Sonnenstrahlung. *Naturwissenschaften* 41: 535-536.

Vielmetter, W. 1958. Physiologie des Verhaltens zur Sonnenstrahlung bei dem Tagfalter *Argynnis paphia* L. - 1. Untersuchungen im Freiland. *Journal of Insect Physiology* 2: 13-37 + 2 pls.

Viette, P. 1949. Contribution à l'étude des Hepialidae (12e note). Genres et synonymie. *Lambillionea* 1949: 101-104.

Viette, P. 1980. Insectes Lépidoptères Limacodidae. *Faune de Madagascar* 53: 162 pp.

Vogel, R. 1912. Über die Chordotonalorgane in der Wurzel der Schmetterlingsflügel. *Zeitschrift für wissenschaftliche Zoologie* 100: 210-244 + pls 7-8.

Vuillaume, M. 1975. Pigments des insectes. *In* Grassé, P.-P. (Ed.). *Traité de Zoologie* 8: 77-184.

Waage, J.K. & Montgomery, G.G. 1976. *Cryptoses choloepi*: a coprophagous moth that lives on a sloth. *Science* 193: 157-158.

Wagner, D.L. & Powell, J.A. 1988. A new *Prodoxus* from *Yucca baccata*: first report of a leaf-mining prodoxine (Lepidoptera: Prodoxinae). *Annals of the Entomological Society of America* 81: 547-553.

Walcott, B. 1975. Anatomical changes during light-adaptation in insect compound eyes. *In* Horridge, G.A. (Ed.). *The compound eye and vision in insects.* pp. 20-33. Clarendon Press, Oxford.

Wallace, M.M.H. 1970. The biology of the Jarrah leaf miner *Perthida glyphopa* Common. *Australian Journal of Zoology* 18: 91-104.

Wasserthal, L.T. 1975. The role of butterfly wings in regulation of body temperature. *Journal of Insect Physiology* 21: 1921-1930.

Watson, A. 1965a. A revision of the Ethiopian Drepanidae (Lepidoptera). *Bulletin of the British Museum (Natural History).* Entomology (Supplement) 3: 177 + 18 pls.

Watson, A. 1965b. *Aethiopsestis* gen. nov. (Lepidoptera), first record of Thyatiridae from the Ethiopian Region. *Journal of the Entomological Society* of Southern Africa 27: 257-266 + 1 pl.

Watson, A. 1975. A reclassification of the Arctiidae and Ctenuchidae formerly placed in the thyretid genus *Automolis* Hübner (Lepidoptera). *Bulletin of the British Museum (Natural History)* Entomology (Supplement) 25: 104 pp. + 33 pls.

Watson, A., Fletcher, D.S., & Nye, I.W.B. 1980. Noctuoidea (part). *In* Nye, I.W.B. (Ed.). *The Generic Names of Moths of the World* 2: 228 pp. British Museum (Natural History).

Watson, A. & Goodger, D.T. 1986. Catalogue of the Neotropical Tiger-Moths. *Occasional Papers on Systematic Entomology* 1: 71 pp. British Museum (Natural History).

Watt, W.B., Hoch, P.C. & Mills, S.G. 1974. Nectar resource use by *Colias* butterflies. Chemical and visual aspects. *Oecologia* 14: 353-374.

Weatherstone, J., MacDonald, A., Miller, D., Riere, G., Percy-Cunningham, J.E. & Benn, M.H. 1986. Ultrastructure of exocrine prothoracic gland of *Datana ministra* (Drury) (Lepidoptera: Notodontidae) and the nature of its secretions. *Journal of Chemical Ecology* 12: 2039-2050.

Weatherstone, J., Percy, J.E., MacDonald, L.M., & MacDonald, J.A. 1979. Morphology of the prothoracic defensive gland of *Schizura concinna* (J.E. Smith) (Lepidoptera: Notodontidae) and the nature of its secretion. *Journal of Chemical Ecology* 5: 165-177.

Weidner, H. 1934. Beiträge zur Morphologie und Physiologie des Genital-apparates der Weiblichen Lepidopteren. *Zeitschrift für Angewandte Entomologie* 21: 239-290.

Wells, P.H. & Rogers, S. *In press.* Is multiple mating an adaptive feature of Monarch butterfly winter aggregations? *In* Malcolm, S.B. and Zaluchi, M. (Eds). *Biology and Conservation of the Monarch Butterfly.* Los Angeles County Museum, Los Angeles.

Wenner, A.M. & Harris, A.M. *In press.* Do Californian Monarchs really migrate? *In* Malcolm, S.B. and Zaluchi, M. (Eds). *Biology and Conservation of the Monarch Butterfly.* Los Angeles County Museum, Los Angeles.

Werny, K. 1966. Untersuchungen über die Systematik der Tribus Thyatirini, Macrothyatirini, Habrosynini und Tetheini (Lepidoptera: Thyatiridae) 463 pp. + 436 figs. Werny, Saarbrücken.

Whalley, P.E.S. 1964. Catalogue of the Galleriinae (Lepidoptera, Pyralidae) with descriptions of new genera and species. *Acta Zoologica Cracoviensia* 9: 561-618 + pls 14-44.

Whalley, P.E.S. 1971. The Thyrididae (Lepidoptera) of Africa and its Islands. A taxonomic and zoogeographic study. *Bulletin of the British Museum (Natural History)* Entomology (Supplement) 17: 198 pp. + pls 1-68.

Whalley, P.E.S. 1976. *Tropical Leaf Moths. A monograph of the subfamily Striglininae (Lepidoptera: Thyrididae)* 194 pp. + pls 1-68. British Museum (Natural History), London.

Whalley, P.E.S. 1977. Lower Cretaceous Lepidoptera. *Nature* 266: 526.

Whalley, P.E.S. 1978. New taxa of fossil and recent Micropterigidae with a discussion of their evolution and a comment on the evolution of Lepidoptera (Insecta). *Annals of the Transvaal Museum* 31: 71-86 + 4 pls.

Whalley, P.E.S. 1986. A review of the current fossil evidence of Lepidoptera in the Mesozoic. *Biological Journal of the Linnean Society* 28: 253-271.

White, M.E. 1986. *The Greening of Gondwana* 256 pp. Reed.

White, T., Weaver, J.S. & Agee, H.R. 1983. Responses of *Cerura borealis* (Lepidoptera: Notodontidae) larvae to low-frequency sound. *Annals of the Entomological Society of America* 76: 1-5.

Wigglesworth, V.B. 1965. *The Principles of Insect Physiology.* 6ᵗʰ Edn. viii+741 pp. Methuen, London.

Williams, C.B. 1958. *Insect Migration.* xii+235 pp. Collins, London.

Williams, G.C. 1966. *Adaptation and Natural Selection.* Princeton University Press, Princeton.

Willis, M.A. & Birch, M.C. 1982. Male lek formation and female calling in a population of the arctiid moth *Estigmene acrea. Science* 218: 168-170.

Wirtz, R.A. 1984. Allergic and toxic reactions to non-stinging arthropods. *Annual Review of Entomology* 29: 47-69.

Wootton, R.J. 1979. Function, homology and terminology in insect wings. *Systematic Entomology* 4: 81-93.

Wootton, R.J. 1981. Support and deformability in insect wings. *Journal of Zoology.* 193: 447-468.

Wootton, R.J. 1987. Insects: the ultimate sailing machines. *The Central Association of Bee-Keepers.* 8 pp. Ilford, Essex.

Wu, J.-T. & Chou, T.-J. 1985. Studies on the cephalic endoskeleton, musculature and proboscis of citrus fruit piercing noctuid moths in relation to their feeding habits. *Acta Entomologica Sinica* 28: 165-172. [In Chinese with English summary.]

Yagi, N. & Koyama, N. 1963. *The compound eye of Lepidoptera.* 319 pp. Shinkyo Press, Tokyo.

Yano, K. 1963. Taxonomic and biological studies of Pterophoridae of Japan (Lepidoptera). *Pacific Insects* 5: 65-209.

Yasuda, T. 1962. On the larva and pupa of *Neomicropteryx nipponensis* Issiki, with its biological notes (Lepidoptera, Micropterygidae). *Kontyû* 30: 130-136 + pls 6-8.

Young, A.M. 1982. *Population Biology of Tropical Insects.* xii+511 pp. Plenum Press, New York and London.

Zacharuk, R.Y. 1985. Antennae and sensilla. *In* Kerkut, G.A. and Gilbert, L.I. (Eds). *Comprehensive insect physiology, biochemistry and pharmacology* 6: 1-69. Pergamon Press.

Zagatti, P. 1981. Comportement sexuel de la Pyrale de la Canne à sucre *Eldana saccharina* (Wlk.) lié à deux phéromones émises par le mâle. *Behaviour* 78: 81-82.

Zagulajev, A.K. & Sinev, S.Y. 1988. Catapterigidae fam. n. - a new family of lower Lepidoptera (Lepidoptera, Dacnonypha). *Entomologicheskoe Obozrenie* 3: 593-601. [In Russian.]

Zerny, H. & Beier, M. 1936. Ordnung der Pterygogenea: Lepidoptera = Schmetterlinge. *In* Kükenthal, W. (Ed.). *Handbuch der Zoologie* 4: 1554-1728. Berlin.

Zimmerman, E.C. 1978. Microlepidoptera II: Gelechioidea. *In* Zimmerman, E.C. (Ed.). *Insects of Hawaii* 9: 883-1903. The University Press of Hawaii, Honolulu.

INDEX

Page numbers to illustrations are indicated in **bold** type.
defn refers to a definition or description of a term.

391